A Practical Guide to Inversely Optimized Treatment Planning

A Practical Guide to Inversely Optimized Treatment Planning

E. Ishmael Parsai
and Ana I. Bejarano B.

Medical Physics Publishing
Madison, Wisconsin

First printing January 2024

Library of Congress Control Number: 2023937520

ISBN hard cover: 978-1-951134-20-4
ISBN eBook: 978-1-951134-21-1

Published by:
Medical Physics Publishing
4555 Helgesen Drive
Madison, WI 53718
800-442-5778
mpp@medicalphysics.org
medicalphysics.org

Printed in the United States of America

Contents

Preface

Since the birth of Intensity-Modulated Radiation Therapy (IMRT), significant progress has been made in enhancing both software and hardware, resulting in the optimization of planning, quality assurance, and delivery procedures. While IMRT is now a standard practice in some regions, many clinical facilities worldwide have yet to integrate these techniques. Once the initial implementation of IMRT is accomplished, one of the most formidable challenges faced by radiation oncology teams is becoming proficient in the planning process for various anatomical sites.

This book has been designed to provide a comprehensive overview for those adopting IMRT for the first time. It outlines the technical prerequisites and offers recommendations for inverse planning, with a specific focus on the planning procedures. The book also includes sample plans from three of the most widely used Treatment Planning Systems (TPS), aiding practitioners in acclimating to IMRT effectively.

Clinical sites that are considering the implementation of IMRT for the first time are highly encouraged to enlist the help of a team of medical physicists, dosimetrists, and other clinical staff with previous experience in intensity-modulated techniques. Contacting vendors for education on hardware and software to be used for the new chapter of the clinic is highly encouraged as well. Sound knowledge of dosimetry, workflow, and optimization techniques in producing treatment plans is essential for radiotherapeutic success. Readers are also encouraged to reach out to medical physics organizations that specialize in international clinical education, such as Medical Physics for World Benefit (mpwb.org) and the American Association of Physicists in Medicine–International Council–Global Medical Physics Education Training Committee (AAPM-IC-GMPETC), and also take advantage of the available references published by the International Atomic Energy Agency (IAEA). This book can be one of the many companions for physicists and dosimetrists during the different phases of first-time IMRT implementation. For further information on the different topics in this book, the reader is redirected to more specific topic-related references. While the references follow mainly the AAPM guidelines and the authors reside in the United States, this book references universal safety.

This book contains nine chapters that fall in the following layout. Some introductory definitions and abbreviations are offered in Chapter one. Chapter two summarizes the resource requirements for IMRT, including machine characteristics, the treatment planning system and commissioning, quality assurance procedures, and other recommendations. Chapter three covers all aspects of patient simulation, including scanner requirements, quality assurance, patient immobilization, and the process of verification simulation. Chapter four covers the beginning of the planning process and optimization specifics for each of the three main treatment planning systems. Chapter five presents quality assurance (QA) in intensity-modulated plans and a discussion on patient-specific QA. Chapter six is a detour to an old technique known as spatially fractionated radiation therapy, also known as grid therapy, using volumetric modulated arc therapy (VMAT) planning and delivery. Chapters seven through nine show sample IMRT cases in three different treatment planning systems to illustrate the concepts learned in Chapter four. The reader should be advised that these chapters cover three different treatment planning systems. The LINAC hardware is limited to C-arm Varian LINACs.

We hope our readers find this book a useful tool in the face of ever-changing clinical cases. We also aim for this book to open meaningful team discussions about the many aspects that need to be considered during a first-time implementation of IMRT. The techniques illustrated in this work have been refined through years of experience in clinical practice. As authors, we want our readers to take these samples into their own understanding in order to develop their own planning style, supported by a solid conceptual understanding of their planning and delivery system as a whole.

Disclaimers: all materials being used for illustrating the planning cases—including dose schemes, case descriptions, and targets—are meant only for educational purposes and are not suggestions for patient treatment. The information pertaining to optimization goals and other plan parameters are meant to illustrate the steps of the planning process and should not be replicated in clinical patient plans without consideration.

For first-time implementations of inverse-modulated techniques, it is highly recommended to have the presence and guidance of a medical physicist with extensive experience in inverse-modulated techniques to advise through the commissioning and implementation process.

E. Ishmael Parsai and
Ana I. Bejarano B.

Acknowledgments

We appreciate the fruitful and stimulating discussions with our colleagues and trainees, in particular, the contribution of Mr. Jonathan N. Bogue, MS, DABR. His review of sections of the manuscript and Eclipse treatment plans is much appreciated.

Ana (Anita) I. Bejarano B. wishes to thank her husband, Zaid Aldoulah, for his support through her medical physics journey. She expresses gratitude to relatives and friends for their support. She also wishes to thank her residency clinical staff and medical physics mentors—Jess McKay, Travis Denton, Bonnie Velasco, and Sammie Hedrick—for their remarkable and joyful dedication to her medical physics education, training, and dosimetric amusement.

Dr. E. Parsai extends his gratitude to the numerous students he has taught throughout his academic journey, from whom he has gained invaluable knowledge. He also expresses appreciation to his family members for their enduring patience, unwavering support, and wholehearted dedication during his scholarly pursuits. In the context of finalizing this book, he specifically wishes to acknowledge his wife and children for their constant presence and understanding in light of the countless late nights and weekends invested in completing this endeavor.

1. Definitions and Abbreviations

3D-CRT (Three-Dimensional Conformal Radiotherapy)—Forward-planned radiotherapy based on a CT dataset where field shape conforms to the shape of the target.

4D-CT (Four-Dimensional Computed Tomography)—Computed tomography dataset with data binned throughout the entire respiratory cycle to capture motion. It is a key tool for measurement and management of respiratory motion.

Amplitude Gating—Gating based on the magnitude of motion measured.

BEV (Beam's-Eye View)—Image from the perspective of the gantry head to help ensure proper patient alignment.

CBCT (Cone-Beam Computed Tomography)—Imaging procedure using cone-beam-shaped kV radiation, typically mounted on a linear accelerator, to obtain three-dimensional image sets for patient alignment prior to treatment.

Compression—The use of abdominal compressors to potentially reduce intra-fraction motion in abdominal or lung tumors.

CT (Computed Tomography)—Imaging procedure used to obtain three-dimensional image sets for treatment planning or diagnostic purposes.

CTV (Clinical Target Volume)—Contour containing the GTV and sub-clinical malignant disease with possibility of recurrence relevant to the treatment.

DIBH (Deep Inspiration Breath-Hold)—A form of amplitude gating where patients voluntarily inspire and hold their breath at a fixed level. This is often used to decrease exposure to the heart for breast plans or other potential toxicities for abdomen or thorax treatments.

DRR (Digitally Reconstructed Radiograph)—Radiograph generated from a computed tomography image set.

DVH (Dose-Volume Histogram)—Histogram relating the radiation dose to tissue volume.

FB (Free Breathing)—No respiratory gating used for the given application.

FFF (Flattening Filter-Free)—A treatment beam without a flattening filter that enables a higher dose rate.

FIF (Field-in-field)—Forward-planned radiation therapy with control points or sub-fields designed by the planner.

Fx (Fraction)—A single treatment of radiation.

Gating—The use of a respiratory monitoring system to trigger the radiation beam. This can be used to reduce the treatment volume and dose to organs at risk for treatments in the abdomen or thorax.

GTV (Gross Tumor Volume)—Contour for the gross palpable and demonstrable extent of the tumor.

Gy (Gray)—Standard unit of measure for radiation dose, equivalent to 1 joule per kilogram (J/kg) or 100 rad (1 cGy = 1 rad).

IGRT (Image-Guided Radiation Therapy)—Use of on-board imaging systems to align the patient prior to radiation treatment.

IMRT (Intensity-Modulated Radiation Therapy)—Inversely optimized radiation therapy treatment method produced by setting constraints and allowing the treatment planning system to generate the optimal fluence and plan delivered with multiple static gantry angles.

Isocenter—Point through which the central rays of the radiation beam pass.

Isodose—A line or color wash that connects all points of an equal dose value, used to visualize the dose distribution.

ITV (Internal Tumor Volume)—Contour containing the CTV with margin to account for uncertainties in size, shape, and position of the CTV within the patient.

kV (kilovolt)—1,000 volts.

MaxIP (Maximum Intensity Projection)—Derived image series from a 4D-CT that retains the highest Hounsfield value for that voxel, representing the most hyperdense structures of the volume.

MeanIP (Mean Intensity Projection)—Derived image series from a 4D-CT that retains the average Hounsfield value for that voxel, representing the average of the volume.

MinIP (Minimum Intensity Projection)—Derived image series from a 4D-CT that retains the lowest Hounsfield value for that voxel, representing the most hypodense structures of the volume.

MLC (Multileaf collimator)—Collimation system using small sets of high-density leaves to shape the dose distribution. A representative diagram is shown in Figure 2.2.

MLC Delivery Methods—The two most common methods of MLC IMRT delivery are "step and shoot," also known as static MLC (SMLC), and "dynamic" (DMLC). The SMLC typically involves 5–15 segments per gantry angle, and the beam is switched off between each segment. The dynamic mode involves continuous movement of the MLC, generally in the form of a "window" sliding from one side of the field to the other. Varying the speeds of the leading and trailing leaves modulates the intensity of the beam.

MV (megavolt)—1,000,000 volts.

OAR (Organ at Risk)—Normal tissue contour.

OBI (On-Board Imaging)—kV or MV imaging systems attached to the linear accelerator to enable IGRT or during treatment imaging.

Phase Gating—Gating based on the periodic phase of the respiratory cycle.

POI (Point of Interest)—A point on an image dataset.

PTV (Planning Target Volume)—Contour containing the ITV with margin to account for uncertainties in internal motion and setup uncertainties.

ROI (Region of Interest)—Normal tissue or critical structure contours.

RPM (Real-Time Patient Management)—A video-based system for gating developed by Varian.

Rx (Prescription)—The physician-prescribed dose and fractionation scheme.

SABR/SBRT (Stereotactic Ablative Radiotherapy/ Stereotactic Body Radiation Therapy)—Highly focused radiation therapy delivered in ≤5 fractions.

Simulation—Procedure for obtaining a computed tomography scan of the patient in the treatment position.

SRS (Stereotactic Radiosurgery)—Highly focused radiation therapy delivered to volumes in the brain in ≤5 fractions.

TPS (Treatment Planning System)—Software used for design and calculation of radiation therapy plans.

VMAT (Volumetric Modulated Arc Therapy)—Inversely optimized radiation therapy treatment method produced by setting constraints and allowing the treatment planning system to generate the optimal fluence and plan delivered with arcs while the gantry and MLCs are continuously moving.

2. Resource Requirements for IMRT

Intensity-modulated techniques like IMRT and VMAT require a particular set of machine characteristics and tools for their safe implementation. The three main requirements include particular features in the delivery system, the planning software, and the patient-specific quality assurance devices.

1. ***The treatment delivery system—typically a linear accelerator (linac)—must be equipped with a working multileaf collimator and applicable software and hardware licenses for the use of IMRT/VMAT***. In the past, for the static delivery of a photon beam with a compensator, IMRT was possible without the use of an MLC. These custom-made devices were made of brass or other similar materials and were being manufactured either in-house or through outside vendors. In the early days of IMRT, these devices were mostly used for IMRT delivery on machines that were not capable of being upgraded or retrofitted with MLC systems.
2. ***The treatment planning software must have a goal-based, inverse-planning optimizer***.
3. ***There must be a detailed machine- and patient-specific quality assurance system***.

A brief overview of the clinical requirements for intensity-modulated techniques is detailed in this book. Further guidance is offered by the American Association of Physicists in Medicine (AAPM) and the International Atomic Energy Agency. Relevant reports include but are not limited to:

- AAPM Task Group Report 82: Guidance document on delivery, treatment planning, and clinical Implementation of IMRT,
- AAPM Medical Physics Practice Guideline 5.b: Commissioning and QA of Treatment Planning Dose Calculations—Megavoltage Photon and Electron Beams,
- AAPM Task Group Report 119: IMRT commissioning: Multiple institution planning and dosimetry comparisons,
- AAPM Task Group Report 120: Dosimetry tools and techniques for IMRT,
- AAPM Task Group Report 219: Independent calculation-based dose/MU verification for IMRT,
- AAPM Task Group Report 218: Tolerance Limits and Methodologies for IMRT Measurement-Based Verification QA,
- AAPM Medical Physics Practice Guideline 8.a.: Linear accelerator performance tests, and
- IAEA-TECDOC-1588 report: "Transition from 2-D Radiotherapy to 3-D Conformal and Intensity Modulated Radiotherapy."

In clinical sites where the implementation of intensity-modulated techniques are being considered, it is imperative that the team has a well-established 3D conformal radiation therapy program prior to adopting intensity-modulated techniques. IAEA-TECDOC-1588 recommends a minimum of 1 to 2 years of experience. This period allows enough time for the team to become familiar with the departmental work flow and the clinical irradiation tools. The staff must be trained in the new delivery method, including radiation oncologists, medical physicists, dosimetrists, and radiation technologists.

Additional quality assurance methods and protocols must be prepared for IMRT implementation. Intensity-modulated techniques also require more extensive quality control and engineering maintenance. A more detailed review of requirements and milestones for IMRT implementation can be found in IAEA-TEC-DOC-1588.

2.1 Machine Characteristics

The use of intensity-modulated techniques requires the linac to have an advanced collimation system and an on-board imaging system with their respective licenses from vendors. Prior to any type of activity, the output of the machine must be calibrated and verified by a second party if possible. The physicist might use the guidelines of AAPM's TG-51 protocol for clinical reference dosimetry of high-energy photon and electron beams (and its addendum) or IAEA TRS reports 398, 381, and 277. Special attention must be placed on small fields, for which the physicist can count on the guidance of Report No. TRS-48—"Dosimetry of small static fields used in external photon beam radiotherapy."

Multileaf collimators (MLCs) (Figure 2.1) typically consist of two banks, or rows, of several high-density

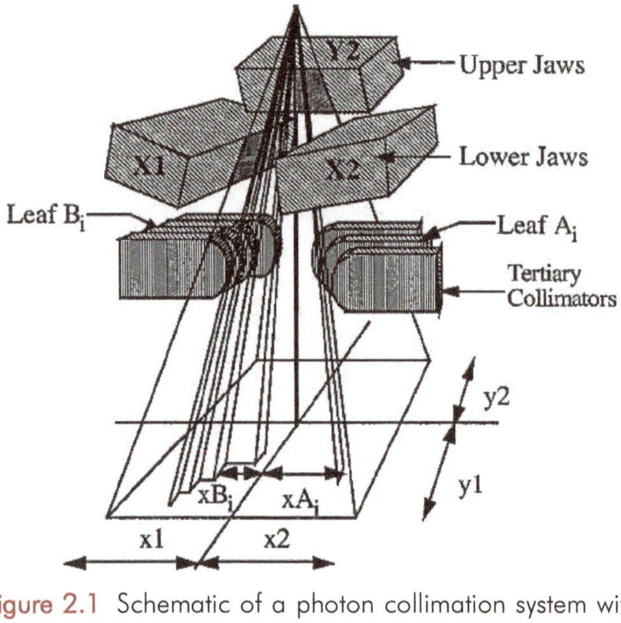

Figure 2.1 Schematic of a photon collimation system with upper and lower jaws and a tertiary multileaf collimator (AAPM TG-50).

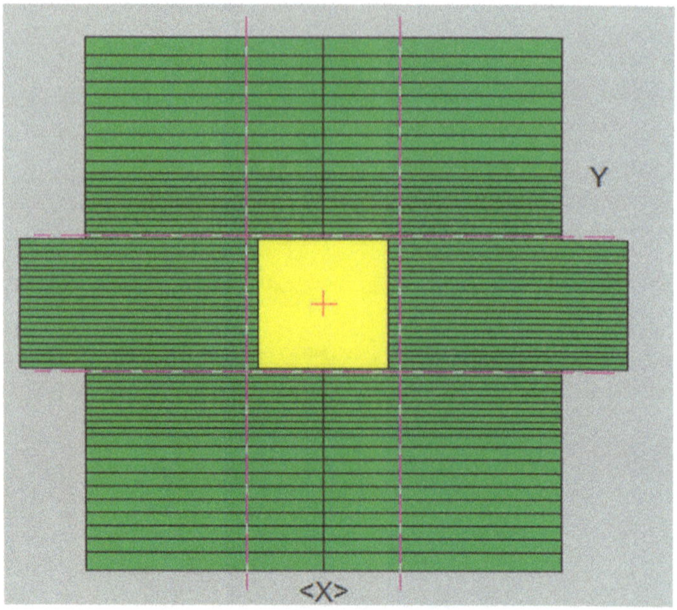

Figure 2.2 Representation of MLC collimation with a Millenium 120 MLC. (Courtesy of Varian Medical Systems.)

(typically tungsten) leaves. These independent leaves are capable of moving in and out of the radiation field, creating different shapes. This modulates the radiation field to customize the delivery for each individual patient without the need for physical attenuators or compensator blocks. The use of MLCs for intensity-modulated techniques allows for the treatment of complex target shapes and the creation of dose distributions that are impossible with conventional 3D conformal plans.

MLCs are often limited in their maximum field size based on the travel distance of their leaves and the size of the MLC's banks (Figure 2.2). The minimal size of treatable fields can be limited by the transverse thickness of the leaves. For example, the maximum MLC field size for a Varian Millennium 120 MLC is 40 × 40 cm, whereas it is 22 cm (y axis) × 32 cm (x axis) on Varian's HD 120 MLC.

- The field size for modulation on the x axis is dependent on the travel distance (the maximum distance a leaf can go in or out of the bank) of the MLC leaves.
- The field size on the y axis is dependent on the size of the banks.

In Varian machines, the leaf travel distance is approximately 14 cm. The minimal size of a treatable field is dependent on:

- the width of the individual MLC leaves,

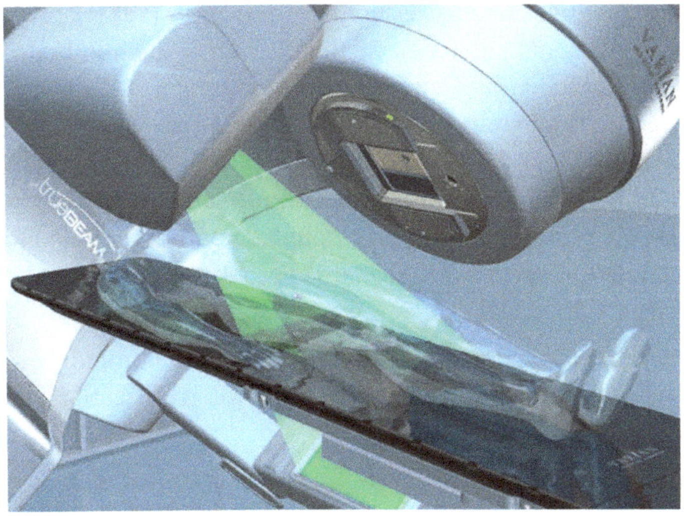

Figure 2.3 On-board imaging system on a linac. (Courtesy of Varian Medical Systems.)

- the minimum field size commissioned (jaw and MLC), and
- the mechanical stability of the machine.

A good practice would be to limit the application of MLCs to treatment volumes larger than three times the leaf thickness or the minimum jaw field size commissioned. Therefore, if 1-cm-thick leaves are being used, the target should have a 3-centimeter diameter for the MLC to adequately modulate the dose within said volume.

The on-board imaging system can include an MV or a kV imaging system (Figure 2.3). Both offer the capability of taking planar images and, additionally, kV systems offer the capability of cone-beam CT (CBCT). The purpose of imaging is to verify the position of the patient before irradiation by allowing the user to acquire images and match them to the planned anatomy. Detailed images, such as CBCTs, can help assess patient anatomy before treatment, such as a full bladder during each fraction of a pelvic treatment. The frequency and type of imaging sessions can change according to case complexity and a physician's request.

For 3D techniques, it is common practice to use high-energy beams (15 MV or 18 MV) for deep-seated tumors. This is not recommended for inverse-modulated techniques. Machine energy for modulated techniques should be set at 10 MV and below. Energies higher than 10 MV produce neutrons, especially when the modulation is increasing, such as with inversely optimized plans. High-density materials in the accelerator, treatment room, and patient can become radioactive at energies higher than 10 MV through photonuclear reactions and neutron-capture reactions. Therefore, it is recommended to keep the energy at 10 MV or less for IMRT or VMAT plans. Beams of 10 MV can be used to treat deep-seated tumors, while 6 MV is ideal for brain, lung, or superficial targets. More information about this issue can be found on the AAPM Report of task group 136: Potential hazard due to induced radioactivity secondary to radiotherapy.

2.2 Treatment Planning System (TPS) Capabilities and Commissioning

Medical physicists are responsible for the commissioning of the treatment planning system. The measurements vary according to the TPS, but they normally require numerous measurements of jaw and MLC-defined fields using a water phantom and detector in addition to measurements of transmission through the leaves and from inter-leaf leakage. The most important aspects of dosimetric machine commissioning for an accurate IMRT beam modeling include:

- accurate dosimetric data collection and entry;
- MLC characterization, including DLG (dosimetric leaf gap) and leaf transmission measurements; and
- correct output factor measurements, with special consideration on small field outputs, among other aspects.

Guidelines for commissioning and validating the TPS can be found in AAPM Medical Physics Practice Guideline 5.b: Commissioning and QA of Treatment Planning Dose Calculations—Megavoltage Photon and Electron Beams. The validation of a TPS consists of an iterative process in which the physicist inserts the data obtained through commissioning into the TPS, creates a series of fields and plans in the TPS, and delivers them to compare the measured doses and profiles to the expectations from the TPS. It is also highly recommended that the beam model be reviewed and tested to work on a variety of cases with different degrees of modulation. Moreover, the physicist should have a good understanding of the mechanisms, capabilities, and limitations of the dose calculation engines used by the different treatment planning systems.

After successfully commissioning and validating the TPS beam model, verification measurements should be performed using a phantom case for an end-to-end test. It is recommended that the final testing includes data transfer from and to the record and verify system and the accelerator. The AAPM's TG-119 titled "IMRT Commissioning Tests: Instructions for Planning, Measurement and Analysis" shows a variety of tests and benchmark cases for implementing intensity-modulated techniques. Other options for end-to-end testing include the use of test phantoms.

2.3 Quality Assurance Procedures

The physics staff should verify the safe operation of the MLC by testing leaf positional accuracy, linear accelerator performance for small fields and small monitor unit delivery, and MLC control accuracy, including position and speed.

The staff should also quantify other physical characteristics—such as the MLC speed against the dose rate and the gantry speed and position—prior to making it available for clinical use. These tests should continue to be performed on daily, monthly, and annual intervals based on the AAPM's TG-142 and AAPM Medical Physics Practice Guideline 8.a. The tests should cover all aspects of the system, such as dosimetry, mechanical tests, and imaging tests (Figure 2.4).

Plan quality assurance should be performed for every treatment plan developed using inverse planning. Pretreatment quality assurance starts with the use of an independent calculation method to verify the number of monitor units and absolute dose using input from the treatment planning system. There is currently a variety of secondary MU check software packages in the market. The secondary MU check software must support IMRT beams and must go through a commissioning and validation process. Relevant reports on secondary MU checks include:

- the Report of AAPM Task Group 219 on independent calculation-based dose/MU verification for IMRT and
- the report of AAPM Task Group 114 on verification of monitor unit calculations for non-IMRT clinical radiotherapy.

Dose verification measurements should also be performed. The treatment plan can be delivered onto a phantom and compared to the planning system's calculated dose. Further information can be found in Chapter 5. Film can be placed around a cylindrical phantom with an ion chamber placed in the center to verify the relative dose distribution and absolute point dose, or detector arrays can be used to verify both at once. It is important that the physics staff understands the limitations of all quality assurance methods and be aware of acceptable uncertainties.

2.4 Other Considerations

The use of intensity-modulated techniques can increase the treatment time and the monitor units necessary to

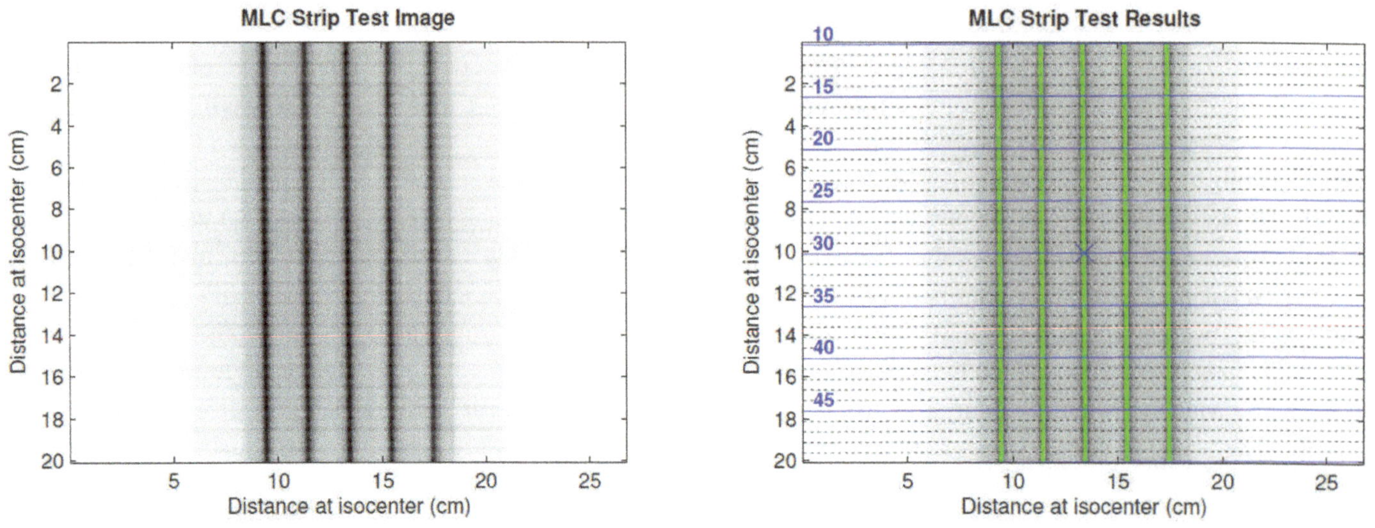

Figure 2.4 MLC picket fence test sample.

deliver a plan. When using step-and-shoot IMRT beams, the user must take into consideration the number of segments used per beam, as this might increase delivery time. When using arcs, the user must take into consideration the number or arcs, as this will also increase the treatment time. An increase in treatment time adds uncertainty in terms of patient workload, patient comfort, patient motion, and physiological processes within the patient (bladder filling, breathing motion, etc.)

The delivery time can be calculated by taking into consideration the basic work flow:

- patient setup,
- setup imaging,
- setup confirmation, and
- beam delivery.

Setup imaging for inverse-planned treatments can include planar imaging (kV or MV) and CBCT (kV). After positioning couch shifts are applied, re-verification images are often acquired prior to start of the treatment.

One way to optimize the delivery time is by arranging the gantry rotation of the beams in such a way that they follow gantry rotation in one direction. For VMAT plans, this consideration can be modified to setting the start rotation angle at the stop angle of the previous arc.

If the plan includes couch rotations, these should be taken into account when optimizing the machine movements as well.

Another tool to decrease treatment times is the dose rate. Use of the maximum dose rate is acceptable in IMRT plans. Dose rate in VMAT plans can be set to maximum, but the final speed of treatment depends on whether the machine is capable of doing *constant dose rate* (CDR) arcs or *variable dose rate* (VDR) arcs. CDR machines maintain the same gantry rotation speed, while VDR can vary gantry rotation delivery speed.

After successful implementation of inverse-planned techniques, experienced clinics can see treatment times of 15 to 20 minutes, counting time from when the patient enters the room to the end of the treatment.

Other factors—such as patient immobilization and target localization—must be reviewed and enhanced prior to starting the clinical delivery of intensity-modulated plans.

2.4.1 Shielding Considerations

It is estimated that the required number of *monitor units* (MU) for an IMRT treatment increase by a factor of 2 to 5 over conventional techniques. This increase in MUs often increases the workload and treatment head leakage for shielding considerations. The physicist should ensure that the facility's shielding is capable of safely accommodating this increased modulation factor, which often is a few times greater than conventional 3D plans.

Common shielding concerns are addressed in NCRP Report No. 151 if working conditions are found to be unsafe, but recommendations in this report do not guarantee to provide safe barriers for linac vaults that are intended to use IMRT and VMAT techniques. A qualified expert should use VMAT-specific use factors when considering the shielding design. Moreover, a record and verify system should also be in place that is capable of handling the IMRT workload and securely holding all critical IMRT parameters.

2.4.2 Staff Requirements

Radiation oncology departments may follow ASTRO's "Safety Is No Accident" recommendations, as well as benchmarking standards from the Association of Community Cancer Centers (ACCC-Cancer.org). Based on the recommended staffing levels from Battista et al.[1] and ASTRO's publication, there is a minimum number of physicist and dosimetrists per number of available treatment delivery systems, number of patients treated per day, and available modalities. The far right two columns in **Table 2.1** show the required number of physicists and dosimetrists for a typical clinic performing HDR brachytherapy and IMRT planning with two linacs.

Similarly, for a typical mid-size radiation oncology department serving 3500 to 5500 patients annually, **Table 2.2** shows the minimum number of other staff required, including:

- ***Therapists***—1 per 86 patients (ACCC-Cancer.org) or 1 per 90 patients (ASTRO),
- ***Dosimetrists***—1 per 328 patients (ACCC-Cancer.org) or 1 per 250 patients (ASTRO),
- ***Physicists***—1 per 244 patients (ACCC-Cancer.org) or 1 per facility (ASTRO), and
- ***Radiation Oncologists***—1 per 30 patients (ACCC-Cancer.org) or 1 per facility (ASTRO).

1. "Medical physics staffing for radiation oncology: a decade of experience in Ontario, Canada." *J. Appl. Clin. Med. Phys.* 13(1):93–110.

Table 2.1 Required number of physicists and dosimetrists for a typical clinic performing HDR brachytherapy and IMRT planning with two linacs

		Relative FTE Factor		Required FTE	
Services — # of Units or Licenses	**# of Systems**	**Physicists**	**Dosimetrists**	**Physicists**	**Dosimetrists**
Multi-energy accelerators	2	0.25	0.05	0.5	0.1
Single-energy accelerators	0	0.08	0.01	0	0
Tomotherapy, CyberKnife, Gamma Knife	0	0.3	0.03	0	0
Cobalt units; IMRT, PACS, EMR, and contouring	1	0.08	0.03	0.08	0.03
Orthovoltage and superficial units	0	0.02	0.01	0	0
Manual brachytherapy, LD seed implants	0	0.2	0.03	0	0
HDR brachytherapy	1	0.2	0.02	0.2	0.02
Simulator, CT-simulator, PET, MRI fusion	1	0.05	0.02	0.05	0.02
Computer planning system (per 10 work stations)	1	0.05	0.02	0.05	0.02
HDR planning system	1	0.2	0.01	0.2	0.01
Annual # of Patients Undergoing Procedures	**# of Patients**	**Physicists**	**Dosimetrists**	**Physicists**	**Dosimetrists**
External beam RT with 3D planning	66	0.0003	0.003	0.0198	0.198
External beam RT with conventional planning	0	0.0002	0.002	0	0
Sealed source brachytherapy (LDR and HDR)	0	0.008	0.003	0	0
Unsealed source therapy	0	0.008	0.005	0	0
IMRT, IGRT, SRS, TBI, SBRT	276	0.008	0.005	2.208	1.38
Education and training (FTE)	0.1	0.667	0.333	0.0667	0.0333
Generation of internal reports (FTE)	0.1	0.667	0.333	0.0667	0.0333
Committees and meetings, including rad. safety (FTE)	0.1	0.667	0.333	0.0667	0.0333
Administration/management	0.5	0.667	0.333	0.3335	0.1665
			TOTAL	**3.8**	**2.0**

Table 2.2 Minimum number of other staff required for a typical mid-sized radiation oncology department serving 3500 to 5500 patients annually

Category	Staffing Recommendations	Actual Staffing*	# of Patients
Radiation oncologists	30 per MD or 1 per facility	4.0	317/4789
Medical physicists	244 per physicist or 1 per facility	4.0	317/4789
Radiation therapists**	1 per 86–90 patients treated annually	5.0	317/4789
RNs	No recommendation	1.5	317/4789
Front desk	No recommendation	1.5	317/4789
Manager	1 per facility	1.0	317/4789

* These numbers could be higher or lower depending on the complexity of the treatments and the complexity of the patient undergoing treatment. Staffing of the linear accelerators will be managed by the lead therapist, and it is such that there will be two people at each accelerator for any routine external beam treatments.
** It is recommended that two therapists be present for any routine radiation therapy treatment.

2.4.3 Additional Staffing Considerations

Here are some other staffing considerations:

- For scheduled time off for therapists, the scheduling will be two therapists per machine per patient, and scheduling of patients is to reflect this rule.
- During times of unexpected time off, the schedules should be arranged so that the center will still comply with two therapists per machine—even if this requires rescheduling patient times—to maintain safety.
- With respect to staffing of the medical physicists and dosimetrists, planned time off is managed so that there will always be one *qualified medical physicist* (QMP) on site. Unplanned absences should be treated the same way to assure a QMP is on-site during all clinical hours. In extreme circumstances, locum tenens should be employed.
- In physician staffing, the recommendation is that there should always be a radiation oncologist present during treatment times. In case of unplanned absences or extreme circumstances, locum tenens should be employed.
- Nursing schedules should be such that one RN or one MA be staffed on a daily basis. Unplanned absences or time away should be managed by the use of contingent RNs or MAs who have been trained in radiation oncology.
- As patient volume increases, the staffing of the radiation oncology department on a contingent or permanent basis should be evaluated by management with patient safety and correct staffing levels as a top priority.

3. Simulation

The simulation process starts the work flow of treatment planning in radiation therapy. The main objectives of the simulation process are to:

- acquire images to identify normal and diseased structures,
- secure reproducible setups for upcoming treatment/s, and
- acquire volumetric representations of patients for planning, dose calculation, and beam arrangement.

Simulation techniques can include x-ray orthogonal films obtained with a simulator and volumetric datasets from *computed tomography* (CT). For inverse-modulated techniques, the use of 3D CT datasets is necessary for target localization and plan calculation. Other datasets, such as 4D CT datasets, can be acquired during simulation for motion assessment and management. Additional imaging datasets from previous studies could be fused for target definition.

All subsequent processes depend on the quality of the simulation of the patient. The following is a short introduction to the simulation process for inverse planned techniques (IMRT, VMAT). The reader is encouraged to find additional information when implementing and upgrading features for clinical use.

3.1 Simulation Imaging: CT

It is strongly recommended that simulation for inverse planning is done with 3D patient image datasets obtained with CT. The 2D images from a conventional simulator do not provide enough information about the target, organs at risk, or other required information to perform dose modulation and calculation. All TPS require computed tomography datasets for dose calculation. The planner can also obtain *digitally reconstructed radiographs* (DRRs) from the CT dataset at the selected isocenter point for other purposes, such as imaging during treatment and verification of beam aperture.

The process begins with the immobilization of the patient to the CT table using an immobilization device. This will be the position of treatment and must be comfortable for the patient. If this position is not comfortable or stable for the patient, motion might occur during treatment. More information on immobilization devices can be found later in this chapter.

All immobilization devices are attached or "indexed" to the CT table, which replicates a treatment table. The physician has the option to add radio-opaque markers to an area to be irradiated for target localization or definition. If the system comes with a fixed laser system, radio-opaque reference markers can be placed on the skin to mark the isocenter (based on room lasers). If the CT has movable lasers, then the isocenter can be set at the end of simulation.

The image is acquired and reviewed for quality purposes. DICOM images and isocenter (if using a movable laser) can be sent from the CT console to the TPS to start the contouring and planning process.

3.2 Scanner Requirements and Parameters

The couch surface must be flat and similar to that used for the accelerator treatment couch. The CT system or room must have its own laser system for localization purposes. These lasers can be either fixed or movable. The actual position of the patient—either *head first supine* (HFS) or *feet first supine* (FFS)—must match that indicated in the scanner position to avoid confusion during treatment imaging or errors in patient orientation during treatment.

Avoid the use of truncated (cut) CT datasets for IMRT and VMAT planning, especially if the cut area is potentially used as a beam entrance. Cut tissue is missing tissue from the calculation, which can result in significant differences in beam path length and miscalculation of dose.

It is recommended that image acquisition protocols with their respective settings are saved in the CT console. Acquisition parameters, such as kV and mAs, can be set up with the help of the CT manufacturer and physicists. The average slice thickness used across many institutions is 2.5 mm for simulation purposes, although it is possible to increase resolution for select anatomic regions such as the central nervous system. The areas to be scanned depend on the extent of the disease, preference, and institutional protocols. It is important to remember that simulation images are not meant to be used for diagnostic purposes.

Some physicians might consider it useful to include contrast in their CT simulation images. Contrast increases the density of the enhanced anatomy and can represent a source of uncertainty in dose calculation. It is possible to have a treatment planning CT without contrast, followed by a CT performed with contrast for target assessment. In cases in which it is not possible to obtain a treatment planning CT without contrast, it is recommended that physicists review these situations and apply the correct density overrides for dose calculation.

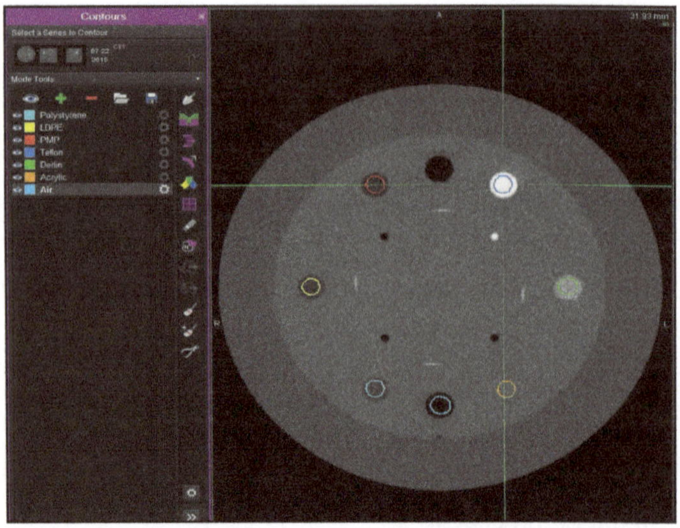

Figure 3.1 HU constancy tested on a Catphan 504 phantom.

3.3 Quality Assurance and Additional Information

CT scanners used in radiation oncology should pass the manufacturer's acceptance testing as well as tests specific to their purpose in the department. AAPM Task Group 39 "Specification and Acceptance Testing of Computed Tomography Scanners" specifies tests for acceptance of CT equipment in radiation therapy.

Medical physicists should perform continuous QA procedures according to the requirements of the local and international regulatory agencies and device manufacturers. One protocol to be followed is the Report of the AAPM Radiation Therapy Committee Task Group No. 66: "Quality Assurance for Computed-Tomography Simulators and the Computed Tomography-Simulation Process." This report does not address QA requirements specific to the scanner or software vendors. Instead, the procedures detailed in the document apply only to CT simulator units for external beam radiation therapy.

The CT scanner used in simulation must be characterized in the TPS through a density vs CT number (or Hounsfield units) curve performed by a medical physicist. This curve is required by the TPS in order to perform dose calculation. This can be done by scanning a phantom with inserts of various known densities and recording the HU values of such materials (Figure 3.1). It is preferred that the phantom be scanned with the clinical protocols. Different protocols can differ in mAs, but not in kV unless there are different curves characterizing different kV.

3.4 Immobilization and Localization

For treatment delivery, IMRT often creates highly conformal plans with steep dose gradients that warrant highly accurate delivery to ensure adequate coverage of the target and sparing of critical structures. Therefore, patient positioning, immobilization procedures, and pretreatment imaging must be reproducible and accurate to ensure safe delivery for the duration of the treatment. All members of the staff—including radiation therapists, physicians, dosimetrists, administration, and physics or physics assistants—should be aware of the changes that inverse planning will bring to their workload. Staff must also communicate expectations of treatment times and immobilization techniques to their patients to prepare them for the process.

During simulation, the patient must be immobilized in a device that will ensure a reproducible and comfortable position every time it is used in treatment. The importance of immobilization greatly increases for

Figure 3.2 Indexed holder (left) and bean bag (right). The vacuum cushion is positioned on top of the indexed holder.

inverse planning techniques like IMRT and VMAT. The immobilization device must not cause artifacts on the image, and it must be attached or indexed to the treatment couch. Popular immobilization devices include:

- ***Styrofoam™ casts.*** These casts are formed with a foaming agent that takes the shape of the patient. The patient is placed on the Styrofoam mold encased in a plastic bag. The foaming agent is poured into the bag, where it takes the shape of the patient's anatomy as it solidifies. These bags cannot be reused.
- ***Vacuum cushions.*** These bags perform similarly to foam casts with the advantage that they are reusable. The patient is placed on top of the bag during simulation, and a vacuum is attached to the bag. As the air exits the bag, it takes the shape of the patient (Figure 3.2).
- ***Carbon fiber boards.*** These boards can be used in addition to a vacuum cushion to achieve inclination. They do not provide effective immobilization needed for intensity-modulated techniques if used alone (Figure 3.3).
- ***Thermoplastic material.*** Masks or other covers made of thermoplastic material can be shaped to the patient's regional anatomy and attached to the table for reproducible shape. The mask is placed in warm water and then applied to the patient, retaining its new shape as it cools down (Figure 3.4). Thermoplastic material cannot be reused. For increased immobilization, some systems attach bite blocks to their thermoplastic masks.

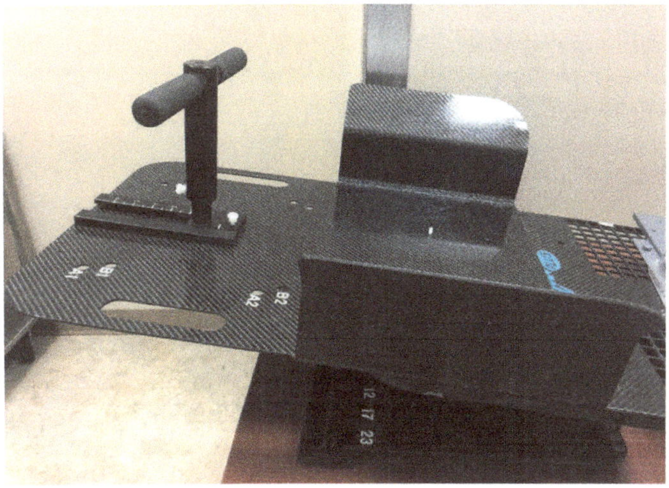

Figure 3.3 Butterfly board.

After acquiring the simulation image, small marks are made on the patient's skin for localization and positioning purposes during treatment. The anatomy to be treated is aligned to the room lasers, and these marks are placed on the lateral and anterior aspects of the patient. It is important that the laser system in the CT, as well as the one located inside the room, have their own quality assurance procedures.

3.5 Verification Simulation

Verification simulation happens during or before the first radiation therapy treatment session. The extent of

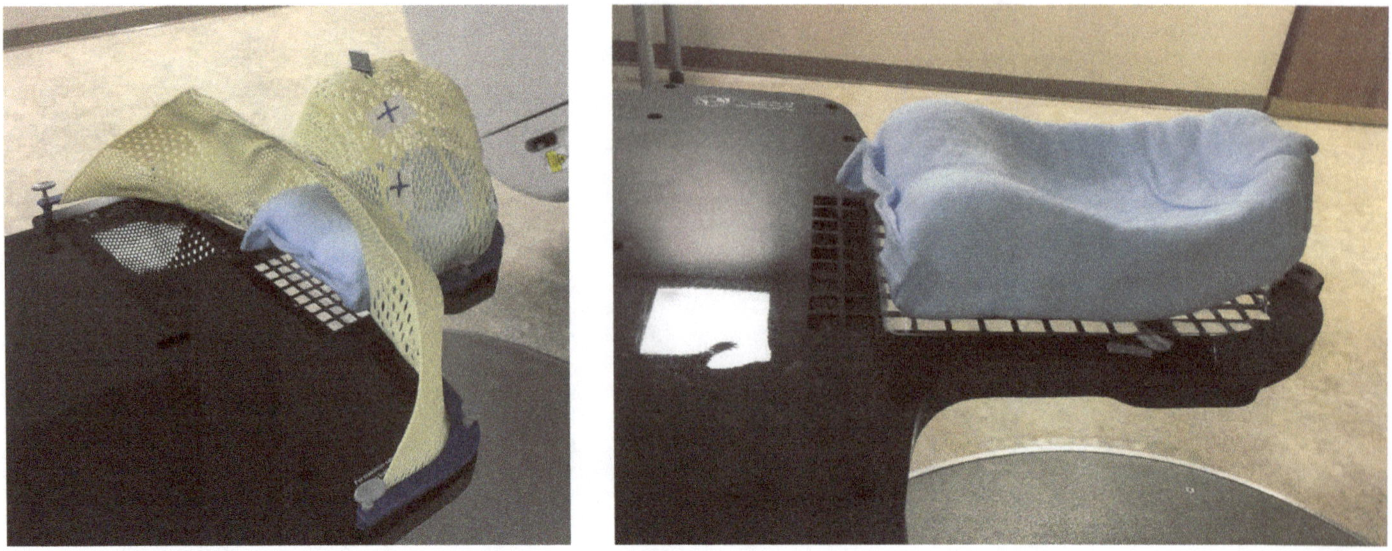

Figure 3.4 Pillow mold (right) and thermoplastic mask (left) for head and neck immobilization.

this verification varies from institution to institution. It is a dry run of the treatment after it is planned and approved for delivery. It takes place in the accelerator vault with the patient on the table. The radiation therapist sets up the patient for delivery. Setup MV or kV images including CBCT can be acquired during verification simulation. Light field and SSD are checked when applicable. The therapist runs through all of the gantry, collimator, and couch configurations for each beam of delivery, if applicable.

4. Planning

This chapter encompasses inverse planning techniques across different treatment planning systems. The different cases presented here should not be taken as step-by-step instructions, but rather as a source of reference for the many different cases that the planner may encounter in the clinical setting.

4.1 Contouring

Contouring consists of drawing volumes of particular structures on a CT dataset, and it is essential to any planning technique, especially with the arrival of sophisticated planning systems. Contours are important since the planner assigns names and characteristics to a volume in the CT dataset that enables the optimization process and dose quantification to each of the structures. These contours can be targets, organs at risk, or optimization contours. They simply identify the enclosed volume so it can be used for a variety of purposes.

4.1.1 Organs at Risk

Organ-at-risk (OAR) contours represent the patient's healthy tissue. These contours can be drawn according to consensus data, such as the RTOG contouring guides (Figure 4.1), and they must be contoured as accurately as possible.

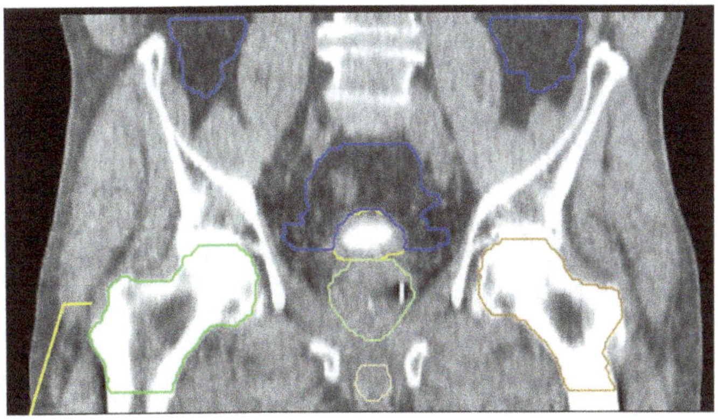

Figure 4.1 Sample contouring for male pelvis. (RTOG contouring atlas).

There are reputable websites that offer guidance on contouring with free registration:

- ***"ProKnow" and "Contouring Accuracy" by Elekta (proknowsystems.com).*** This Elekta website contains how-to presentations as well as practice online datasets with scoring capabilities. Users can learn to contour challenging anatomy (such as brachial plexus, head and neck structures, and others) and have their practice contours compared to expert datasets with feedback.
- ***E-Contour (e-contour.org).*** This physician-led website contains a variety of CT datasets, giving the user the experience of having a 3D image encyclopedia of contours for quick reference.

It is also important to know the difference between *serial organs* and *parallel organs* in terms of sensitivity to radiation. The dose and dose distribution can affect the sensitivity of an organ. One might think of all organs as divided into subunits, which can be functional or nonfunctional after radiation damage. Serial organs can fail entirely if any subunit has been damaged (examples include the spinal cord, optic nerves, and the brainstem). Parallel organs can fail entirely if many or all subunits are damaged (examples include the liver, lungs, and kidneys).

All plans must be approved and signed by the radiation oncologist after reviewing target coverage and dose to OARs.

4.1.2 Optimization Contours

Optimization contours are drawn and used with the sole purpose of helping enhance the dose distribution during inverse planning. Overlapping optimization contours can cause conflicts in the optimizer. To avoid unnecessarily overlapping structures for optimization, here is some guidance:

- ***Separating dose levels into targets that have more than one dose level is important.*** Overlaps create conflicts for most optimizers. Create one contour for each individual dose level. Targets (GTV, CTV, ITV, and PTV) are drawn by the radiation oncologist.
- ***In cases where the radiation oncologist wishes to spare healthy tissue at the cost of target coverage, the planner can use Boolean operators to create a special optimization structure without modifying the original target structures.*** A subtraction of the target minus the OAR (typically with a 2-mm margin) can become an optimization target structure that can be used in the optimizer for planning.
- ***In cases where the radiation oncologist wishes to cover the target at the expense of healthy tissue, the planner can use Boolean operators to create an OAR optimization structure that avoids the target with a margin of typically 2 mm.*** This structure can be used in the optimizer for planning.
- ***Planners can choose to leave margins in their optimization target contours or to draw "rings" around them.*** Rings are structures that act like "barriers" between dose levels to achieve the desired dose falloff or to avoid *dose spills* (high dose going in regions of low dose). In cases in which the TPS system does not have an optimization option to spare normal tissue, the user can use a series of concentric rings around the target. Another option is to make a 2- to 3-cm "rind" from the external or body contour.

4.2 Isocenter Placement

The beam isocenter is the point at which the central axis of the beams will be directed during treatment. In the case of having a moving laser system or a static laser system, the placement of a point within the patient can be achievable during the simulation process or the planning process. Labeling these points correctly is imperative to avoid mistakes through planning, quality assurance, and delivery. There are usually two points of interest at which beams can be directed: the isocenter from CT simulation or the beam isocenter.

4.2.1 Isocenter from CT Simulation

This is the point at which the patient is marked during the simulation process. If the institution counts with a moving laser system, during the simulation process it is possible to "drop" a point in a particular part of the body (generally the middle of the target) and make the lasers move to that point so that tattoos and marks can be drawn at the actual site of alignment for treatment. If the location of this isocenter is optimal, then it can also be used as an isocenter of the beams, instead of just using it as a point to set the lasers. If the location is not optimal, then another point can be placed on the planning system (i.e., the beam isocenter).

4.2.2 Beam Isocenter

Throughout this book, "beam isocenter" is the point at which all beams converge (Figure 4.2). Prior to treat-

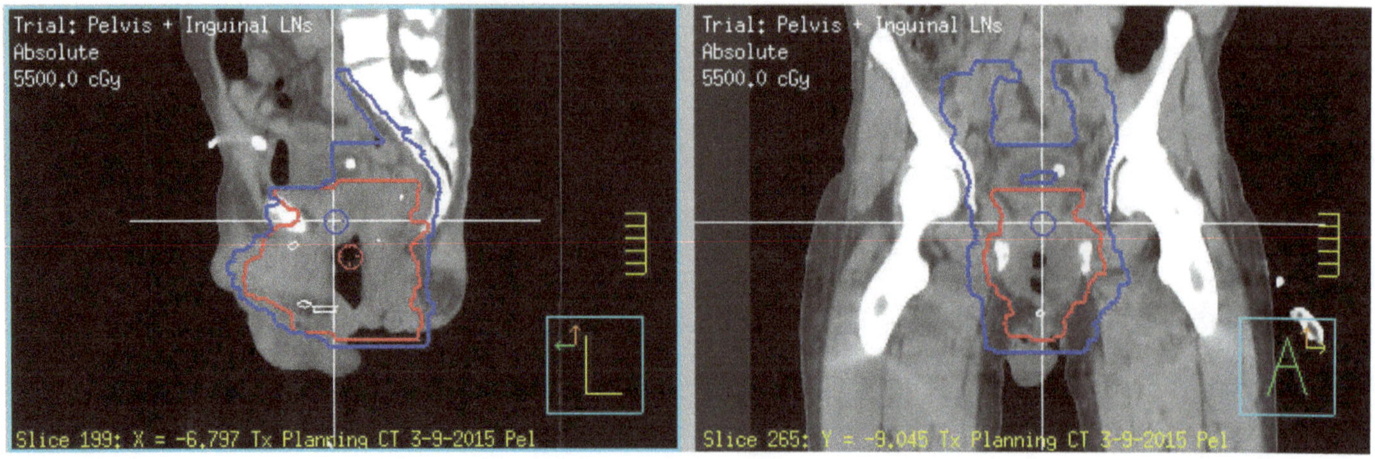

Figure 4.2 Isocenter position (blue POI) in the central aspect of the target.

ment, therapists will receive shifts from the marked isocenter in simulation to the true beam isocenter. In this case, for treatment, the patient will be set to the tattoos of the original isocenter and then moved to the beam isocenter by shifts provided by the planning system.

A good general placement of a beam isocenter is one that is in the middle of the target, with relatively easy-to-make shifts from the original simulation isocenter (if applicable). Not placing the isocenter for the beams in the middle of the target could compromise conformity of the planned dose since most accelerators have high-definition MLCs only in the middle of their field.

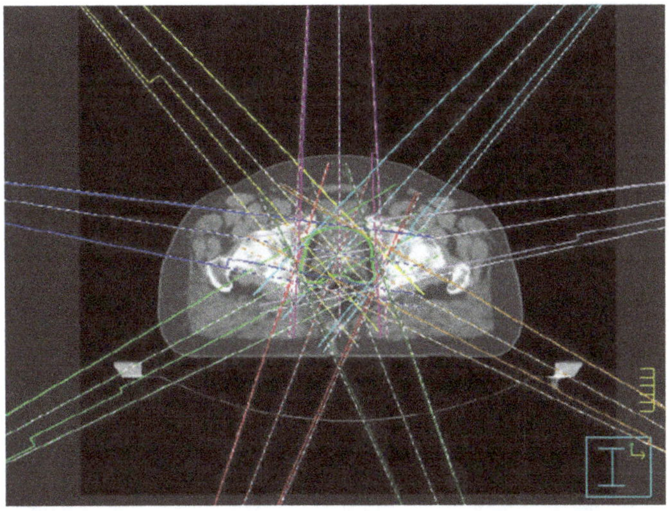

Figure 4.3 A typical 9-beam IMRT beam arrangement.

4.3 Beam Arrangement

4.3.1 IMRT Beam Arrangement

IMRT can have a variety of beam arrangements depending on the target location and coverage needed. A traditional plan will involve 7 to 9 beams (Figure 4.3) located around the patient at fixed or variable gantry angle intervals. Collimator rotations are not needed, but the planner is encouraged to use them in cases where they are valuable to spare an organ, cover a target, etc.

For fixed-jaw machines, the use of split beams can help cover wide targets. Other options include:

- ***manual setup of split fields*** (have one beam divided into two beams with different MLC carriage positions), and
- ***vendor-specific packages***, such as the large-field IMRT provided by Varian Eclipse.

The beam arrangement can be one-sided if the lesion to be treated is not located centrally in the body (Figure 4.4). The planner must avoid irradiating healthy structures as much as possible, unless coverage of the lesion is greatly improved and desired and the dose to healthy tissue is under reasonable tolerance.

4.3.2 VMAT Beam Arrangement

The most common VMAT beam arrangement encompasses two or more arcs around the patient. The collima-

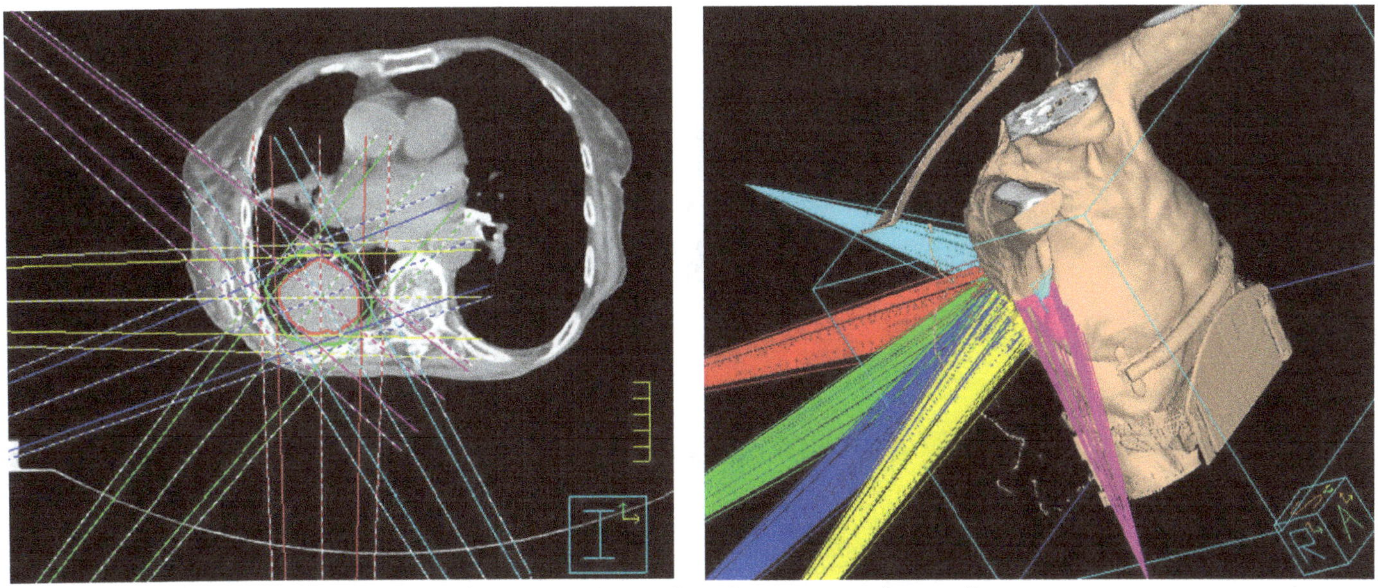

Figure 4.4 A typical beam arrangement for a lesion located on the posterior chest wall.

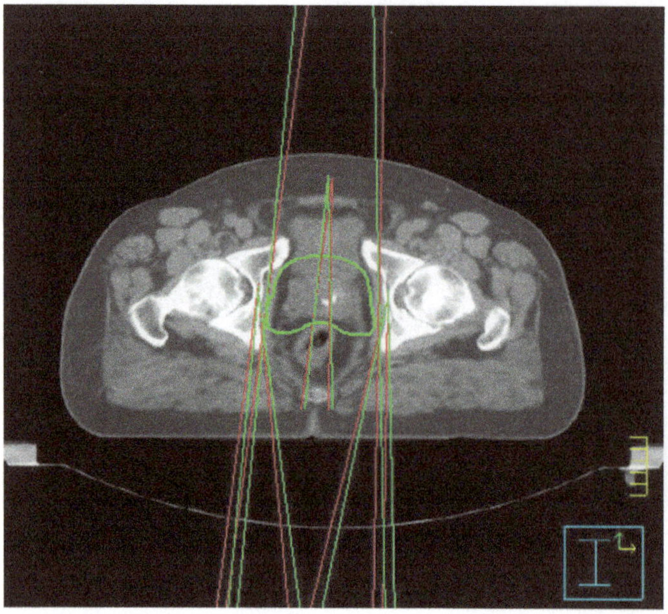

Figure 4.5 Two VMAT arcs with a 2-degree offset.

tor can be placed at an angle (30 to 90 degrees) to decrease the inter-leaf leakage. It is important to know the machine and system limitations in terms of start and stop angles for an arc.

Another parameter to keep in mind is the gantry spacing or space between control points. In some planning systems, the user has the option to set control points every 5, 4, 3, or 2 degrees. The increased resolution would take more system resources to optimize and calculate, which might make the calculation long and impractical. One way to offset the disadvantages of this scenario is to offset the start and stop angles of each arc by the gantry spacing divided by two. For example, if the gantry spacing is 4 degrees, the planner can create a plan with two VMAT arcs offset by 2 degrees from each other's start and stopping angles (Figure 4.5).

The planner must be aware of the capabilities of the accelerator and R&V system, especially in terms of delivery, gantry rotation, and collision. For example, some systems might allow a 180- to 179.9-degree gantry rotation (almost 360 degrees), while others allow a 181- to 179-degree gantry rotation. Since the gantry will be going around the patient, the planner must be careful to not set a beam arrangement that could cause a collision.

The arcs can be set to cover 360 degrees, as well as partial arcs when needed, for one-sided lesions (Figure 4.6).

4.4 Optimization

Inverse planning refers to planning done by providing the computer with a description of the desired solution for which to develop a plan. The solution the user provides to the computer is the desired dose distribution, and the development of a plan refers to the creation of control points or beamlets to achieve the solution. Different planning systems have different nomenclature for the optimization process, but the IMRT/VMAT basis is still the same.

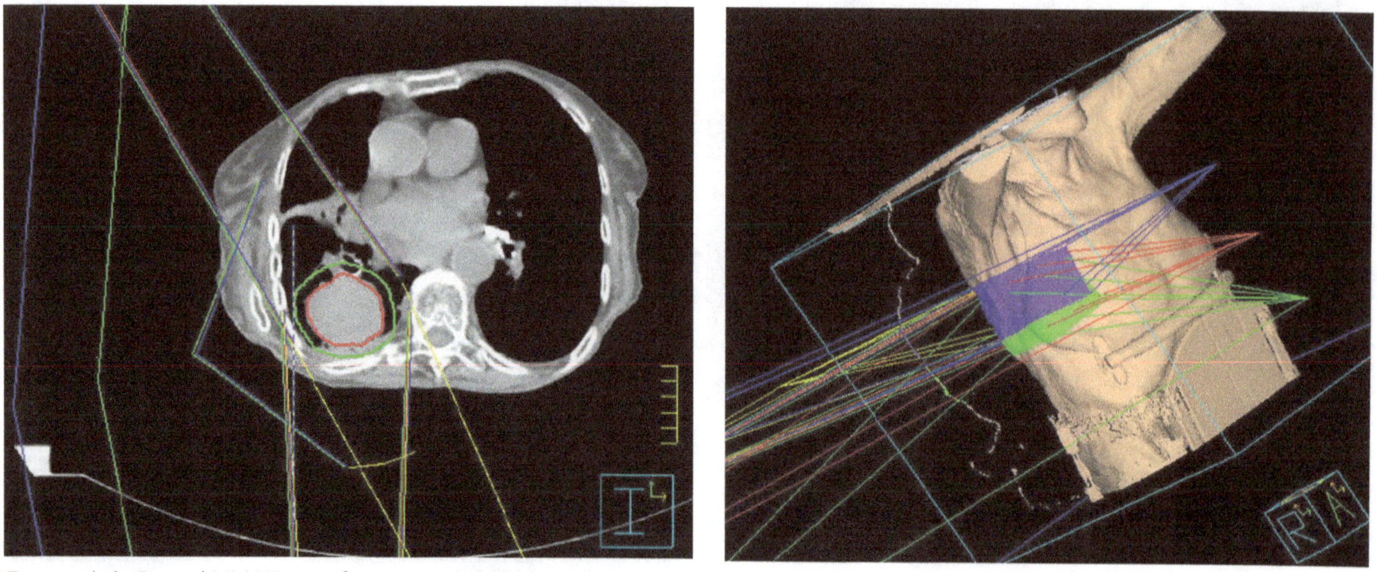

Figure 4.6 Partial VMAT arcs for a one-sided lesion.

For each beam placed on the patient, the IMRT optimization will try to find the best control points/MLC positions to achieve the desired dose distribution (Figure 4.7). *Control points* or *beamlets* are individual parts of a dynamic IMRT beam to regulate dose to a certain area.

To start the optimization, one needs to have targets defined by a physician, the beams set up, and all necessary contours defined. For the TPS to achieve an optimal dose distribution, it is important to set appropriate objectives, constraints, and beam parameters.

Some planning systems require the user to convert the dose fluence map to a deliverable plan, while others do this automatically after the optimization process. Several optimizations can be necessary to achieve the desired dose distribution.

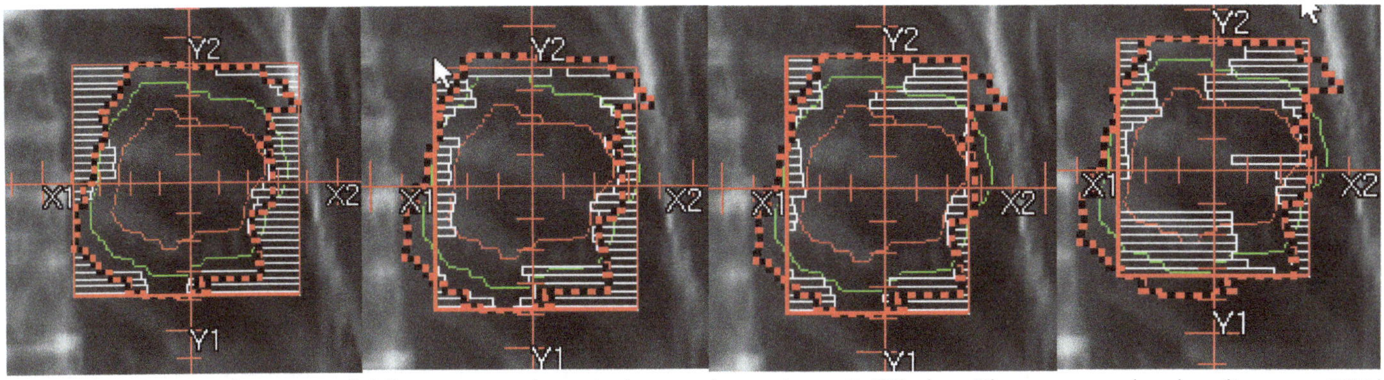

Figure 4.7 Series of images of different control points for one beam in an IMRT plan. The green and red outlines are PTV/GTV target contours.

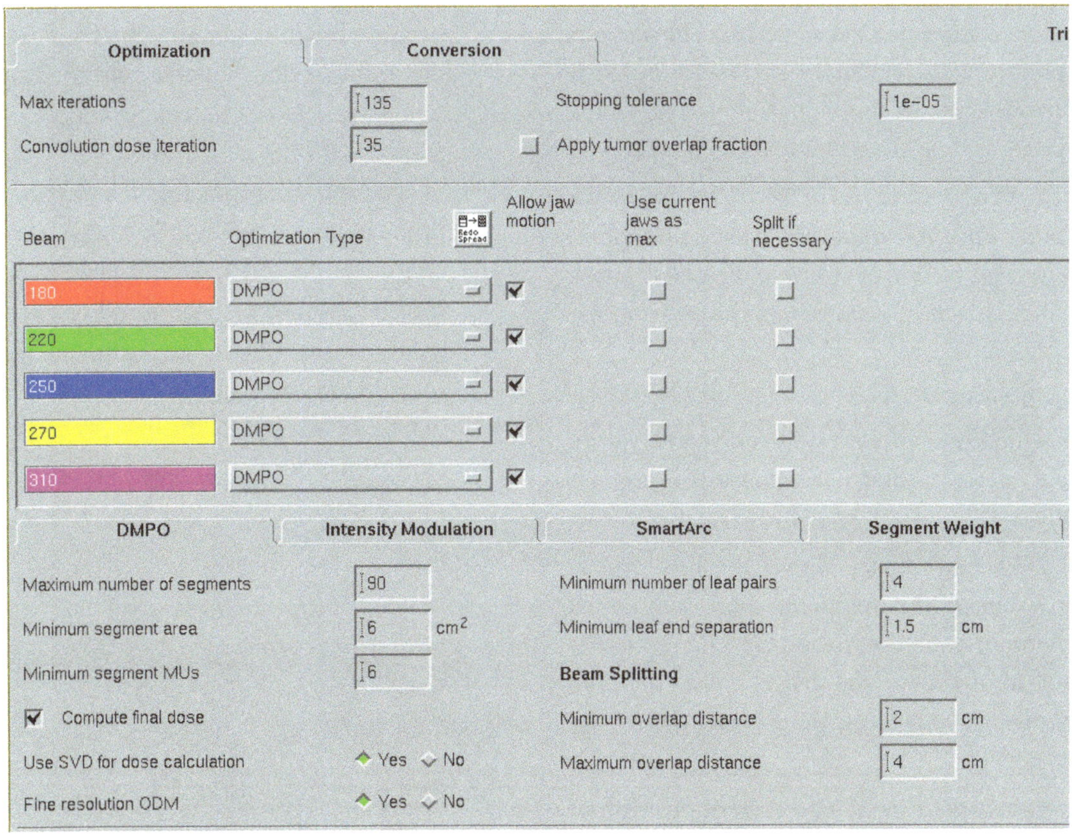

Figure 4.8 Sample IMRT Parameters menu under the IMRT tab in Pinnacle 9.8. Similar options can be found in other versions of the TPS.

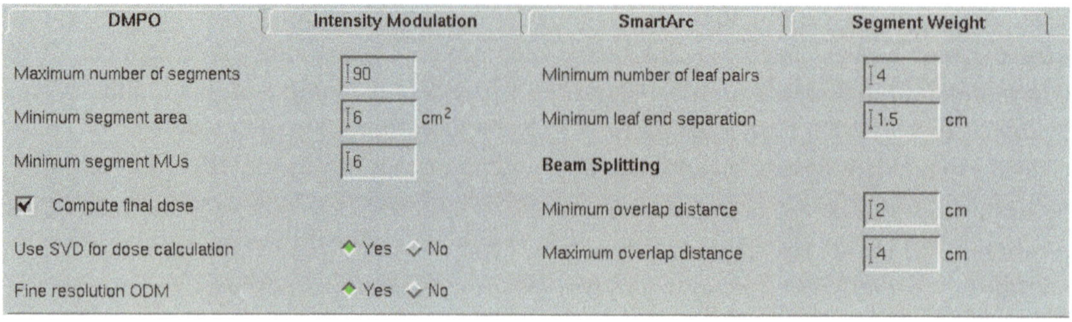

Figure 4.9 DMPO options in Pinnacle 9.8.

4.4.1 Pinnacle Optimization

4.4.1.1 Pinnacle IMRT/VMAT Parameters

Pinnacle IMRT parameters are shown in Figure 4.8. The following settings are found in Pinnacle version 9.8. However, they are still applicable for newer versions of the software.

For IMRT, the following beam parameters are of importance:

- ***Maximum iterations.*** The basic rule is to set the convolution dose iteration to one quarter to one third of the maximum iteration number. For example, one can set up 75 maximum iterations with convolution dose calculation at 25. The user is free to choose any number of iterations, but it is important to keep in mind that choosing a low value does not give the system enough iterations to achieve the best optimization results.
- ***Convolution dose iteration.*** This sets up the number of iterations before computing the convolution dose, adding machine parameters (close to a real deliverable plan).
- ***Stopping tolerance.*** The *objective value* (defined later) at which the software will stop working on achieving a certain objective because it is considered achieved or it is not possible to be achieved.
- ***Moving jaws.*** If the user allows the machine to move jaws during the treatment, these will adjust to the shape of each control point. This option is important to decrease the MLC scattering and leakage, but it depends on the capabilities of the machine.
- ***Use current jaws as max.*** The current position of the jaws will be the maximum limit during the optimization. Though rarely used, this can be selected only when the warning of the "field size too large" shows up during optimization.
- ***Split if necessary.*** This automatically splits beams by making another one.

4.4.1.2 Pinnacle Optimization Types

- ***DMPO.*** The acronym DMPO stands for *direct machine parameter optimization*. Instead of just providing an intensity modulation dose, it also includes machine characteristics, MLC position, and segment/control point weights (Figure 4.9). DMPO produces plans that are delivered more efficiently than an intensity modulation plan because machine parameters have been used in the optimization. The number of segments refers to the number of control points per plan (step and shoot IMRT).

  The more complicated the shape of the target is or the closest a target is to organs at risk, the more control points the plan will need to achieve the desired isodose distribution. For example, in this example, the number of segments has a default of 50, which can be increased according to how many control points the system ends up using to control the dose. Too many control points can increase the delivery time, so a good rule is to set

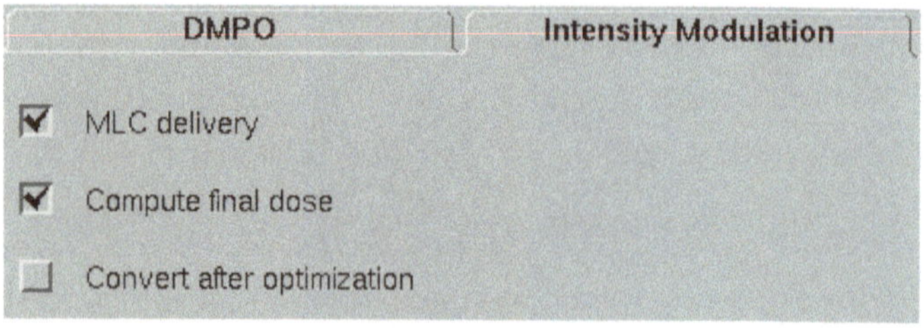

Figure 4.10 Intensity modulation options in Pinnacle 9.8.

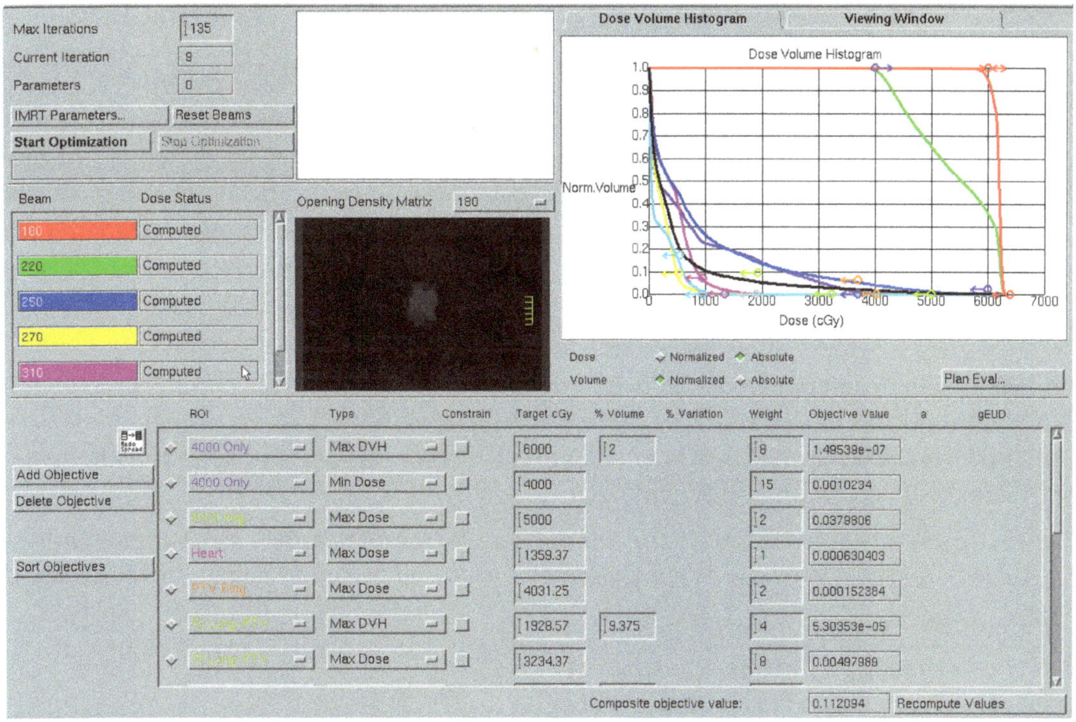

Figure 4.11 IMRT optimizer screen in Pinnacle.

a maximum number of segments equal to 10 times the field number.

- ***Intensity modulation.*** "MLC delivery" is selected if the machine will use MLC for delivery (Figure 4.10). Leave this option unchecked if using a compensator (less common nowadays).

4.4.1.3 Pinnacle Optimization Process

Pinnacle's IMRT optimization process screen is shown in Figure 4.11 and features the following:

- ***ODM*** is a transmission filter used by the software to represent a matrix with the relative fluence of a modulated beam of radiation with respect to the intensity of an open field.
- ***Objectives:***

  Max dose—accomplished when maximum dose = 0% of the target

  Min dose—accomplished when minimum dose = 100% of target

  Max DVH— accomplished when maximum DVH = specified percentage

  Min DVH—accomplished when minimum DVH= specified percentage

  Uniform dose—accomplished when the dose is uniform at specified dose

- ***Constrain.*** By checking this box, the system tries to accomplish the objective regardless of other objectives being met. One must be careful in using constraints since it might make the plan unable to go through optimization since it cannot meet too many conflicting constraints. Figure 4.12 shows a representation of minimum/maximum dose and minimum/maximum DVH point objectives.
- ***Target (cGy).*** The dose value in cGy for the objective for that particular ROI.

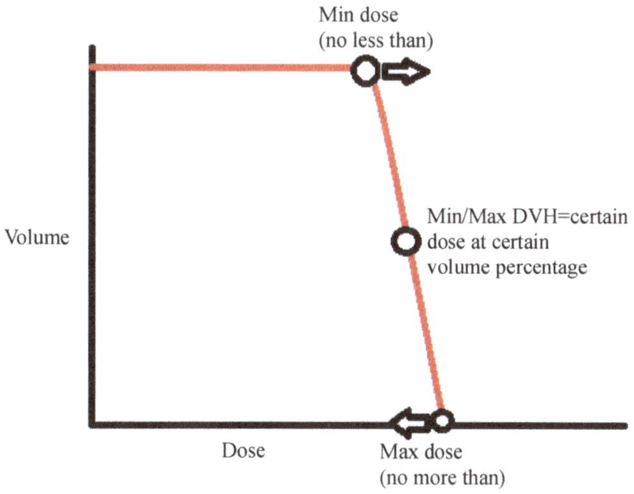

Figure 4.12 IMRT objectives.

- ***% Volume.*** Applicable to certain objectives only, usually "Max and Min DVH."
- ***% Variation.*** Applicable to certain objectives only, usually "Uniformity."
- ***Weight.*** On a scale from 1 to 100, this determines how much priority the planner would like the system to put into achieving a particular objective or constraint. Once set, they are all relative to each other.
- ***Objective value.*** This is the value of each objective function multiplied by its weight. The higher it is (compared with the others), the harder the system is working to accomplish that particular objective.
- ***gEUD (available in some systems).*** This stands for *generalized equivalent uniform dose* value.
- ***a-value (or a-parameter).*** This is available in some systems. This is a parameter used by the EUD model that controls hot or cold spots. Less than 1 is appropriate for targets. The smaller the a-value, the more sensitive to cold spots the computation becomes. The higher the value, the more sensitive to hot spots the calculation becomes, with a = 1 corresponding to the mean dose to the target.

4.4.2 Eclipse Optimization

4.4.2.1 Eclipse IMRT Parameters

Eclipse version 13 uses a separate optimizer for IMRT and VMAT (Figure 4.13) Both optimizers are accessible from the planning menu. The user can hover over the buttons to see a description of their function. Eclipse offers two main optimizers: PRO (Progressive Resolution Optimizer) and PO (Photon Optimizer). The options on the user interface are similar. There are a variety of studies comparing the two optimization algorithms, and users are encouraged to read Varian's documentation on both optimizers. This book uses samples of the PRO optimizer only.

- ***Load templates.*** This option allows the planner to load, save, or manage optimization objective templates. These templates can be created from an ideal optimized plan to be used in future, similar cases.
- ***Exclude structures.*** This allows the planner to remove structures from the optimization window to simplify the display.
- ***Use Normal Tissue Objective (NTO).*** This enables the built-in dose fall-off objective in Eclipse. The distance from the target border defines the distance in cm from the target that the

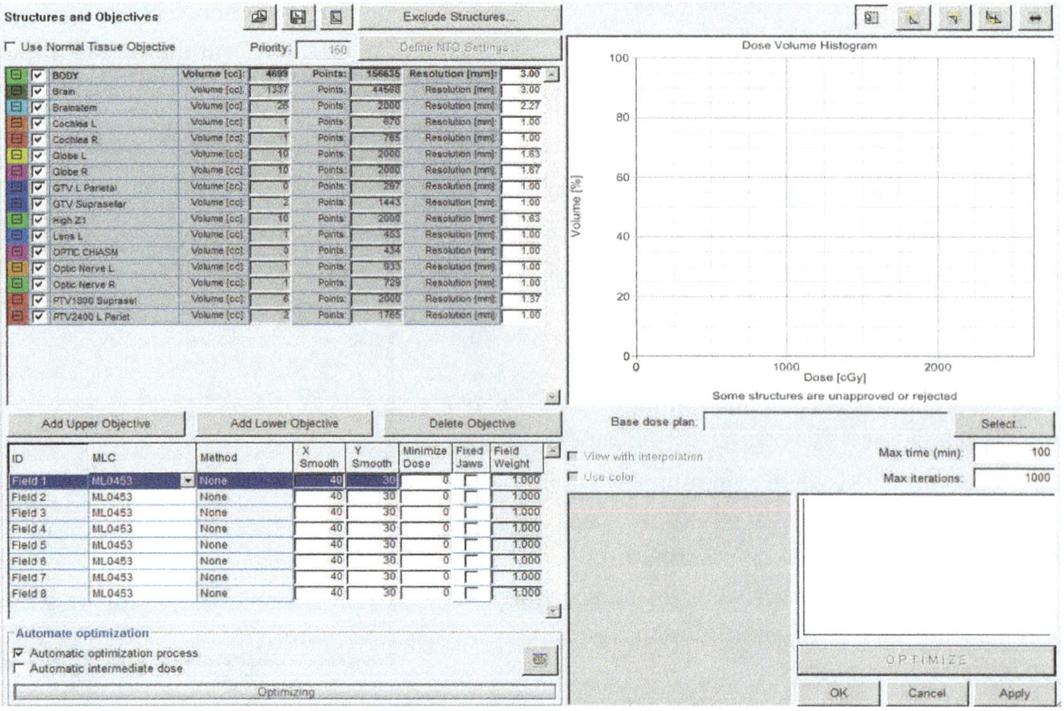

Figure 4.13 IMRT dose volume optimizer in Eclipse.

optimizer will start trying to achieve the objective (Figure 4.14). A common setting is 0.1 cm from the target with a start dose of 100% and an end dose of 60% to 50%, with a fall-off of 0.05 for standard fractionated treatments, or 0.15 to 0.2 for stereotactic cases where a steeper dose gradient is required.

NTO works very well for spherical or cylindrical targets and can frequently eliminate the need for ring optimization structures. In cases of irregular geometry or if NTO does not create an acceptable plan, rings can be created and used as an alternative without NTO enabled. The priority sets how hard the optimizer should work on a particular objective (defaulted to 150).

- ***Objectives.*** Upper or lower objectives can be added from the buttons in the top right of the optimizer window or the large labeled buttons below the structure list.

4.4.2.2 *Eclipse VMAT Parameters*

The VMAT optimizer window (Figure 4.15) is similar to the IMRT window with a few differences. This section will focus on explaining these settings.

- ***Avoidance sectors.*** This feature allows the user to define up to two lengths of arc per field where the

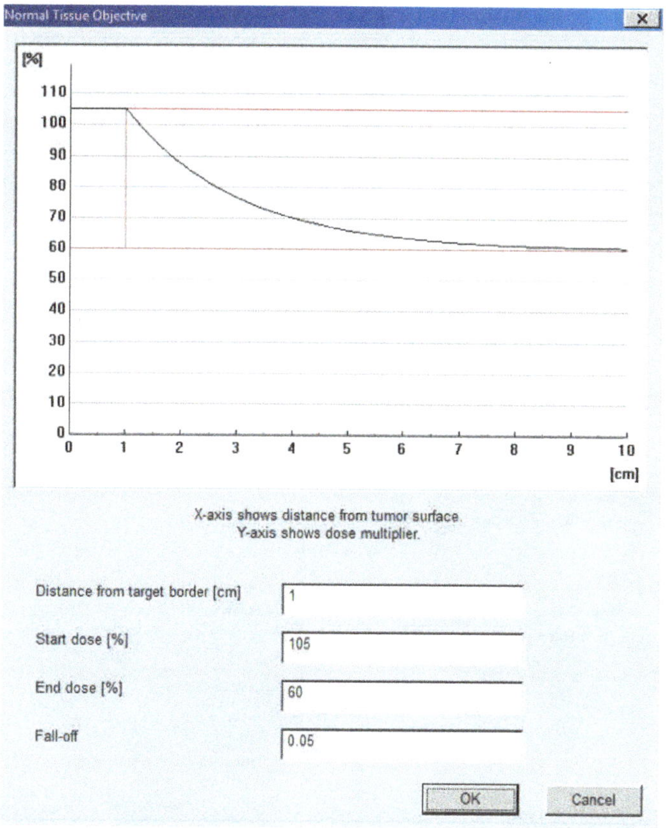

Figure 4.14 NTO settings. In this case, Eclipse will reduce dose starting at 1 cm out from the target and starting at 105% of the lowest maximum dose objective at a fall-off of 0.05.

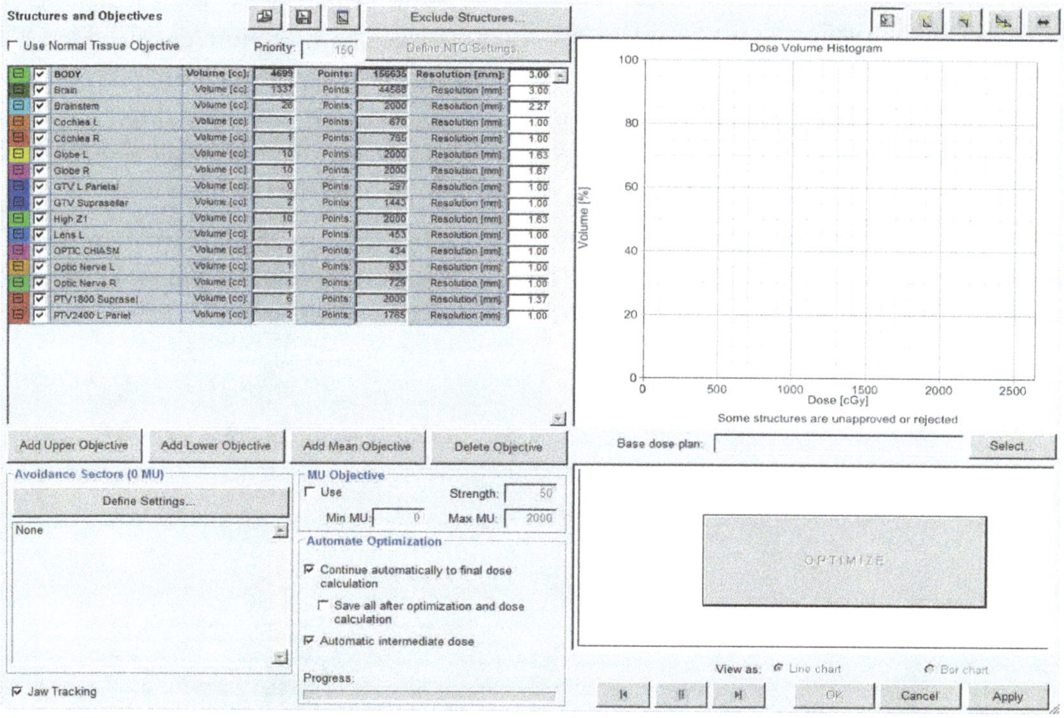

Figure 4.15 VMAT dose volume optimizer in Eclipse.

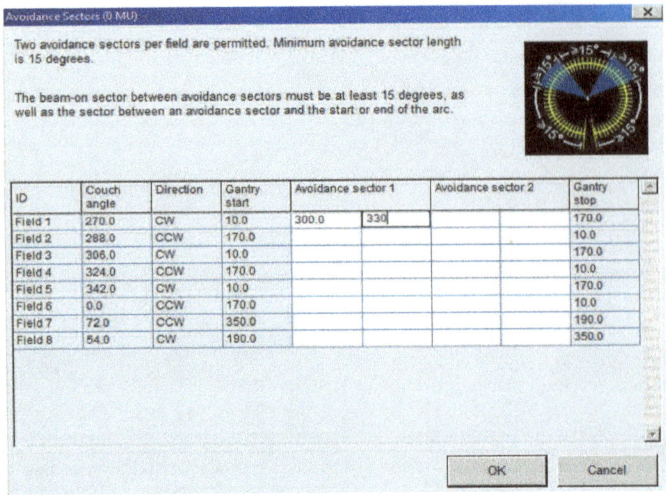

Figure 4.16 Settings for avoidance sectors.

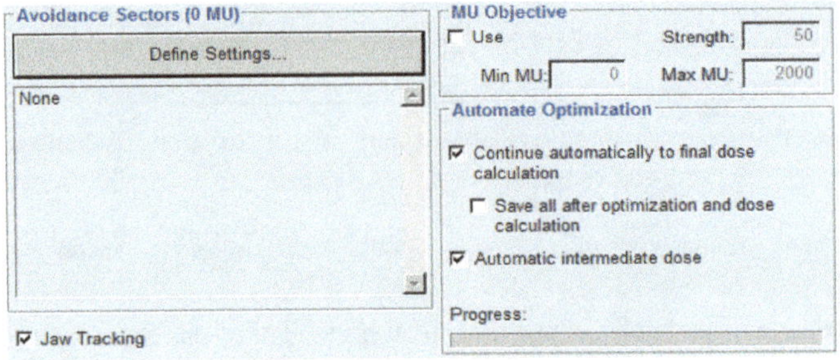

Figure 4.17 Additional VMAT settings.

treatment machine will hold the beam and deliver zero monitor units (Figure 4.16). This can be particularly useful if the planner is trying to avoid dose entering through a certain region, such as the arms or eyes of the patient.

- ***MU objective.*** This setting allows the user to assign a minimum and maximum number of monitor units to limit modulation for the plan. The strength sets how hard the planning system should work to achieve the limit MU.

- ***Automate optimization.*** All check boxes for the automate optimization options should also be checked to allow the optimizer to complete the process and save the plan without need for input from the user (Figure 4.17).

- ***Optimization progress.*** The bottom right-hand corner of the optimizer window can show either a line or bar chart during the optimization for the cost function, depicting how hard the optimizer is working on objectives for a given structure and other plan settings (Figure 4.18). The bottom options allow the user to progress through the optimizer and *machine resolution* (MR) levels or go backwards, if necessary, along with other basic optimizer controls.

4.4.2.3 Eclipse Optimization Process

Upon launching the IMRT optimizer in Eclipse version 13, the user can see the window shown in Figure 4.19 where the following options are offered:

- ***Objectives.*** Upper or lower objectives can be added from the buttons in the top right of the optimizer window or the large labeled buttons below the structure list. Additional options for displaying contours on the DVH can be displayed by right clicking in this window. The volume, number of points, and resolution (determined from the contour volume) are shown to the right of the contour name. Most of the options should be set based on user preference and to allow automation of the optimization process. While the IMRT and VMAT have slightly different options, including avoidance sectors, they share common optimization objectives (Figure 4.20):

  Upper Objective—used to minimize dose to a relative volume; a volume of 0% indicates a maximum dose

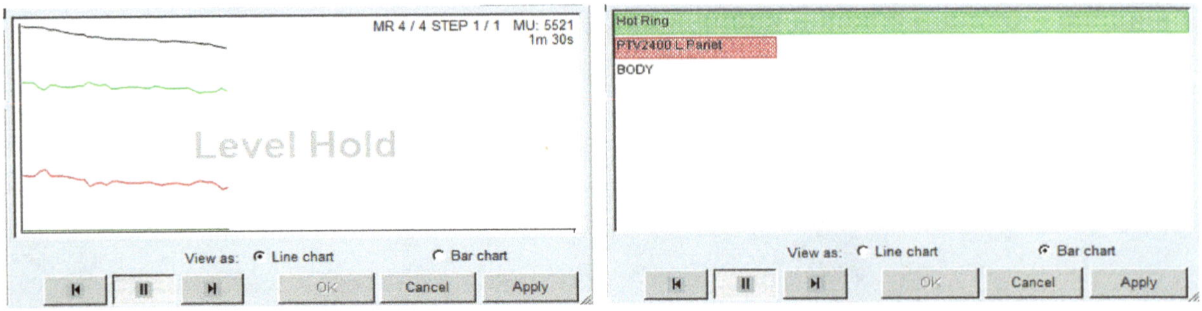

Figure 4.18 Optimization process view options.

Lower Objective—used to deliver a minimum dose to a relative volume; a volume of 100% indicates a minimum dose

Mean Objective—used to keep the mean dose value to that structure below the set value, only available on VMAT optimizations

- ***Priority.*** Assigns an importance to the associated objective. Values can be adjusted by the user from 1–1000, with a value of 1000 indicating the highest importance and 1 being the lowest. The NTO is set to 150 by default, so objectives are often scaled relative to NTO. Objectives with high importance can routinely be set to 250 and those with low importance to 50.
- ***DVH display.*** The sample objectives of Figure 4.20 can be seen in Figure 4.21 with the corresponding DVH display. The directions of the arrows indicate upper or lower objectives, and vertical lines appear when the relative volume is set to 0% or 100%, respectively. Mean objectives are displayed as small diamonds, corresponding to the dose value.
- ***Field parameters.*** The selectable MLC drop-down (Figure 4.22) is predetermined based on the treatment machine. The X and Y smoothing gives a degree of smoothing to the dose in those directions with acceptable default values. Minimize dose is how much the planning system tries to limit the dose, with 0 being the ideal and default value. Fixed jaws can be selected if excessively large fields are causing an issue with the optimization process. Both options for automatic optimization should be checked to enable the planning system to proceed through the entire optimization

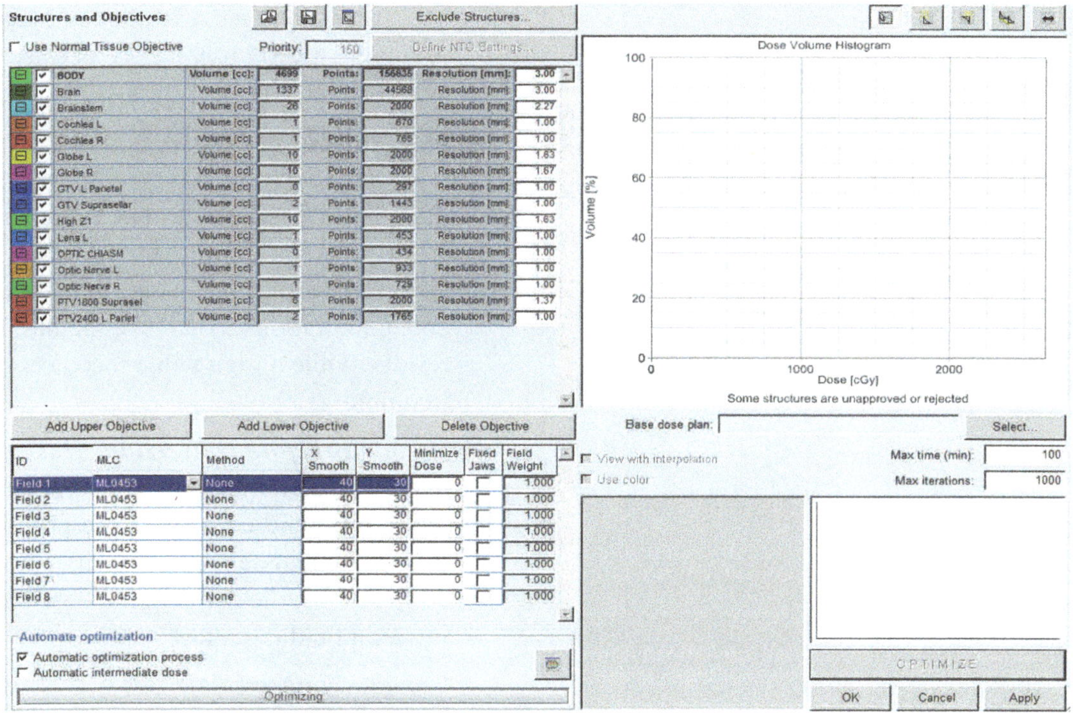

Figure 4.19 Eclipse optimization window.

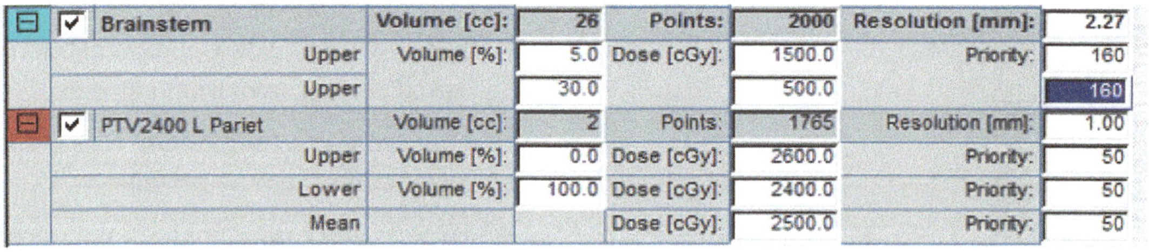

Figure 4.20 Objective sample in Eclipse.

process without the need for user input, and the bottom bar shows the optimization progress.

- ***Optimization process and Fluence Display.*** The maximum time is expressed in minutes and the number of iterations can be set here based on user preference (Figure 4.23). The default values of 100 minutes or 1000 iterations tends to be sufficient as it is an upper maximum, and the optimizer typically does not run for the full length of time. However, if the user wants to complete the optimization within a given time frame, this setting can be reduced. The lower window shows how hard the optimizer is working on the colored structure, and the general options for the optimization process are below.

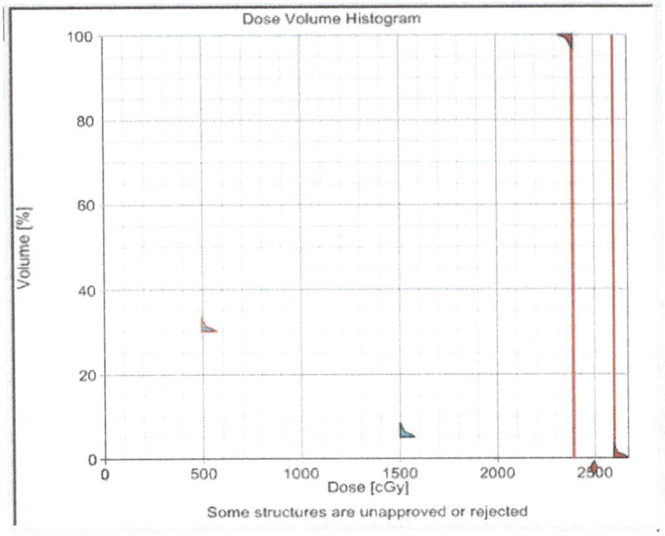

Figure 4.21 DVH display with objective points.

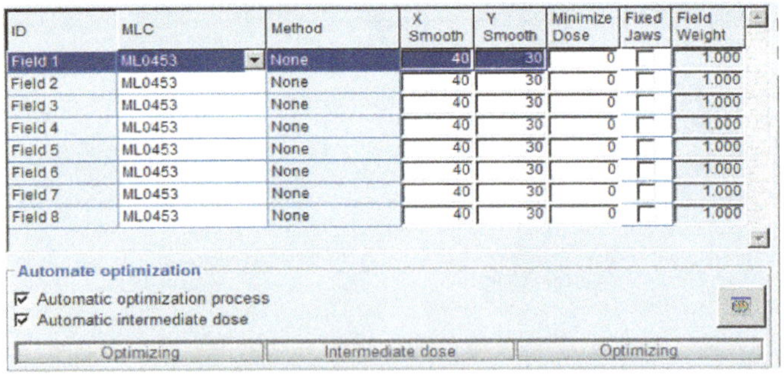

Figure 4.22 Field parameters section.

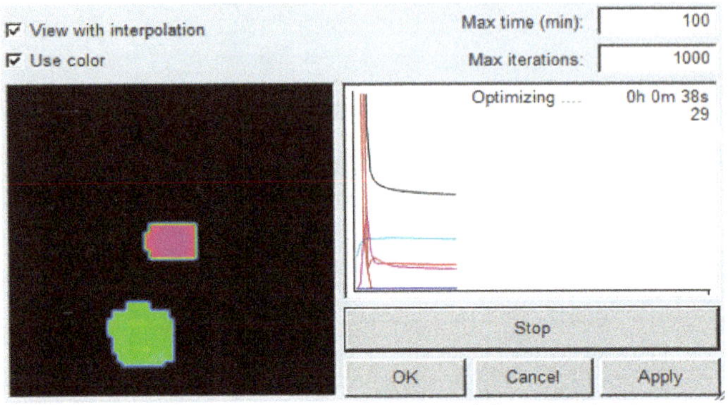

Figure 4.23 Optimization process and fluence display.

4.4.3 RayStation Optimization

4.4.3.1 RayStation IMRT Parameters

The IMRT optimization type can be selected for RayStation version 6 in the Plan Design window under the Photons modality (Figure 4.24). The TPS is able to optimize step-and-shoot (SMLC) and sliding window plans (DMLC), but the delivery depends on machine capabilities. The planner can review the Pinnacle optimization guide, since most parameters are similar.

- ***Optimization tolerance.*** This is the objective value at which the software will stop working on achieving a certain objective because it is considered achieved or it is not possible to achieve.
- ***Max number of iterations.*** This sets up the maximum number of times the software refines the plan. The user is free to choose the number of iterations, but it is important to keep in mind that choosing a low value may not give the system enough iterations to achieve the best optimization results, while a high value increases the computation time.
- ***Iterations before conversion.*** This establishes the number of iterations before computing the convolution dose and converting the plan to MLC segments (the real deliverable plan). For example, for 75 iterations, the convolution dose iteration in the iteration field is set to 15–20.
- ***Compute intermediate dose.*** This box should be checked to allow automatic conversion and for most of the optimization process to proceed automatically.
- ***Compute final dose.*** This box should be checked to allow final dose computation using the collapsed cone convolution algorithm.
- ***Max number of segments.*** This refers to the number of control points per plan.
- ***Min segment area.*** This is the open area of control points or beamlets. It can be used to prevent

the optimizer from creating very small, unpractical control points. The default of 4 cm^2 is often sufficient.

- ***Min segment MU per fraction.*** The minimum number of monitor units a control point can deliver.
- ***Min number of open leaf pairs.*** The minimum number of leaf pairs that must be open for a control point.
- ***Min leaf end separation.*** The minimum distance between two leaf ends.
- ***Max number of control points per beam (DMLC only).*** This is similar to the max number of segments, but a sliding window needs many more to create an acceptable plan. The user can edit this field as necessary, but the default of 100 tends to be sufficient.
- ***Exclude.*** Allows the user to exclude select beams from the optimization process (Figure 4.25).
- ***Segment shapes.*** Allows the system to determine the optimal segment shapes.
- ***Segment MU.*** Allows the system to determine the optimal weight for the segment MU.
- ***Beam MU.*** If control points and segment MU have already been created, the Beam MU box can be checked to optimize beam weights.

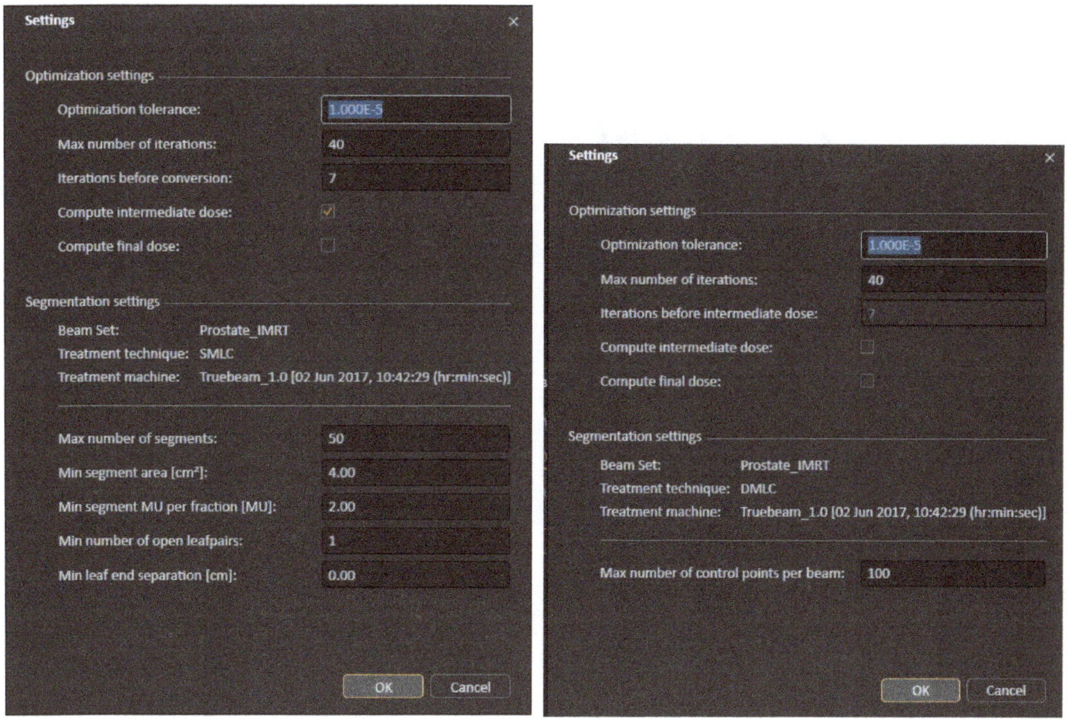

Figure 4.24 Optimization settings for SMLC (left) and DMLC (right).

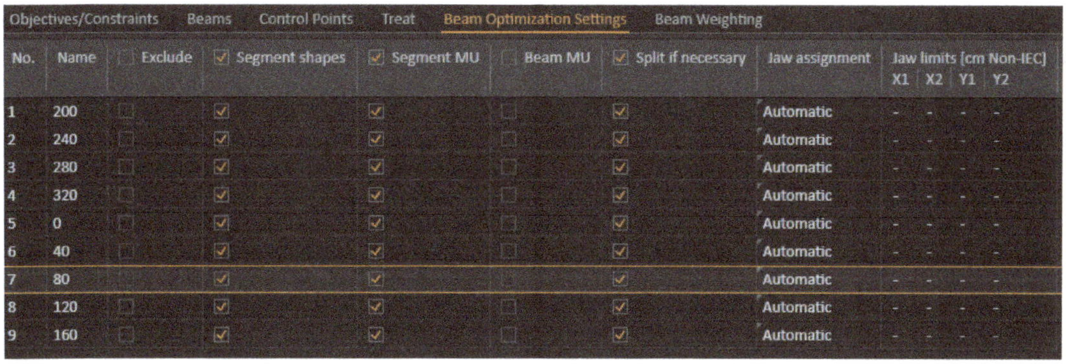

No.	Name	Exclude	Segment shapes	Segment MU	Beam MU	Split if necessary	Jaw assignment	Jaw limits [cm Non-IEC] X1	X2	Y1	Y2
1	200	☐	☑	☑	☐	☑	Automatic	-	-	-	-
2	240	☐	☑	☑	☐	☑	Automatic	-	-	-	-
3	280	☐	☑	☑	☐	☑	Automatic	-	-	-	-
4	320	☐	☑	☑	☐	☑	Automatic	-	-	-	-
5	0	☐	☑	☑	☐	☑	Automatic	-	-	-	-
6	40	☐	☑	☑	☐	☑	Automatic	-	-	-	-
7	80	☐	☑	☑	☐	☑	Automatic	-	-	-	-
8	120	☐	☑	☑	☐	☑	Automatic	-	-	-	-
9	160	☐	☑	☑	☐	☑	Automatic	-	-	-	-

Figure 4.25 Beam optimization settings for SMLC.

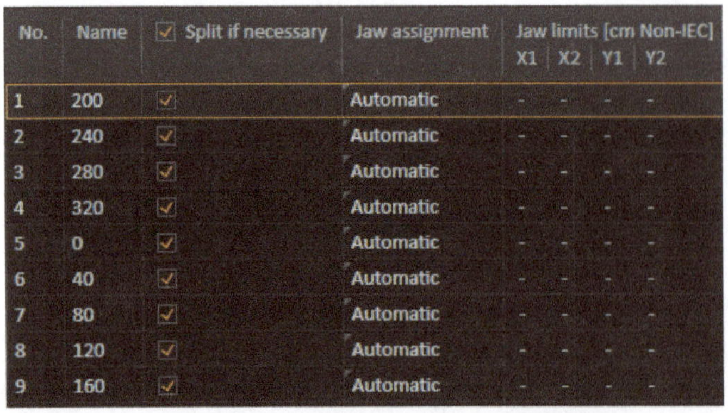

Figure 4.26 Beam optimization settings for DMLC.

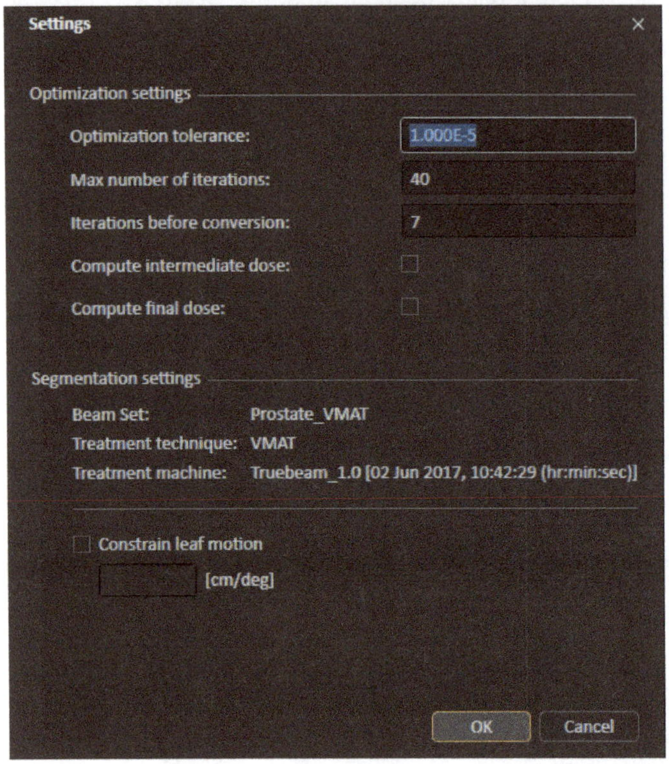

Figure 4.27 VMAT optimization settings.

- ***Split if necessary.*** If the optimizer created control points that are too wide to avoid open leaves in the field, "split if necessary" can be selected to split the control points and jaw assignment to allow sufficient modulation of the entire field width (Figure 4.26).
- ***Jaw assignment.*** This can be set to automatic, which allows the optimizer to determine the best jaw positions to limit leakage from the leaves, or to use limits as max if there is a wide field and the user does not want jaw positions larger than a fixed limit.

4.4.3.2 RayStation VMAT Parameters

The VMAT optimization type can be selected for RayStation version 6 in the Plan Design window under the Photons modality (Figure 4.27).

Most options work identically to those found in the IMRT settings window (refer to Section 4.4.3.1).

- ***Constrain leaf motion.*** This value should be entered based on the physical limitations of the treatment machine. It can also be set by default in the physics module of the planning system.
- ***Create Dual Arcs.*** If this option is checked, the optimizer will automatically create another arc of the same length going the opposite direction. This can be checked to minimize set-up by the planner and allow for more modulation of the field (Figure 4.28).
- ***Gantry spacing.*** This is the spacing in degrees between control points. Entering a gantry spacing of 4 degrees will create a new control point every 4 degrees.
- ***Max deliver time.*** This is the maximum delivery time for a particular arc.
- ***Limit MU and Max MU/fx.*** This is used to limit the total number of monitor units to either decrease modulation or to help decrease treatment time. This option can also be used to create an approximate conformal arc plan using VMAT.

4.4.3.3 RayStation Optimization Process

Options and buttons for controlling or evaluating the optimization are located at the top of the optimizer window screen (Figure 4.29). The user can see the DVH, the clinical goals with physician-set dose objectives, the dose statistics for the visible contours, or a 2D view of the dose distribution. The top right can be set as a 2D view of the dose distribution, BEV (beam's-eye view), or 3D dose cloud or visualization of the structure set. The bottom left is currently selected to show the objectives/constraints, but it can also be set to any of the other options for editing beam parameters. During the optimization process, the progress can be viewed in the bottom right hand corner with the function values versus iteration number.

Optimization objectives (Figure 4.30) can be selected and configured.

- ***Optimization objectives:***

  Min dose—used to deliver a minimum dose, dose to 100% of contour

Figure 4.28 Beam optimization settings for VMAT.

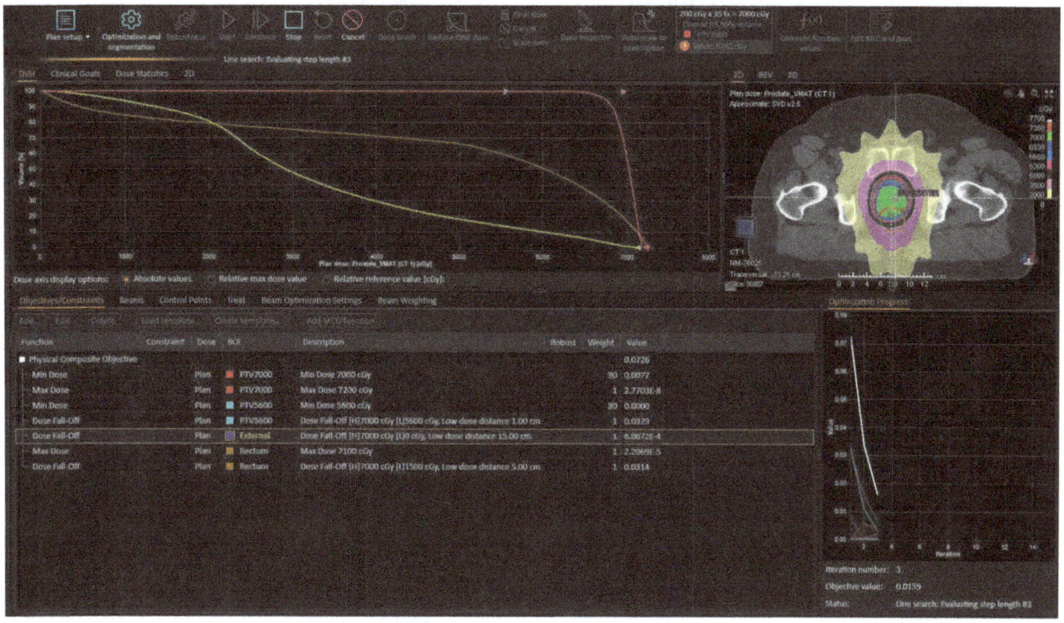

Figure 4.29 Optimizer window in RayStation.

Max dose—used to limit maximum dose, dose to 0% of the contour

Min DVH—used to deliver a minimum dose to a relative volume of a contour

Max DVH—used to minimize a set dose to a relative volume of the contour

Uniform dose—sets a min and max dose for a given contour to deliver a uniform dose

Min EUD—Uses an “a” value of less than or equal to 1. a=1 defines a mean dose, giving cold and hot spots equal weighting; a<1 gives lower doses a higher weight and more importance to cold spots, approaching a min dose as a approaches negative infinity.

Max EUD—Uses an “a” value of greater than or equal to 1. a=1 defines a mean dose, giving cold and hot spots equal weighting; a>1 gives higher dose more importance and approaches a max dose as a approaches infinity.

Target EUD—Pushes to achieve an EUD for the given A value.

Dose Fall-Off—Can be applied to OARs or targets. Allows the max dose level to fall-off linearly at the dose distance from target. Can be selected to adapt to target dose levels for the dose level of each target volume.

Uniformity Constraint—Can be applied to OARs or targets. Sets the variation of dose within the ROI.

***a-parameter* (for min and max EUD)**—This is a biological parameter used by the EUD model that controls hot or cold spots. Less than 1 is appropriate for targets. The smaller the “a” value, the more sensitive to cold spots the computation becomes, approaching a minimum dose. An “a” value equal to or more

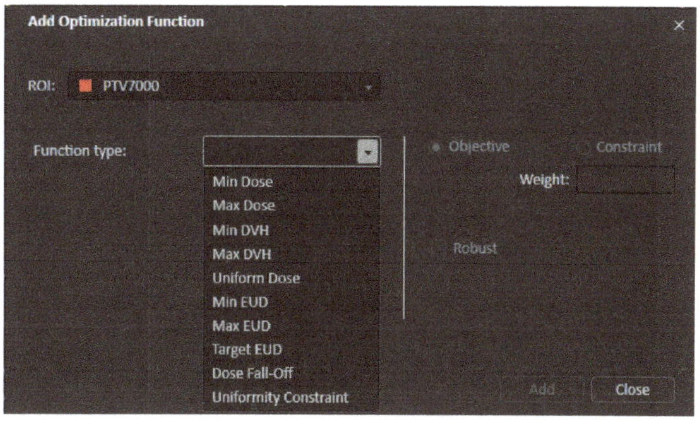

Figure 4.30 RayStation's Add Optimization window.

than 1 is more applicable to critical structures. The higher the value, the more sensitive to hot spots the calculation becomes, approaching a maximum dose. a=1 corresponds to the mean dose to the target.

- ***Robustness.*** This can be set to allow for variability in setup or tumor motion. Robustness allows the user to set a direction and magnitude for an ROI's positional uncertainty and optimize the plan to account for it.

RayStation also includes functions such as Reduce OAR Dose and the Dose Brush to further edit isodose volumes or automatically reduce dose to OAR structures. These advanced functions and Multi-Criteria Optimization can be explored using the user manual and trial and error (if those licenses are purchased) to allow the user to gain experience.

4.4.4 Basic Guidelines for Optimization

The following is an overview of how to start the optimization process. The user can opt for an initial optimization for target coverage purposes only and then include other structures during subsequent optimizations. The weight indicates how much the system will work on a particular objective, which is indicated on the objective value. The higher the value, the harder the work to obtain what the parameter asks for.

After the first optimization, the planner can increase the number of parameters and change existing parameters having an idea of what the DVH looks like. This process might take several repetitions until the desired isodose distribution is achieved.

4.4.4.1 The First Optimization

These are general guidelines on the objectives to introduce into the planning system for the first optimization. After the first dose distribution, these objectives can be modified, and more can be added to limit the doses in each area (contour).

Generally, check all beam parameters that can affect optimization or the final result:

- No collimators are at 0 degrees in VMAT arcs.
- No IMRT fields are directly opposing each other.
- In cases of step-and-shoot IMRT cases, make sure that there is an acceptable number of control points set up for the optimization to take place.

Plans that require more modulation (irregular or unusual target shapes, targets close to critical organs, or targets with several dose levels) can be set up to have more control points than what is set up by the institution. As an example, one could set up 50 control points per plan. This number can be increased to higher values, such as 75, 80, 90, 100, 110, 120, etc. Too many control points can result in very long treatment times with little or no dosimetric benefit.

General checks before optimizing could include:

- The correct beam isocenter location is selected.
- The correct machine is selected for optimization in case the center has more than one machine.
- The appropriate beam energy and beam parameters are entered, such as gantry and collimator settings.
- The appropriate jaw is specified.
- The dose grid is drawn correctly with the right resolution for the case (usually matching CT slice thickness).
- The correct prescription is entered on the system.
- The target, organs at risk, and optimization contours are drawn correctly.

Sample objectives for the first optimization are:

- Targets = minimum dose set to prescription, min DVH set to prescription covering 95% of target, and maximum dose set to 110% as a starting point.
- Use of NTO or rings (described below).
- Ring (separating dose levels) = maximum dose of a value halfway between the two dose levels it separates.
- PTV ring (the ring immediately outside PTV) = maximum dose value equivalent of slightly less than the PTV prescription.
- Body ring (a ring made of the body or external contour and its contraction by 2 to 3 cm) = maximum dose of 50% of high dose region prescription.
- Organs at risk = work according to dose tolerances for the case.
- The weights could be left at 1 for the first optimization so the system can work with the inherent objective values. The user can also opt for changing the weights as well. One could place more weight for coverage on min dose or min DVH objectives for targets at first optimization.

After the first optimization, make sure the dose is normalized correctly. Additional and current objectives may be added or modified.

4.4.4.2 Objectives to be Added After the First Optimization

Look at the DVH and dose distribution on the CT dataset, and ask the following questions:

- Are all OARs being taken into consideration?
- Does a dose in the target need to be uniform?
- Is the target meeting the right hot spot?
- Is the hot spot inside the GTV contour?
- Is significant dose leaking from the PTV into healthy tissue? Add more weight into the max dose objective of the PTV ring as well as decrease the dose or modify the NTO objective.
- Are there streaks going on the outside of the body? If so, check the NTO or the body ring and add additional max DVH objectives.

4.4.5 Introductory Case: A Lung Lesion IMRT

Now let's put all of our optimization tools and understandings to work on a sample case. The lesion to treat is on the right lung (Figure 4.31). There are two PTVs:

1. PTV 6000 (inner red PTV) and
2. PTV 4000 (green outer PTV).

The PTV 6000 is prescribed 6000 cGy in 5 fractions, while the PTV 4000 is prescribed 4000 cGy in 5 fractions.

The PTV 4000 contour contains the PTV 6000 contour, so they will need to be subtracted from one another for optimization. For this case, organs at risk include the lungs, the spinal cord, and the heart.

Since the lesion is located at one side of the thoracic cavity only, the placement of the beams is such that minimum irradiation is delivered to healthy tissue. A typical beam placement of 360 degrees around the patient is not recommended in this case since the lesion is not in a central location. This even beam placement would not yield a good dose distribution, and it might also lead to gantry table collisions during treatment given the position of the isocenter and machine limitations.

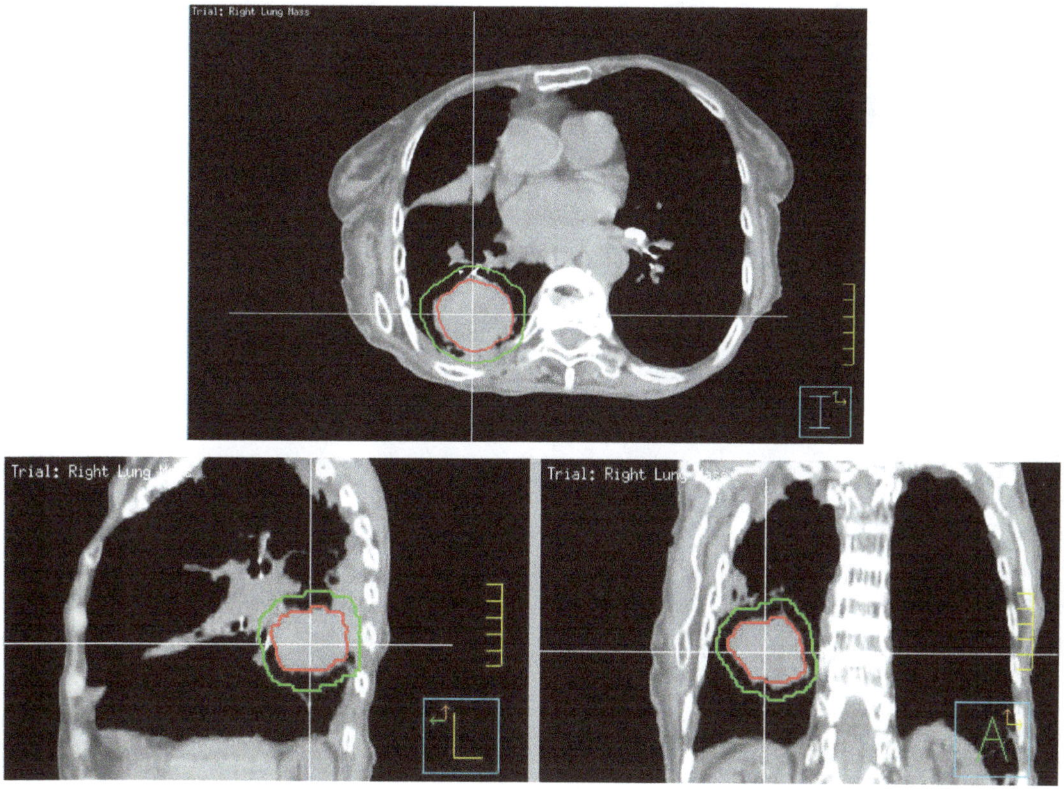

Figure 4.31 Contours of PTV 6000 (red) and PTV 4000 (green).

For this lesion, a nonstandard six-beam placement has been chosen (Figure 4.32). The beams have been placed in a way such that the lesion can receive dose from almost all possible sides, while trying to avoid the organs at risk (spinal cord, heart, and lungs). A sufficient number of control points per plan is also an important parameter to assure a good dose distribution. In this case, the final plan contains 90 control points.

The "dose painting" process starts with a series of contours to control the dose distribution.

1. ***Separating dose levels.*** In this case, the contour PTV 4000 contains PTV 6000 inside. If one optimizes PTV 4000 to have a dose distribution of 4000 cGy, the planning system will not be able to achieve this objective since PTV 6000 will try to achieve 6000 cGy within the same PTV 4000 contour. In this case, the purple contour is "4000 cGy only" (Figure 4.33), which excludes the PTV 6000 contour. The PTV 6000 contour is left alone since it does not have any conflicts with other dose levels. The ring can also have a 2-mm margin around the high-region PTV to help with dose fall-off, or the user can opt for a thin ring between the two dose levels.

2. ***Rings.*** A ring around the target can separate the lesion to be treated from healthy tissue. It provides the sufficient dose fall-off between these structures. These rings can have a thickness of 1 to 2 cm. In this case, the PTV ring (orange in Figure 4.34) has a thickness of 1 cm. An additional 2-mm margin from the target can help in dose fall-off from the target to the ring without significant optimizer penalties.

Inner PTV rings are necessary when dealing with several dose levels in a target. In this case, a ring for the PTV 6000 with a thickness of 0.25 cm (the yellow contour in Figure 4.35) was made to help the dose change from 6000 cGy to 4000 cGy. To facilitate optimization, this structure should not overlap other structures. In this

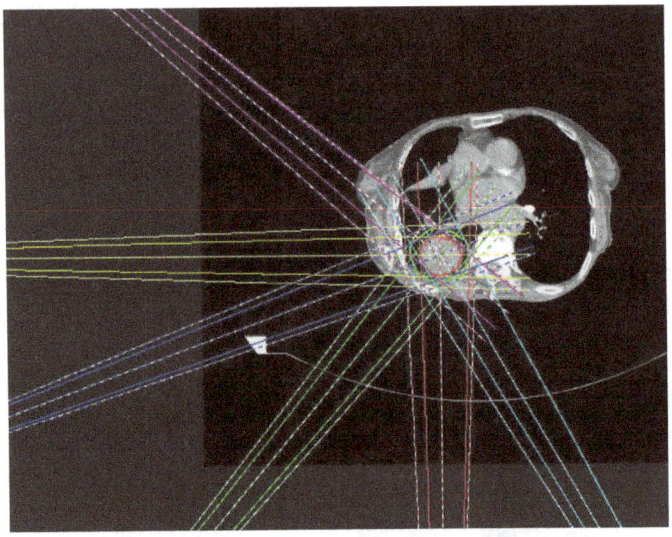

Figure 4.32 IMRT beam arrangement.

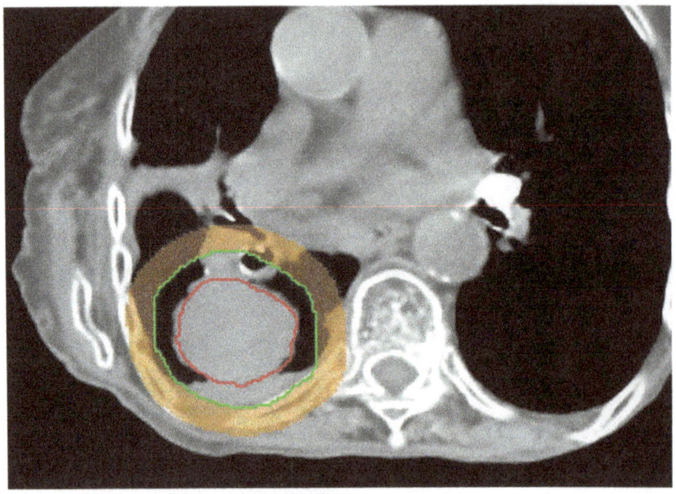

Figure 4.34 PTV ring (orange colorwash).

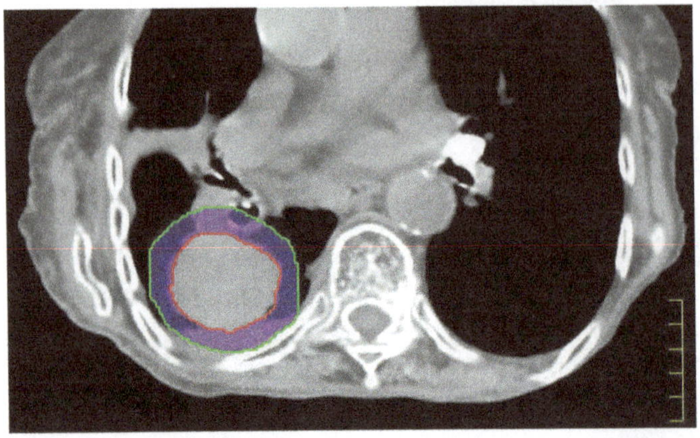

Figure 4.33 Contours of PTV that contain 4000 cGy (purple colorwash).

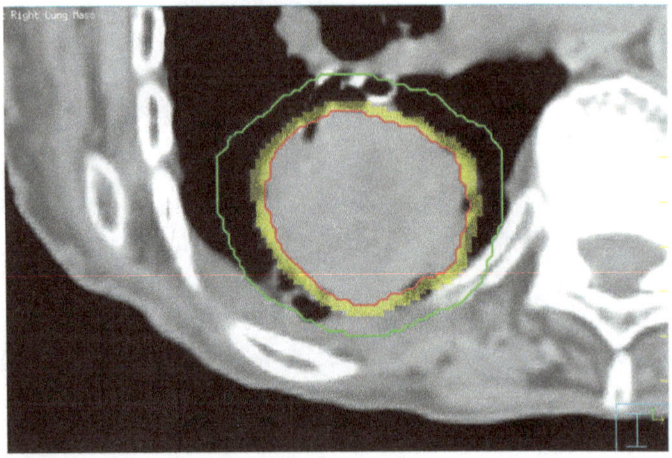

Figure 4.35 Inner PTV ring (yellow colorwash).

case, the PTV 4000 optimization structure could have a margin away from the PTV 6000 structure.

A ring for fall-off dose must be used during optimization to avoid low-dose streaks at the outer borders of the body in case the optimizer does not have an NTO option. This type of structure can be 2 to 3 cm thick and should only cover a few centimeters above and below the target (Figure 4.36). This structure is made from the external or body contour.

3. ***Contours made while optimizing.*** During the optimization process, if modifying and adding constraints to the organs at risk and the contours previously reviewed do not deliver an appropriate dose distribution, one of the possible solutions is to contour the problem

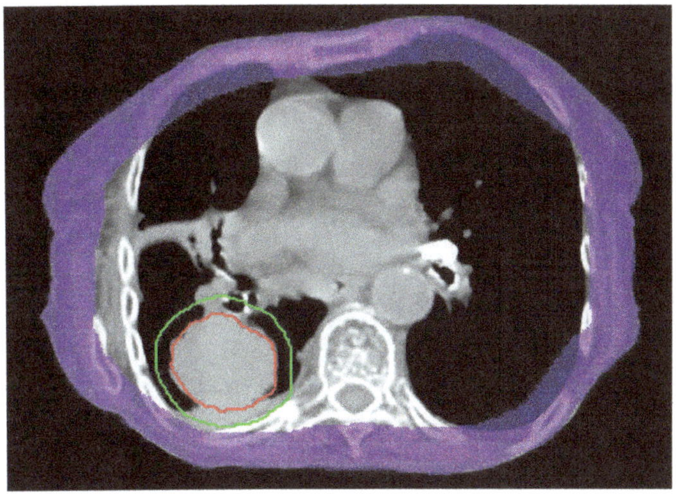

Figure 4.36 Body ring (purple colorwash).

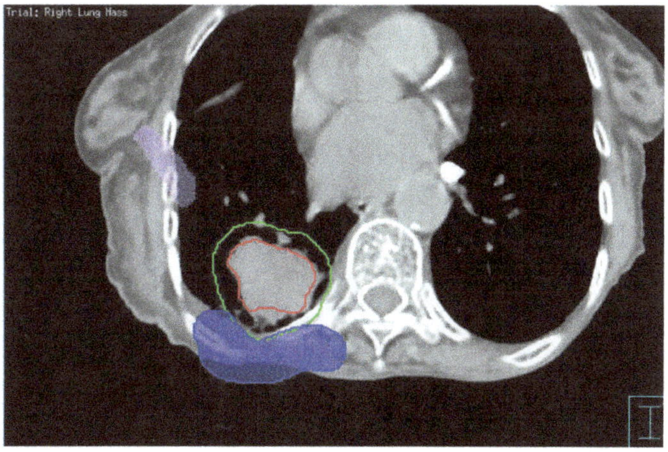

Figure 4.37 Additional optimization contours.

ROI	Type	Constrain	Target cGy	% Volume	% Variation	Weight	Objective Value	a	gEUD
4000 Only	Max DVH		6000	2		8	1.49539e-07		
4000 Only	Min Dose		4000			15	0.0010234		
6000 ring	Max Dose		5000			2	0.0379806		
Chest wall	Max Dose		3700			25	0.000787747		
Heart	Max DVH		953.125	7.0312		5	0		
Heart	Max Dose		1359.37			1	0.000630403		
Lt Lung	Max DVH		531.25	8.5938		2	0		
PTV Ring	Max DVH		3703.12	5.8594		10	0.00109122		
PTV Ring	Max Dose		4031.25			2	0.000152384		

ROI	Type	Constrain	Target cGy	% Volume	% Variation	Weight	Objective Value	a	gEUD
PTV Ring	Max Dose		4031.25			2	0.000152384		
Rt Lung-PTV	Max DVH		1928.57	9.375		4	5.30353e-05		
Rt Lung-PTV	Max Dose		3234.37			8	0.00497989		
SC	Max DVH		546.875	17.187!		1	0		
SC	Max Dose		953.125			1	0.000176379		
	Max Dose		1900			7	0.000521762		
lung PTV6000	Max Dose		6400			3	0		
lung PTV6000	Min Dose		6000			15	0.000557886		
lung PTV6000	Uniform Dose		6000			55	0.0599015		
spill	Max Dose		3700			15	0.00423808		

Figure 4.38 Optimization parameters for this sample

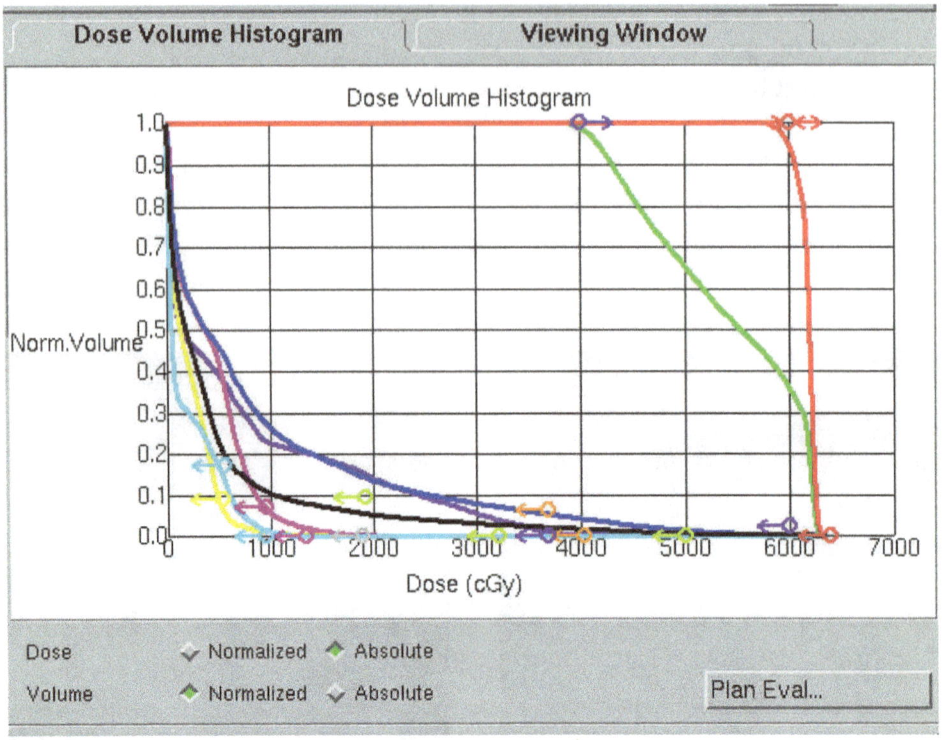

Figure 4.39 DVH in optimization window.

areas and assign them to the optimizer for dose modulation.

In this example, the light purple contour (Figure 4.37) represents a low-dose spill, while the blue contour attempts to fix a dose spill (4000 cGy out of PTV 4000).

4. ***Optimization.*** For the FIRST optimization, one might choose the following parameters:

PTV 6000 = minimum dose 6000 cGy for good target coverage

PTV 6000 = minimum DVH of 6000 cGy to 95% of the target with greater weight than the minimum dose objective

PTV 6000 = maximum dose to 108% or 110% to reduce hot spots within the target

PTV 4000 only = minimum dose 4000 cGy for that contour coverage

PTV 4000 only = minimum DVH of 4000 cGy to 95% of the target with greater weight than the minimum dose objective

PTV 4000 only = maximum dose of 6000 cGy to avoid dose from the high dose region

PTV 6000 RING = maximum dose of 5000 cGy (usually, a value half way between the two dose levels)

PTV 6000 RING = Instead of maximum dose, one can limit the dose to the desired value to 5% of the volume

PTV RING = maximum dose of 3900 cGy or 4000 cGy to start

Body RING = maximum dose of 50% of high dose region prescription

Organs at risk = maximum dose 50% of high-dose region prescription or work according to dose tolerances used

For example, the optimization process in Pinnacle 9.8 can be found under the IMRT tab. Figure 4.38 shows a sample list of all final objectives.

The "weight" is a way to prioritize certain objectives. The higher the value, the more the system will work on achieving the objective. "Objective Value" shows how much the system will actually work on the objective to meet its criteria. Higher values represent more work toward the objective, while a value close to 0 represents accomplishment or simply an unrealistic objective the system cannot meet (not a high priority on weight, too).

Each constraint can be visualized on the DVH (Figure 4.39) as an arrow that indicates where dose will be directed at that point.

5. Quality Assurance of Intensity-Modulated Plans

Intensity-modulated plans contain several subfields with different jaw and MLC positions. Due to their complexity, it is necessary to verify their safe delivery. This task falls on the medical physics team, and it is not only part of system commissioning for intensity-modulated treatments, but a continuous process of quality assurance.

Current IMRT quality assurance programs include pretreatment measurements, secondary MU checks, correct commissioning of IMRT, third-party phantom irradiation verification (such as the Imaging and Radiation Oncology Core (IROC) at M.D. Anderson in the United States), MLC log review, and 3D DVH calculation based on log files or exit dosimetry, among other methods.

Patient-specific quality assurance can take place before the start of every patient treatment course. Routine machine quality assurance, including MLC tests, can be performed at recommended intervals. Strong clinical work flows and staff communication can mark a difference in the quality of treatments in clinic. Continuous quality assurance also involves weekly checks of treatment records, trending of MLC delivery records, and all processes that ensure the treatment is being delivered correctly every day.

Pretreatment IMRT quality assurance is important, but it only captures the plan once before treatment. IMRT QA measurements can detect commissioning errors in MLC modeling, small field modeling, as well as work flow errors, such as errors in exporting data, wrong datasets, etc.

However, physicists should not rely on pretreatment IMRT QA as their only quality check. Other ways to successfully prevent errors come in the form of work flow reviews, and machine or dose monitoring during treatment delivery. Work flow reviews can include physics chart reviews, physics weekly chart checks, therapists chart review, check lists, and pretreatment check lists. Machine or dose monitoring during treatment delivery includes machine MLC log review, 3D DVH reconstruction based on MLC logs or exit dosimetry, transit dosimetry, and, more recently, the use of artificial intelligence.

Treatment plans should go through at least two types of verification prior to delivery:

1) a secondary Monitor Unit (MU) verification and

2) the delivery verification.

The medical physicist is encouraged to consult various AAPM reports such as

- AAPM Task Group 71: "Monitor Unit Calculations for External Photon and Electron Beams,"
- AAPM Task Group Report 219: "Independent Calculation-based Dose/MU Verification for IMRT,"
- Report No. 114: Verification of Monitor Unit Calculations for Non-IMRT Clinical Radiotherapy," and
- other reports mentioned in other sections of this book.

Each physicist should determine appropriate acceptance criteria per treatment site, objectives, and the institutional policies, based on consensus criteria.

5.1 Secondary Monitor Unit Verification

Monitor units for each field of the treatment plan can be verified independently of the TPS used for the plan. The need for this verification was born when calculations were performed by hand using lookup tables and having a second person repeat the calculation. A modern TPS uses sophisticated algorithms to calculate dose in 3D geometrical patient models with heterogeneous tissue densities. The goal of secondary verification is to ensure that the monitor unit calculation of the TPS has adequate parameters for the treatment. This takes place at a larger scale during commissioning and validation of the TPS, but it must also take place for each patient plan.

The monitor unit verification process is straightforward for 3D conformal plans, but intensity-modulated plans require a more in-depth calculation because of their modulation of beam intensity. In 3D conformal therapy, the central axis dose and off-axis dose can be easily calculated. IMRT/VMAT fields are modulated and include regions of high dose gradients and heterogeneities. The calculation at central axis is useful, however verification at other points is necessary for inverse-planning validation. The medical physicist can use a variety of software and tools to perform monitor unit verification, including commercial software or software tools developed in-house.

Most of the secondary MU calculation software makes an automated hand calculation based on parameters from the exported plan file from the TPS, meaning that the dose calculated is dose to water as obtained in the commissioning data. There might be larger-than-usual discrepancies for small fields, regions of high gradient, fields with few MUs, split fields, and low-density areas. Moving to a 3D algorithm approach could solve these uncertainties.

In recent years, with more accurate and Monte Carlo-based TPS calculation algorithms, there has been a need for these secondary MU checks to become a true verification on the calculation algorithm (Figure 5.1). AAPM Task Group Report 219 sets the tone to a transition from automated hand calculations to actual recomputation of dose with heterogeneities, including software packages with Monte Carlo modules that can second check the TPS better than standard hand calculations. Another solution would be to have a secondary TPS recalculate and compare the plan to the main TPS.

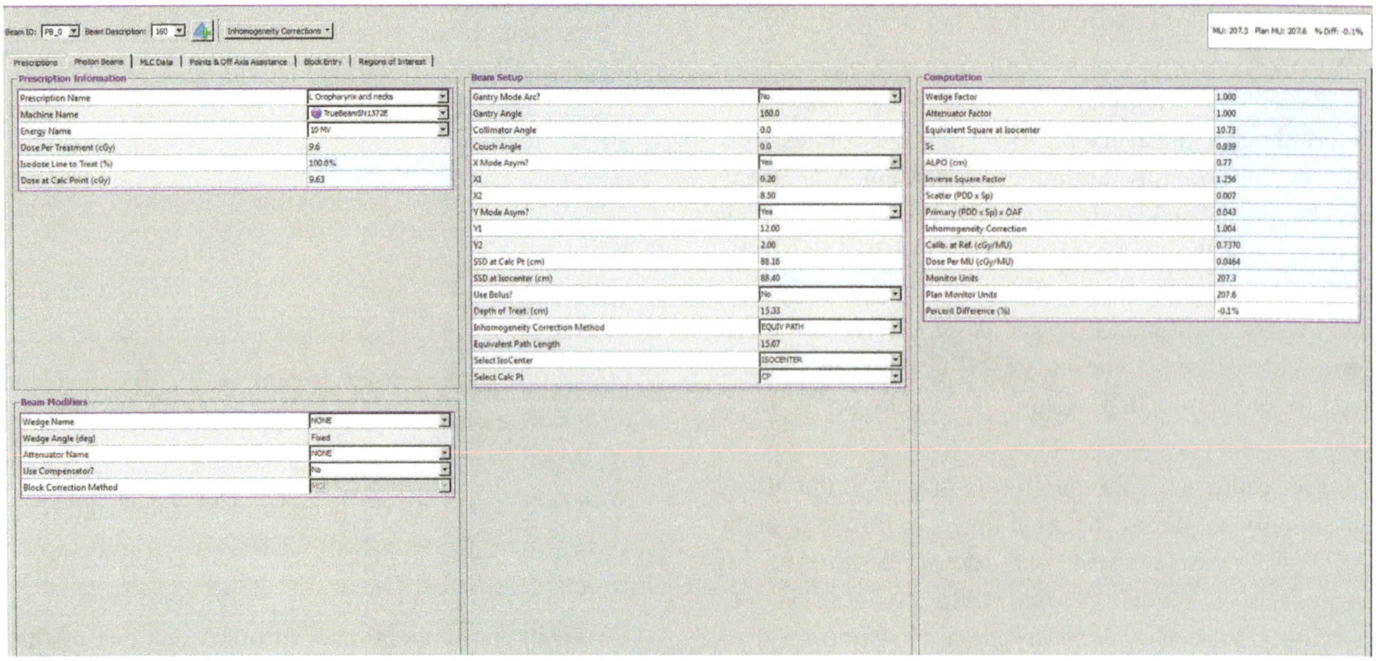

Figure 5.1 Sample of a secondary MU check software (Radcalc).

5.2 Delivery Verification (Pretreatment Measurements)

There are several types of quality assurance systems that can be implemented for treatment delivery verification for pretreatment quality assurance and checks during the course of treatment. These systems focus on testing the dose and the MLC/jaw positions on which the fluence map is dependent. Some systems can offer dose checks only and some others will allow the user to check both parameters. It is important to remember that dose is dependent on the beam parameters, but small deviations in MLC/jaw positions can be detected separately from dose measurement results.

Tolerances for the quality assurance program should be created based on the limitations of the system used, as well as consensus data from task group reports. Acceptance thresholds for quality assurance measurements should take those factors into consideration. The literature on the system in place will also inform your acceptance thresholds.

This chapter will quickly summarize several patient-specific QA methods. Before implementing any quality assurance program, it is recommended that the user analyze each method in terms of efficiency, accuracy, precision, availability, and cost-effectiveness.

5.2.1 Point Dose Measurements

For this method, the physicist compares a measurement taken with an ion chamber in a known medium at a known depth with the dose computed by the TPS (Figure 5.2). The medium of measurement or phantom could be water, acrylic, or plastic water. The term "point-dose" does not reflect the nature of this measurement, which is truly a small-volume measurement.

The ion chamber type and size should be appropriate to what is being measured and should be placed in a dose region that does not have steep gradients. A cylindrical geometry minimizes the sensitivity response variation as a function of beam entry angle, and a small volume delivers measurements with high spatial resolution. Another alternative to using ion chambers for this measurement is the use of solid-state dosimeters or plastic scintillators.

The treatment fields are recalculated on a phantom dataset CT of the phantom in the TPS, and the measured point dose is compared to the calculated point dose. This method is relatively easy to implement, but it does not accurately test parameters such as MLC/jaw positions.

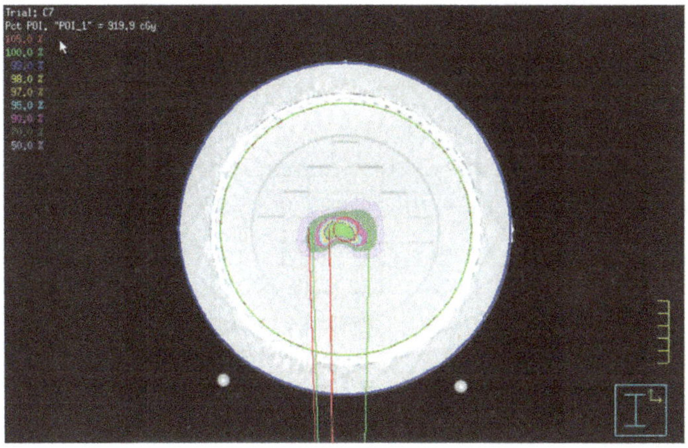

Figure 5.2 A CT scan of a circular phantom with calculated dose.

For further verification, film measurements, portal imaging, or detector array measurements can accompany this method.

5.2.2 Radiochromic Film

Radiochromic film can be placed at a known depth in slabs of plastic water or acrylic. After field delivery, the films can be scanned and compared with planar dose files from the TPS with the help of QA software. Film requires a linearity calibration before QA of the plan and high-quality processing for the film (including using a high-quality scanner). Films can be inserted into flat or cylindrical phantoms and can be accompanied by other testing methods. The medical physicist is encouraged to read AAPM Task Group Report 69: "Radiographic Film for Megavoltage Beam Dosimetry."

There are various considerations for the use of silver halide film. Silver halide film sensitivity can be affected by variations in film batch, processor, variations in photon beam spectrum, as well as densitometry artifacts. On the other hand, radiochromic film does not require a processor, but dose sensitivity can be low.

Gafchromic EBT, a high-sensitivity radiochromic film, has a good response to radiation, but it requires a high-quality scanner for processing. Radiochromic film is further discussed in AAPM TG-55: "Radiochromic Film Dosimetry" and AAPM TG-235: "Radiochromic Film Dosimetry," which is an update to TG-55.

5.2.3 Portal Imaging

The portal imager across from the gantry head can also be used to acquire a dosimetric integrated image of the fields. The resulting images can then be compared to planar dose files from the TPS using appropriate software. Portal imagers offer better spatial resolution than detector arrays. The response of the *electronic portal imaging device* (EPID) is converted to dose and compared to the predicted dose image. In terms of absolute dose measurements, the portal imager needs to be configured, calibrated, and cross-checked with another QA measurement device. The characterization of this calibration must be done in a range of clinical conditions, including correct energies, dose rate, field size, and leaf speed. Quality control of the dose detected must be done to keep the integrity of the system.

5.2.4 Detector Arrays

Detector arrays can be made of ion chambers or diodes. They need to be calibrated for absolute dose measurements according to the manufacturer's instructions. These arrays usually come with their own software and accept planar dose files from the TPS for comparison purposes. Figure 5.3 shows an example of TPS calculation of dose fluence of an IMRT plan. This dose plane can be used for comparison with measured data from detector arrays.

Detector arrays are highly efficient in terms of optimization of measurements and clinical work flow. After cross calibration, the medical physicist can obtain the dose fluence per beam and total plan. Some detector arrays are sensitive enough to provide data on absolute dose and other factors, such as field size and MLC positioning errors.

The spatial resolution of these devices must be taken into account when analyzing acquired data. There are commercial detector arrays designed for small-field dosimetry in stereotactic plans.

5.3 Measurement Acquisition Methods

The AAPM Task Group Report 218 has identified common methods of measuring data for pretreatment IMRT QA. These methods are: true composite, perpendicular field-by-field, and perpendicular composite. This section offers an overview of the definitions and variations of each method, as well as the advantages and disadvantages for clinical implementation. Medical physicists might use one or a combination of these acquisition techniques.

- ***True Composite.*** Here the plan is being delivered to the device with actual treatment parameters (gantry, collimator). The continuous data acquisi-

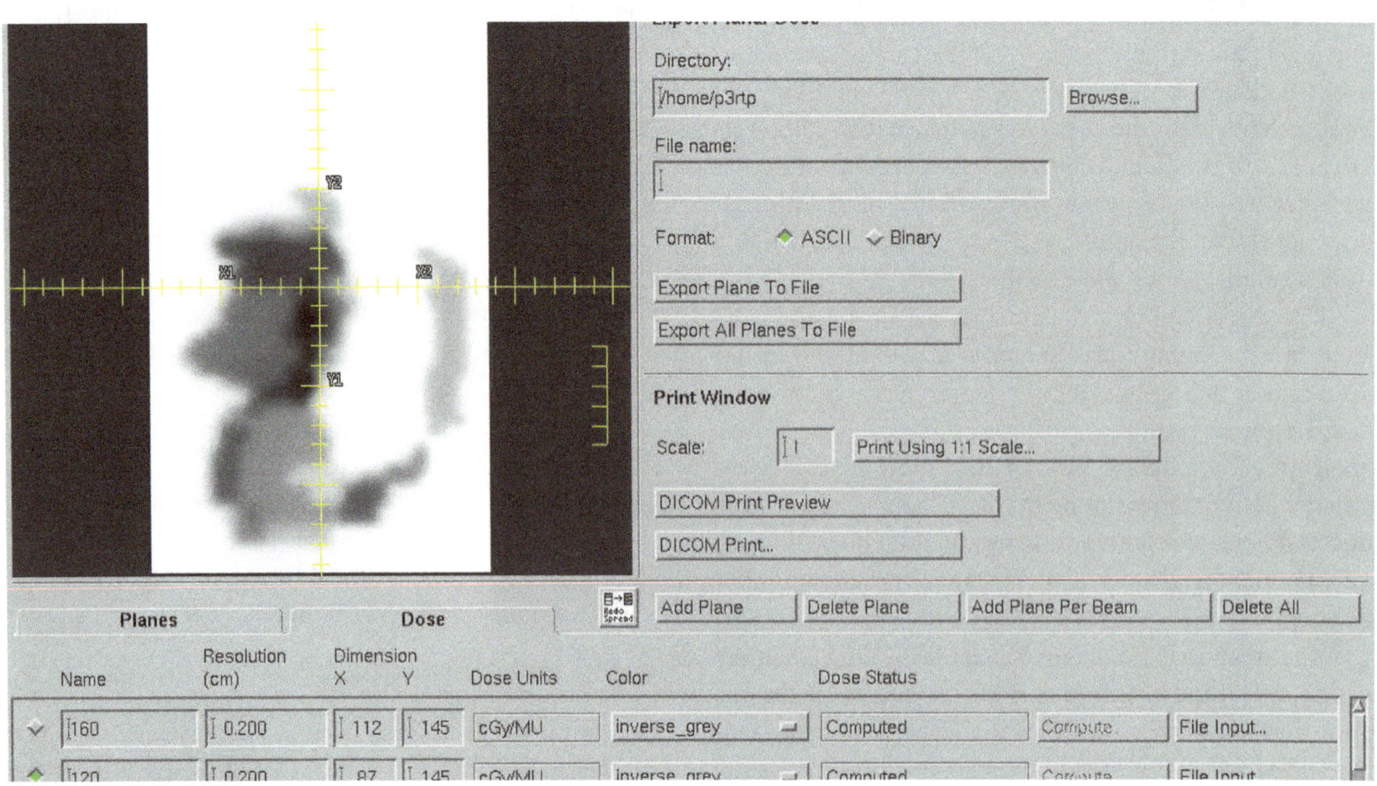

Figure 5.3 Planar dose file generation in Pinnacle.

tion starts before the first beam is delivered and ends after the last beam is delivered. While this method allows the physicist to have a composite dose of all beams being delivered with actual treatment parameters, it is impossible to isolate the cause of errors or evaluate beam doses individually.

- ***Perpendicular Field-by-Field.*** The measurement device here is perpendicular to the direction of the beam. In order to deliver the plan at the correct gantry angles, the measurement device must be fixed or mounted to the gantry. The gantry could also stay fixed in an upright position while facing the device, with the disadvantage that the effects of gravity on the gantry (MLCs especially) will not be tested. The acquisition of data for analysis is broken down to a dataset per field. The advantage of this method is the potential to detect errors more easily and track them to specific fields.
- ***Perpendicular Composite.*** This measurement device is perpendicular to the beam, and the same conditions from the perpendicular field-by-field method can be replicated. The difference from the previous method is that the data acquisition encompasses all fields without pause for individual fields. As with the true composite method, this method might have the potential to hide delivery errors.

5.4 Measurement Analysis Methods

A common method to evaluate measured dose distributions is to compare the measurement to another dose plane computed in the TPS for the same field. In most cases, the physicist must recalculate the patient plan into a CT dataset of the phantom or measurement device. The dose planes of the fields at a chosen depth are exported into the analysis software. The reference dose is the dose plane calculated by the planning system, while the evaluated dose is the measurement. Figure 5.4 is an example of calculated dose by the TPS (top left window) compared to measured dose (top right window) by the detector array. The figure's lower right window shows a profile comparing normalized or absolute dose data using distance-to-agreement, gamma, and gradient.

There are three main methods of analyzing the datasets for dose and its distribution: dose difference, *distance-to-agreement* (DTA), and gamma analysis.

- ***Dose difference.*** Here the evaluation criteria is focused on the difference in dose (cGy or Gy). This method is overly sensitive to spatial alignment of the dose planes to be compared as well as to steep dose gradients. A dose difference test could easily fail in regions of steep dose gradients, a reason why it is more suited to dose maps with more homogeneous dose.

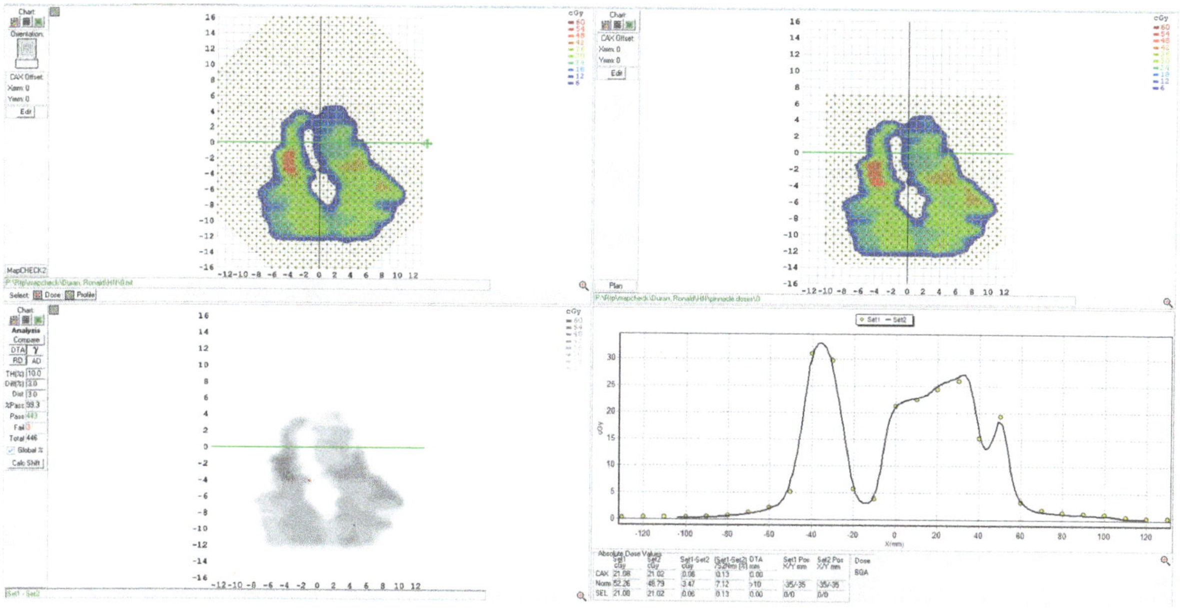

Figure 5.4 Gamma analysis performed on a Mapcheck (Sun Nuclear Corporation).

- ***Distance-to-Agreement (DTA).*** This analysis method was created in response to the feedback of the dose difference method. This method evaluates the differences between two dose distributions. In contrast to the dose difference method, this method is suitable for dose maps with high gradients. Each point in the reference dose map is evaluated according to the DTA parameter (mm), which is the minimum distance to an evaluated point that shares the dose value. The DTA test is sensitive as to which dose map is set as reference or measurement, since this comparison is not a simple profile difference operation. Unlike the dose difference test, the DTA test depends on which set is selected as the reference distribution. This is because a search is required for the points on the reference set such that, if one were to exchange the reference and evaluated dose distributions, a unique DTA result would be determined.
- ***Gamma Analysis.*** This analysis method was created in response to the previous two methods. It combines both dose difference and DTA methods into a single passing-score metric. For a single point in the planar dose map, a search function looks at dose and distribution parameters. If the gamma value is less than 1, the point is given a passing score. If the gamma value is greater than 1, the point is given a failing score. A recommended criterion is 3%/2 mm, 10% threshold, with an agreement score of 95%. The physicist can use a global dose normalization for routine comparisons and a local dose normalization for a stricter analysis or troubleshooting. The resolution of the evaluated dose should not be greater than one-third of the DTA criterion.

6. Application of VMAT for Spatially Fractionated Radiation Therapy (Grid Therapy)

The focus of this book is on how modern *treatment planning systems* (TPS) are used in conjunction with linacs employing *multileaf collimator* (MLC) systems and inverse treatment planning algorithms to produce highly accurate treatments that excel at sparing healthy tissues. These new tools were not designed with a 100-year-old cancer treatment technique in mind, but sometimes old techniques and new tools find each other after many years. Therefore, before moving to this book's final three chapters showing sample treatment plans developed using Pinnacle, Eclipse, and RayStation, we take a slight detour to describe how today's new tools have brought about a resurgence in *grid therapy,* also known as *spatially fractionated radiotherapy* (SFRT).

Grid therapy has been around for nearly 100 years (1–3). Significant literature exists describing the application of this technique and its outcomes in treating bulky tumors. Conventionally, grid therapy refers to a delivery technique of megavoltage x-rays, where a high dose of radiation is given to a tumor site in a single fraction by converting a large treatment field into many smaller fields.

6.1 Early Applications of Grid Therapy

Prior to the availability of advanced linear accelerators equipped with high-definition MLC systems and the *volumetric modulated arc therapy* (VMAT) delivery technique, grid therapy was accomplished through the placement of a device typically manufactured of Cerrobend block with diverging cylindrical holes in a large attenuator mounted on an accessory tray placed in the path of an open radiation beam.

Figure 6.1 shows an in-house fabricated grid made at the University of Toledo Medical Center in 1995 with 77 diverging holes. Using this compensator block, bulky tumors of the head and neck, liver, sarcoma, etc., were treated on a linear accelerator using a 6-MV or 10-MV x-ray beam through a single static field. The prescription dose was typically 15 Gy to 20 Gy to the depth of

Figure 6.1 Side view (left) and top view (right) images of a typical in-house fabricated grid manufactured with Cerrobend block.

maximum dose (D_{max}) delivered through this sieve-like radiation pattern to the target volume with nonuniform dose distribution. The patient is set up at 100 cm SSD with the beam isocenter located in the center of the center open hole of the grid. This SFRT treatment would be followed by 20 Gy to 50 Gy of fractionated conventional radiotherapy.

Other indications for grid therapy were locally advanced cancers that would not be amenable to conventional radiation or that had proved refractory to chemoradiation.

The grid device was primarily used during the era of orthovoltage therapy where photon beams in the range of 120 keV to 300 keV were being used to treat cancer. In this energy range, the maximum dose will be deposited at the skin level, which is a major concern when the radiation dose needs to be delivered to deeply seated tumors. High dose to the skin often produced skin breaks, resulting in non-healing wounds or fistulas. One solution at the time was to use a grid where a desired level of radiation dose could be given to the patient without encountering expected radiation-induced complications (3–6).

Overall, the concept of grid therapy, in simple terms, is the delivery of nonuniform, high-dose, pencil-beam radiation through a collimated grid with a 50:50 open-to-closed blocked area ratio. It was found in the early years of radiotherapy through clinical experience that spatial fractionation minimizes the toxicity of high-dose radiation by limiting the volume of normal tissue receiving radiation. This is an effect typically known as the *dose-volume effect*, which allows nearby normal cells to migrate and mediate the repair of damaged areas (7,8).

Another clinical rationale presented in explaining grid therapy is that high-dose radiotherapy in bulky tumors creates a *radiation-induced bystander effect* (RIBE), which results in cell death in neighboring cancer cells through apoptosis. SFRT is able to destroy portions of the tumor receiving the highest dose of radiation, while inducing RIBE in peripheral cells within the target but far enough away from adjacent healthy structures (4,8).

6.2 Grid Therapy Makes a Return

With the advent of linear accelerators in the 1950s, when higher energies became available, grid therapy lost its usefulness because of the skin-sparing effect, an inherent feature of megavoltage beams. By the mid-1980s, however, grid therapy was back again, being used for the treatment of large-area bulky tumors. Using a linear accelerator, one can produce an external beam radiation plan with dose distributions similar to that of an interstitial brachytherapy procedure.

Dosimetry of grid therapy was not a straightforward solution, and different authors had slightly different approaches. A simple method was explained by Mitev and Suntharalingamin in 1986 (9) whereby a measured, single-hole dose distribution was used to calculate dose profiles at a given depth when grid was implemented. Their modeled result agreed fairly well with measured data, but was limited to two dimensions only. Komanduri et al. (10) developed a semiempirical computer method in 1996 to model the three-dimensional dose

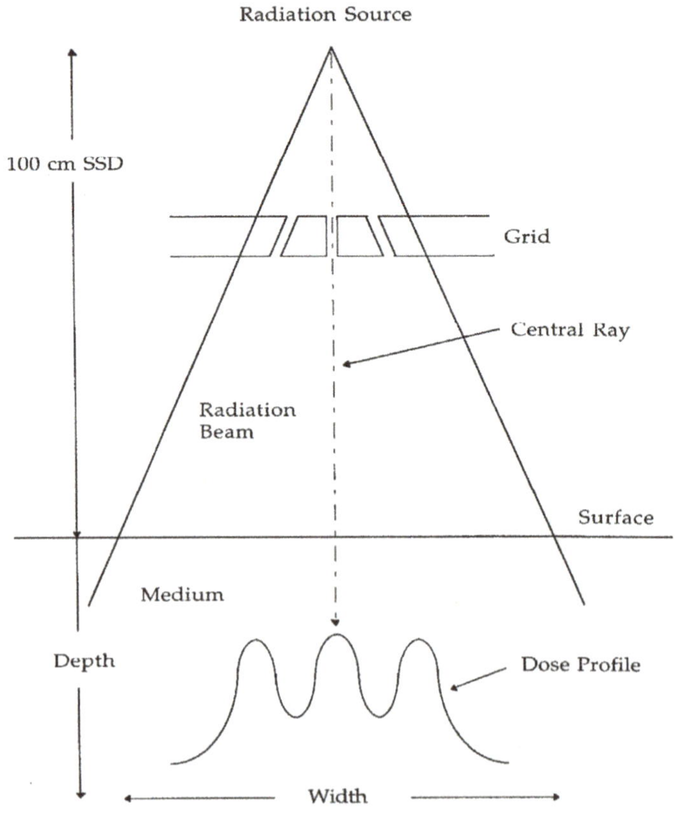

Figure 6.2 Schematic of grid therapy setup. The radiation source is provided by a linear accelerator and originates from a point that is located 100 cm from the surface of the medium (i.e., the patient's skin). Note the one-dimensional dose profile resulting from the open and blocked portions of the grid. (Reprinted from Komanduri M. Ayyangar and T. Pawlicki; Computers and Biomedical Research 29:247–58 (1996), with permission.)

distribution from a high-energy photon beam using measured data from a single cylindrical grid hole, as shown in Figure 6.2. In their model, dose could be graphically displayed as a series of isodose curves in two dimensions (e.g., depth and width), as a beam profile at a given depth in one dimension, or as isodose lines as shown in Figure 6.3.

The lines indicated by 90, 70, 50, 30, and 10 in Figure 6.3 are isodose line percentages of the maximum dose. Converting this 2D graph to a table, one can see the numerical distribution as a function of distance from the center of the hole, as shown in **Table 6.1**.

A 3D isosurface dose distribution with the grid installed at the head of the gantry using a 6-MV x-ray is shown in Figure 6.4.

Gholami et al. (11) reported that hole sizes in the range of 1 to 1.25 cm and with spacing of 1.7 to 1.8 cm

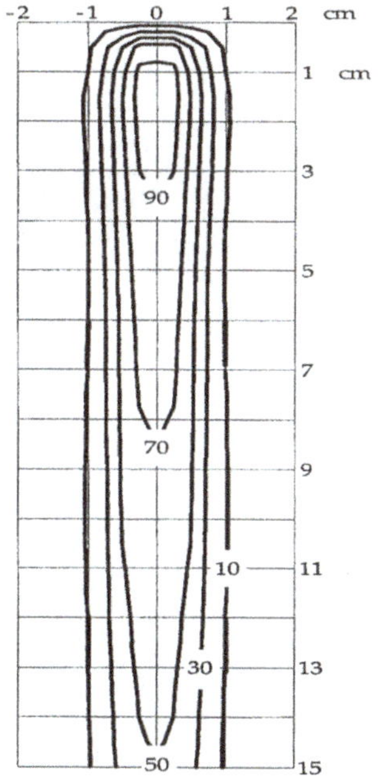

Figure 6.3 A partial display of the measured 2D isodose distribution of a single 1-cm-diameter hole that is symmetric about the central ray. The transmission component of dose has been subtracted. (Reprinted from M. Komanduri et al., Computers and Biomedical Research 29, 247–58 (1996), with permission.)

Table 6.1 A partial table of the 2D isodose distribution for a single hole of 1 cm diameter shown in Figure 6.3 using 6-MV x-rays. (Reprinted from Komanduri M. Ayyangar and T. Pawlicki; Computers and Biomedical Research 29:247–58 (1996), with permission.)

Depth (cm)	0.0	0.05	0.10	0.15	0.20	0.25	0.30	0.35	0.40
0.5	80.3	80.4	80.1	79.7	79.3	78.1	75.3	72.0	68.2
1.0	97.8	98.1	97.9	97.5	96.7	95.5	92.5	88.4	84.1
1.5	100.1	100.6	100.6	100.3	99.3	98.3	95.4	91.1	87.1
2.0	98.4	98.7	98.7	98.2	97.1	96.1	93.4	89.2	84.9
2.5	96.2	96.4	96.3	95.7	94.6	93.5	91.0	87.0	82.4
3.0	96.0	94.2	94.1	93.3	92.1	90.9	88.7	84.8	80.0
3.5	91.2	91.3	91.1	90.2	89.0	87.8	85.8	82.0	77.0
4.0	89.1	89.0	88.9	87.8	86.6	85.3	83.5	79.8	74.7
4.5	87.0	86.9	86.7	85.5	84.2	82.9	81.4	77.7	72.4
5.0	84.8	84.6	84.4	83.1	81.8	80.4	79.0	75.4	70.0
5.5	82.6	82.4	82.3	81.1	79.9	78.6	77.3	73.9	68.8
6.0	80.3	80.1	80.0	79.0	77.8	76.6	75.4	72.1	67.4
6.5	78.2	78.1	78.0	77.1	76.0	74.9	73.7	70.6	66.2
7.0	76.4	76.2	76.1	75.4	74.4	73.4	72.2	69.2	65.1
7.5	74.2	74.1	74.0	73.4	72.4	71.5	70.4	67.6	63.7
8.0	72.3	72.3	72.2	71.7	70.8	70.0	68.9	66.2	62.6
8.5	70.5	70.5	70.5	70.1	69.2	68.4	67.4	64.8	61.5
9.0	68.5	68.5	68.5	68.2	67.4	66.7	65.7	63.3	60.2
9.5	67.0	67.0	67.0	66.8	66.1	65.4	64.5	62.1	59.3
10.0	65.1	65.1	65.1	65.1	64.4	63.8	62.9	60.7	58.0
10.5	63.5	63.5	63.5	63.4	62.8	62.1	61.2	59.1	56.5
11.0	61.6	61.5	61.5	61.4	60.8	60.0	59.1	57.2	54.6

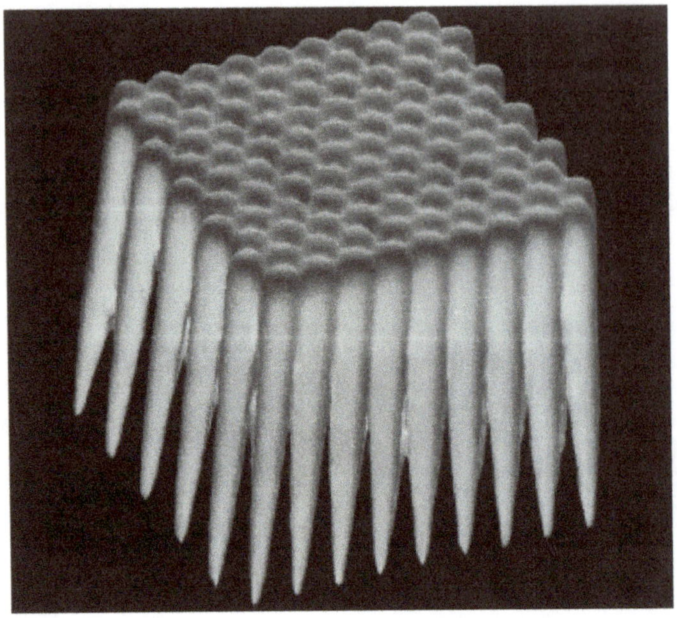

Figure 6.4 Three-dimensional image of 50% isosurface dose shown for a grid with 104 1-cm-diameter holes. (Reprinted from Komanduri M. Ayyangar and T. Pawlicki; Computers and Biomedical Research 29:247–58 (1996), with permission.)

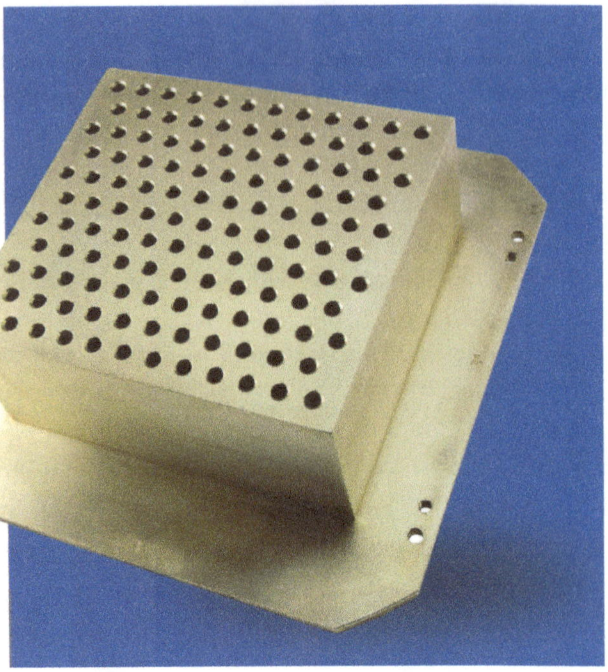

Figure 6.5 Grid compensator. (Courtesy of Dot Decimal Inc., Stanford, Florida.)

in the grid block optimize the therapeutic ratio and normal tissue sparing. In recent years, the company Dot Decimal, Inc., began offering a commercially produced grid manufactured from brass material (Figure 6.5) allowing spatial fractionation with different grid sizes, ranging from 120 to 250 rays of beamlet entries. These compensators can be ordered with customizable aperture sizes, center-to-center spacing, and holes that follow beam divergence, but are typically designed as 1-cm holes equidistantly spaced. Other specifics of the Dot Decimal Grid are that it is approximately 7.62 cm thick, weighs about 15 to 20 kg, and has hole centers 2.1 cm from center to center and 1.4 cm in diameter at the isocenter.

To make the implementation of grid therapy feasible for more institutions, Ha et al. (12) showed that an MLC can be used to deliver SFRT with dosimetric properties as good as the ones obtained by using a Cerrobend block grid. The typical dose delivered using an MLC is similar to having a physical grid, and in the range of *stereotactic radiosurgery* (SRS) or *stereotactic body radiotherapy* (SBRT) 15 to 20 Gy is delivered in a single fraction. The dose is typically prescribed to D_{max} and is followed by 20 to 50 Gy of conventionally fractionated radiotherapy.

In a retrospective study of 79 patients, Neuner et al. (13) compared the MLC-based SFRT delivery (40 patients) to Cerrobend grid delivery (39 patients) and found comparable results. The MLC-based grid was designed by creating holes of 1 cm per side arranged 1 cm apart in a line with a 1-cm strip between lines and with the next line of holes staggered by 1 cm. When combined, the holes created through the MLC provided a profile similar to that of a Cerrobend grid. Neuner et al. concluded that MLC-based SFRT should allow clinics to adopt this treatment approach more easily for the treatment of bulky tumors in palliative or curative settings.

Another method using an MLC was described by Grams et al. (14). They used a VMAT technique employing the MLC to deliver the dose to 1.5-cm diameter spherical contours placed throughout the *gross tumor volume* (GTV). This is also known as *lattice radiotherapy*, where vertices are strictly contained within the GTV while keeping the dose level lower at the periphery of the tumor to avoid related toxicity.

The placement of spheres may be variable, but they all must be placed within the GTV (or a target volume provided by a radiation oncologist), maintain a 3-cm center-to-center separation, and the edge of any sphere must be at least 1 cm from any *organ at risk* (OAR). Following this guideline, one can place as many spheres as possible within the *clinical target volume* (CTV), an example of which is shown in Figure 6.6.

Typical optimization techniques are then applied through the application of concentric ring structures to

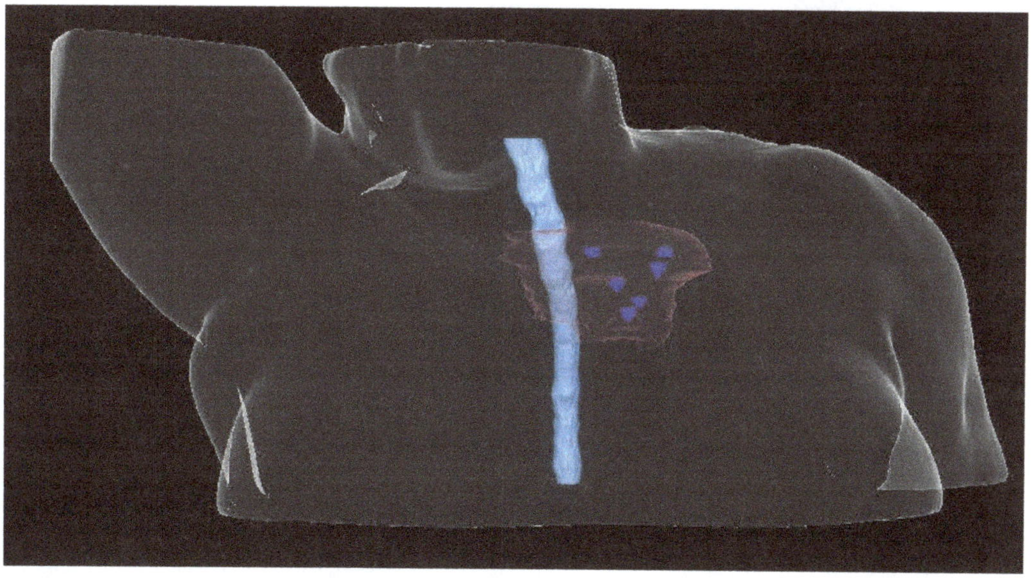

Figure 6.6 Three-dimensional rendering of a patient with a paraspinal mass (CTV) shown in orange. The 1.5-cm-diameter spherical contours shown in blue have been placed throughout the CTV following the guidelines in the text. To spare the spinal cord, no sphere is placed near that structure.

confine the highest level of dose to the center of each sphere, while sparing the tissue between them, similar to the conventional use of the physical grid. This will result in alternating high-dose and low-dose regions throughout the CTV and sparing the OARs. The energy used for this treatment technique is typically a 6-MV (or sometimes 10-MV) *flattening filter-free* (FFF) x-rays beam.

After the placement of spheres within the CTV, a single isocenter is placed at the CTV's geometric center (approximately the position of the center of the mass). This point can be moved if needed to afford proper gantry clearance. The number and angular extent of treatment arcs used for this approach are similar to a standard VMAT plan. Figure 6.7 shows the beam

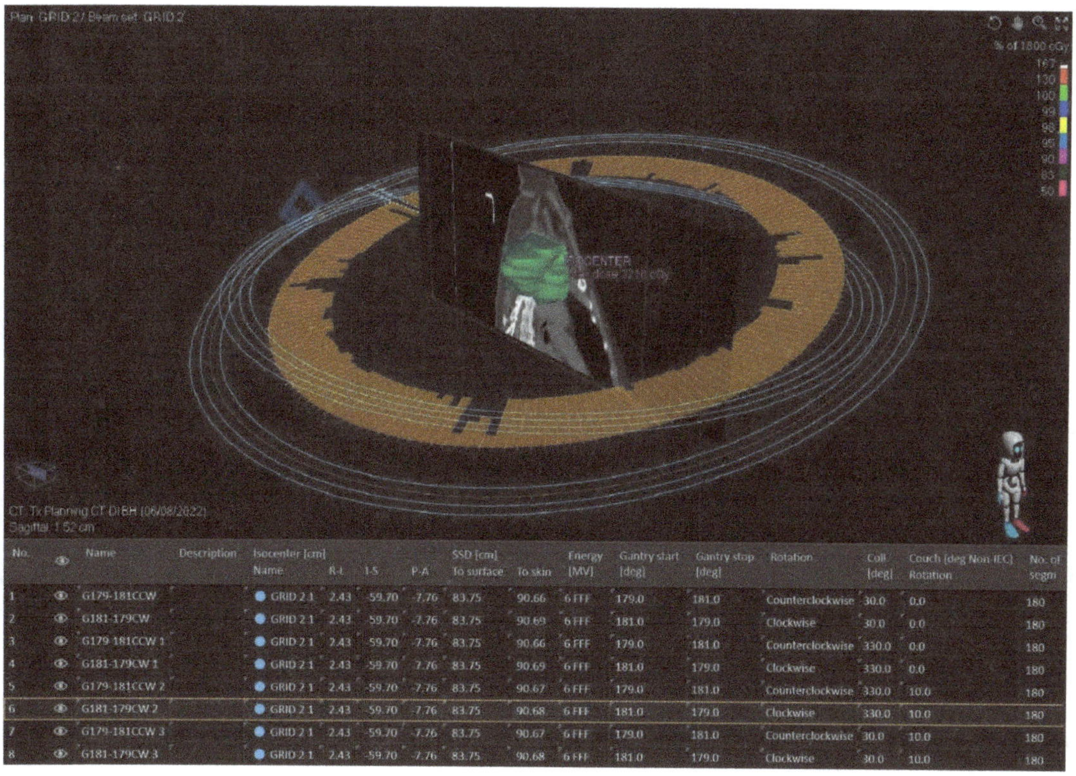

Figure 6.7 Beam arrangement with MLC.

arrangement used for this plan where 8 full 360-degree arcs were used.

In Figure 6.8 transverse plans with a couple of the high-dose spheres are shown with a few optimization rings. Yellow shows the 1.2-cm-diameter contour. Green on the upper sphere and red in the lower sphere are the 1.5-cm-diameter contours. Bright yellow on the upper and blue in the lower sphere are the 3-cm rings designed to keep the high dose within the spheres. The reddish contour surrounding all spheres is the CTV.

Optimization objectives for this plan are displayed in Figure 6.9. As indicated earlier, the plan optimization

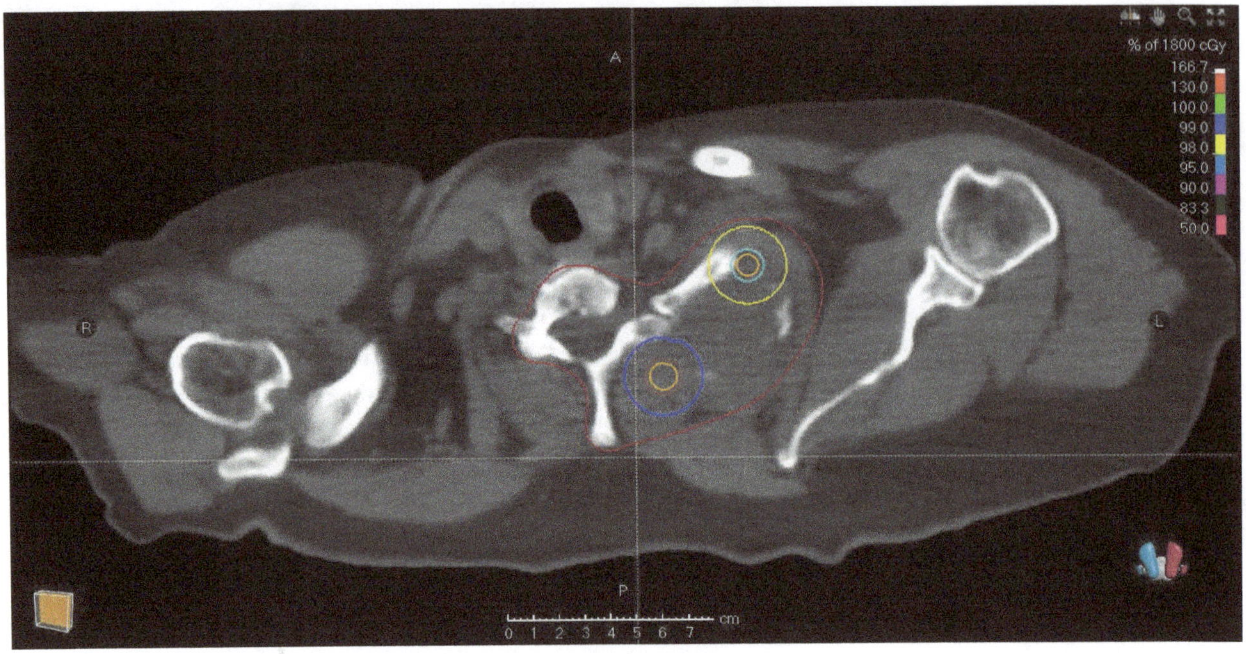

Figure 6.8 Contours showing CTV and various contours of 1.2 cm, 1.5 cm, and 3 cm to manage optimization.

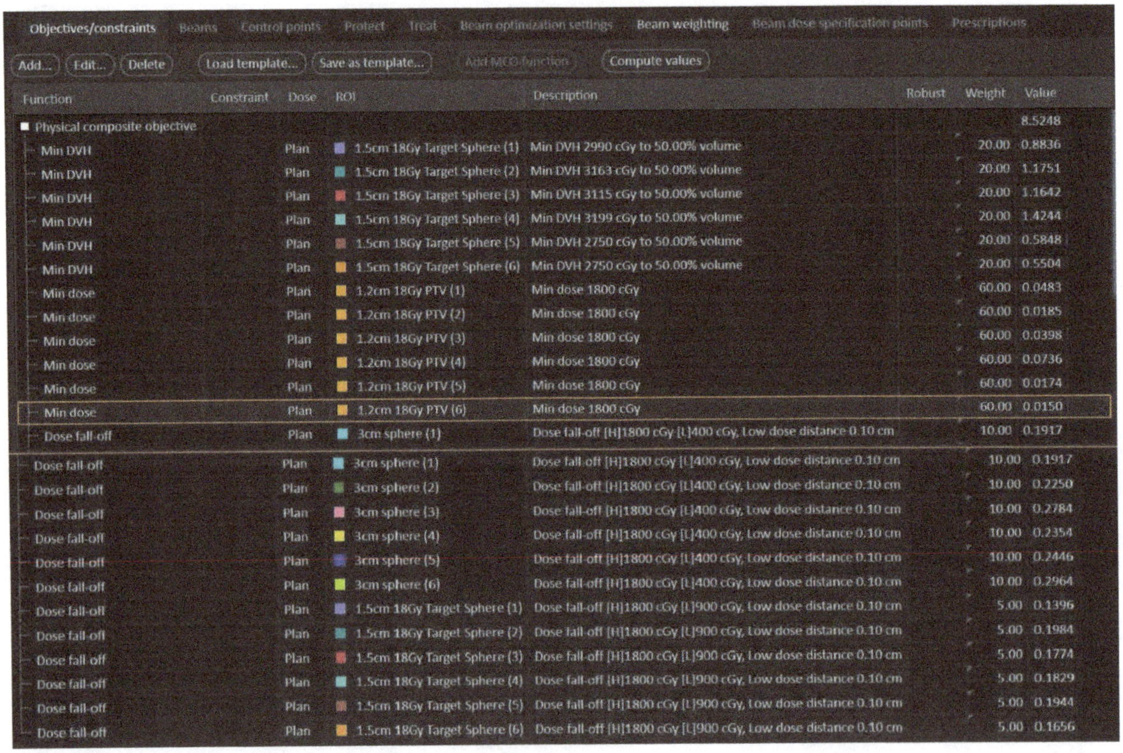

Function	Constraint	Dose	ROI	Description	Robust	Weight	Value
Physical composite objective							8.5248
Min DVH		Plan	1.5cm 18Gy Target Sphere (1)	Min DVH 2990 cGy to 50.00% volume		20.00	0.8836
Min DVH		Plan	1.5cm 18Gy Target Sphere (2)	Min DVH 3163 cGy to 50.00% volume		20.00	1.1751
Min DVH		Plan	1.5cm 18Gy Target Sphere (3)	Min DVH 3115 cGy to 50.00% volume		20.00	1.1642
Min DVH		Plan	1.5cm 18Gy Target Sphere (4)	Min DVH 3199 cGy to 50.00% volume		20.00	1.4244
Min DVH		Plan	1.5cm 18Gy Target Sphere (5)	Min DVH 2750 cGy to 50.00% volume		20.00	0.5848
Min DVH		Plan	1.5cm 18Gy Target Sphere (6)	Min DVH 2750 cGy to 50.00% volume		20.00	0.5504
Min dose		Plan	1.2cm 18Gy PTV (1)	Min dose 1800 cGy		60.00	0.0483
Min dose		Plan	1.2cm 18Gy PTV (2)	Min dose 1800 cGy		60.00	0.0185
Min dose		Plan	1.2cm 18Gy PTV (3)	Min dose 1800 cGy		60.00	0.0398
Min dose		Plan	1.2cm 18Gy PTV (4)	Min dose 1800 cGy		60.00	0.0736
Min dose		Plan	1.2cm 18Gy PTV (5)	Min dose 1800 cGy		60.00	0.0174
Min dose		Plan	1.2cm 18Gy PTV (6)	Min dose 1800 cGy		60.00	0.0150
Dose fall-off		Plan	3cm sphere (1)	Dose fall-off [H]1800 cGy [L]400 cGy, Low dose distance 0.10 cm		10.00	0.1917
Dose fall-off		Plan	3cm sphere (1)	Dose fall-off [H]1800 cGy [L]400 cGy, Low dose distance 0.10 cm		10.00	0.1917
Dose fall-off		Plan	3cm sphere (2)	Dose fall-off [H]1800 cGy [L]400 cGy, Low dose distance 0.10 cm		10.00	0.2250
Dose fall-off		Plan	3cm sphere (3)	Dose fall-off [H]1800 cGy [L]400 cGy, Low dose distance 0.10 cm		10.00	0.2784
Dose fall-off		Plan	3cm sphere (4)	Dose fall-off [H]1800 cGy [L]400 cGy, Low dose distance 0.10 cm		10.00	0.2354
Dose fall-off		Plan	3cm sphere (5)	Dose fall-off [H]1800 cGy [L]400 cGy, Low dose distance 0.10 cm		10.00	0.2446
Dose fall-off		Plan	3cm sphere (6)	Dose fall-off [H]1800 cGy [L]400 cGy, Low dose distance 0.10 cm		10.00	0.2964
Dose fall-off		Plan	1.5cm 18Gy Target Sphere (1)	Dose fall-off [H]1800 cGy [L]900 cGy, Low dose distance 0.10 cm		5.00	0.1396
Dose fall-off		Plan	1.5cm 18Gy Target Sphere (2)	Dose fall-off [H]1800 cGy [L]900 cGy, Low dose distance 0.10 cm		5.00	0.1984
Dose fall-off		Plan	1.5cm 18Gy Target Sphere (3)	Dose fall-off [H]1800 cGy [L]900 cGy, Low dose distance 0.10 cm		5.00	0.1774
Dose fall-off		Plan	1.5cm 18Gy Target Sphere (4)	Dose fall-off [H]1800 cGy [L]900 cGy, Low dose distance 0.10 cm		5.00	0.1829
Dose fall-off		Plan	1.5cm 18Gy Target Sphere (5)	Dose fall-off [H]1800 cGy [L]900 cGy, Low dose distance 0.10 cm		5.00	0.1944
Dose fall-off		Plan	1.5cm 18Gy Target Sphere (6)	Dose fall-off [H]1800 cGy [L]900 cGy, Low dose distance 0.10 cm		5.00	0.1656

Figure 6.9 Optimization objectives for the SFRT paraspinal plan.

is very similar to any VMAT plan with the insertion of rings to manage dose to OARs and dose fall-off.

Figure 6.10 shows the dose distribution in an (a) upper (cephalic), (b) mid-plane, and (c) lower (caudal) cross section in three planes of transverse, sagittal, and coronal views.

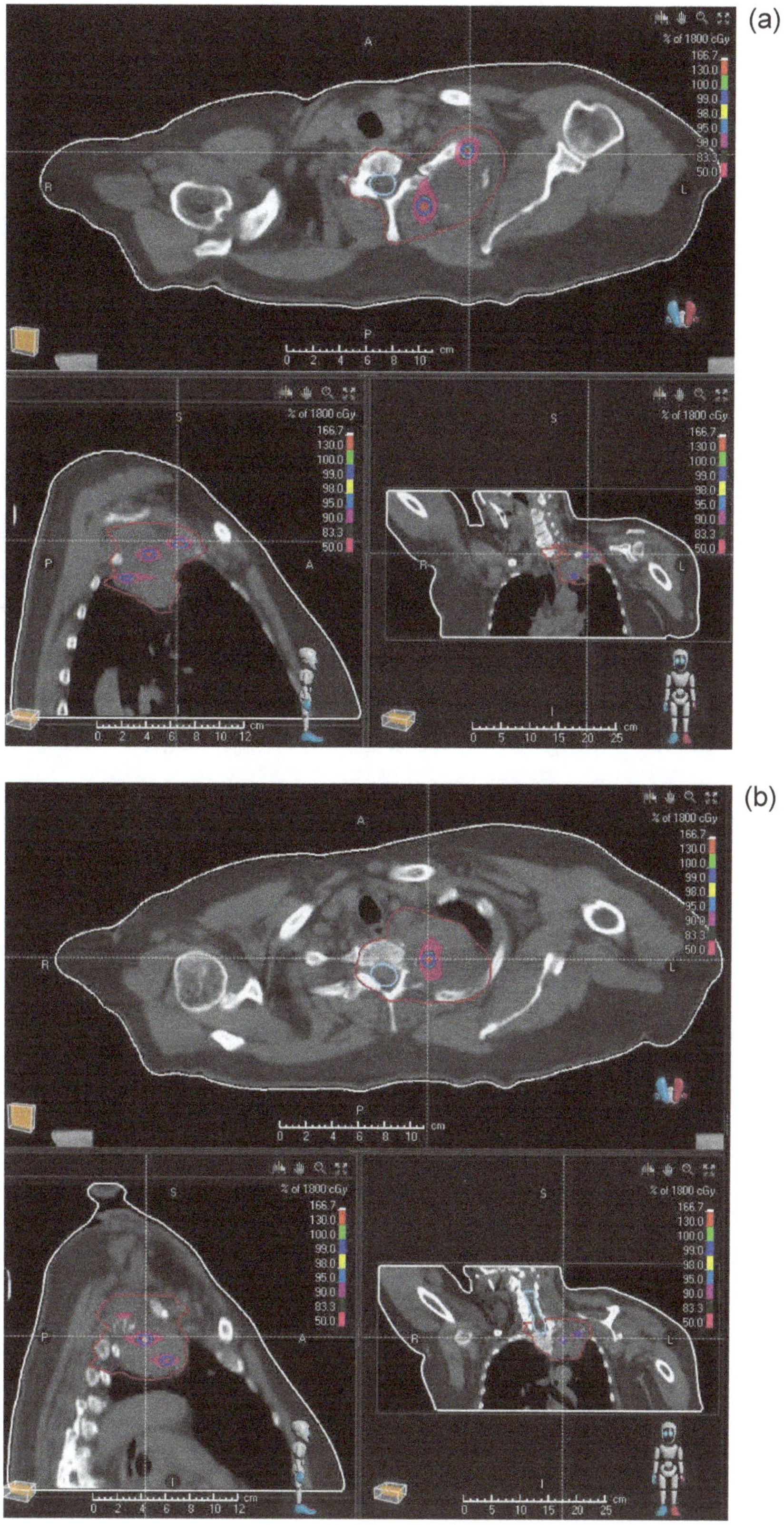

Figure 6.10 Axial, sagittal, and coronal views of high-dose spheres in (a) superior, (b) mid-plane, and (c) inferior sections of the CTV.

Figure 6.11 shows an example of a dose profile through two neighboring high-dose spheres in the treatment plan.

Figure 6.12 shows *digitally reconstructed radiographs* (DRR) of various MLC and jaw positions from different gantry angles.

Figure 6.10 (continued).

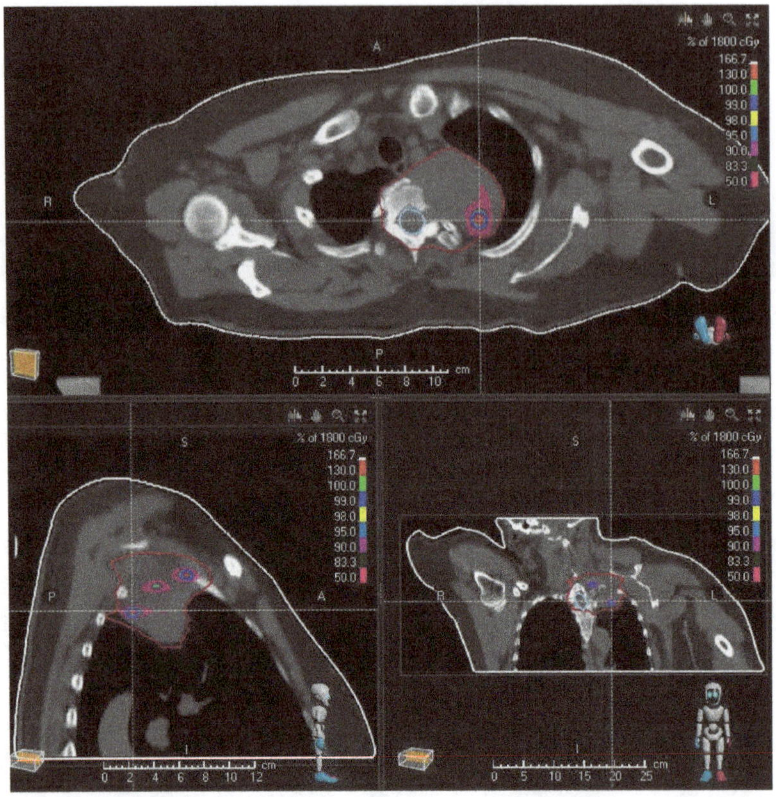

(c)

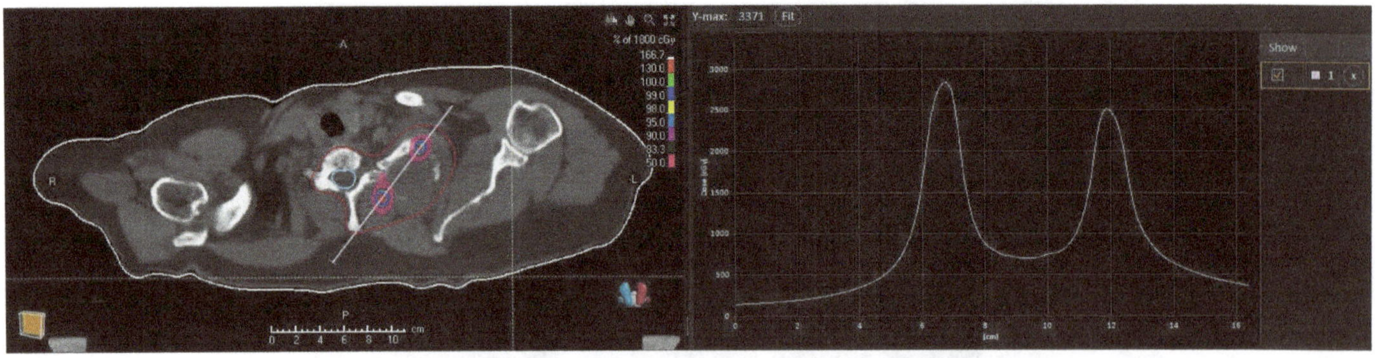

Figure 6.11 Axial view of high-dose spheres within the dose profile across two neighboring high-dose spheres within the CTV.

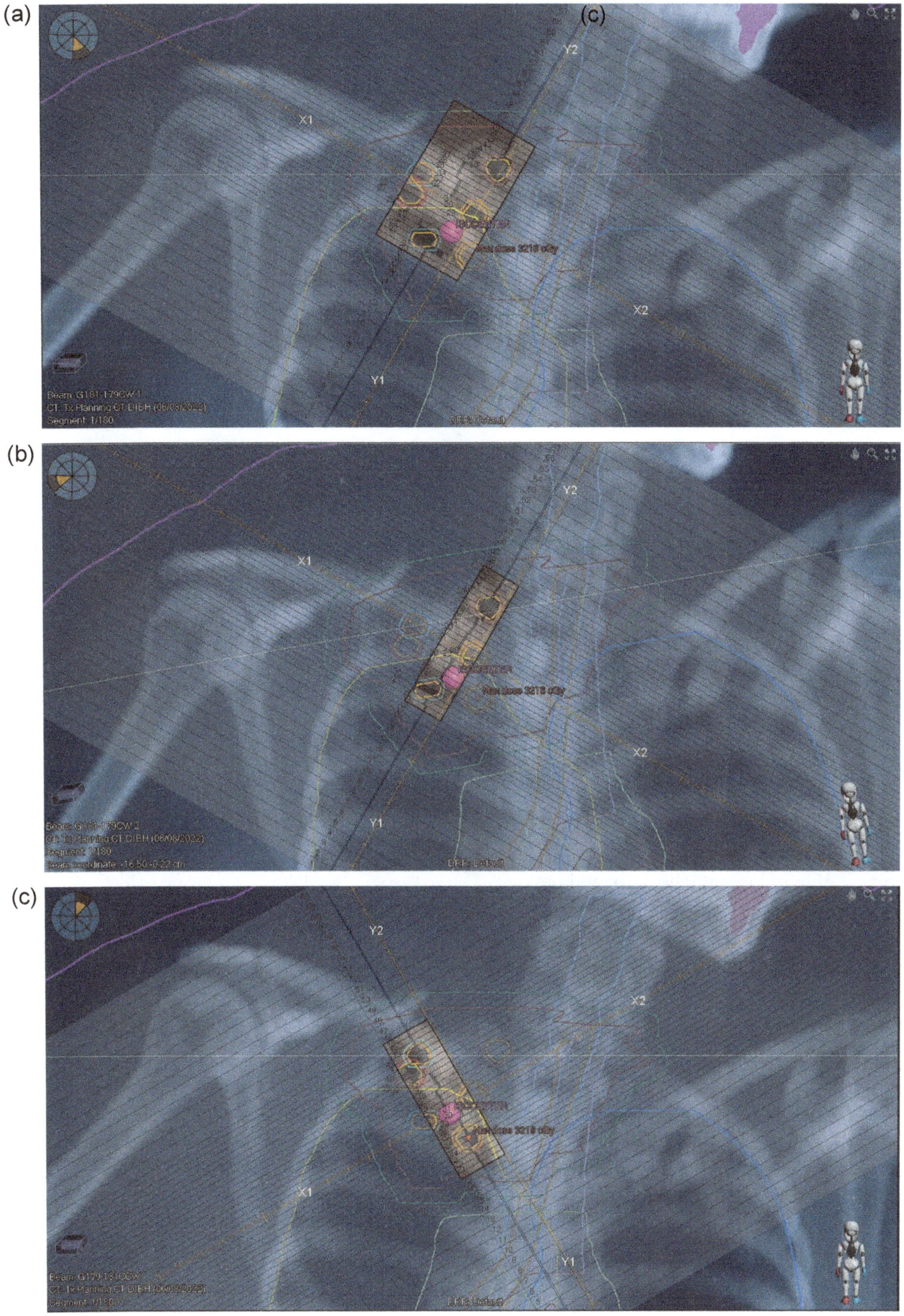

Figure 6.12 Display of MLC and jaw positions from different gantry, collimator, and couch angles. The small orange section in the top left inter-circle in each image shows the gantry position where each image was captured at the first segment. (a) Arc beam moving clockwise from gantry angle 181 to 179; image captured at segment 1 at gantry 181°, collimator 330°, and couch angle 0°. (b) Arc beam moving clockwise from gantry angle 181 to 179; image captured at segment 1 at gantry 181°, collimator 330°, and couch angle 10°. (c) Arc beam moving counterclockwise from gantry angle 179 to 181; image captured at segment 1 at gantry 179°, collimator 30°, and couch 0°.

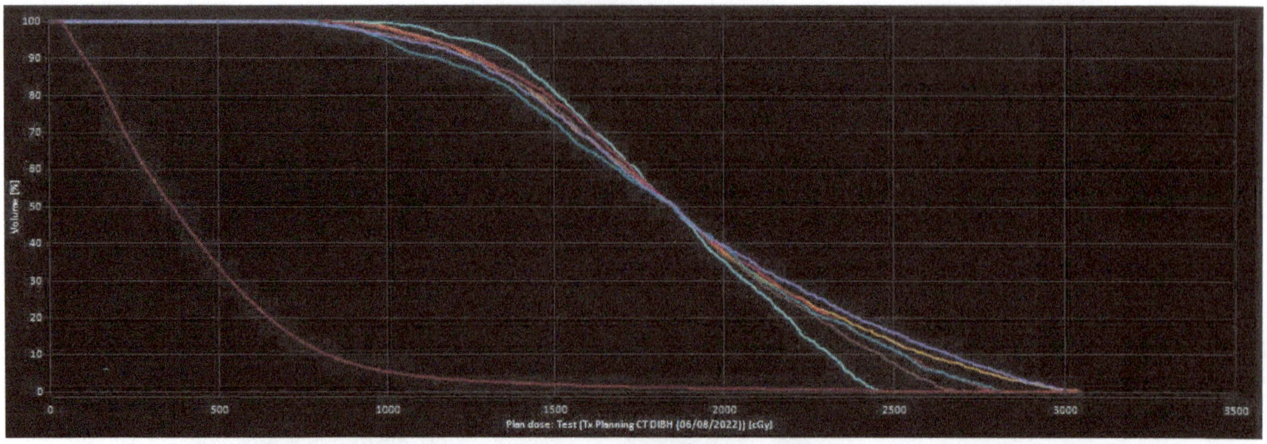

Figure 6.13 DVH curves generated for a VMAT SFRT plan, grouped such that 50% of the volume receives the prescription dose of 18 Gy delivered to 6 spherical regions with diameters of 1.5 cm. The solid red line represents the CTV.

Figure 6.13 shows the *dose volume histogram* (DVH) for this paraspinal plan where 50% of the CTV receives the prescription dose of 18 Gy. The long tail of the DVH curve is specific to grid therapy and indicates that higher doses are delivered to small volumes, while the majority of the CTV receives lower doses indicative of the sparing between spheres.

6.3 Conclusion

In summary, a major drawback of physical grid is that the highest dose is always delivered to D_{max}, which potentially may not be within the tumor. Moreover, since the dose is delivered through a single static field, a significant portion of exit dose may be delivered beyond the target into normal tissue.

Delivering SFRT using a VMAT technique, however, allows delivery of maximum dose to delineated targets and eliminates these problems through continuous gantry rotations and beam modulations specifically designed to spare the OARs. Moreover, using the MLCs for grid therapy gives the planner more flexibility in determining aperture size and center-to-center spacing on a case-by-case basis without having to commission the creation of a new Cerrobend or brass block. With MLCs, the apertures that may fall directly in line with any OARs or other critical structures can easily be closed.

A major limitation with MLCs, however, is that it takes a much longer time to complete patient treatment due to the added modulation factor, resulting in a higher number of MUs. This increases the radiation leakage dose, which may undesirably increase surface doses and valley-to-peak ratios.

Chapter References

1. Köhler, A. (1909). "Theorie einer methode bisher unmöglich anwendbar hohe dosen röntgenstrahlen in def tiefe des geweben zur therapeutischen wirksamkeit zu bringen ohne schwere schädigung des patienten, zugleich eine methode des schutzes gegen röntgenverbrennung überhaupt." *Fortschr. a. d. Geb. d. Röntgenstrahlen* 14:27–29.
2. Köhler, A. (1909). "Zur röntgentiefentherapie mit massendosen." *Müchen med. Wchnschr.* 56:2314–16.
3. Liberson, F. (1933). "The value of a multi-perforated screen in deep x-ray therapy." *Radiology* 20:186–95.
4. Mohiuddin, M., M. Fujita, W. F. Regine, et al. (1999). "High-dose spatially-fractionated radiation (GRID): a new paradigm in the management of advanced cancers." *Int. J. Radiat. Oncol. Biol. Phys.* 45:721–27.
5. Marks, H. (1952). "Clinical experience with irradiation through a Grid." *Radiology* 58:338–42.
6. Loevinger, R. (1952). "Depth dose curves for Grids in x-ray therapy." *Radiology* 58:351–59.
7. Prasanna, A., M. M. Ahmed, M. Mohiuddin, and C. N. Coleman (2014). "Exploiting sensitization windows of opportunity in hyper and hypofractionated radiation therapy." *J. Thorac Dis.* 6:287–302.
8. Hopewell, J. W. and K. R. Trott (2000). "Volume effects in radiobiology as applied to radiotherapy." *Radiother. Oncol.* 56:283–88.
9. Mitev, G. and N. Suntharalingam (1986). "Semi-empirical calculation of dose distributions for high energy photon beam Grid therapy." *Med. Phys.* 13:592.
10. Ayyangar, K. M. and T. Pawlicki (1996). "A computer method to model the dose distribution of high energy photon grid therapy in three dimensions." *Comput. Biomed. Res.* 29:247–58.
11. Gholami, S., H. A. Nedaie, F. Longo, M. R. Ay, S. A. Dini, and A. S. Meigooni (2017). "Grid block design based on Monte Carlo simulated dosimetry, the linear quadratic and Hug-Kellerer radiobiological models." *J. Med Phys.* 42:213–21.
12. Ha, J. K., G. Zhang, S. A. Naqvi, W. F. Regine, and C. X. Yu (2006). "Feasibility of delivering Grid therapy using a multiload collimator." *Med. Phys.* 33(1):76–82.
13. Neuner, G., M. M. Mohiuddin, N. Vander Walde, et al. (2012). "High-dose spatially fractionated GRID radiation therapy (SFGRT): A comparison of treatment outcomes with Cerrobend vs. MLC SFGRT." *Int. J. Radiat. Oncol. Biol. Phys.* 82:1642–49.
14. Grams, M. P., D. Owen, S. S. Park, I. A. Petersen, M. G. Haddock, E. B. Jeans, R. R. Finley, and D. J. Ma (2021). "VMAT Grid Therapy: A Widely Applicable Planning Approach." *Pract. Radiat. Oncol.* 11:e339–47.

7. Pinnacle TPS Sample IMRT and VMAT Plans

Case Number pTA01: Thorax and Abdomen—Gastroesophageal Cancer Radiation Treatment

General Information

Brief Clinical History

54 year-old male with poorly differentiated adenocarcinoma with signet ring features of the cardia/gastroesophageal junction. EGD (esophagogastroduodenoscopy) revealed tumor on the greater curvature side of the cardia as well as gross Barrett's esophagus of the distal 2 to 3 cm of the esophagus. Neoadjuvant chemotherapy followed by restaging and planned surgery.

Prescription

Esophagus/stomach and lymph nodes:
PTV 4500 = 180 cGy × 25 fractions
PTV 5250 = 210 cGy × 25 fractions

Simulation

The patient was simulated in the supine position with a vacuum cushion device as immobilization device. The targets were drawn with the help of a PET-CT/Planning CT image fusion.

> **NOTE:** All sample cases in this chapter are for educational purposes only. Hypofractionated regimens, SRS and SBRT procedures must be tested and implemented clinically in a separate commissioning process from basic IMRT/VMAT implementation. All dose distributions are not suggestions for clinical treatments. Objectives in the optimizer as well as beam arrangements are for illustration/educational purposes only.

Target

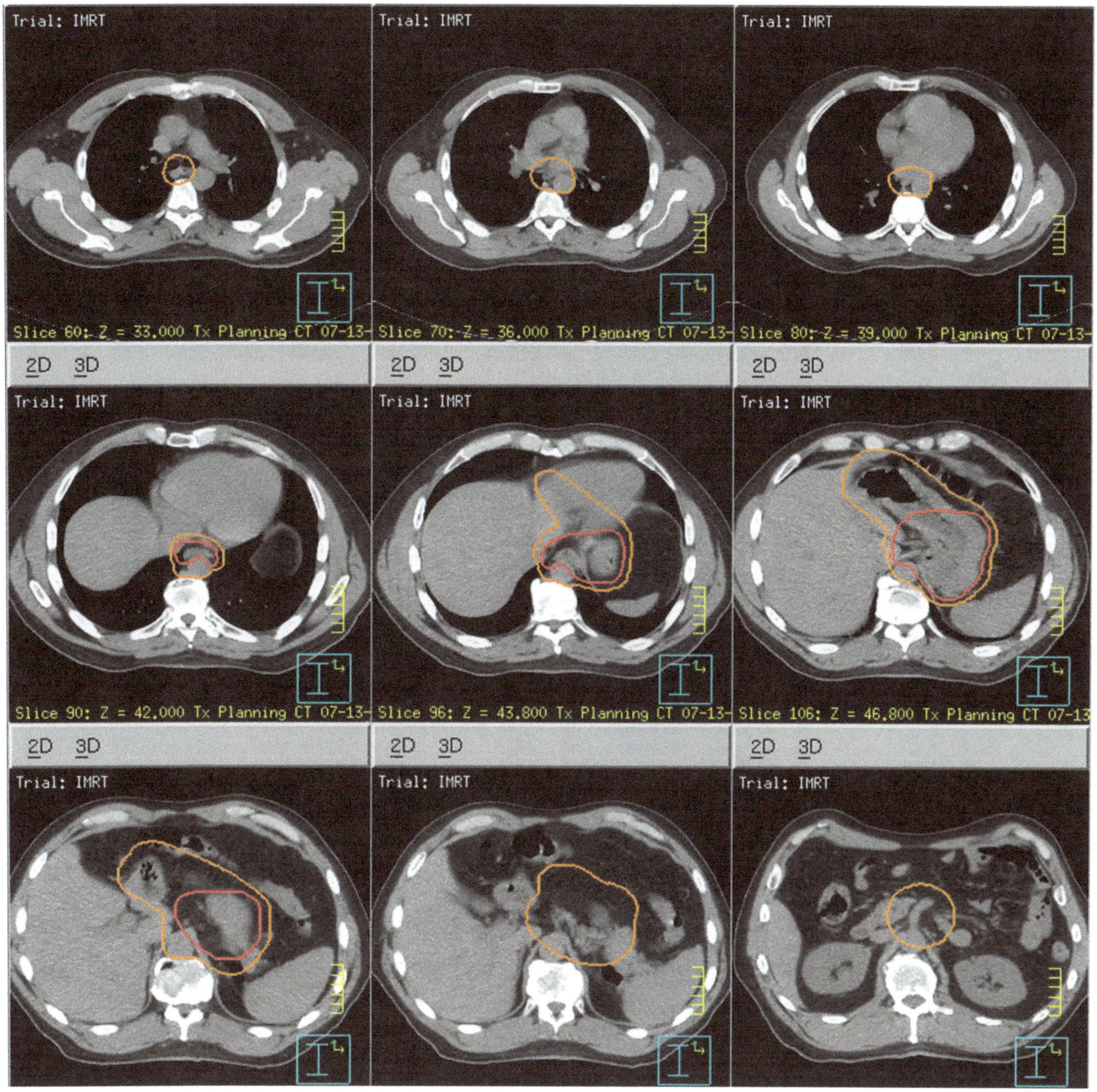

Figure 7.1 Contours of PTV 4500 (orange) and PTV 5250 (red).

Figure 7.1 shows the target contours drawn by the radiation oncologist. Contours of normal structures are not shown on this figure. The use of inversely optimized modulated techniques can be beneficial to the patient due to the unusual target shape and the OAR proximity to the target. Inversely optimized techniques would make it possible to spare healthy tissue, such as the bowel (small intestine), liver, spinal cord, and lungs.

It is important to accurately contour the OAR, especially in cases in which the target volume is likely to include functional organs. Dose constraints in these cases are set at the discretion of the prescribing physician. Two types of contours could be used to evaluate the DVH: the original OAR and the OAR minus the target volume overlap.

Beam Arrangement

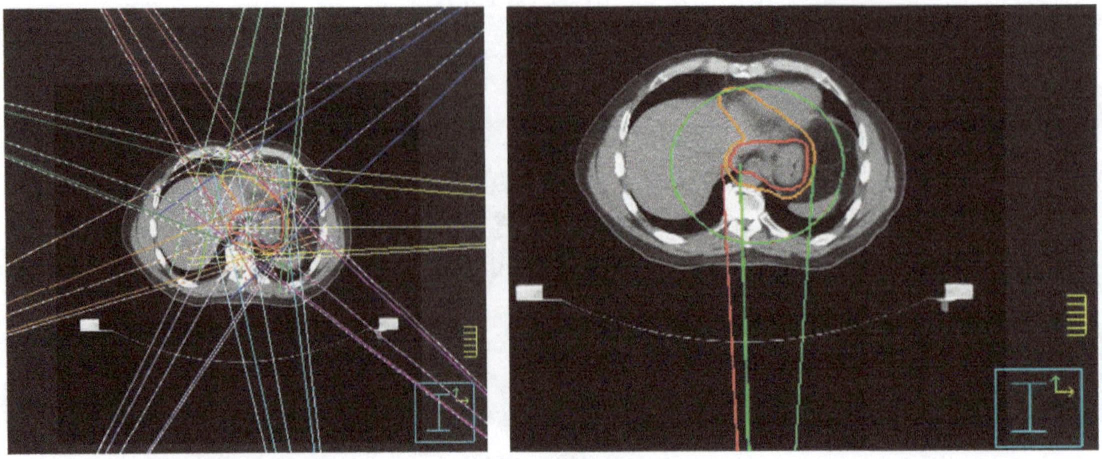

Figure 7.2 Beam arrangement for IMRT (left) and VMAT (right).

Figure 7.3 Beam arrangement for IMRT (left) and VMAT (right) on a skin surface reconstruction.

Nine beams were distributed around the patient for the IMRT plan. The target is located in the middle of the axial plane, and this type of arrangement provides dose coverage to the lesion from all sides, decreasing the chances of a nonhomogeneous dose distribution.

The planner can certainly start with a 7-field IMRT and switch to having more beams or having a VMAT plan. In case of a VMAT plan, collimator rotations are necessary to reduce the effect of the inter-leaf leakage. This is at the discretion of the planner and the complexity of the target. More complex dose distributions require more control points, which can be translated into more beams. Avoid excessive numbers of beamlets for static-gantry IMRT as they are time-consuming during delivery. Use only them if they are completely necessary to obtain a superior dose distribution.

Gantry Angle	Couch	Collimator	Energy (MV)
10	0	0	10
50	0	0	10
90	0	0	10
130	0	0	10
170	0	0	10
210	0	0	10
250	0	0	10
290	0	0	10
330	0	0	10

Optimization

Dose levels within target: It is important to separate regions of different dose prescriptions within the target before running an optimization. For example, in this case we have two dose levels: 4500 cGy and 5250 cGy.

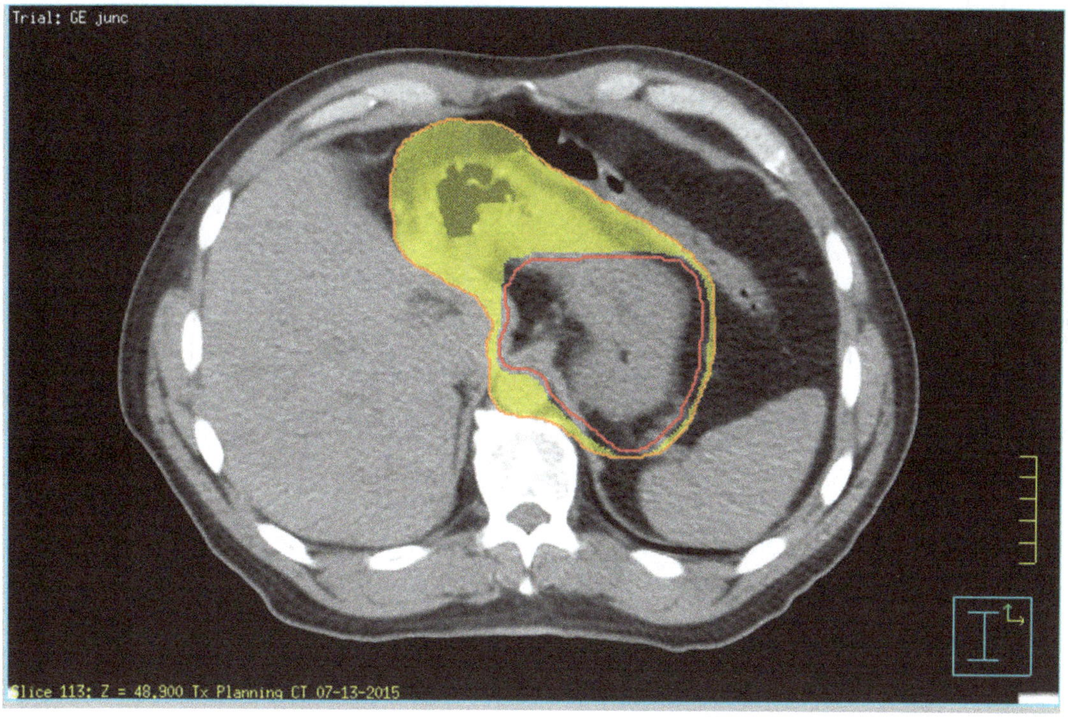

Figure 7.4 PTV 4500–PTV 5250 (subtraction) contour in yellow (colorwash).

The yellow contour (PTV 4500–PTV 5250) in Figure 7.4 is used to optimize only the area that is receiving 4500 cGy. This is done to avoid conflicts in the optimization objectives since the PTV 4500 also contains PTV 5250. Notice the 2-mm margin around the region of high dose in the yellow contour. This helps the planner reach the desired dose gradient. These margins can be used alone or in combination with ring structures.

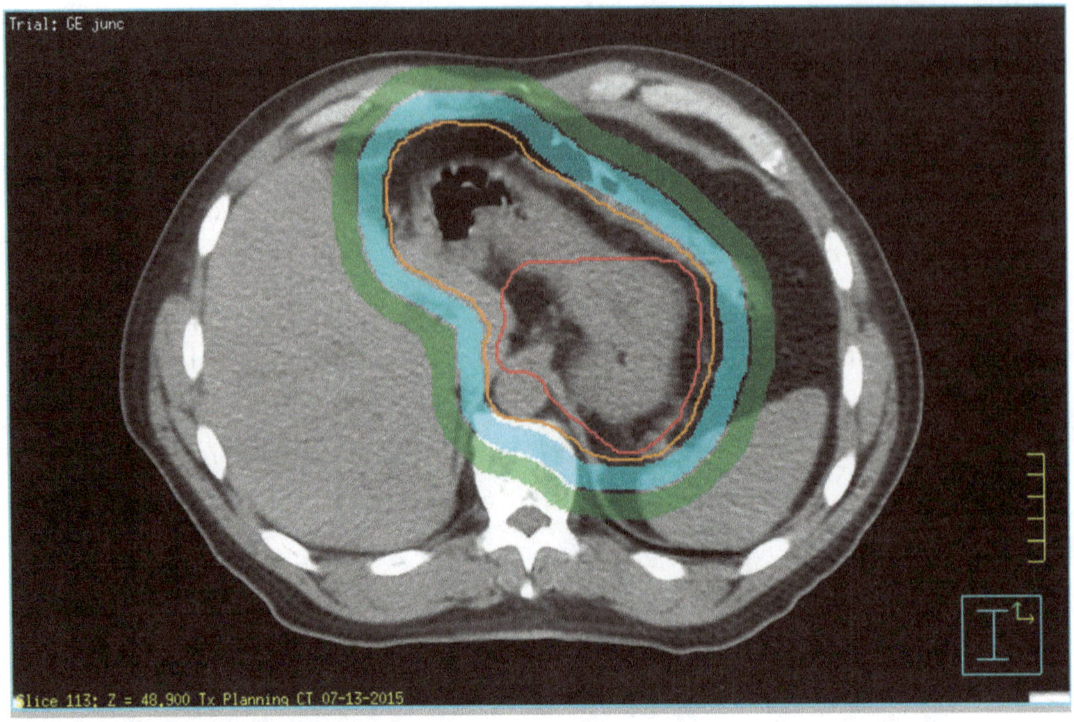

Figure 7.5 Optimization structures in color wash display.

Optimization contours: PTV rings can be used to make the dose conformal to the target as well as to prevent streaks of low dose.

In Figure 7.5, the blue ring is a PTV ring. Instead of a body ring, the planner has used the green ring in this case. The objectives for maximum dose of the green ring are lower than those of the PTV ring. For example, the blue ring might be set up to a maximum dose of 4500 cGy, while the green ring might be set up to a maximum dose of 3800 cGy.

Optimizer: Most optimizations can be started with basic objectives, covering the target and avoiding PTV rings and other optimization structures.

For example, one might start with an optimization along the following criteria:

PTV 5250	→ Minimum dose = 5250 cGy
PTV 5250	→ Maximum dose = 5500 cGy
PTV 4500–PTV 5250	→ Minimum dose = 4250 cGy
PTV 4500–PTV 5250	→ Maximum dose = 5000 cGy
PTV ring	→ Maximum dose = 4500 cGy
Outer ring (green)	→ Maximum dose = 3800 cGy

The dose distribution in the target is meant to be uniform since functional organs are being irradiated. The PTV ring can be optimized to get a dose less than 4500 cGy, and the outer ring can be optimized to a maximum dose of about 70% of the high dose prescription. If the other ring is pushed down to a dramatically lower dose with respect to the target prescription, the dose in the target will become less uniform.

After optimizing, the planner has a better understanding of the coverage that can be achieved and the dose that the OARs are receiving. In further optimizations, the planner can introduce other criteria that lower dose to OARs while conserving coverage. In this case, OARs included spinal canal, kidneys, bowels, heart, and lungs. For coverage, the planner can put minimum DVH objectives in the target, such as 100% of dose covering 95% of the target.

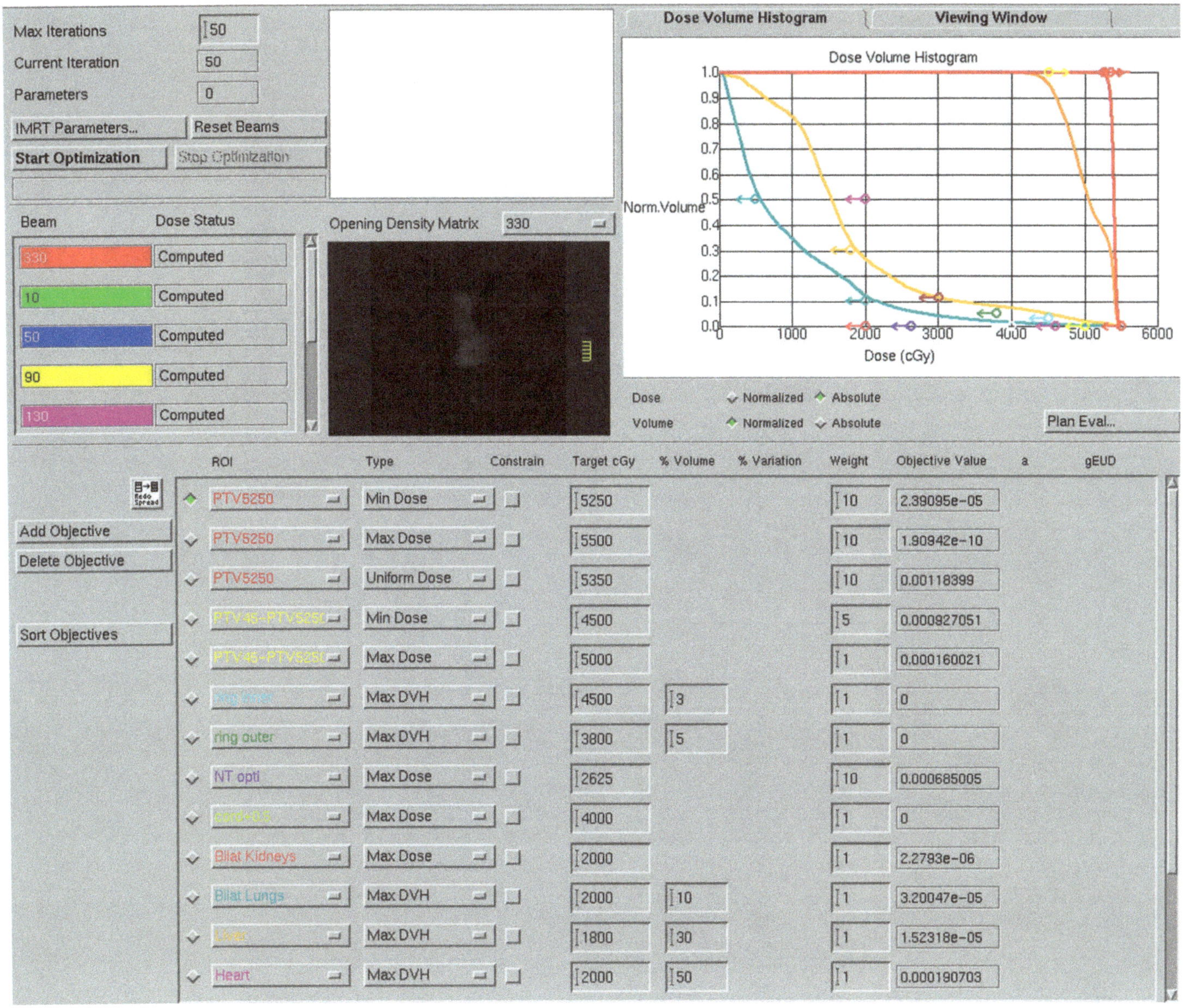

Figure 7.6 Optimizer window of Pinnacle 9.8.

Dose Distribution

Figures 7.7 and 7.8 show the final dose distribution for IMRT and VMAT plans. A measure of a good dose distribution is not only the conformality of the dose to the target, but also the lower isodose lines.

This 50% isodose line should be present in a small, reasonable volume around the target, and it should not form streaks going to the surface of the patient. The final DVH obtained is one of the most important tools to assess a radiation treatment plan, along with the dose distribution seen on the CT dataset. Notice the 5250-cGy isodose line is contained to PTV 5250 only, and it is not a significant portion of the PTV 4500 contour. This sharp fall-off of PTV 5250 is an indication of a good dose conformity.

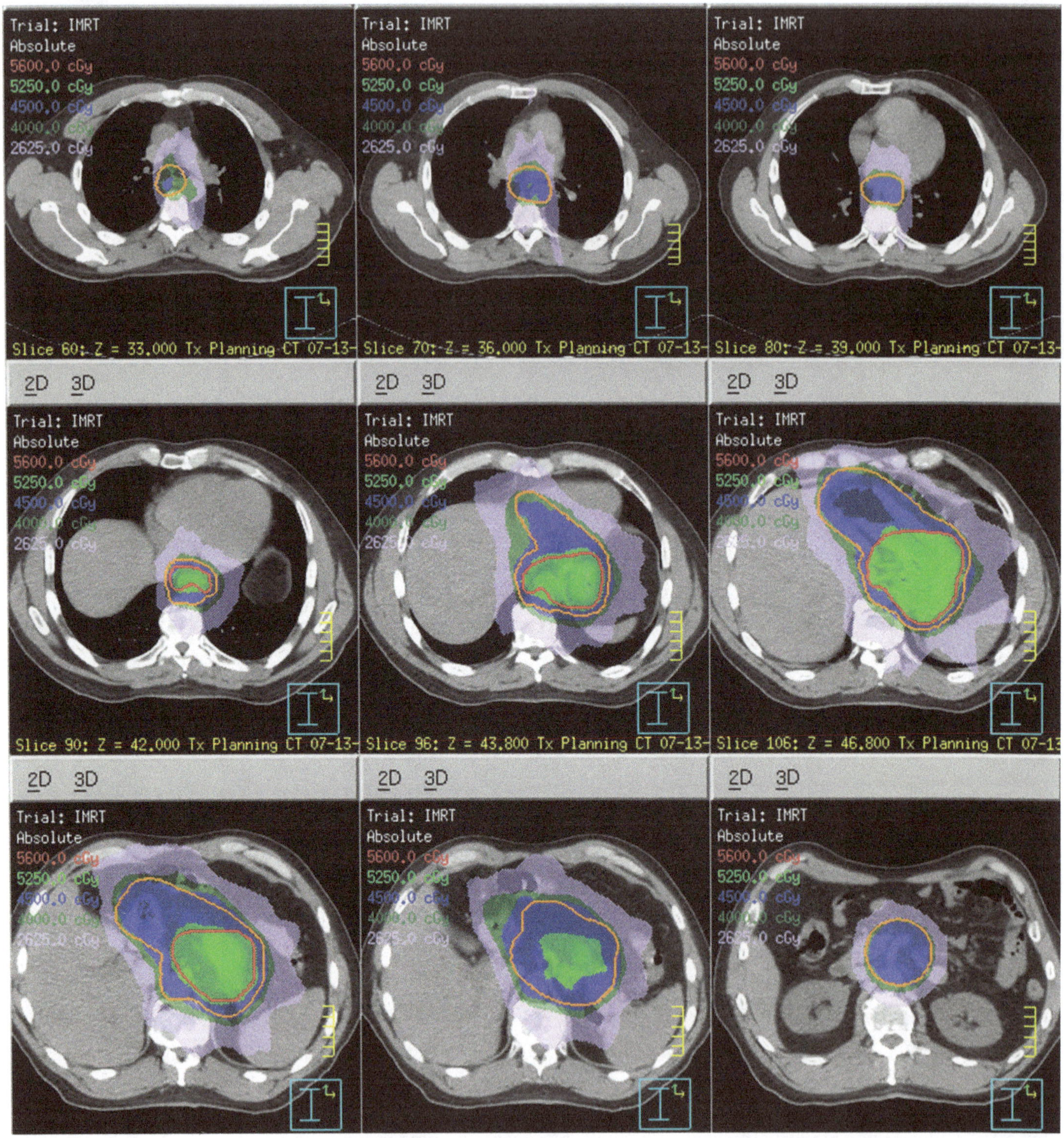

Figure 7.7 Final IMRT dose distribution: 5250 cGy (green), 4500 cGy (blue), 4000 cGy (dark green), 2625 cGy (50% IDL, purple).

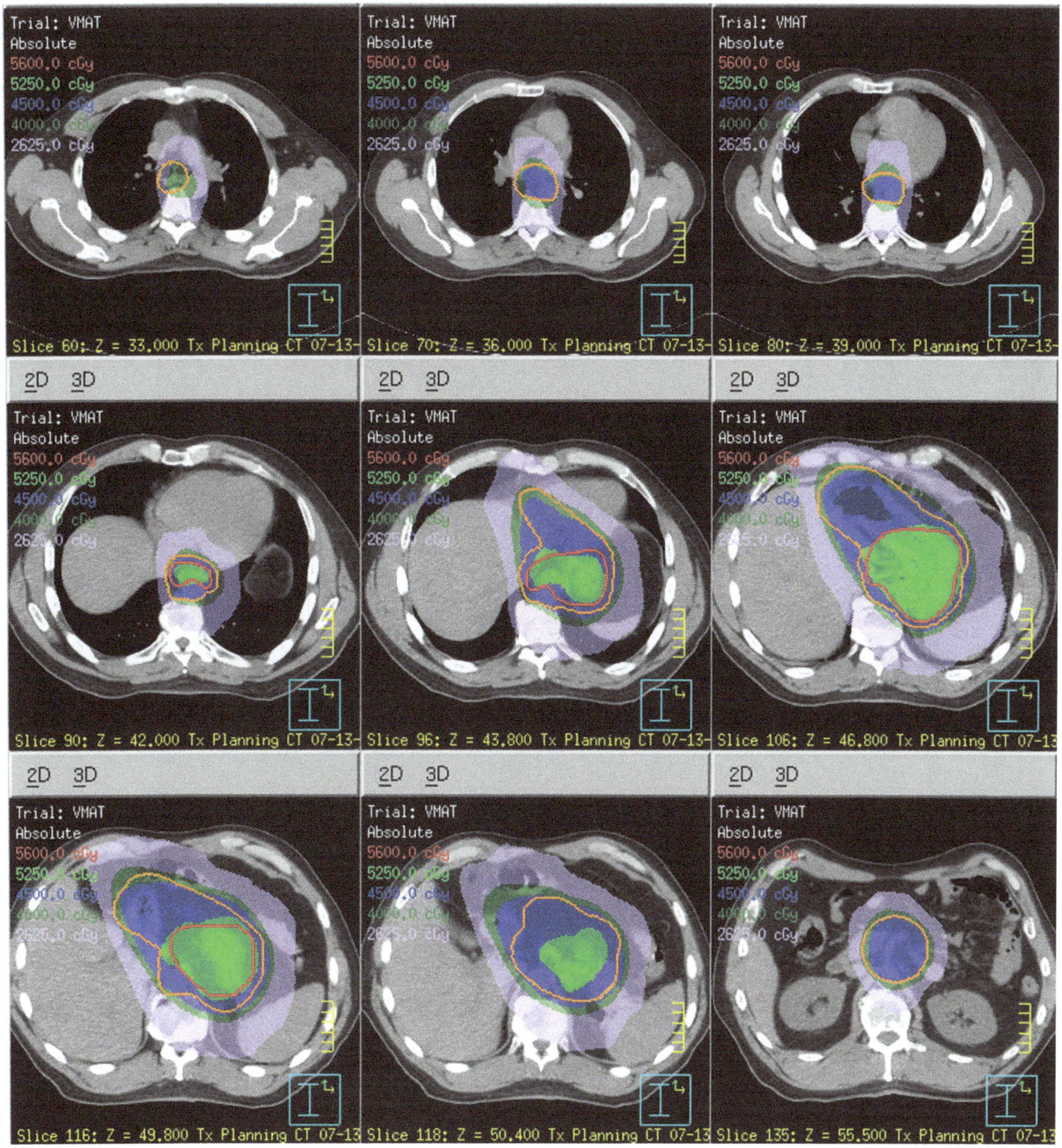

Figure 7.8 Final VMAT dose distribution: 5250 cGy (green), 4500 cGy (blue), 4000 cGy (dark green), 2625 cGy (50% IDL, purple).

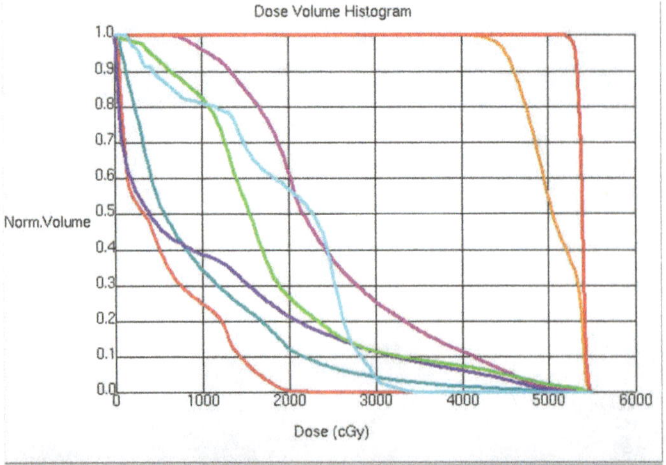

ROI Statistics

Line Type	ROI	Trial or Record	Min.	Max.	Mean	Std. Dev.
	Bilat Kidneys	IMRT	37.7	2203.7	554.6	550.6
	Bilat Lungs	IMRT	31.5	5477.8	938.7	925.6
	Bowel	IMRT	3.6	5397.6	1049.8	1321.6
	Heart	IMRT	466.7	5458.4	2447.3	1022.0
	Liver	IMRT	42.4	5482.6	1777.1	1081.7
	PTV4500	IMRT	3318.0	5484.1	5041.5	317.9
	PTV5250	IMRT	4939.8	5484.1	5388.0	38.8
	cord	IMRT	149.0	3429.0	1912.6	866.8

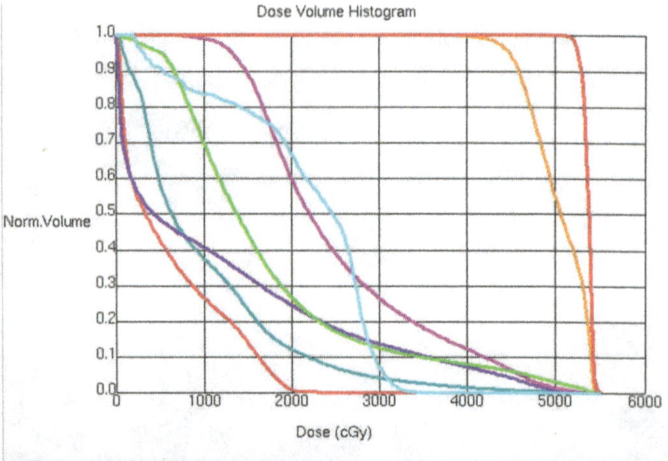

ROI Statistics

Line Type	ROI	Trial or Record	Min.	Max.	Mean	Std. Dev.
	Bilat Kidneys	VMAT	37.9	2274.8	607.8	604.7
	Bilat Lungs	VMAT	31.1	5321.4	985.0	892.6
	Bowel	VMAT	3.7	5424.3	1145.2	1401.5
	Heart	VMAT	621.1	5397.2	2501.9	1004.7
	Liver	VMAT	44.0	5466.3	1706.0	1169.8
	PTV4500	VMAT	3178.5	5514.2	5023.5	326.6
	PTV5250	VMAT	4680.7	5514.2	5362.0	57.1
	cord	VMAT	198.8	3433.5	2126.1	864.3

Figure 7.9 Dose volume histograms (DVHs) for IMRT (left) and VMAT (right) plans.

Case Number pTA02: Thorax and Abdomen—Left Upper Lung Cancer Radiation Treatment

General Information

Brief Clinical History

70-year-old female diagnosed with keratinizing squamous cell carcinoma of the left upper lung with extension into the left perihilar region. Concurrent chemotherapy and radiation therapy.

Prescription

Left upper lung:
PTV 6250 = 250 cGy × 25 fractions
PTV 4500 = 180 cGy × 25 fractions

Simulation/Comments

The simulation was accomplished with a vacuum cushion immobilization setup. Fusion of images was done with a recent PET/CT for evaluation of the extent of the lesion and targeting by the radiation oncologist. A 4D CT was acquired to verify the extension of the PTV volume as well as the respiratory motion.

Target

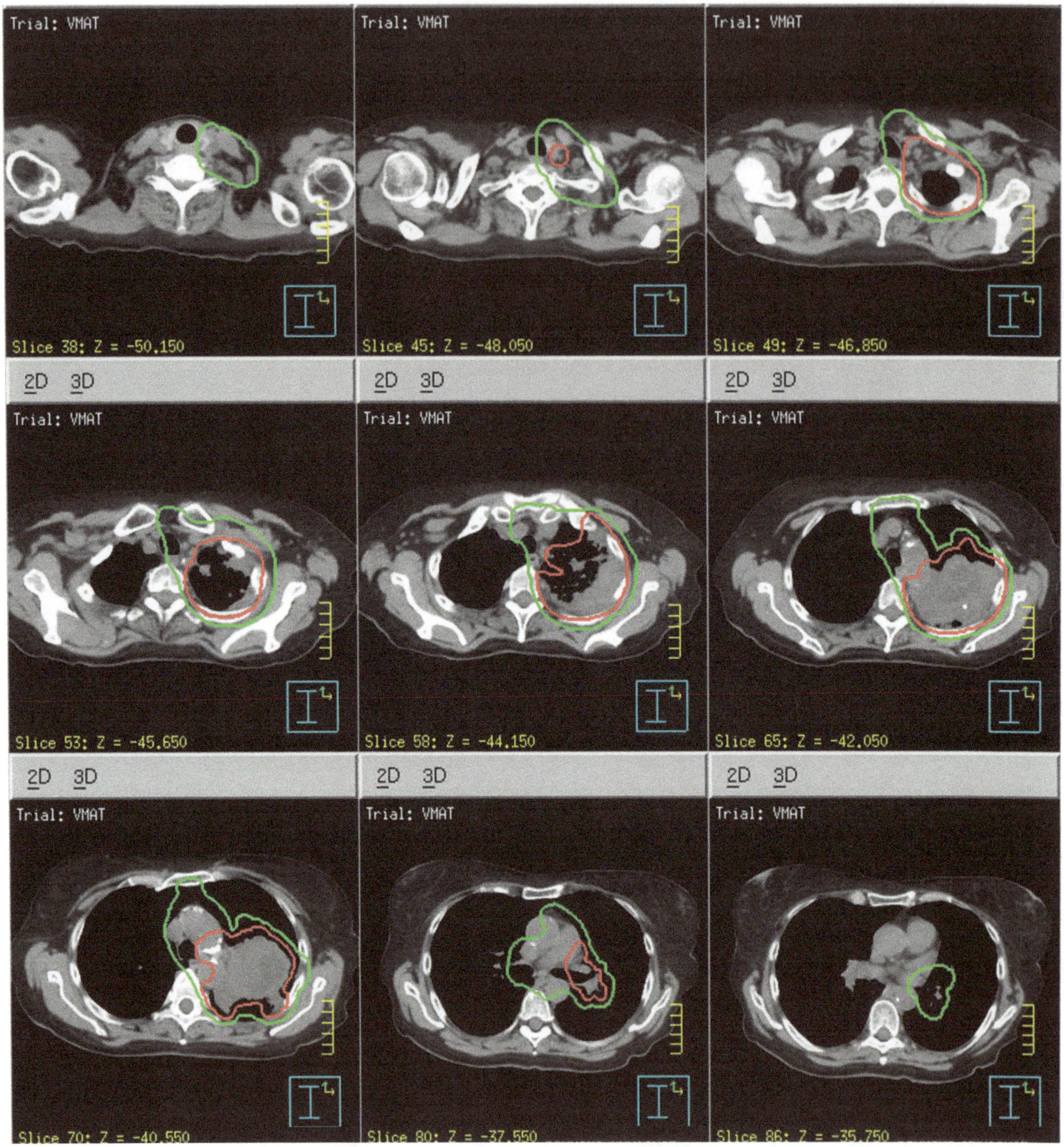

Figure 7.10 PTV contours for case pTA02.

There are two PTV contours drawn by the radiation oncologist: PTV 6250 and PTV 4500. It is important to accurately contour healthy tissue, especially when treatment volumes overlap the organs at risk. The planner can evaluate the healthy tissue on the DVH, representing it as a raw lung contour and by making a new contour by subtracting the GTV or PTV volume.

Beam Arrangement

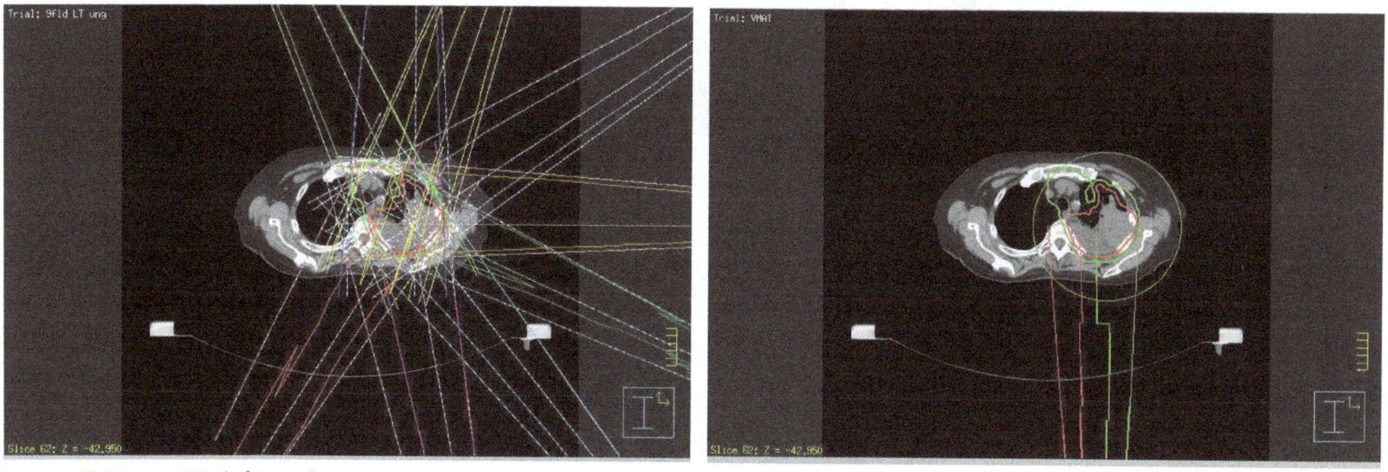

Figure 7.11 IMRT (left) and VMAT (right) beam arrangements for case pTA02.

A nine-beam arrangement has been chosen for case pTA02. Due to the location of the isocenter and the one-sided nature of the treatment volume, the beams can be placed in such a way that they do not enter the patient through the healthy right lung.

Gantry Angle	Couch	Collimator	Energy (MV)
170	0	0	10
140	0	0	10
115	0	0	10
90	0	0	10
56	0	0	10
18	0	0	10
0	0	0	10
335	0	0	10
212	0	0	10

The VMAT beam arrangement takes two arcs that go around the patient. In this case, the delivery machine has variable gantry speed and dose rate, which means that the plan can be optimized in such a way that there is very small (insignificant) dose delivered through the right lung while the gantry accelerates its way in the rotation. Machines with constant dose rates will also offer sparing, but at a lesser degree. The planner has to verify possible collisions between the gantry and the couch because of the location of the isocenter. Another arc arrangement to be considered is that of partial arcs encompassing only the left side of the patient.

Figure 7.12 IMRT and VMAT beam arrangement for case pTA02 on a surface rendering.

Optimization

Dose levels within target: The different contours for dose prescriptions have to be separated prior optimization, as shown in Figure 7.13. Notice the gap left between dose levels. This helps achieve the dose gradient needed between the two PTV contours.

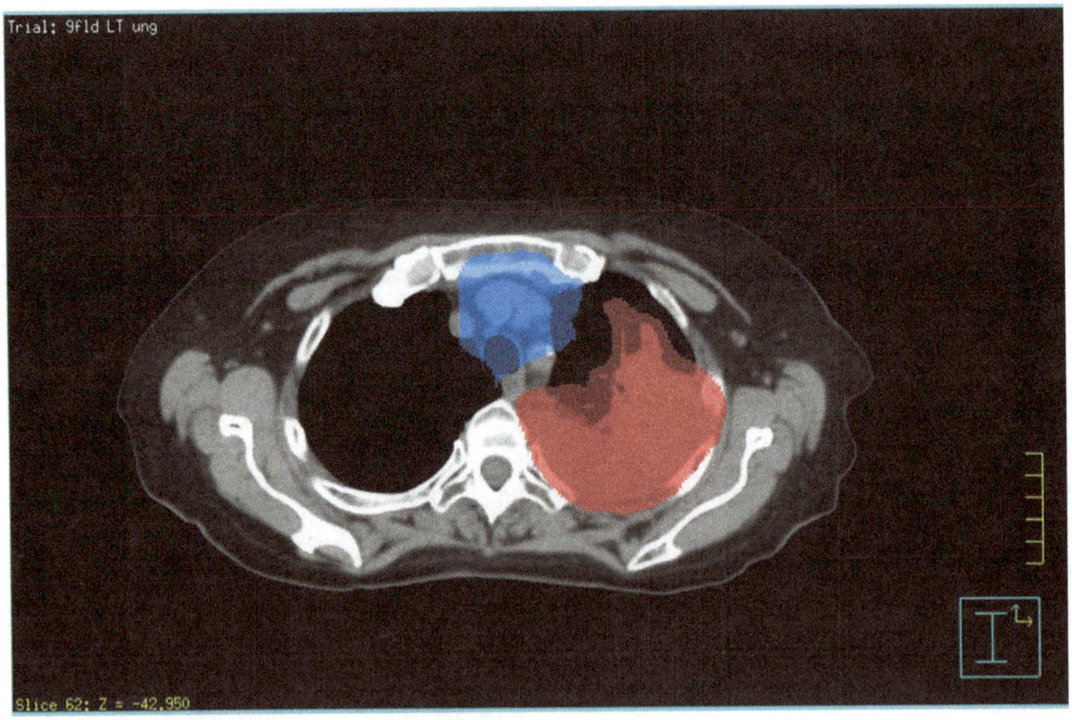

Figure 7.13 Optimization contours for different level PTV.

Optimization contours: PTV rings are used to make the dose conformal to the target as well as to prevent streaks of low dose. The body ring helps pull in low regions of dose from the body surface. These contours can be seen in Figure 7.14. The planner can decide to not have the body ring overlap with other optimization structures.

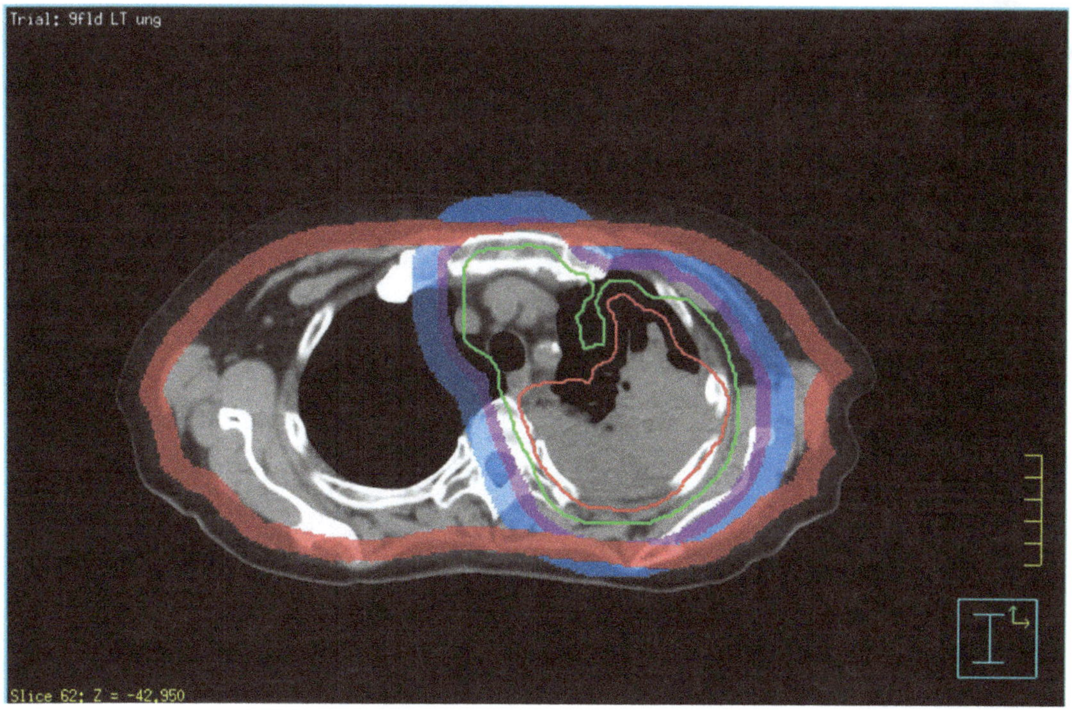

Figure 7.14 Optimization contours: PTV rings (purple and blue colorwash) and body ring (red colorwash).

Optimizer: The optimization process can be started with some basic objectives that cover the target and avoid normal structures. For example, the planner might start with an optimization along the following criteria

PTV 6250	→ Minimum dose = 6250 cGy
PTV 6250	→ Maximum dose = 6600 cGy
PTV 4500–PTV 6250	→ Minimum dose = 4500 cGy
PTV 4500–PTV 6250	→ Maximum dose = 6250 cGy
PTV inner ring	→ Maximum dose = 4500 cGy
Outer PTV ring	→ Maximum dose = 3800 cGy
Body ring	→ Maximum dose = 3000 cGy

The planner can review the DVH, target coverage and dose constraints after the first optimization and add objectives and contours accordingly in further optimizations.

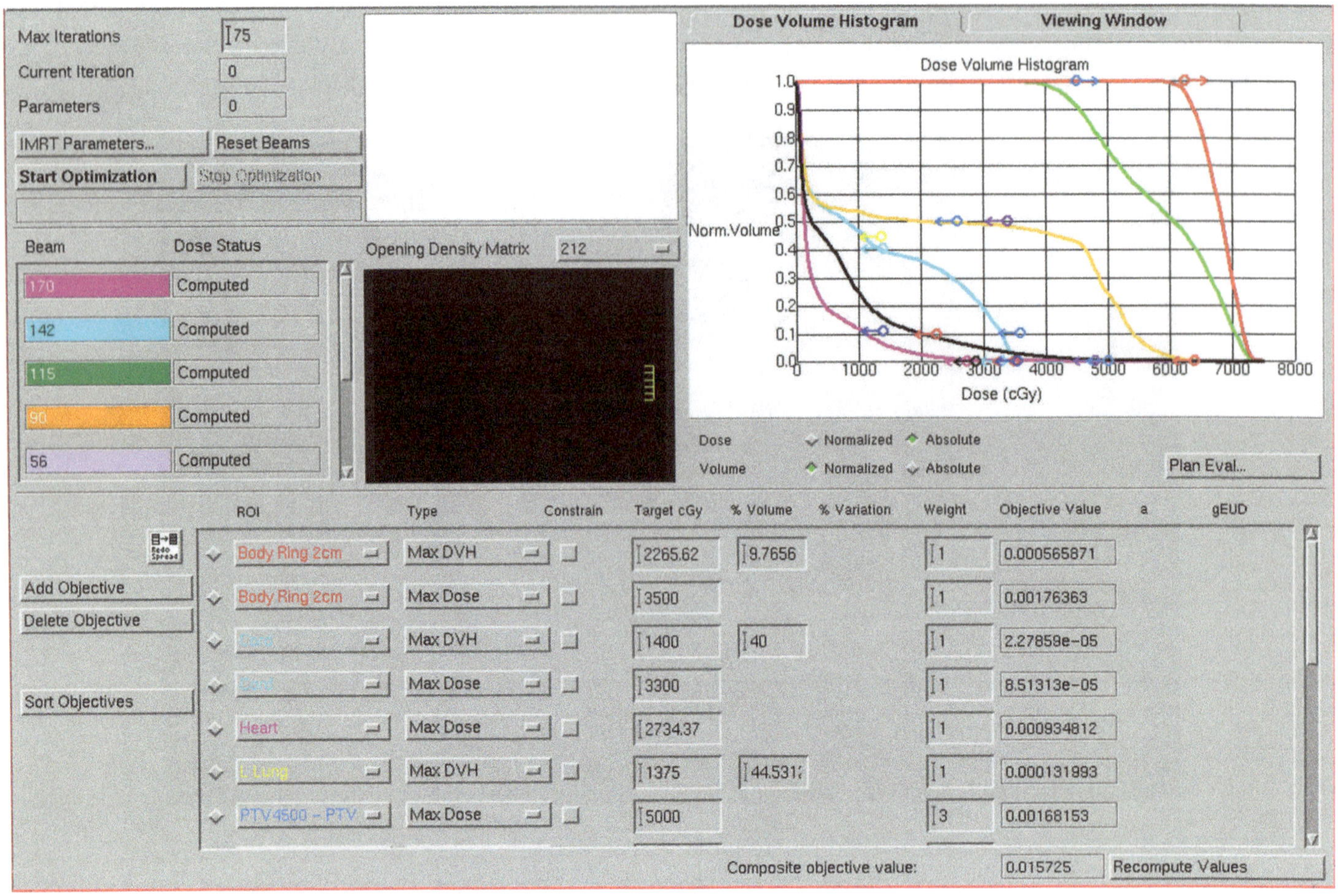

Figure 7.15 IMRT Optimizer window for case pTA02.

Dose Distribution

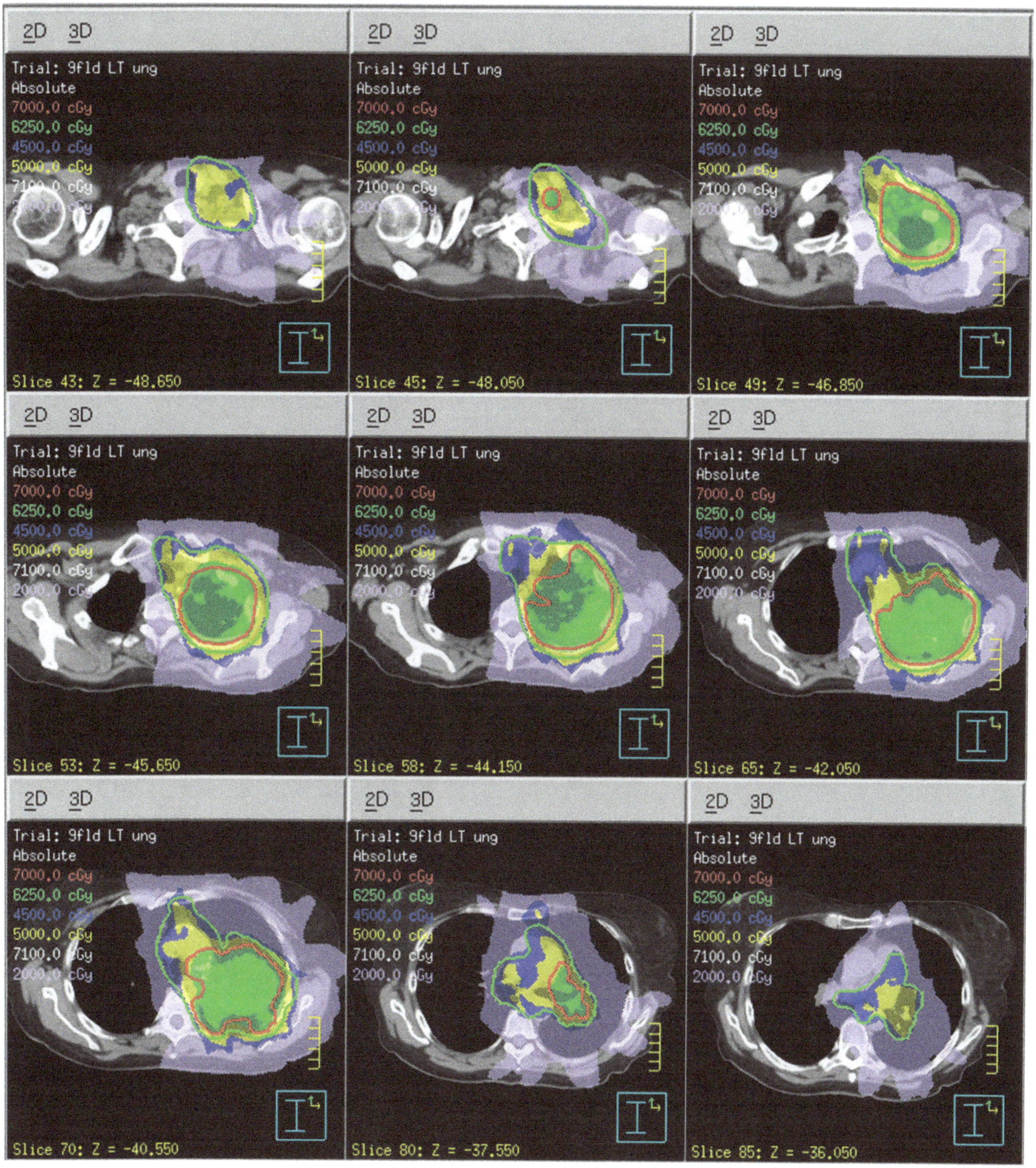

Figure 7.16 IMRT final dose distribution.

Figures 7.16 and 7.17 show the final dose distribution for IMRT and VMAT plans respectively. Both plans show similar target coverage. It is important that the areas of high dose or "hotspots" are within the CTV. In the IMRT plan, the 2000 cGy isodose line can be seen at the dose entrance points in the IMRT plan.

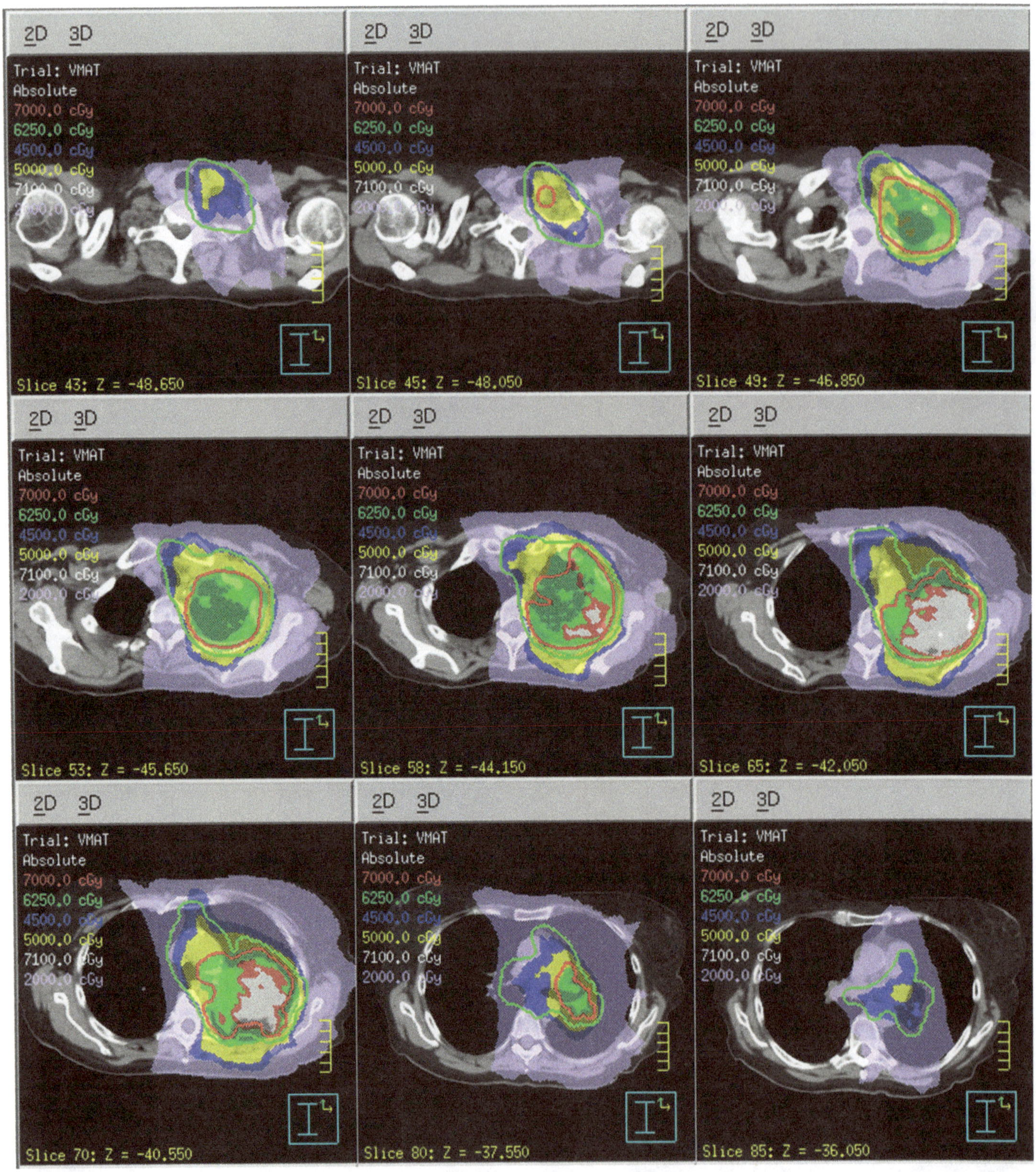

Figure 7.17 VMAT final dose distribution.

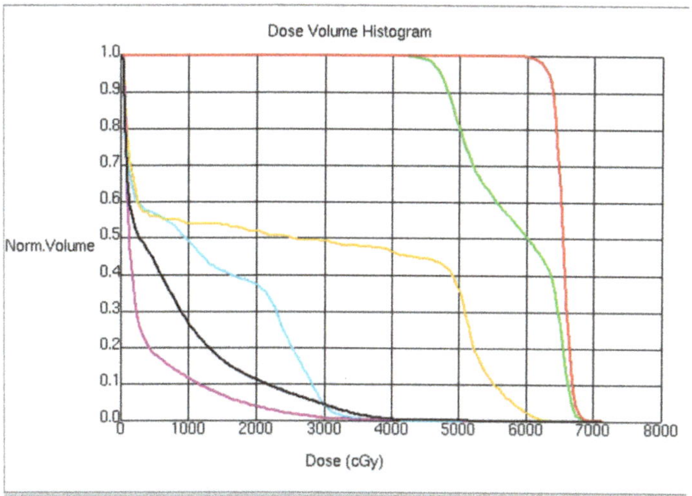

ROI Statistics

Line Type	ROI	Trial or Record	Min.	Max.	Mean	Std. Dev.
	Cord	9fld LT ung	32.2	4136.9	758.9	1082.0
	Esophagus	9fld LT ung	40.3	6319.0	2686.5	2467.8
	Heart	9fld LT ung	44.0	5067.7	375.1	593.6
	PTV4500	9fld LT ung	--	7114.0	5822.0	733.9
	PTV6250	9fld LT ung	5309.2	7114.0	6535.8	135.6
	combined lung	9fld LT ung	27.0	5138.5	730.5	909.4

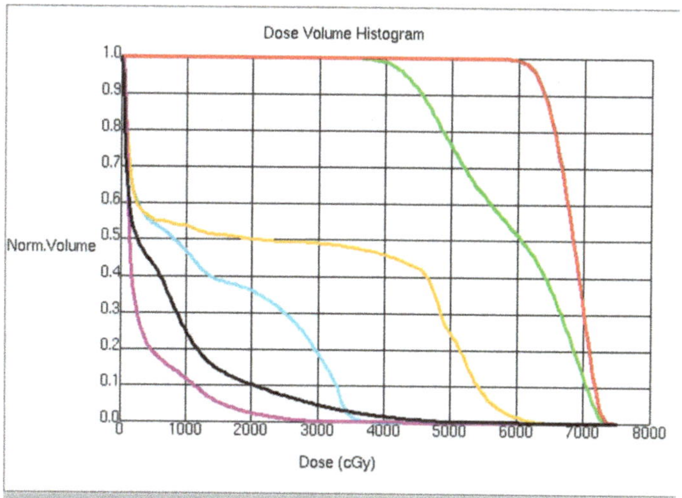

ROI Statistics

Line Type	ROI	Trial or Record	Min.	Max.	Mean	Std. Dev.
	Cord	VMAT	22.1	3828.5	812.6	1210.9
	Esophagus	VMAT	43.2	6406.8	2580.2	2406.6
	Heart	VMAT	52.8	4310.2	359.5	513.6
	PTV4500	VMAT	33.1	7460.3	5883.4	966.8
	PTV6250	VMAT	5344.3	7460.3	6808.9	299.5
	combined lung	VMAT	12.5	6049.2	712.6	947.2

Figure 7.18 Final DVH for case pTA02 for IMRT (left) and VMAT (right).

Case Number pTA03: Thorax and Abdomen—Mediastinum and Lung Cancer Radiation Treatment

General Information

Brief Clinical History

46-year-old male diagnosed with non-small-cell lung cancer (NSCLC), presenting with right hilar mass. Concurrent chemotherapy and radiation therapy prior to surgery.

Prescription

Mediastinum and lung:
PTV 6250 = 250 cGy × 25 fractions
PTV 4500 = 180 cGy × 25 fractions

Simulation/Comments

The simulation was accomplished with a vacuum cushion as an immobilization device. Fusion of images was done with a recent PET/CT for evaluation of the extent of the lesion and targeting by the radiation oncologist. A 4D CT was acquired to assess the motion of the lesion during the respiratory cycle.

Target

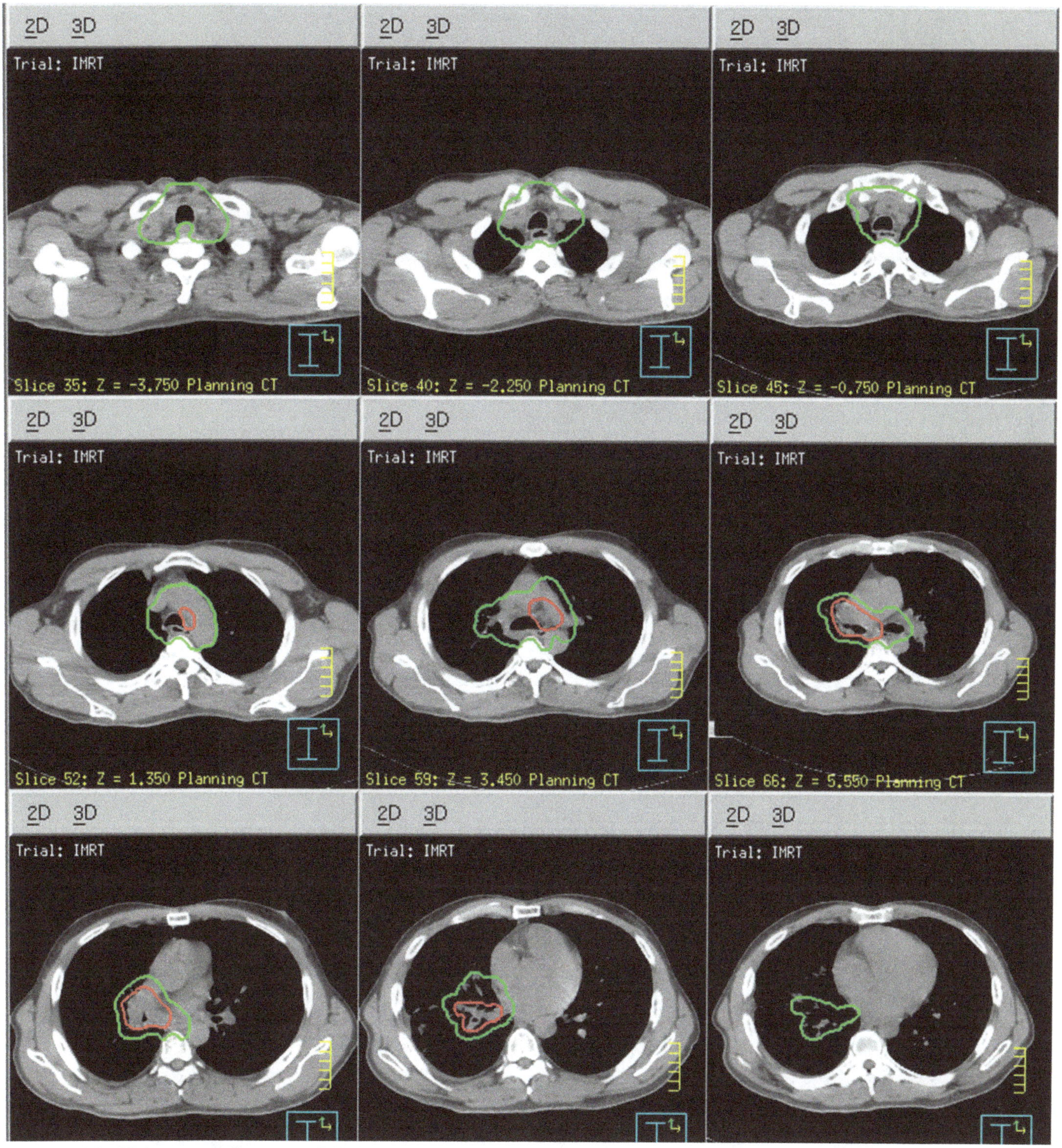

Figure 7.19 PTV contours for case pTA03.

Figure 7.19 shows the targets drawn by the radiation oncologist: PTV 6250 (red) and PTV 4500 (green). The target is overlapping with some normal structures, such as the trachea, the esophagus, great vessels, and the heart. The highly irregular shape of the target makes this case an excellent case for the use of inverse modulated techniques.

Beam Arrangement

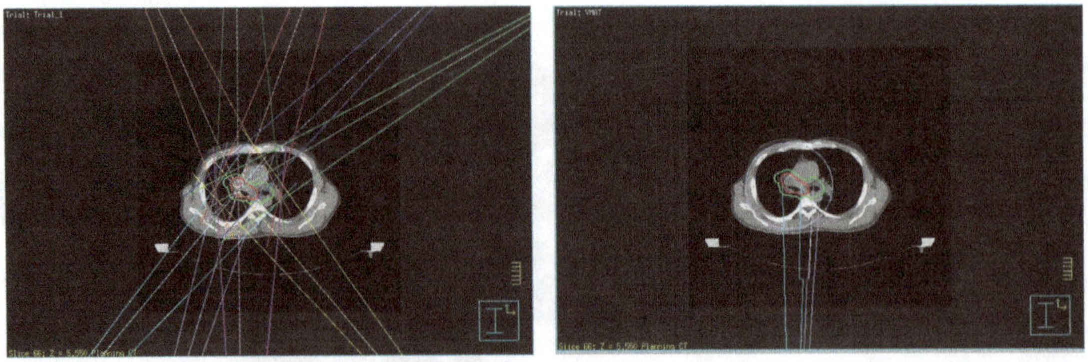

Figure 7.20 IMRT (left) and VMAT (right) beam arrangements for case pTA03.

The IMRT treatment plan had nine beams distributed in the anterior and posterior aspects of the patient. The anterior beams favor the superior aspect of the target (heart), and the posterior beams favor the inferior aspect of the target (lung lesion). The following is a summary of IMRT field arrangement.

Gantry Angle	Couch	Collimator	Energy (MV)
185	0	0	10
205	0	0	10
225	0	0	10
345	0	0	10
0	0	0	10
15	0	0	10
30	0	0	10
45	0	0	10
155	0	0	10

The VMAT plan has two arcs going around the patient, as seen in Figures 7.20 and 7.21. Two arcs provide enough control points to achieve conformity in the dose distribution while sparing healthy tissue. It is recommended to add structures to decrease the entrance dose and spare the lungs as much as reasonably achievable.

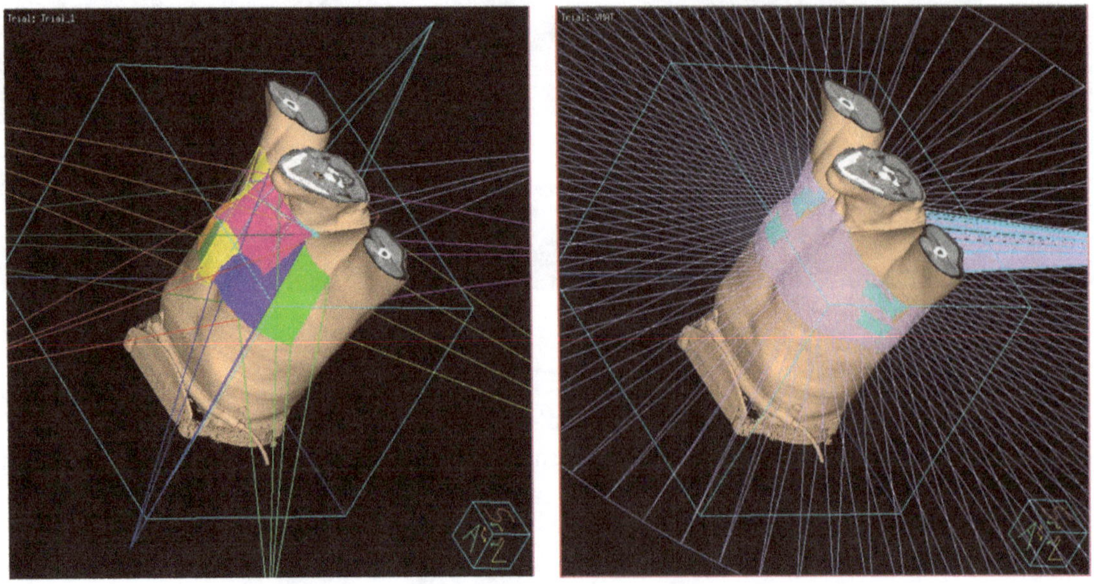

Figure 7.21 IMRT and VMAT beam arrangement for case pTA03 on a surface rendering.

Optimization

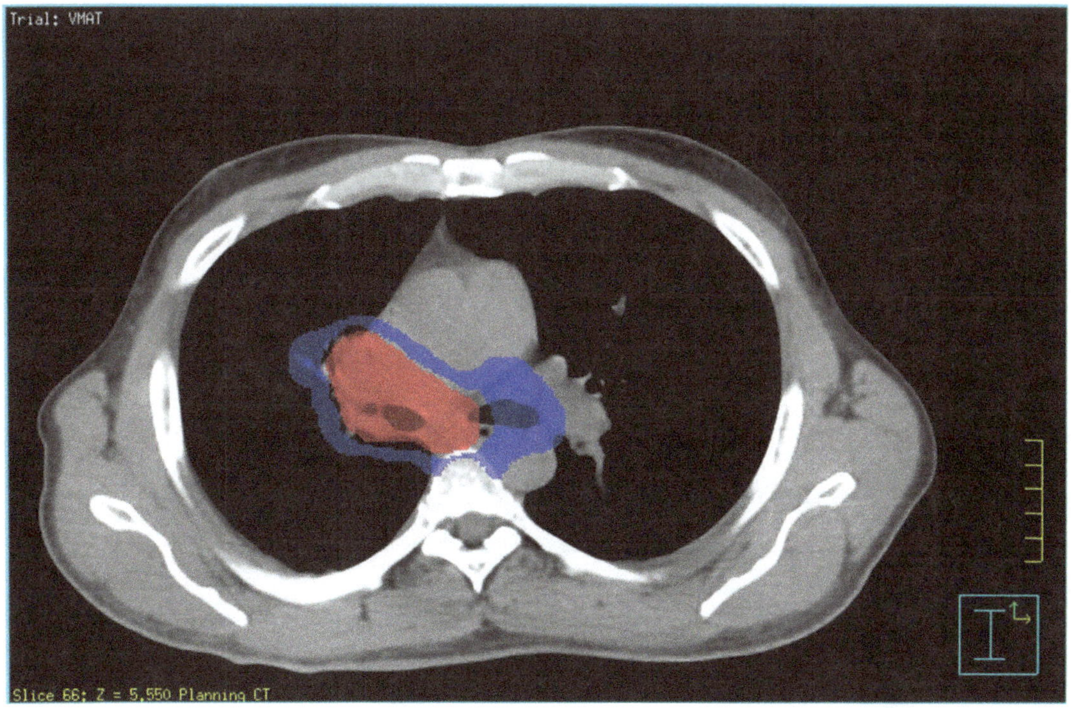

Figure 7.22 Optimization contours for different level PTV.

Dose levels within target: Figure 7.22 shows the separation between the different dose prescriptions, with an additional gap to help dose conformity during the optimization process.

Optimization contours: The main optimization contours for this case are a series of PTV rings to control the dose gradient between the target and the healthy tissue. The body ring avoids low dose streaks (50% IDL) on the surface of the patient. Add a 2 mm margin between structures to facilitate the optimization objectives.

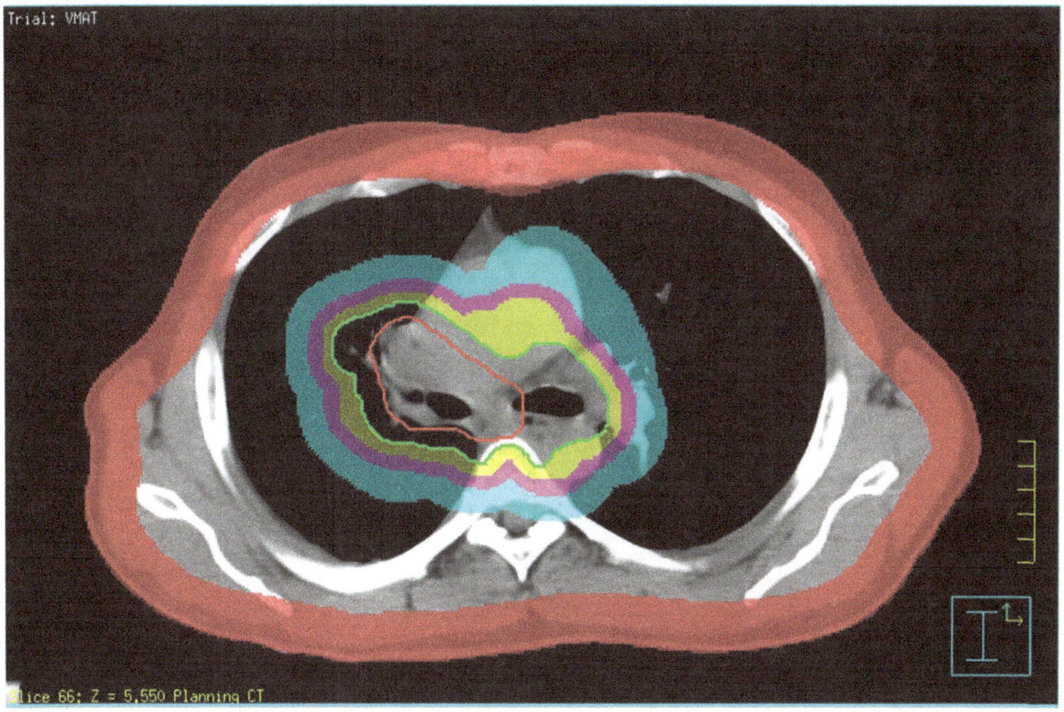

Figure 7.23 Optimization contours: PTV rings (yellow, pink and blue colorwash) and body ring (red colorwash).

Optimizer: The optimization process can be started with basic objectives focusing on covering the target and reaching some dose conformity. For example, one might start with an optimization along the following criteria:

PTV 6250	→ Minimum dose = 6250 cGy
PTV 6250	→ Maximum dose = 6600 cGy
PTV 4500–PTV 6250	→ Minimum dose = 4500 cGy
PTV 4500–PTV 6250	→ Maximum dose = 6250 cGy
PTV rings	→ Maximum dose = decreasing from 4500 cGy to 3000 cGy
Body ring	→ Maximum dose = 3000 cGy

Additional objectives for organs at risk or other contours can be added after evaluating the first DVH and dose distribution. VMAT plans can use the same parameters.

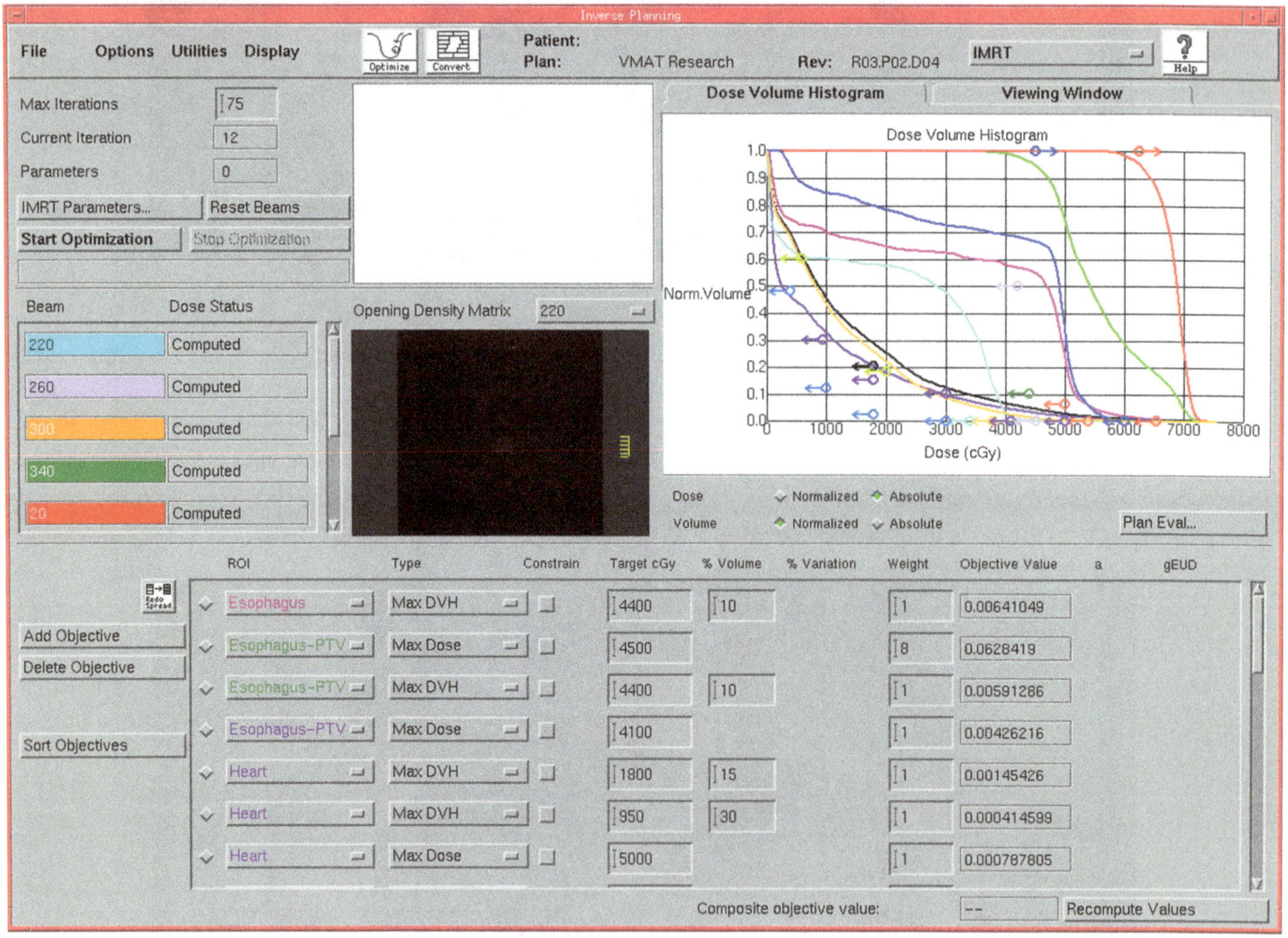

Figure 7.24 IMRT Optimizer window for case pTA03.

Dose Distribution

Figures 7.25 and 7.26 show the final dose distribution for IMRT and VMAT plans. The VMAT plan dose distribution spills in the anterior and posterior aspects of the patient in order to spare the lungs. Notice the sparing of the spinal canal in both plans. All regions of high dose (hot spots) are within the regions of CTV and PTV 6250. The 4500-cGy isodose line spillage in the AP direction can be seen as a consequence of intense sparing of the lung tissue.

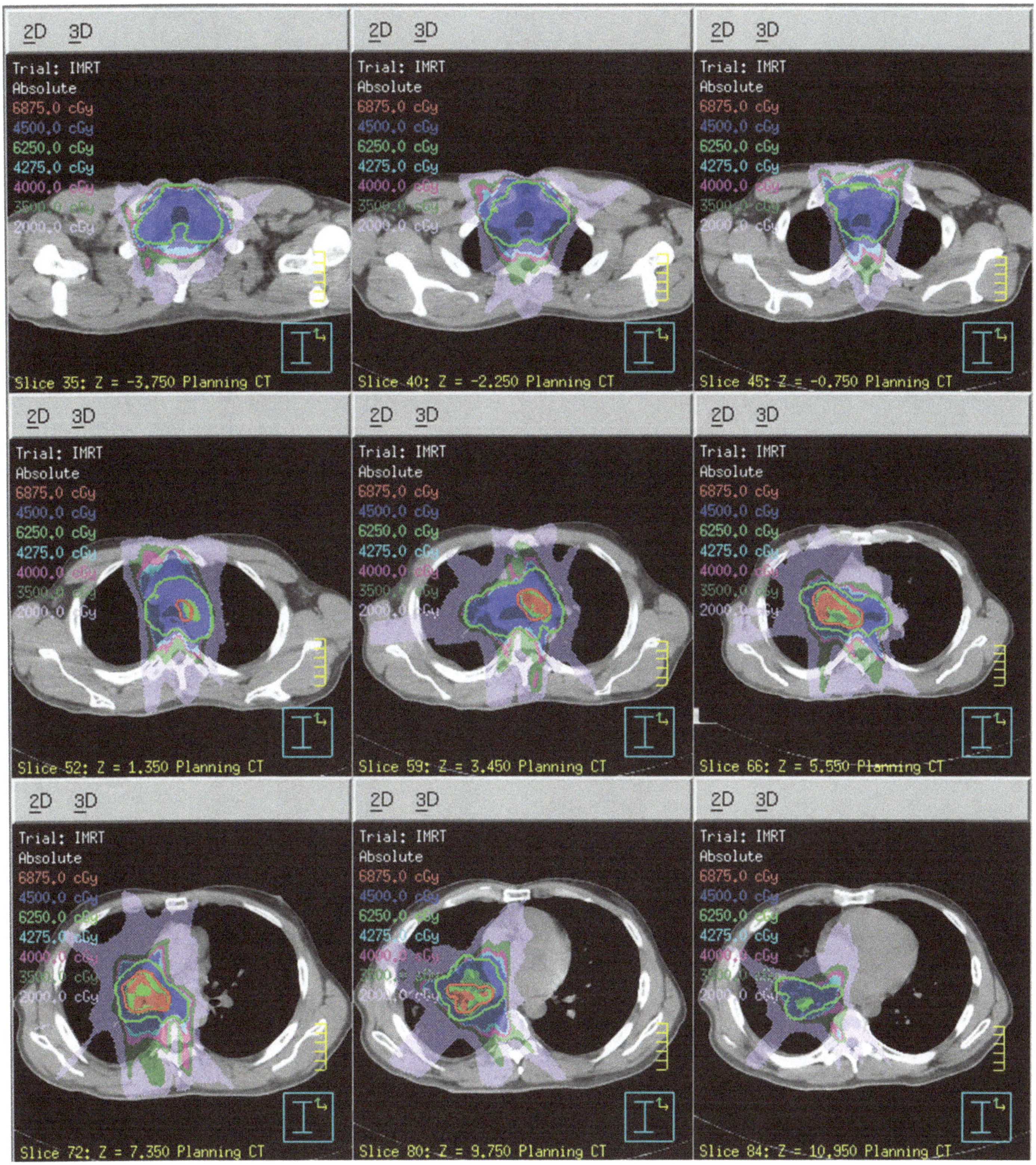

Figure 7.25 IMRT final dose distribution.

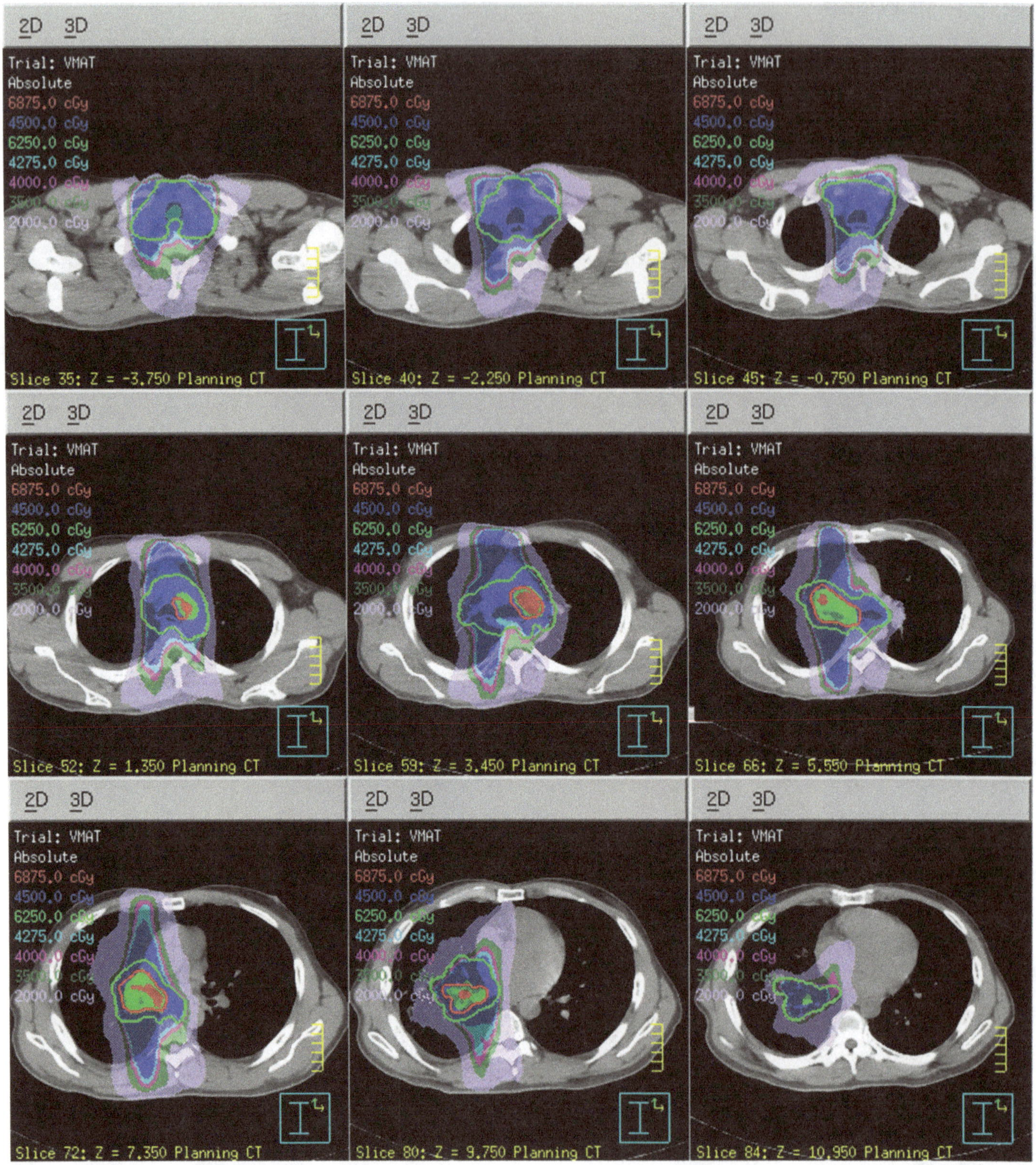

Figure 7.26 VMAT final dose distribution.

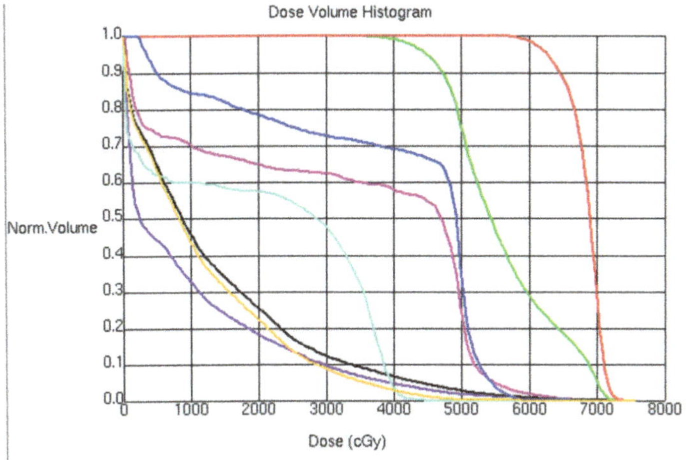

ROI Statistics

Line Type	ROI	Trial or Record	Min.	Max.	Mean	Std. Dev.
	Esophagus	IMRT	46.0	6963.2	3244.3	2191.3
	Heart	IMRT	26.5	7257.1	974.6	1314.5
	Lungs	IMRT	10.9	7214.3	1332.5	1377.0
	PTV4500	IMRT	2891.8	7400.6	5582.3	795.4
	PTV6250	IMRT	5328.3	7400.6	6825.3	275.2
	Spinal Canal	IMRT	18.4	4465.2	1201.3	1623.3
	Trachea bronchi-PTV 62	IMRT	228.2	6580.4	3406.1	2114.0
	lungs-ptv	IMRT	10.9	5862.0	1166.4	1129.4

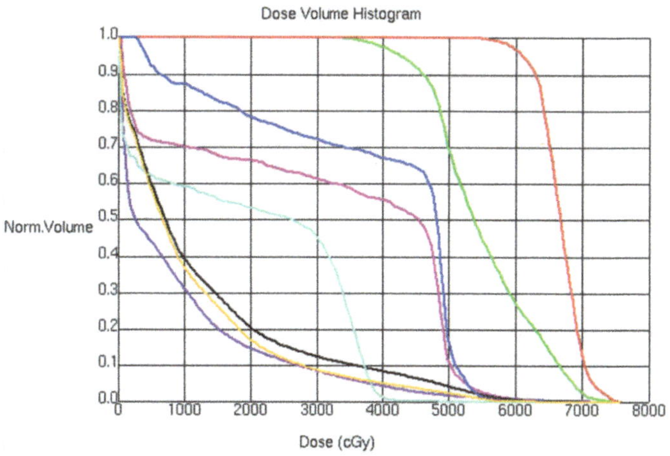

ROI Statistics

Line Type	ROI	Trial or Record	Min.	Max.	Mean	Std. Dev.
	Esophagus	VMAT	42.3	6846.9	3132.7	2121.6
	Heart	VMAT	21.8	7142.4	900.0	1248.7
	Lungs	VMAT	9.8	6990.4	1248.5	1439.9
	PTV4500	VMAT	2792.2	7552.2	5488.0	800.2
	PTV6250	VMAT	5066.4	7552.2	6660.6	322.9
	Spinal Canal	VMAT	14.6	4836.4	1126.9	1549.9
	Trachea bronchi-PTV 62	VMAT	268.4	6436.7	3333.8	2027.2
	lungs-ptv	VMAT	9.8	6330.2	1090.0	1232.1

Figure 7.27 Final DVH for case pTA03 for IMRT (left) and VMAT (right).

Case Number pTA04: Thorax and Abdomen—Gastroesophageal Cancer Radiation Treatment

General Information

Brief Clinical History

74-year-old male with distal esophageal cancer extending into cardia of stomach.

Prescription

Gastroesophageal:
PTV 5250 = 210 cGy × 25 fractions
PTV 4500 = 180 cGy × 25 fractions

Simulation/Comments

The simulation was accomplished with a vacuum cushion as an immobilization device. Fusion of images was done with a recent PET/CT for evaluation of the extent of the lesion and targeting by the radiation oncologist.

Target

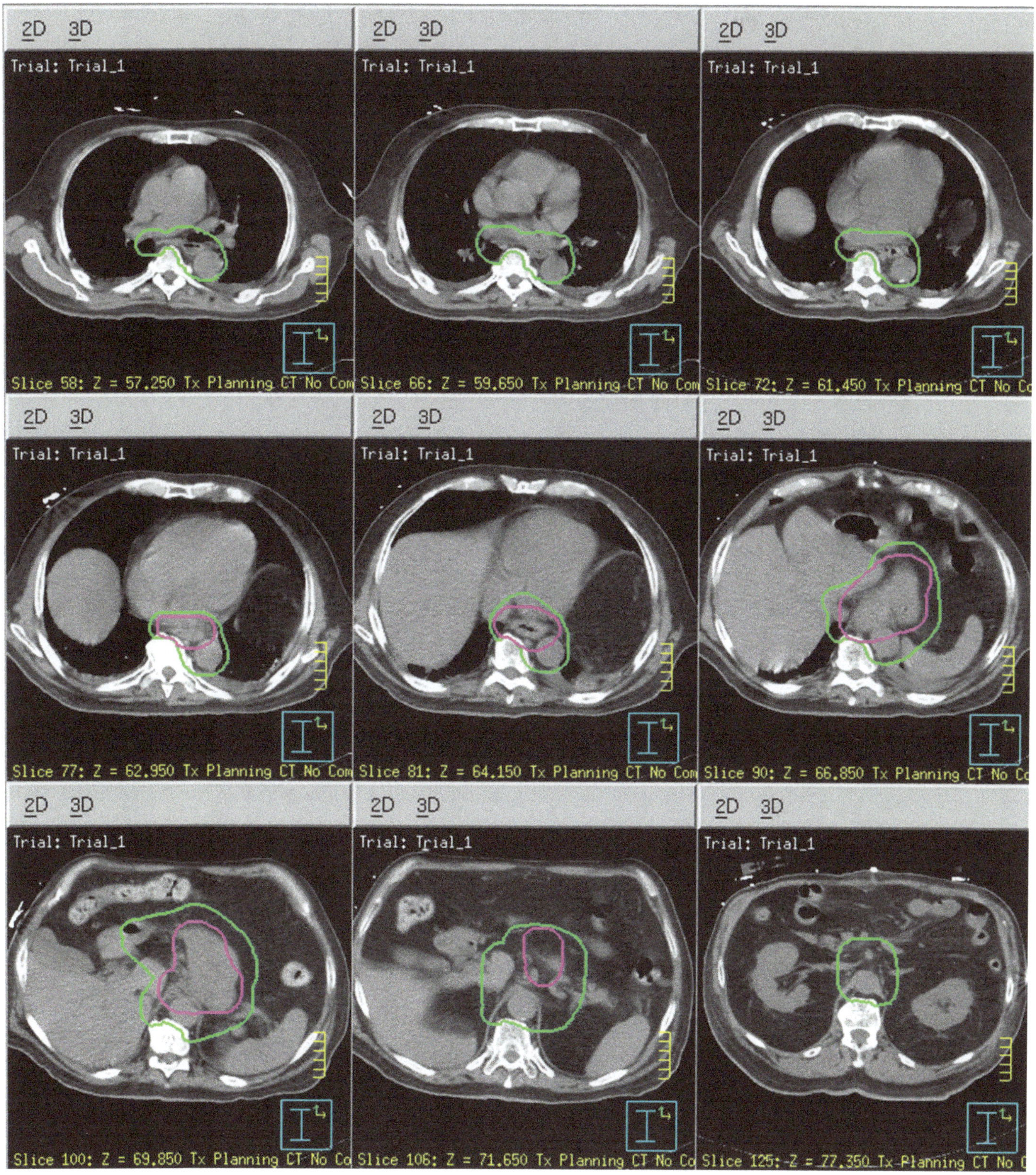

Figure 7.28 PTV contours for case pTA04.

There are two PTV contours for this case: PTV 4500 (green) and PTV 5250 (pink), as seen in Figure 7.28. The target overlaps with normal structures such as the esophagus, heart, stomach, and bowel. The irregular shape of the target makes this case suitable for IMRT and VMAT.

Beam Arrangement

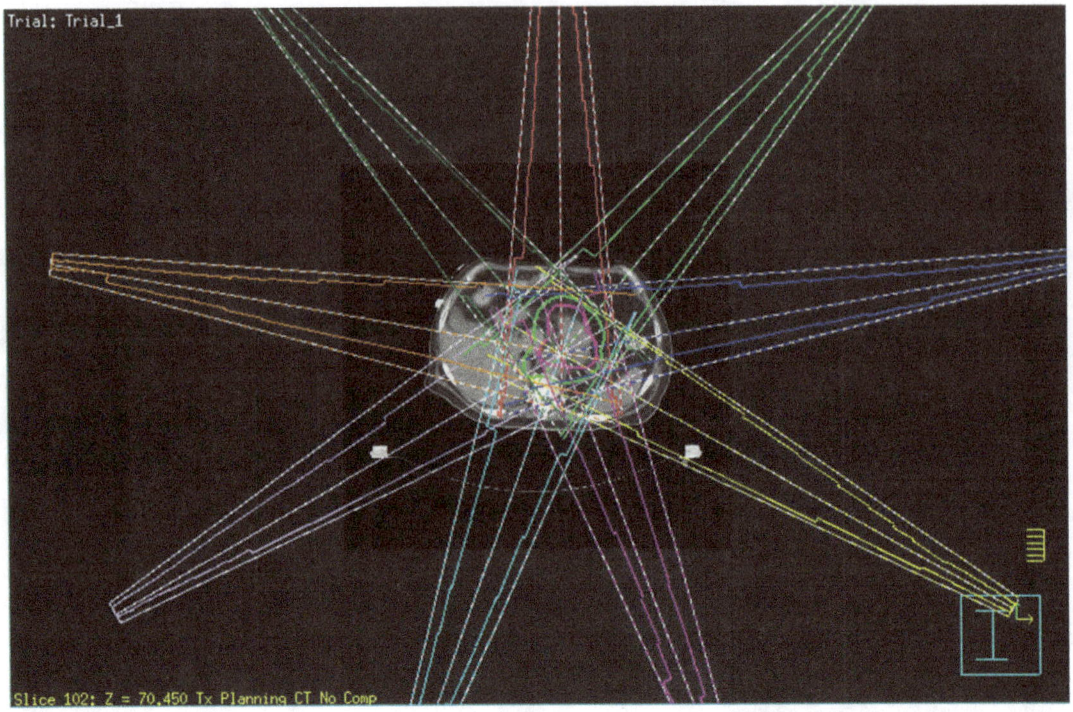

Figure 7.29 IMRT beam arrangement for case pTA04.

Figures 7.29, 7.30, and 7.31 show the beam arrangement for IMRT and VMAT plans. The IMRT plan has nine beams distributed around the patient for the centrally located lesion. The following is a summary of IMRT field arrangement.

Gantry Angle	Couch	Collimator	Energy (MV)
200	0	0	10
240	0	0	10
280	0	0	10
320	0	0	10
0	0	0	10
40	0	0	10
80	0	0	10
120	0	0	10
160	0	0	10

The VMAT plan has two arcs that go around the patient 360 degrees. The advantages of VMAT in terms of beam arrangement would be the shorter delivery time, although planning times might be increased according the computing power of the TPS. It is recommended to add optimization structures to decrease the entrance dose and avoid lung irradiation as much as reasonably achievable.

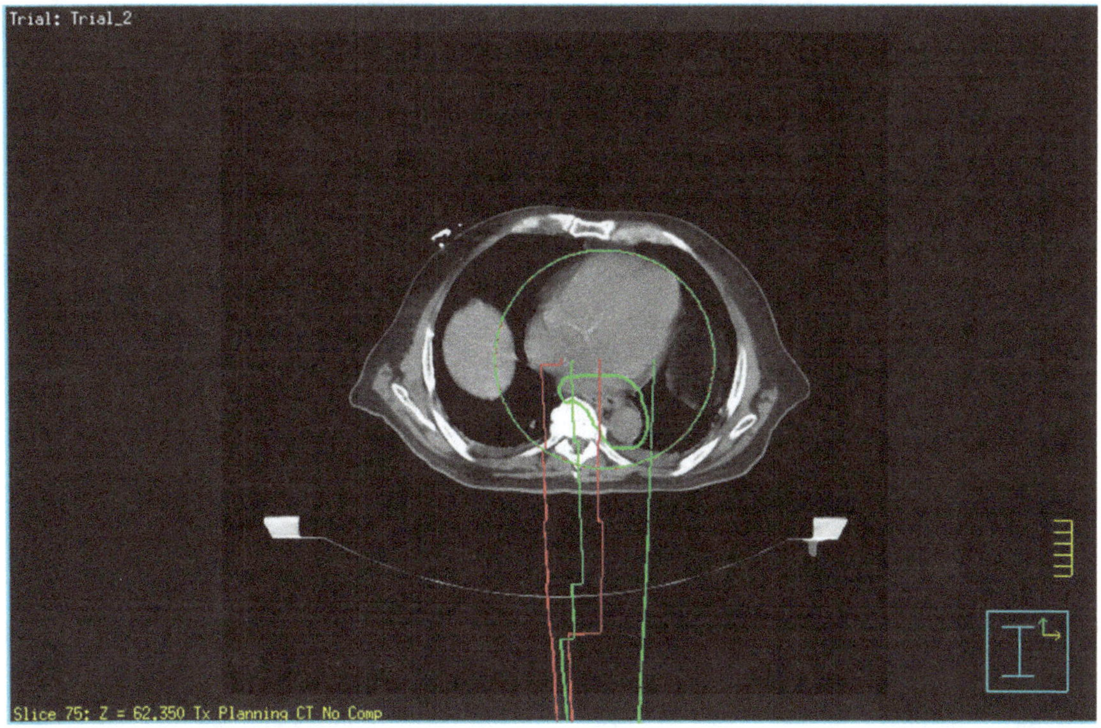

Figure 7.30 VMAT beam arrangement for case pTA04.

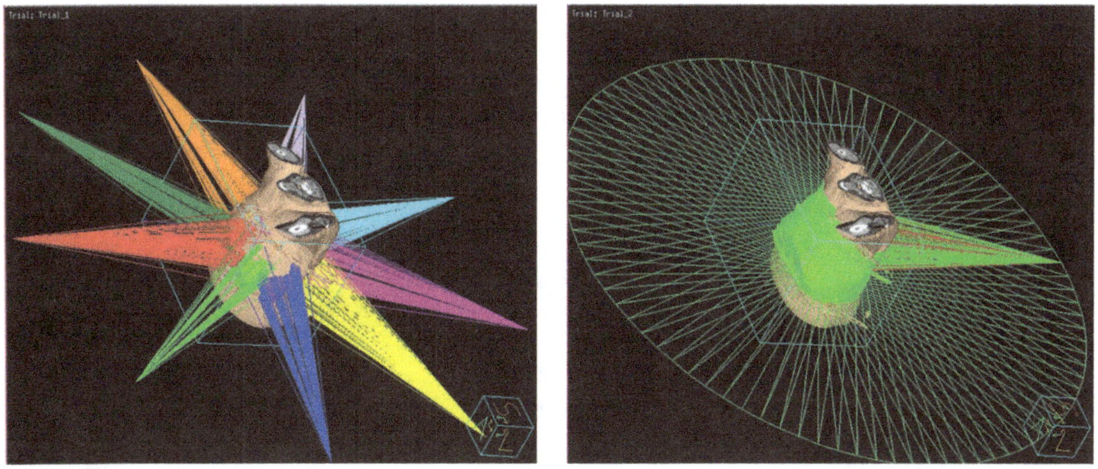

Figure 7.31 IMRT and VMAT beam arrangement for case pTA04 on a surface rendering.

Optimization

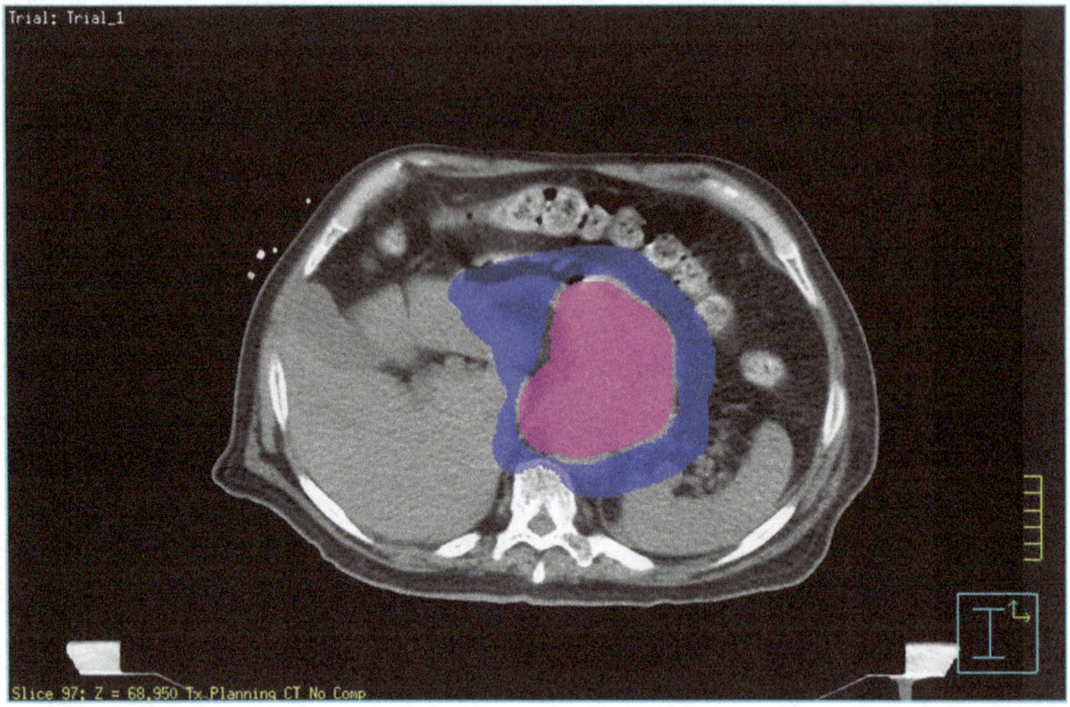

Figure 7.32 Optimization contours for different level PTV.

Dose levels within target: Figure 7.32 shows the individual contours for the two PTV dose levels. Add a 2-mm margin between structures to facilitate the optimization objectives.

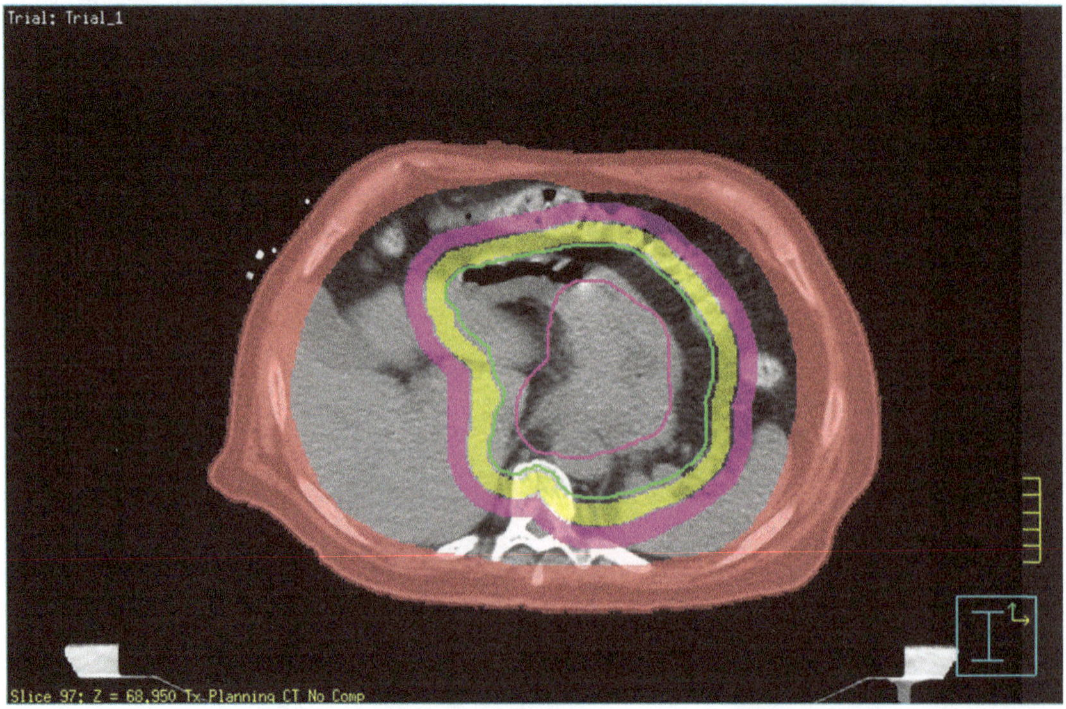

Figure 7.33 Optimization contours: PTV rings (yellow and pink colorwash) and body ring (red colorwash).

Optimization contours: The main optimization contours for this case include two PTV rings to control the dose gradient between the target and the healthy tissue: inner and outer PTV ring. The body ring avoids low dose streaks (50% IDL) on the surface of the patient.

Optimizer: The optimization process can be started with basic objectives focusing on covering the target and reaching some dose conformity.

One might start with an optimization along the following criteria:

PTV 5250	→ Minimum dose = 5250 cGy
PTV 5250	→ Maximum dose = 5500 cGy
PTV 4500–PTV 5250	→ Minimum dose = 4500 cGy
PTV 4500–PTV 5250	→ Maximum dose = 5250 cGy
PTV rings	→ Maximum dose = decreasing from 4500 cGy to 3000 cGy
Body ring	→ Maximum dose = 3000 cGy

Additional objectives for organs at risk or other contours can be added after evaluating the first DVH and dose distribution. VMAT plans can use the same parameters.

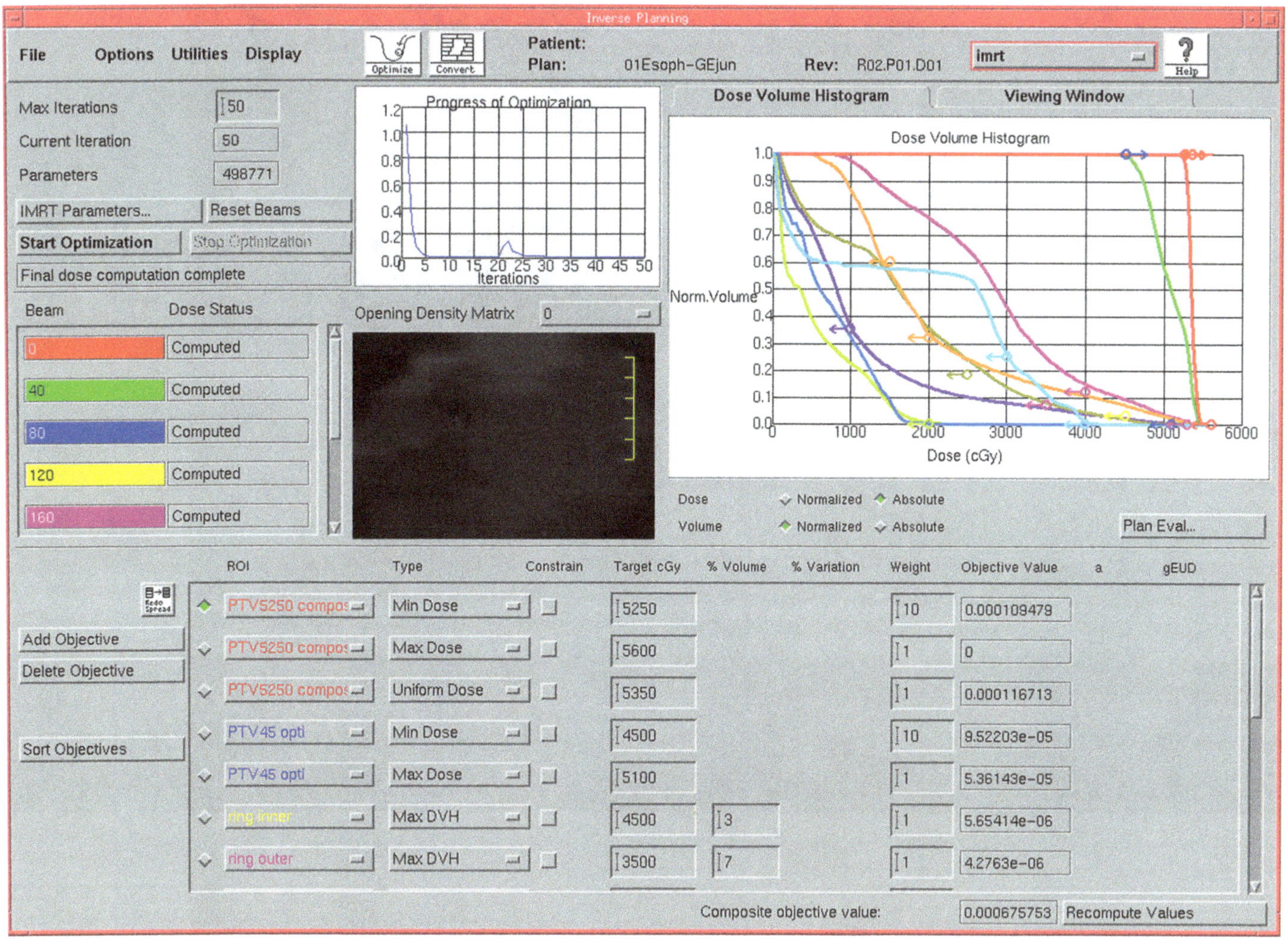

Figure 7.34 IMRT Optimizer window for case pTA04.

Dose Distribution

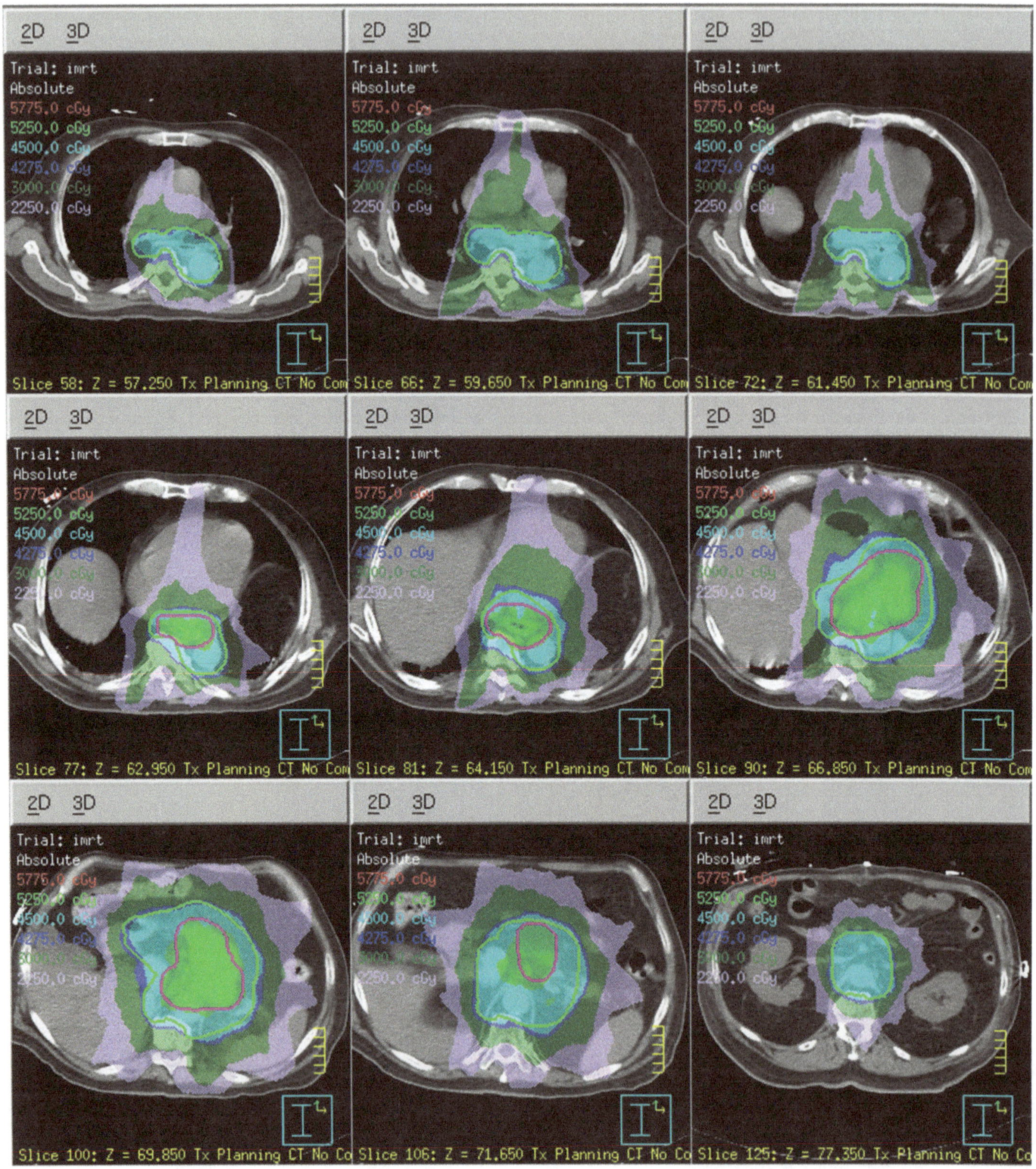

Figure 7.35 IMRT final dose distribution.

Figures 7.35 and 7.36 show the final dose distribution for IMRT and VMAT plans, respectively. All the regions of high dose (hot spots) are within the regions of CTV and PTV 5250.

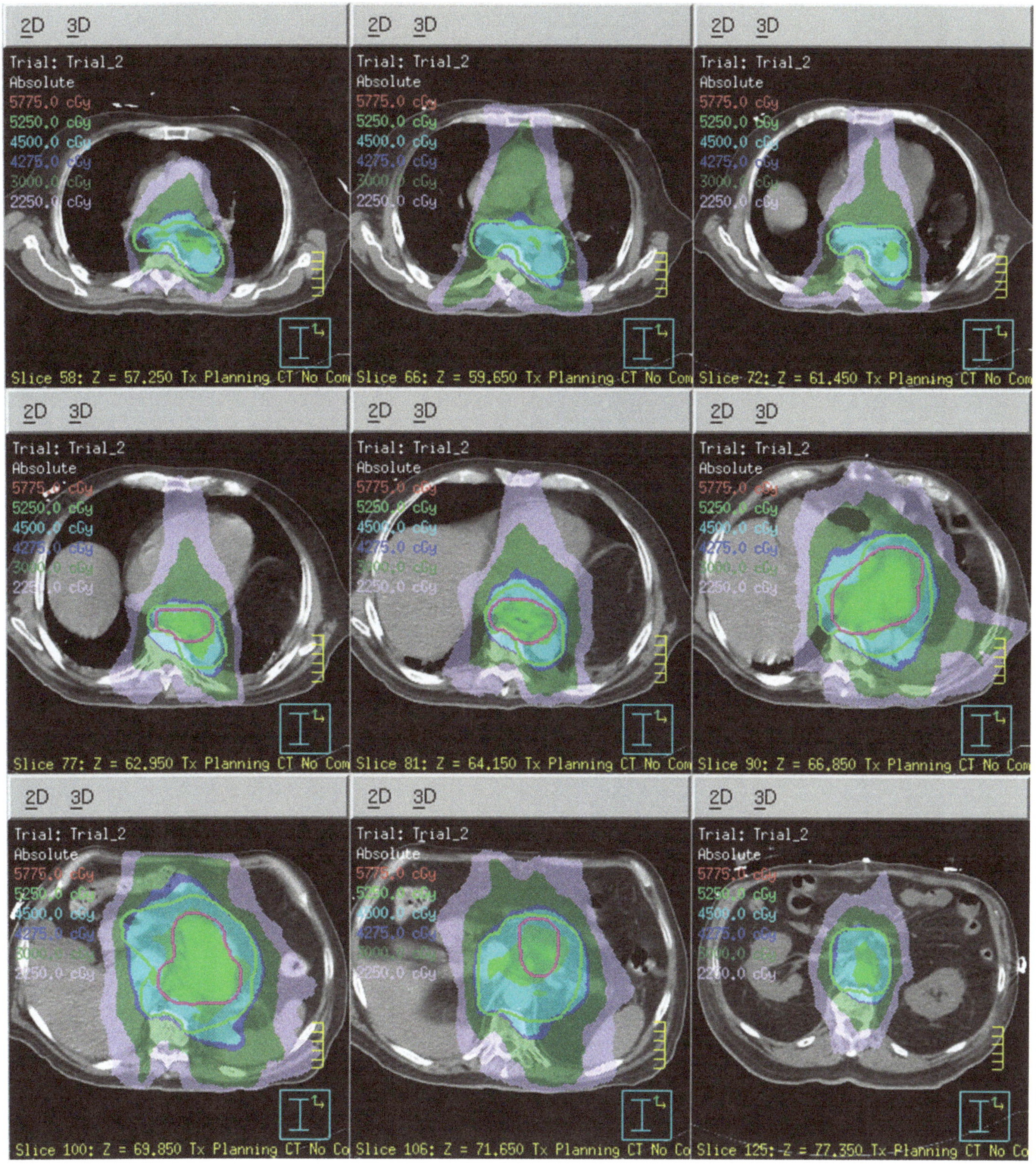

Figure 7.36 VMAT final dose distribution.

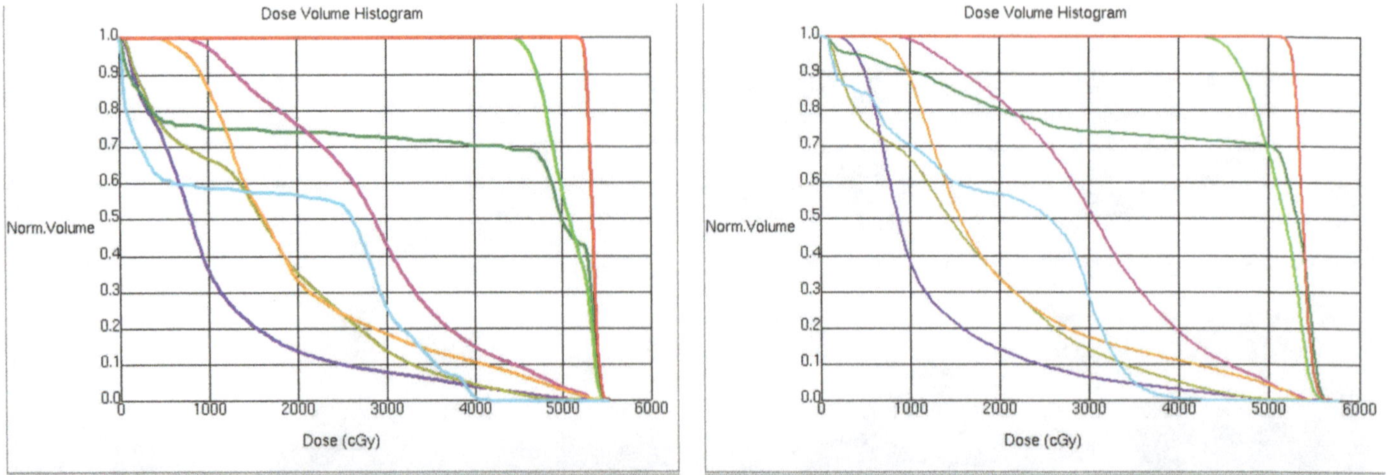

ROI Statistics (IMRT, left)

Line Type	ROI	Trial or Record	Min.	Max.	Mean	Std. Dev.
	Bilat lungs	imrt	44.8	5452.6	1097.5	1041.2
	Bowel	imrt	45.1	5189.6	1661.2	1176.8
	Esophagus	imrt	23.0	5492.1	3821.4	2159.6
	Heart	imrt	731.1	5443.0	2643.5	1086.1
	Liver	imrt	395.7	5518.8	2008.6	1174.4
	PTV4500 final	imrt	4066.9	5523.1	5064.9	254.5
	PTV5250 composite	imrt	76.4	5523.1	5345.5	51.5
	Spinal Canal	imrt	19.0	4162.1	1437.9	1499.8

ROI Statistics (VMAT, right)

Line Type	ROI	Trial or Record	Min.	Max.	Mean	Std. Dev.
	Bilat lungs	Trial_2	182.2	5421.0	1191.2	905.3
	Bowel	Trial_2	58.2	5359.0	1629.5	1185.0
	Esophagus	Trial_2	83.0	5697.7	4265.0	1823.6
	Heart	Trial_2	770.4	5622.4	3066.8	1051.8
	Liver	Trial_2	498.5	5488.4	2015.3	1189.6
	PTV4500 final	Trial_2	3687.1	5721.4	5111.6	275.1
	PTV5250 composite	Trial_2	4360.4	5711.7	5372.7	79.3
	Spinal Canal	Trial_2	78.9	4263.8	1578.6	1355.7

Figure 7.37 Final DVH for case pTA04 for IMRT (left) and VMAT (right).

Case Number pTA05: Thorax and Abdomen—Spinal Cord Cancer Radiation Treatment

General Information

Brief Clinical History

35-year-old woman with a grade II ependymoma in the thoracic spinal cord. History of multiple surgeries with evidence of recurrence. Radiation therapy after surgery.

Prescription

Spinal cord:
PTV 5400 = 180 cGy × 30 fractions

Simulation/Comments

The simulation was accomplished with a vacuum cushion as an immobilization device. Fusion of images was done with a recent PET/CT and MRI for evaluation of the extent of the lesion and targeting by the radiation oncologist.

Target

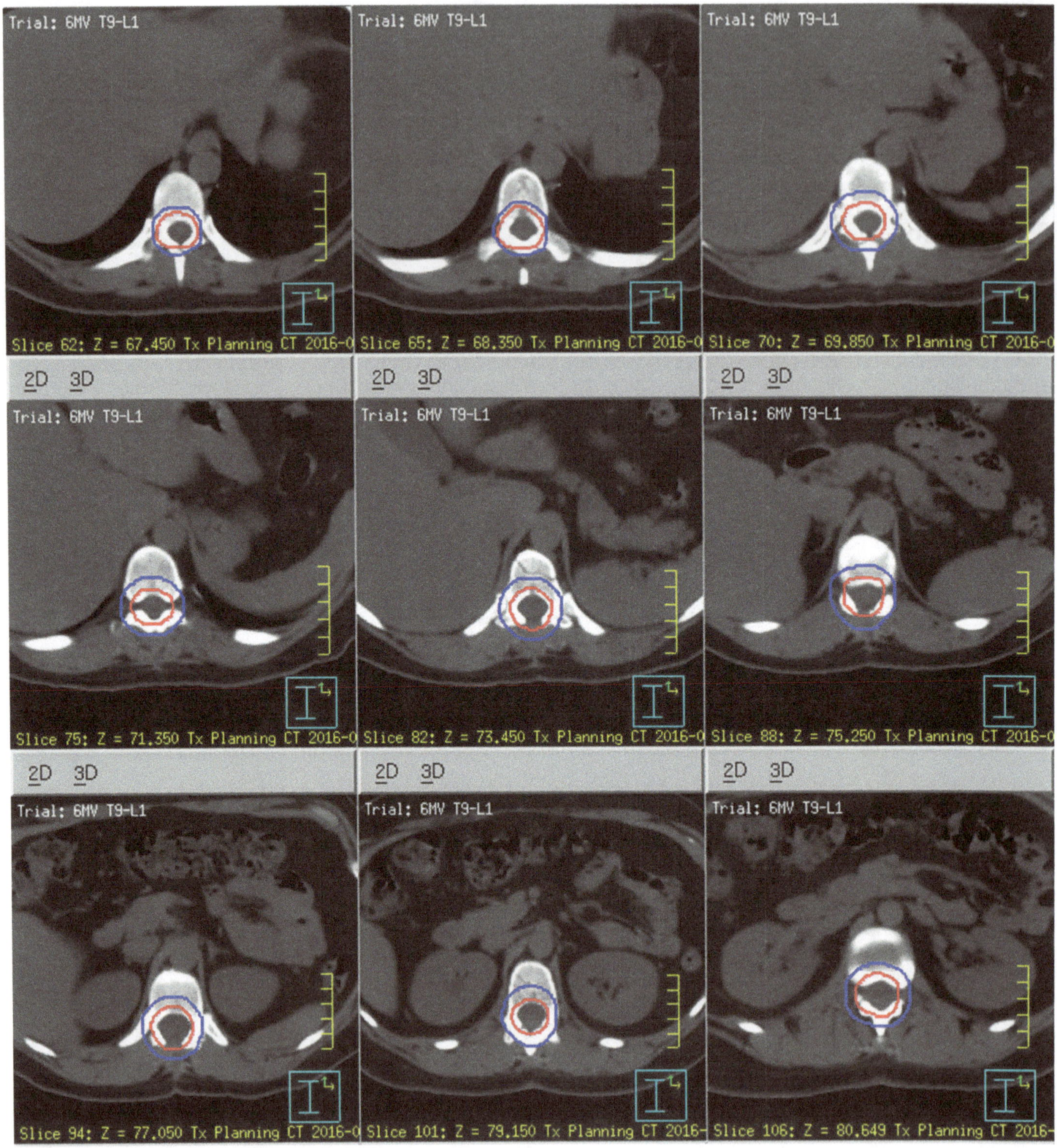

Figure 7.38 PTV contours for case pTA05.

There is only one PTV contour in this case: PTV 5400 (blue contour) as seen in Figure 7.38. The red contour is the CTV. The extent of the lesion covers various anatomical areas including organs at risk in the thorax and abdomen.

Beam Arrangement

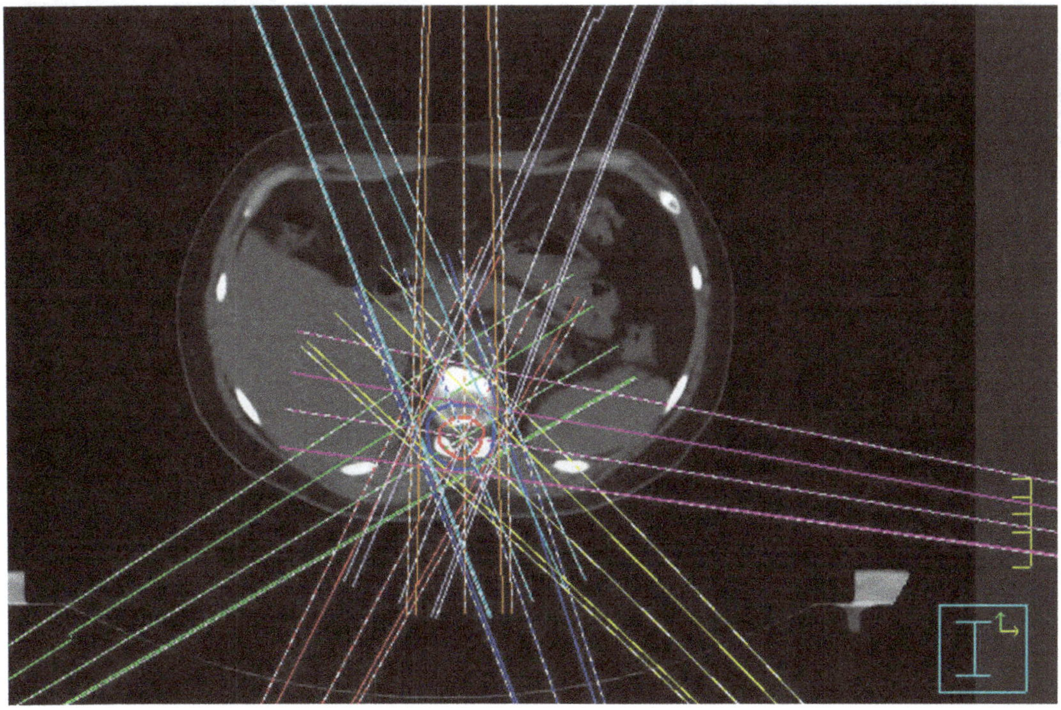

Figure 7.39 IMRT beam arrangement for case pTA05.

Figures 7.39, 7.40, and 7.41 show the beam arrangement for IMRT and VMAT plans. The IMRT plan has eight beams distributed around the patient for the centrally located lesion. The following is a summary of IMRT field arrangement.

Gantry Angle	Couch	Collimator	Energy (MV)
225	0	0	10
250	0	0	10
315	0	0	10
0	0	0	10
45	0	0	10
100	0	0	10
130	0	0	10
160	0	0	10

The VMAT plan has two arcs that go around the patient 360 degrees. The main advantages of VMAT are its shorter delivery time and its fewer number of beams. Planning times might be increased for VMAT optimization according to the computing power of the TPS.

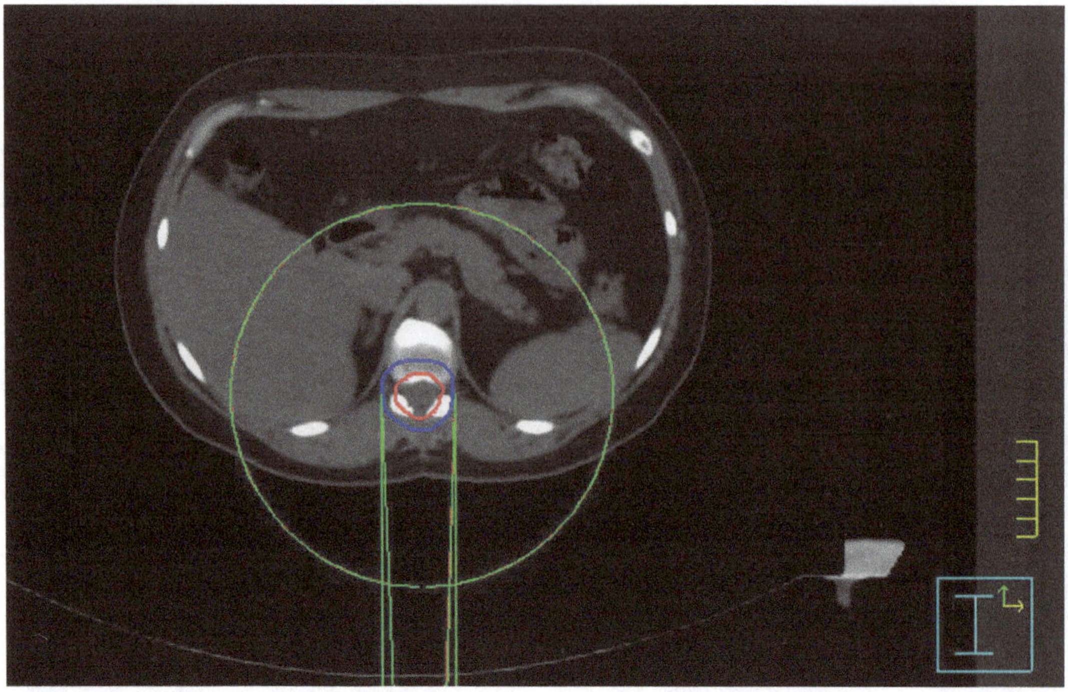

Figure 7.40 VMAT beam arrangement for case pTA05.

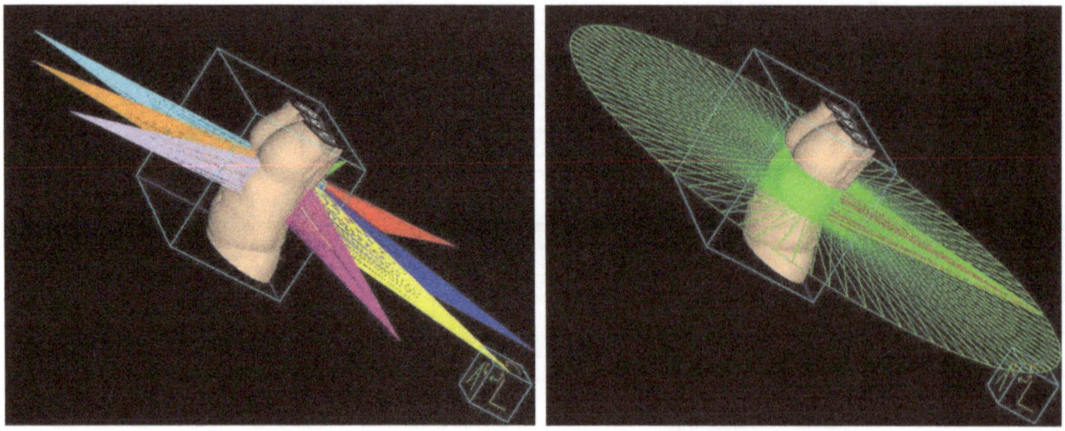

Figure 7.41 IMRT and VMAT beam arrangement for case pTA05 on a surface rendering.

Optimization

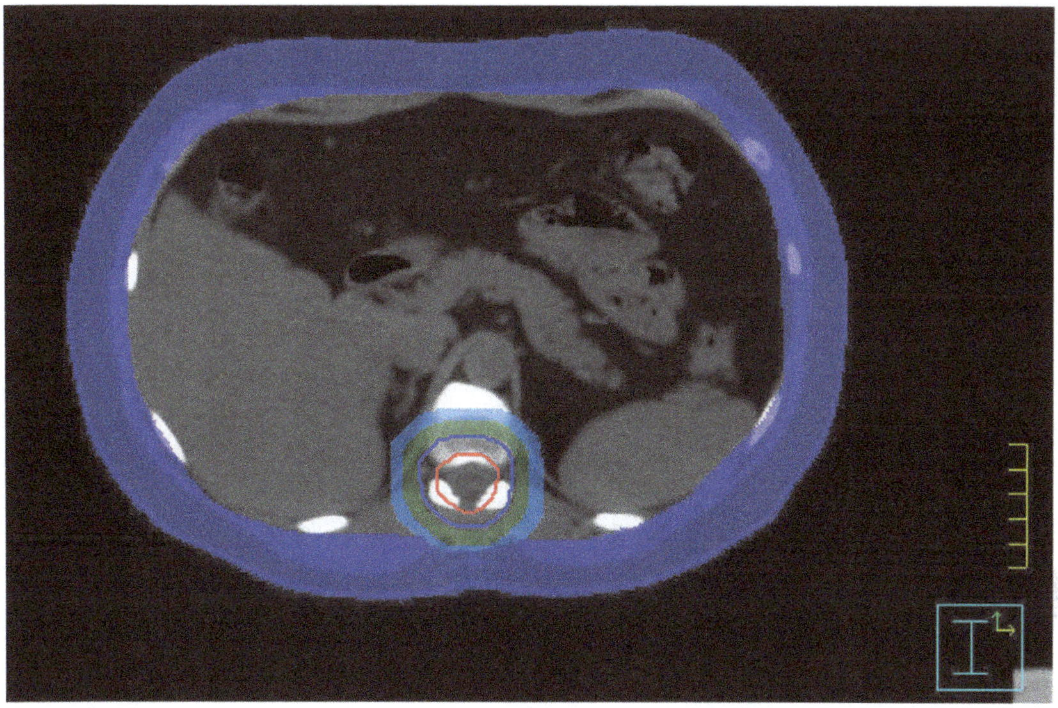

Figure 7.42 Optimization contours: PTV rings (purple and blue colorwash); body ring (red colorwash).

Optimization contours: The main optimization contours for this case include two PTV rings to control the dose gradient between the target and the healthy tissue: inner and outer PTV ring. The body ring avoids low-dose streaks (50% IDL) on the surface of the patient.

Optimizer: The optimization process can be started with basic objectives focusing on covering the target and reaching some dose conformity. One might start with an optimization along the following criteria:

PTV 5400	→ Minimum dose = 5400 cGy
PTV 5400	→ Maximum dose = 5600 cGy
PTV rings	→ Maximum dose = decreasing from 4500 cGy to 3000 cGy
Body ring	→ Maximum dose = 2500 cGy

Additional objectives for organs at risk or other contours can be added after evaluating the first DVH and dose distribution. VMAT plans can use the same parameters.

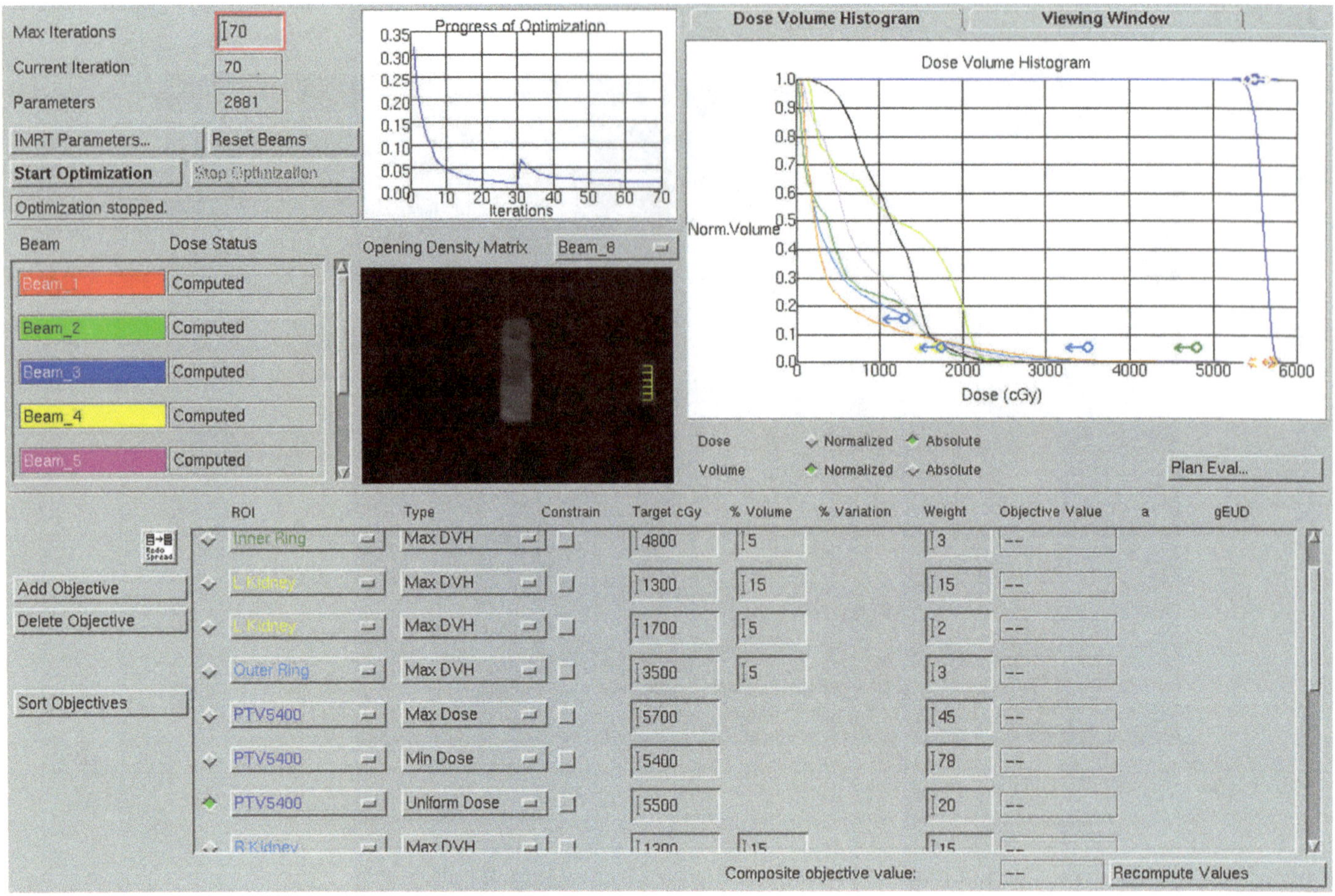

Figure 7.43 IMRT Optimizer window for case pTA05.

Dose Distribution

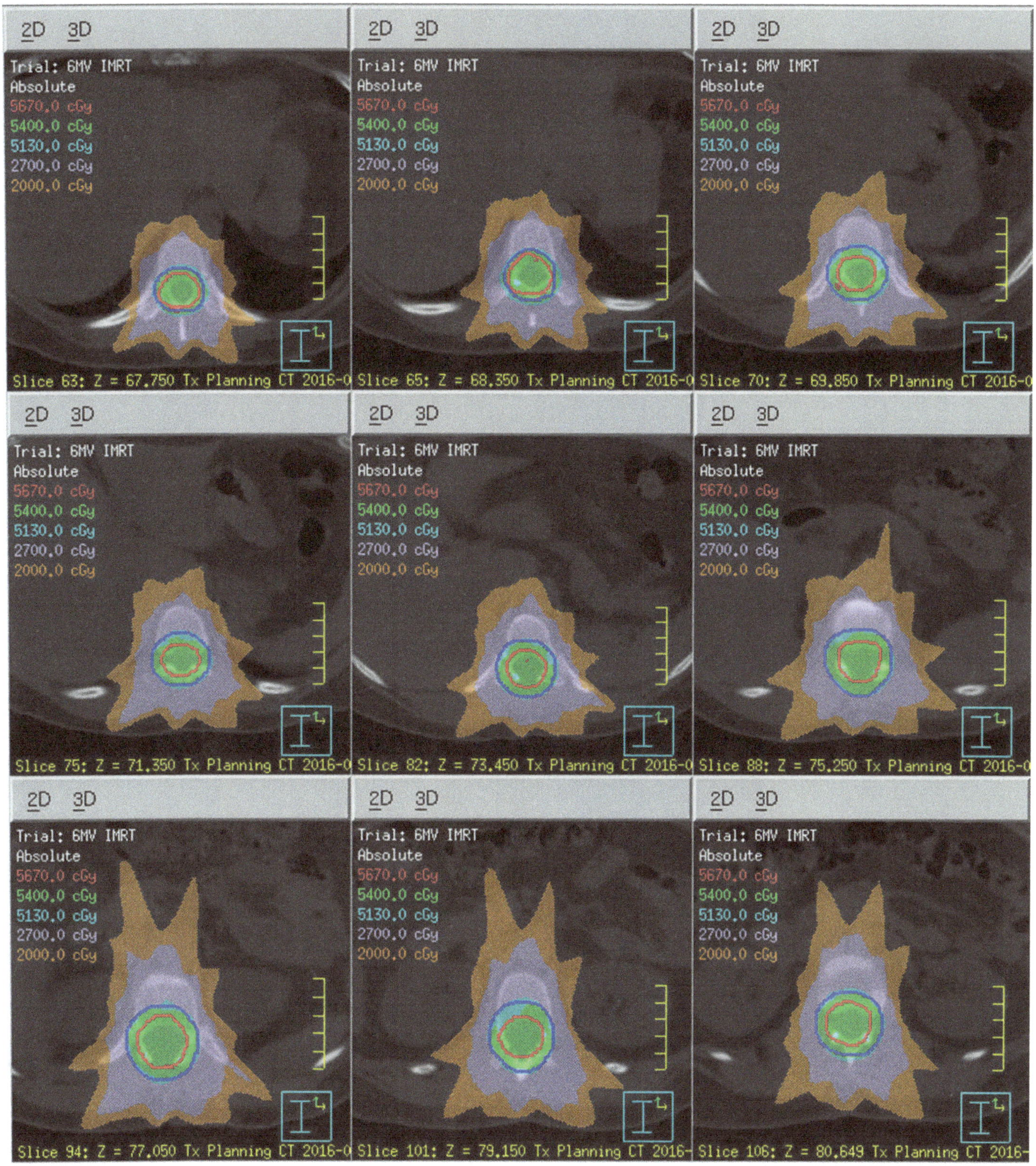

Figure 7.44 IMRT final dose distribution.

Figures 7.44 and 7.45 show the final dose distribution for IMRT and VMAT plans, respectively. Notice the low dose spill in both cases. The increased angles and control points are of advantage in this case for VMAT planning. Both plans have similar coverage of the PTV volume, as well as dose to OARs that are under tolerance levels.

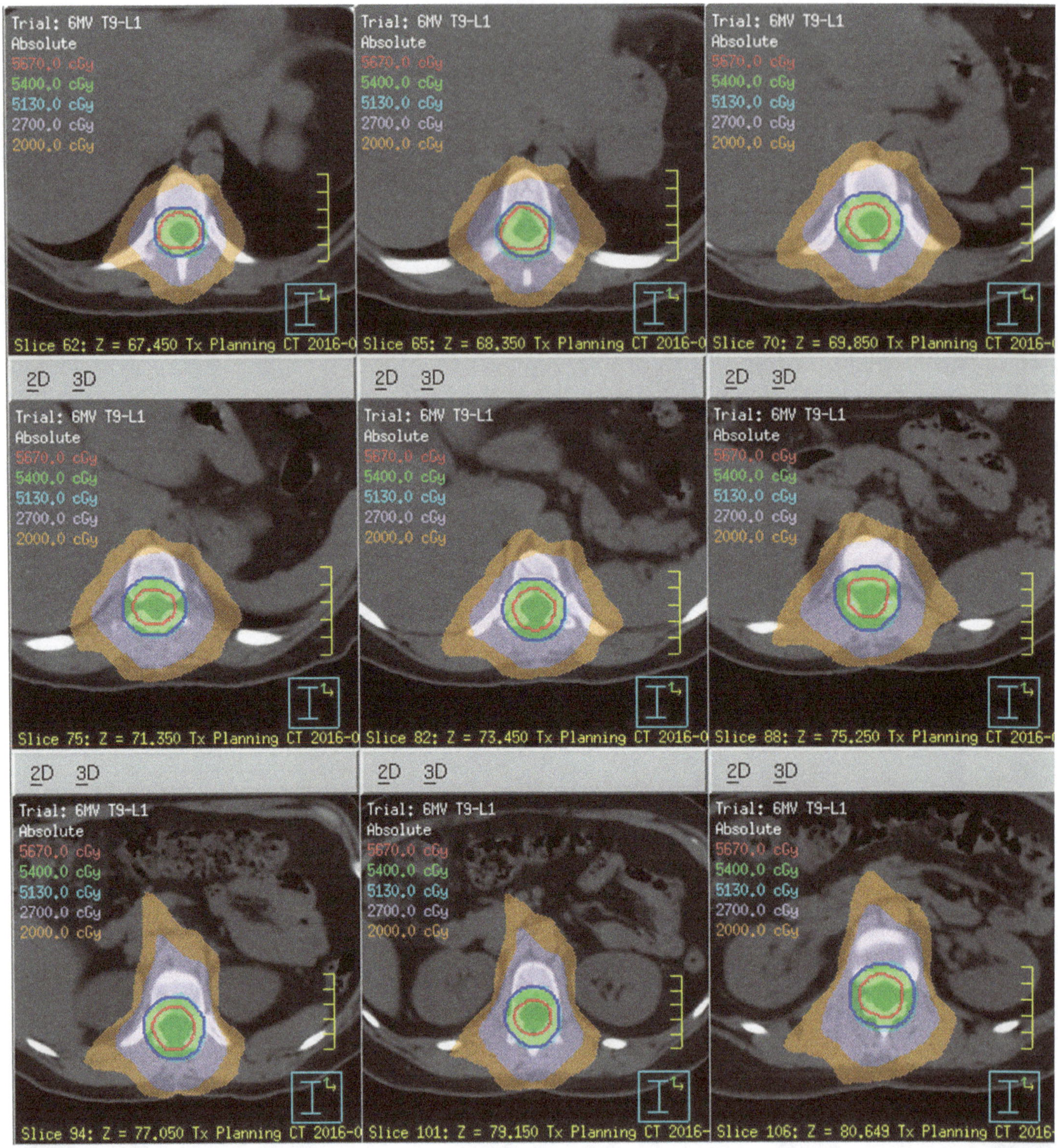

Figure 7.45 VMAT final dose distribution.

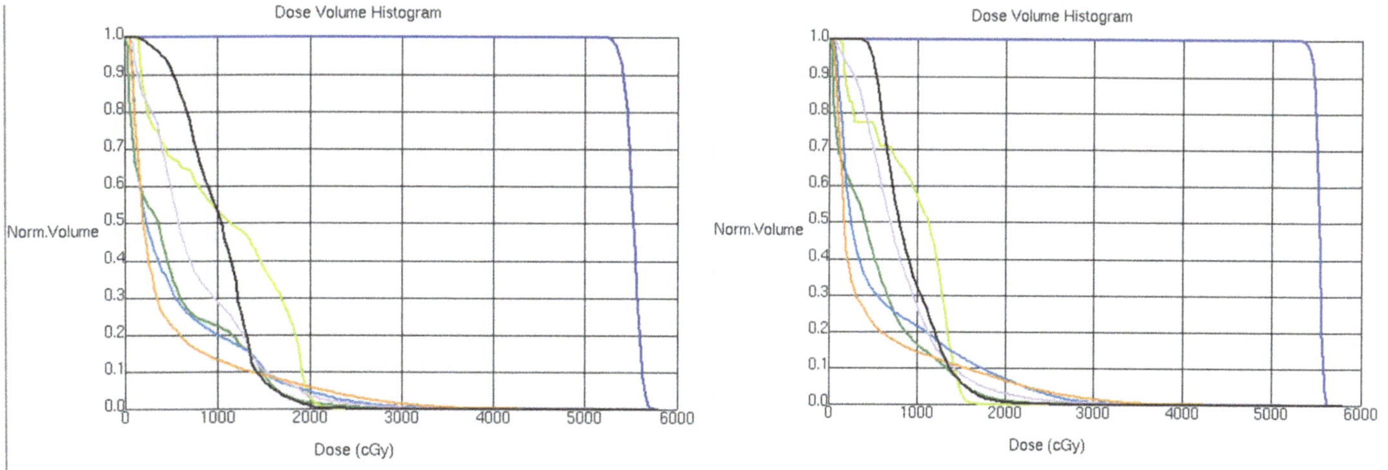

ROI Statistics

Line Type	ROI	Trial or Record	Min.	Max.	Mean	Std. Dev.
	Bowel	6MV IMRT	16.1	2765.7	536.0	568.9
	Esophagus	6MV IMRT	140.7	2256.0	1032.6	700.9
	L Lung	6MV IMRT	57.7	4007.6	138.9	400.8
	Liver	6MV IMRT	62.9	5108.2	736.2	578.5
	PTV5400	6MV IMRT	5077.9	5748.2	5516.0	82.1
	R Lung	6MV IMRT	58.8	4276.9	59.9	283.5
	Stomach	6MV IMRT	146.4	2414.0	1005.1	370.1

ROI Statistics

Line Type	ROI	Trial or Record	Min.	Max.	Mean	Std. Dev.
	Bowel	6MV T9-L1	20.2	2655.4	519.3	499.7
	Esophagus	6MV T9-L1	137.5	1716.6	893.5	496.1
	L Lung	6MV T9-L1	57.0	3976.4	150.1	432.6
	Liver	6MV T9-L1	53.5	5120.7	772.8	506.8
	PTV5400	6MV T9-L1	5251.7	5640.7	5525.6	45.5
	R Lung	6MV T9-L1	48.6	4235.1	60.2	293.5
	Stomach	6MV T9-L1	360.2	2515.2	895.6	339.0

Figure 7.46 Final DVH for case pTA05 for IMRT (left) and VMAT (right).

Case Number pHN01: Head and Neck Radiotherapy Treatment

General Information

Brief Clinical History

66-year-old man with T4 N2cMo squamous cell carcinoma of the larynx. Patient had a tracheostomy.

Prescription

Larynx and lymph nodes:
PTV 6250 = 250 cGy × 25 fractions
PTV 6000 = 240 cGy × 25 fractions
PTV 4500 = 180 cGy × 25 fractions

Simulation/Comments

The simulation was accomplished with a thermoplastic mask and a moldable pillow. Fusion of images was done with a recent PET/CT for evaluation of the extent of the lesion and targeting by the radiation oncologist.

Target

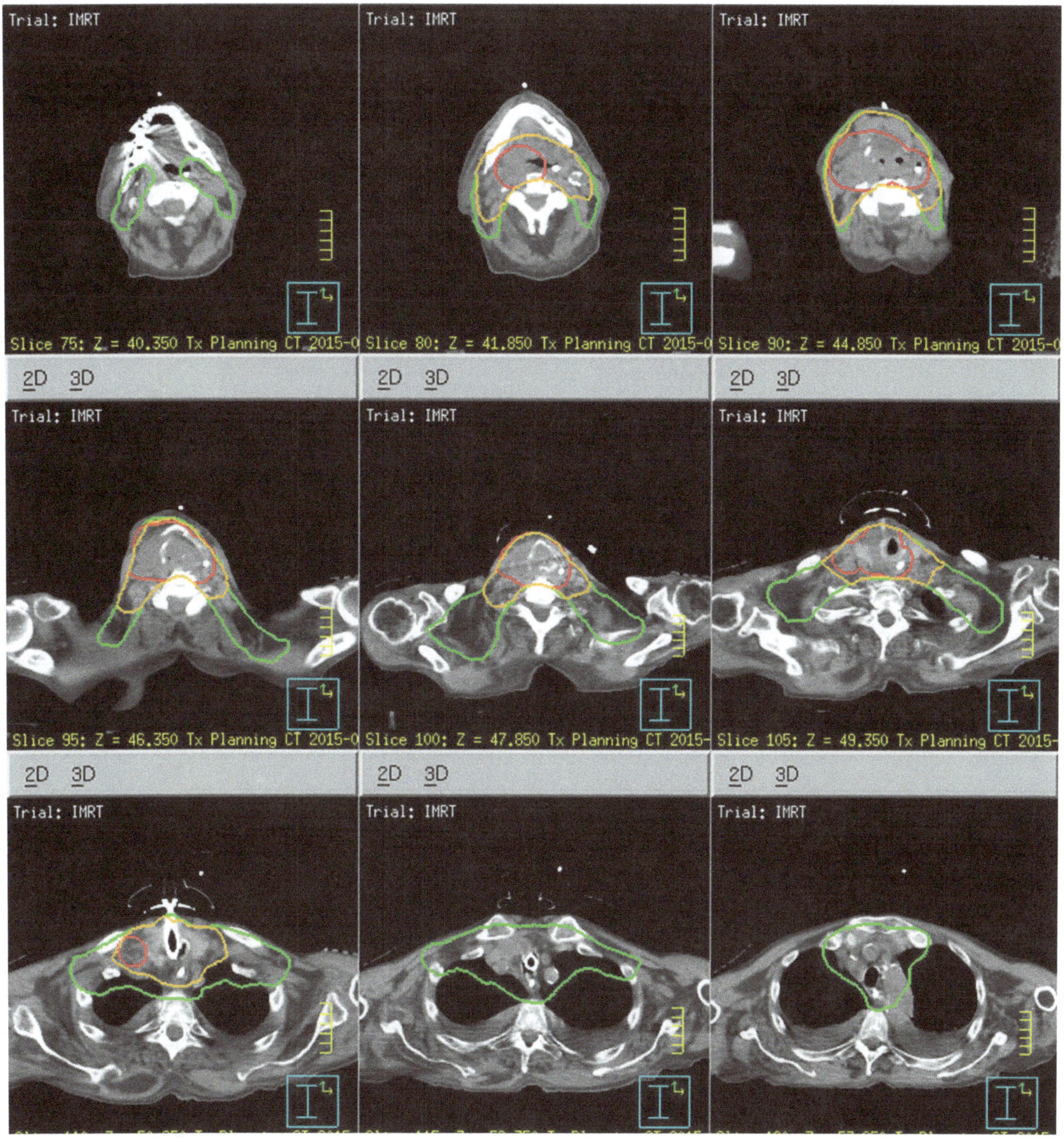

Figure 7.47 PTV contours for case pHN01: PTV 6250 (red), PTV 6000 (orange), and PTV 4500 (green).

Figure 7.47 shows the PTV contours drawn by the radiation oncologist. Reviewing the targets before planning is an important step in evaluating which structures are the organs at risk and which problems will have to be dealt with during planning. In this case, the planner has to be careful with the spinal cord, as well as the salivary glands, mandible, and other structures. The dose gradient from one PTV to the other is another factor to take into account for future optimization, and these different dose levels can be separated for fewer conflicts in the optimization objectives.

Beam Arrangement

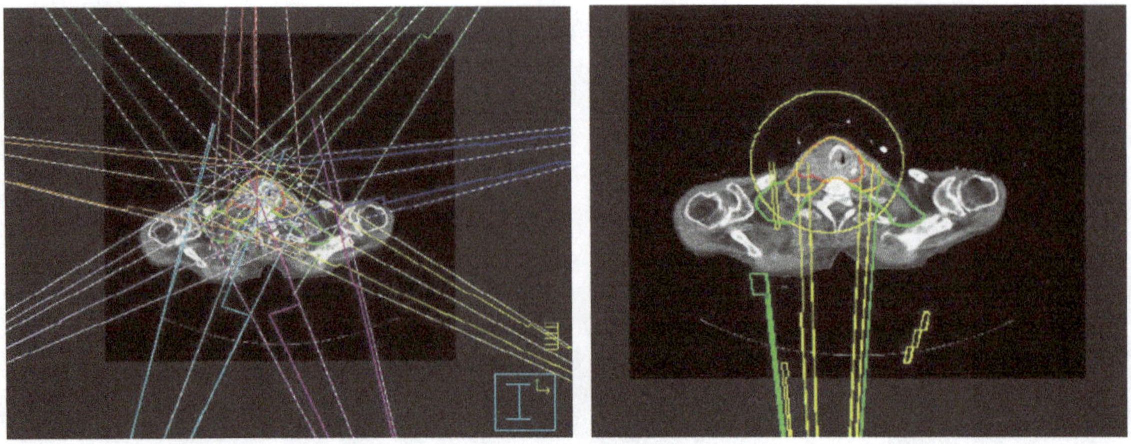

Figure 7.48 Beam arrangement for IMRT (left) and VMAT (right) plans.

The beam arrangement for this case can be seen in Figure 7.48. The target contour is located symmetrically in the body, and it is not favoring any side. The planner opted to distribute the beams evenly every 40 degrees around the patient's body. In other cases, beam distribution can favor certain angles or sites to avoid irradiating organs at risk. When placing beams, avoid having beams facing directly at each other. The following is a summary of the beam arrangement for this case.

Gantry Angle	Couch	Collimator	Energy (MV)
0	0	0	6
40	0	0	6
80	0	0	6
120	0	0	6
160	0	0	6
200	0	0	6
240	0	0	6
280	0	0	6
320	0	0	6

VMAT beam arrangement for this case consists of two arcs going around the patient back and forth.

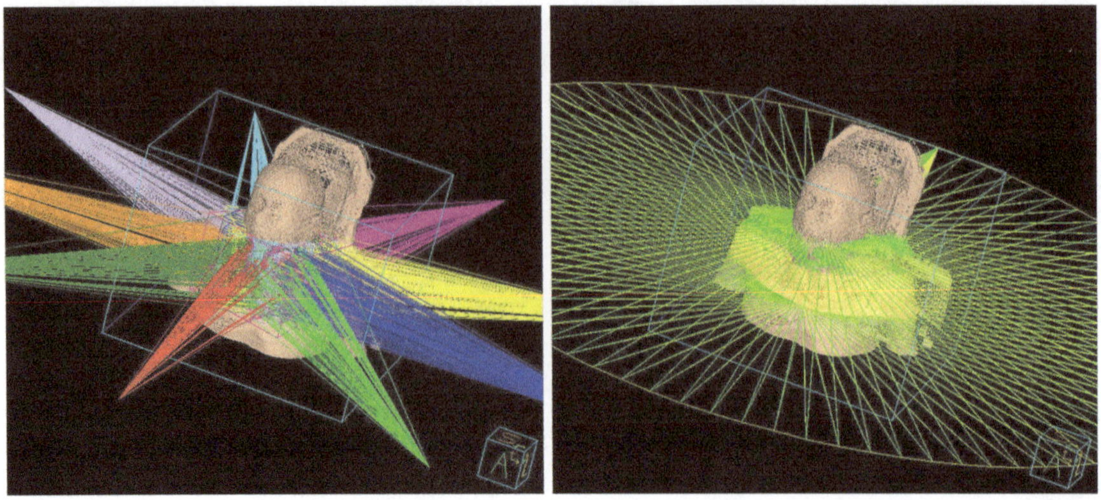

Figure 7.49 Beam arrangement for IMRT (left) and VMAT (right) on a skin surface reconstruction.

Optimization

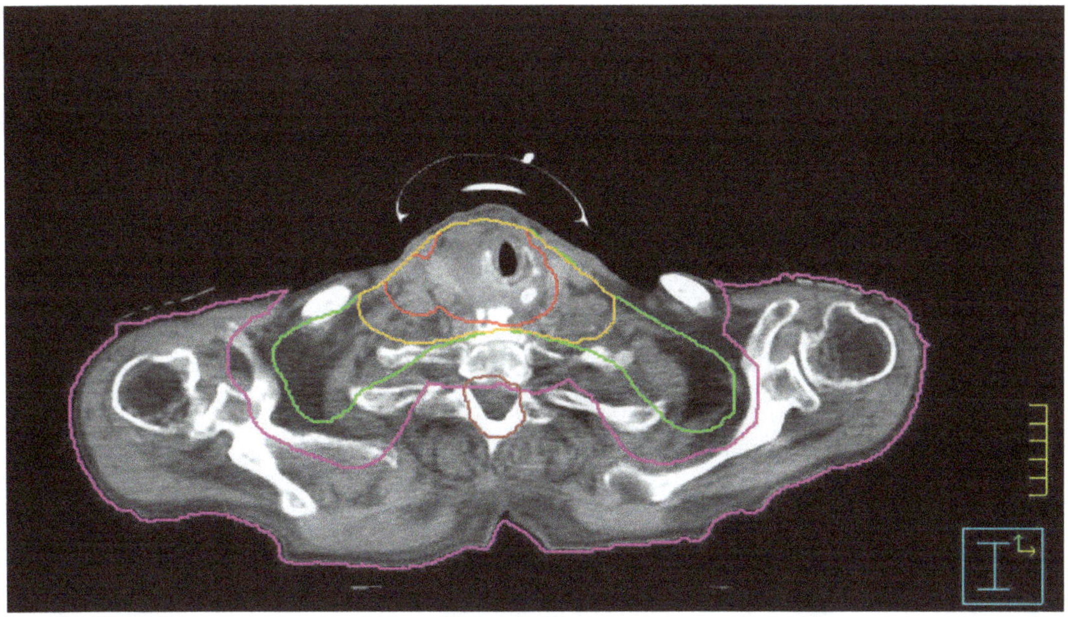

Figure 7.50 Optimization contours: body ring (pink), cord + margin (brown).

Dose levels within target: The first thing the planner should do is to separate the dose levels within the target to avoid objective conflicts in the optimizer. In this case, the planner would have three optimization contours: PTV 6250 (the highest prescription dose), PTV 6000–PTV 6250 (PTV 6000 without the overlap region of the highest prescription dose PTV), and PTV 4500–PTV 6000 (PTV 4500 without the overlap region of the contour PTV 5000). The planner can choose to leave a 2-mm margin in between contours to allow dose fall-off or to create rings for each structure.

Optimization contours: Figure 7.50 contains some important optimization contours for this case. Here we observe the body ring (pink) that the planner has designed for this plan. Since the lesion is close to the anterior surface of the patient, the planner has decided to enforce a body ring covering most of the normal structures (there is another structure for the anterior skin in other slices). The cord plus a margin of 3 mm can also be seen on Figure 7.50 (brown). The margin might have been helpful to the planner to constrain the dose to the cord even farther with more volume to constrain.

Optimizer: To start an IMRT optimization, basic optimization parameters can be configured first: PTV, PTV rings, body rings, and some organs at risk. For the first iteration of this plan, the planner could start with:

PTV 6250	→ Minimum dose = 6250 cGy
PTV 6250	→ Maximum dose = 6750 cGy (108% dose)
PTV 6000–PTV 6 250	→ Minimum dose = 6000 cGy
PTV 6000–PTV 6250	→ Maximum dose = 6250 cGy
PTV 4500–PTV 6000	→ Minimum dose = 4500 cGy
PTV 4500–PTV 6000	→ Maximum dose = 6250 cGy
PTV rings	→ Maximum dose = dose point in the middle of the gradients
Body ring	→ Maximum dose = 3800 cGy

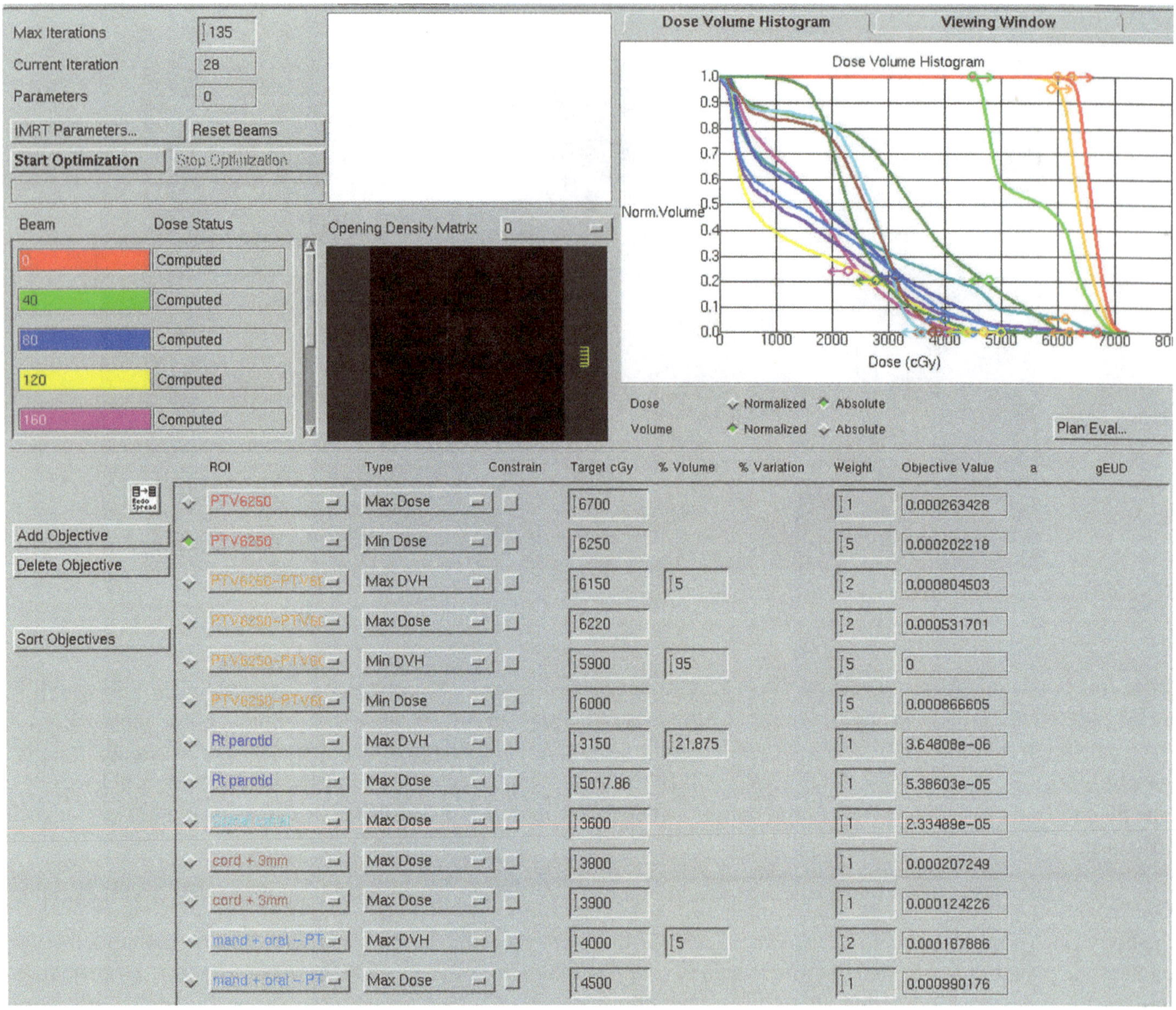

Figure 7.51 Optimizer window for case pHN01.

The planner can later add other critical structures. After the first optimization, the planner can increase the number of parameters and change existing parameters having an idea of what the DVH looks like. This process might take several iterations.

Dose Distribution

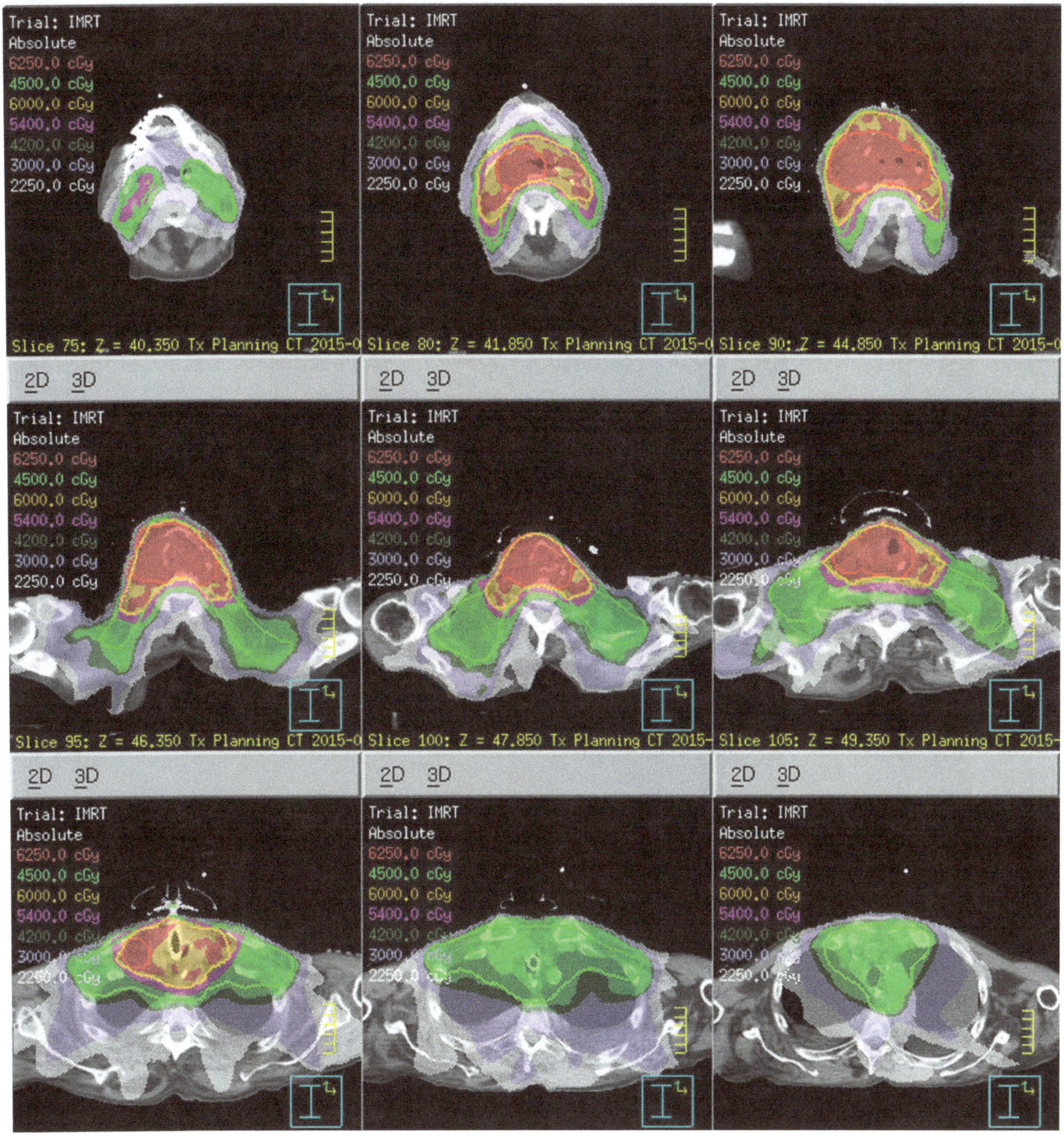

Figure 7.52 Final dose distribution for IMRT plan.

Figures 7.52 and 7.53 show the final dose distribution for the IMRT and VMAT plans, respectively. Both plans are normalized to have the same degree of coverage for the PTV with the highest prescription. The VMAT plan manages to accomplish both coverage and conformity to the targets due to the increased number of control points. Figure 7.54 contains the final DVH for the plan. The coverage and dose limitations for all the structures can be seen here with the help of the planning software tools.

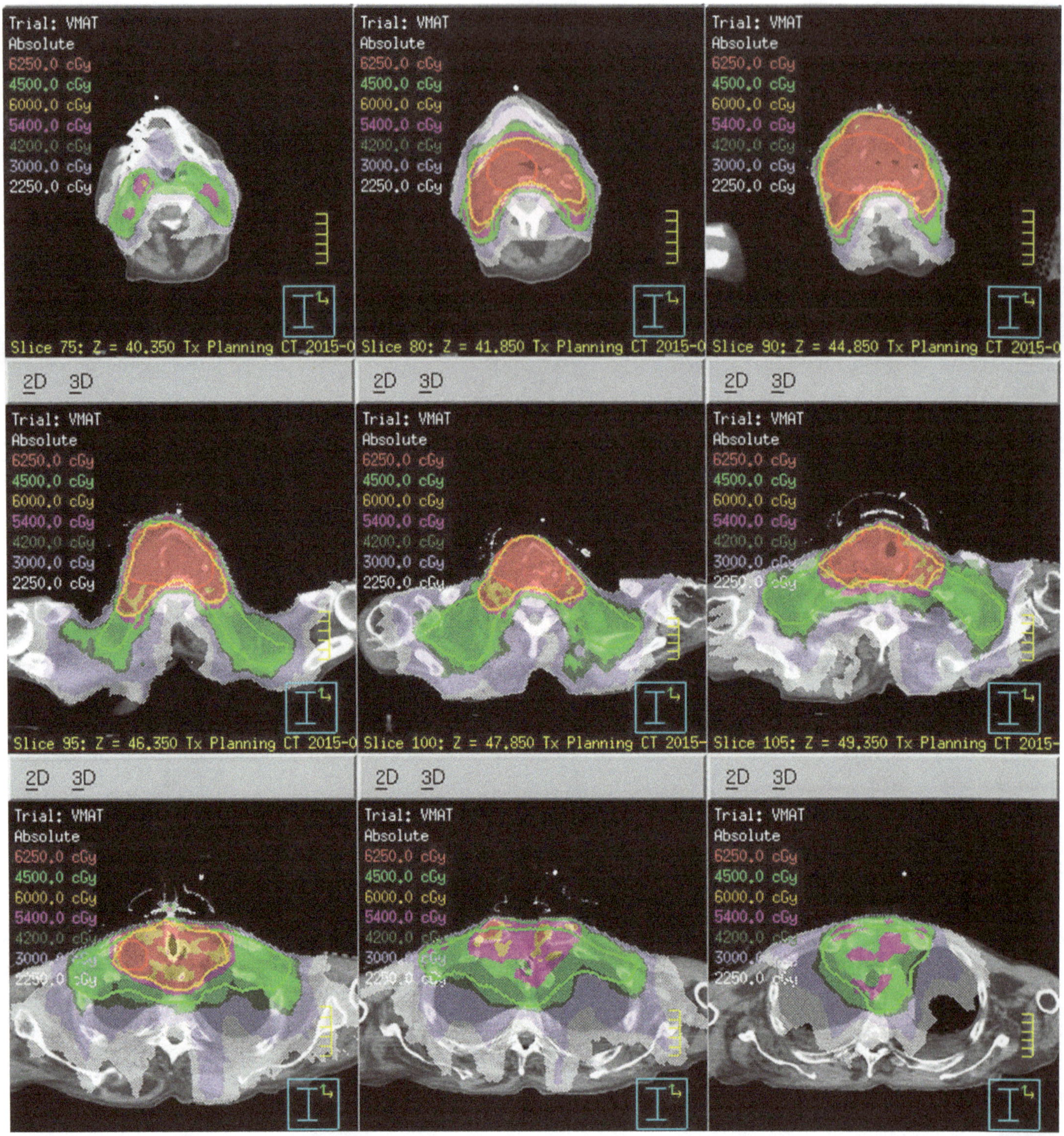

Figure 7.53 Final dose distribution for VMAT plan.

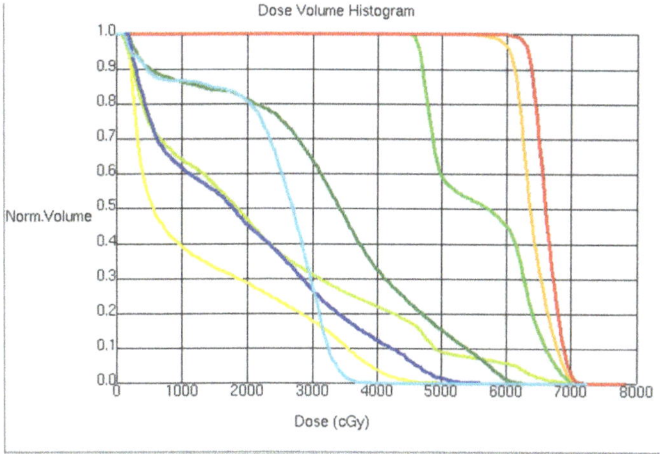

ROI Statistics

Line Type	ROI	Trial or Record	Min.	Max.	Mean	Std. Dev.
	Esophagus	IMRT	0.0	7203.0	1333.3	1819.1
	Lt parotid	IMRT	168.4	4928.9	1284.7	1309.5
	Mandible	IMRT	167.5	6251.2	3162.8	1670.3
	PTV4500	IMRT	4182.3	7203.0	5613.5	789.4
	PTV6000	IMRT	4799.1	7203.0	6404.6	262.3
	PTV6250	IMRT	5272.2	7203.0	6614.1	198.8
	Rt parotid	IMRT	166.3	5595.4	1952.4	1489.1
	Spinal canal	IMRT	117.5	3902.8	1939.7	1264.7

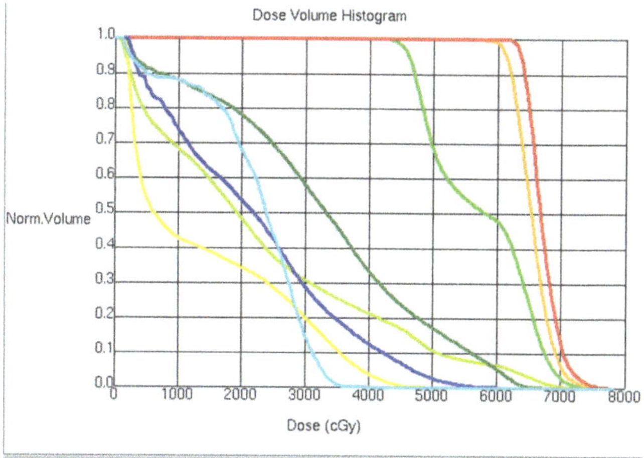

ROI Statistics

Line Type	ROI	Trial or Record	Min.	Max.	Mean	Std. Dev.
	Esophagus	VMAT	2.1	7756.4	1373.5	1837.7
	Lt parotid	VMAT	156.3	4856.5	1400.3	1354.9
	Mandible	VMAT	162.6	6701.3	3143.1	1743.7
	PTV4500	VMAT	3851.5	7756.4	5725.9	832.7
	PTV6000	VMAT	4693.2	7756.4	6544.3	285.7
	PTV6250	VMAT	5318.9	7756.4	6707.5	255.7
	Rt parotid	VMAT	193.1	6011.4	2196.9	1393.3
	Spinal canal	VMAT	119.4	3974.7	1770.3	1163.8

Figure 7.54 Dose volume histogram for case pHN01.

Case Number pHN02: Head and Neck Radiotherapy Treatment

General Information

Brief Clinical History

62-year-old male with a stage IVA, T1N2cM0 SCC of the left of the base of tongue.

Prescription

Left base of tongue:
PTV 6250: 250 cGy × 25 fractions
PTV 6000: 240 cGy × 25 fractions
PTV 4500: 180 cGy × 25 fractions

Simulation/Comments

The simulation was accomplished with a thermoplastic mask and a moldable pillow on a head and neck board. The patient had to take dentures out for both the simulation and treatment. Fusion of images was done with a recent PET/CT for evaluation of the extent of the lesion and targeting by the radiation oncologist.

Target

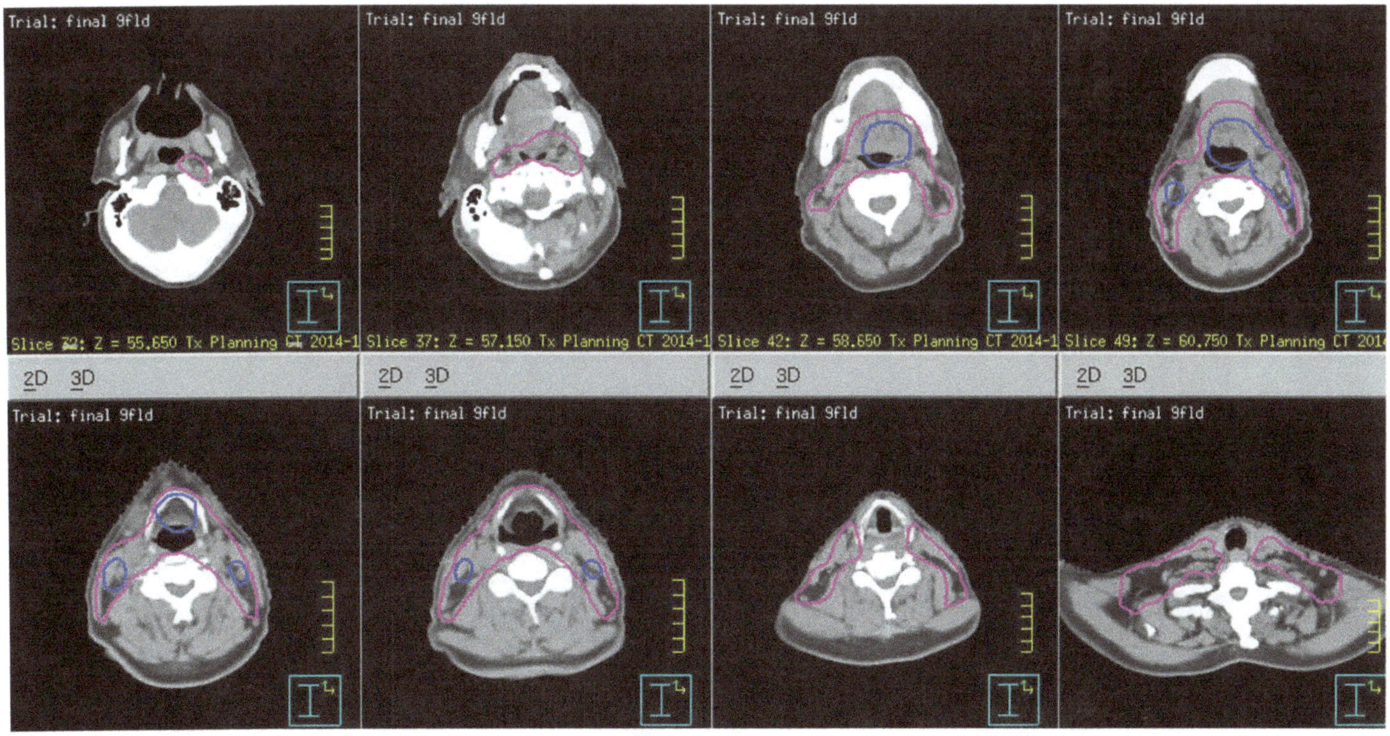

Figure 7.55 PTV contours for case pHN02: PTV 6250 (red), PTV 6000 (blue), and PTV 4500 (pink).

Figure 7.55 shows three PTVs: PTV 6250, PTV 6000, and PTV 4500. There might be overlaps from the PTV structures into the normal tissue structures. The planner must make sure to accurately contour normal structures—such as the spinal canal, the mandible, the esophagus, the trachea, and parotid glands—regardless of how close or overlapping the PTV contour is to these structures.

Beam Arrangement

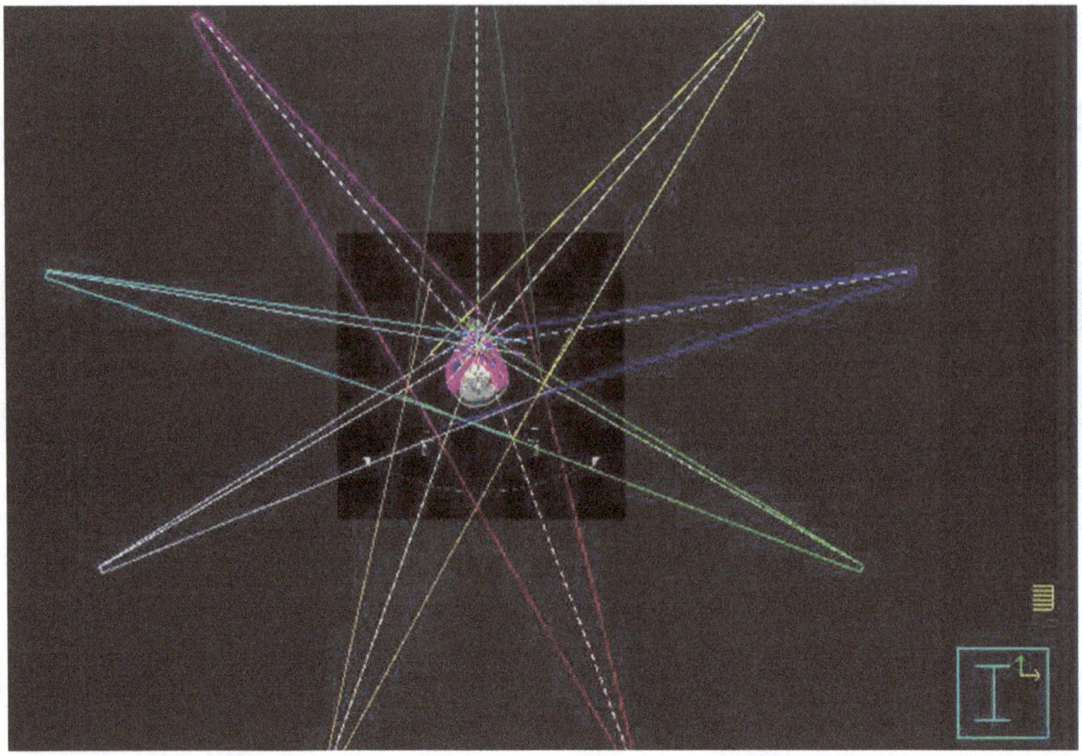

Figure 7.56 IMRT beam arrangement for case pHN02.

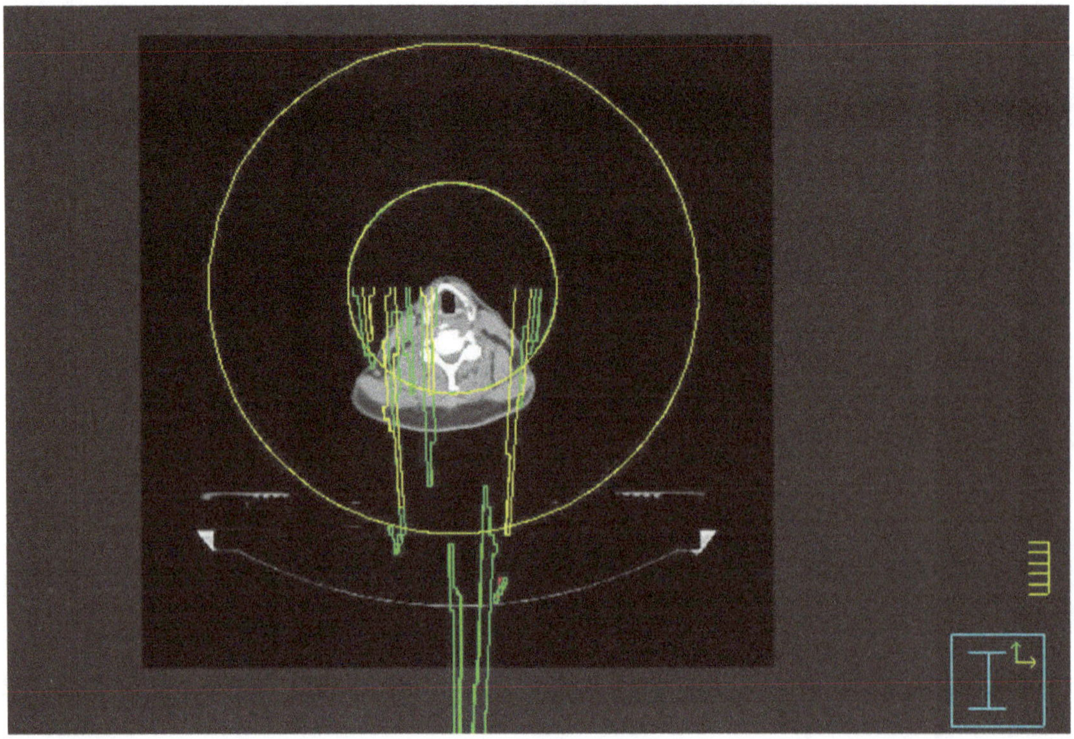

Figure 7.57 VMAT beam arrangement for case pHN02.

The IMRT plan has a 9-field arrangement as shown in Figure 7.56. In this case, the planner has placed the beams every 40 degrees around the patient. Notice that the beams are not facing each other. The following is a summary of main beam parameters used for this treatment plan.

Gantry Angle	Couch	Collimator	Energy (MV)
160	0	0	6
120	0	0	6
80	0	0	6
40	0	0	6
0	0	0	6
320	0	0	6
280	0	0	6
240	0	0	6
200	0	0	6

The VMAT plan has a 2-field arrangement covering 360 degrees around the patient, as seen in Figure 7.57. Two arcs allow for enough modulation to achieve the correct dose distribution. It is important to know machine and system limitations in such beam arrangements. The energy used for the VMAT plan was 6 MV.

Figure 7.58 IMRT and VMAT beam arrangement on a surface rendering.

Optimization

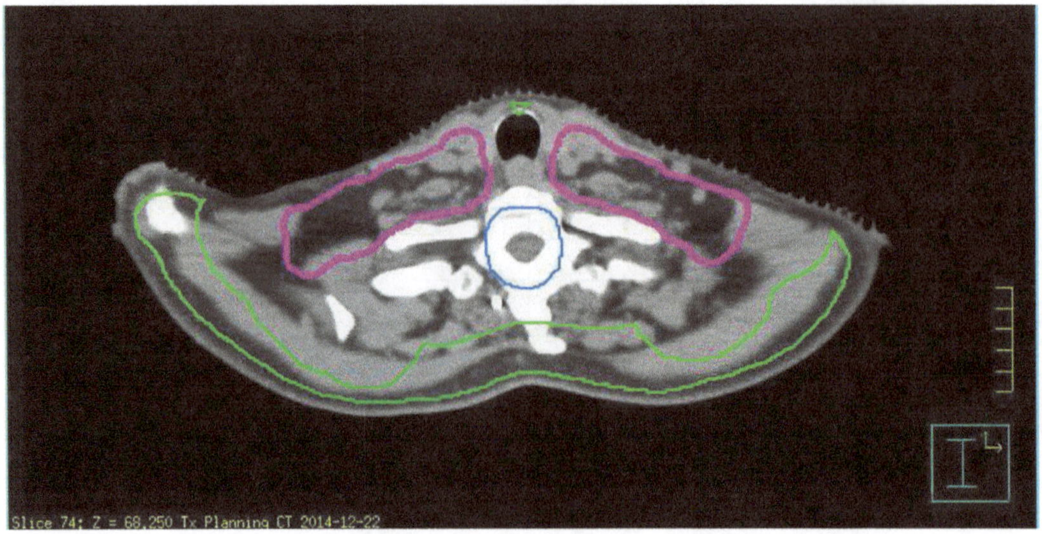

Figure 7.59 Optimization contours: body ring (green), cord + expansion (blue).

Dose levels within target: The planner must separate the dose levels within the target to avoid objective conflicts in the optimization process. In this case, PTV 6250 (the highest prescription dose), PTV 6000–PTV6250 (PTV 6000 without the overlap region of the highest prescription dose PTV), and PTV 4500–PTV 6000 (PTV 4500 without the overlap region of the contour PTV 5000). The planner can create rings for each structure to help with dose conformity, especially in a case like this in which the high-dose gradient region is separated in distinct nodules.

Optimization contours: Figure 7.59 shows the optimization contours. The body ring, if used correctly, can help bring dose inward and prevent spillage or excess entry dose. The cord plus a margin of 10 mm helps spare the spinal cord from prescription dose as well as some high-dose lines.

Optimizer: Figure 7.60 shows a partial list of the optimization parameters used for this plan in the optimizer window. The planner can start by placing basic parameters for main targets and running one optimization. Based on the results of the first optimization, the planner can add more parameters and even contours that control the dose in the target as well as in the OARs so it can reach the desired distribution. The optimization process can be repeated a few times until the system finds a suitable answer for the parameters placed.

For the first iteration of this plan, the planner could start with:

PTV 6250	→ Minimum dose	= 6250 cGy
PTV 6250	→ Maximum dose	= 6750 cGy (108% dose)
PTV 6000–PTV 6250	→ Minimum dose	= 6000 cGy
PTV 6000–PTV 6250	→ Maximum dose	= 6250 cGy
PTV 4500–PTV 6000	→ Minimum dose	= 4500 cGy
PTV 4500–PTV 6000	→ Maximum dose	= 6250 cGy
PTV rings	→ Maximum dose	= dose point in the middle of the gradients
Body ring	→ Maximum dose	= 3800 cGy

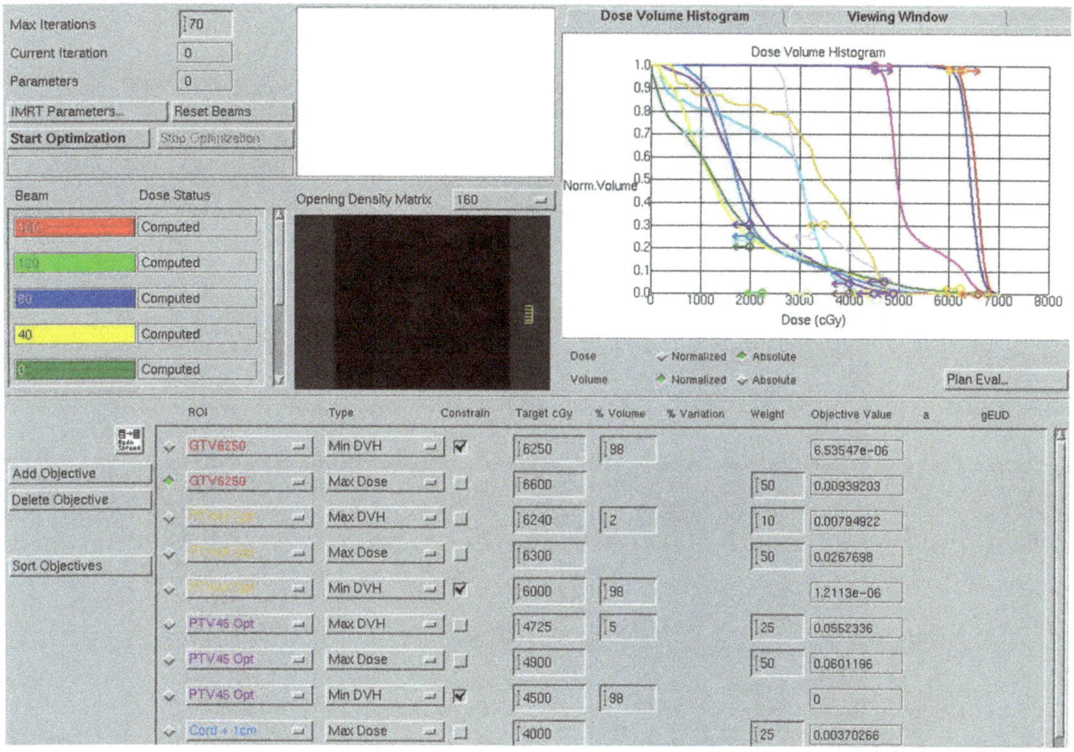

Figure 7.60 IMRT Optimizer window for case pHN02.

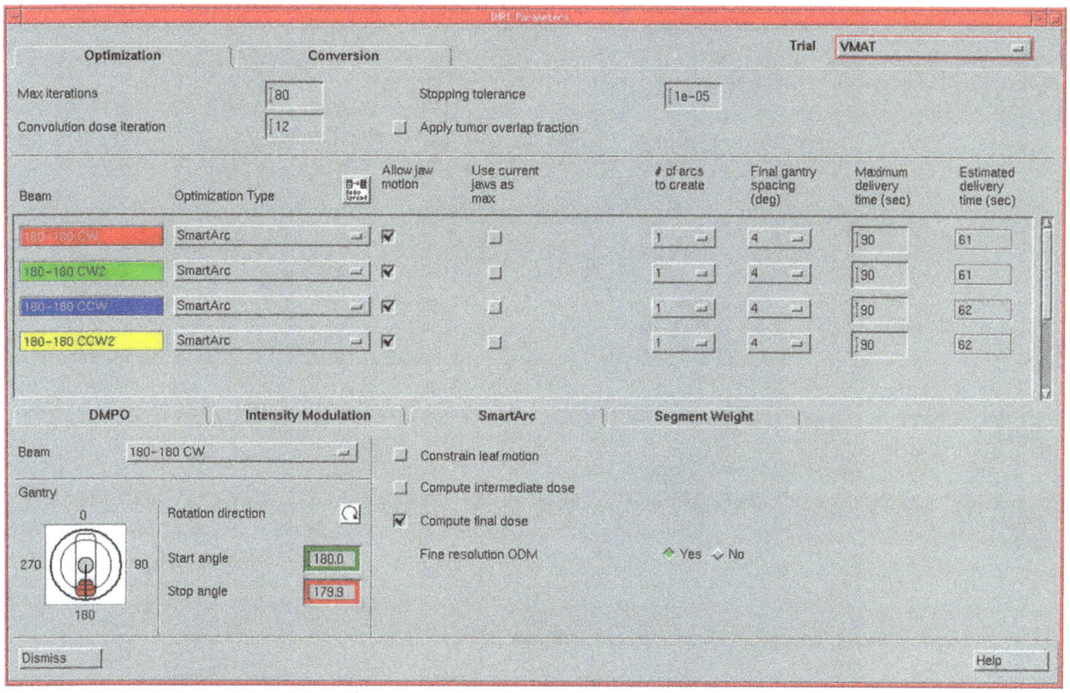

Figure 7.61 VMAT parameters window for case pHN02.

Dose Distribution

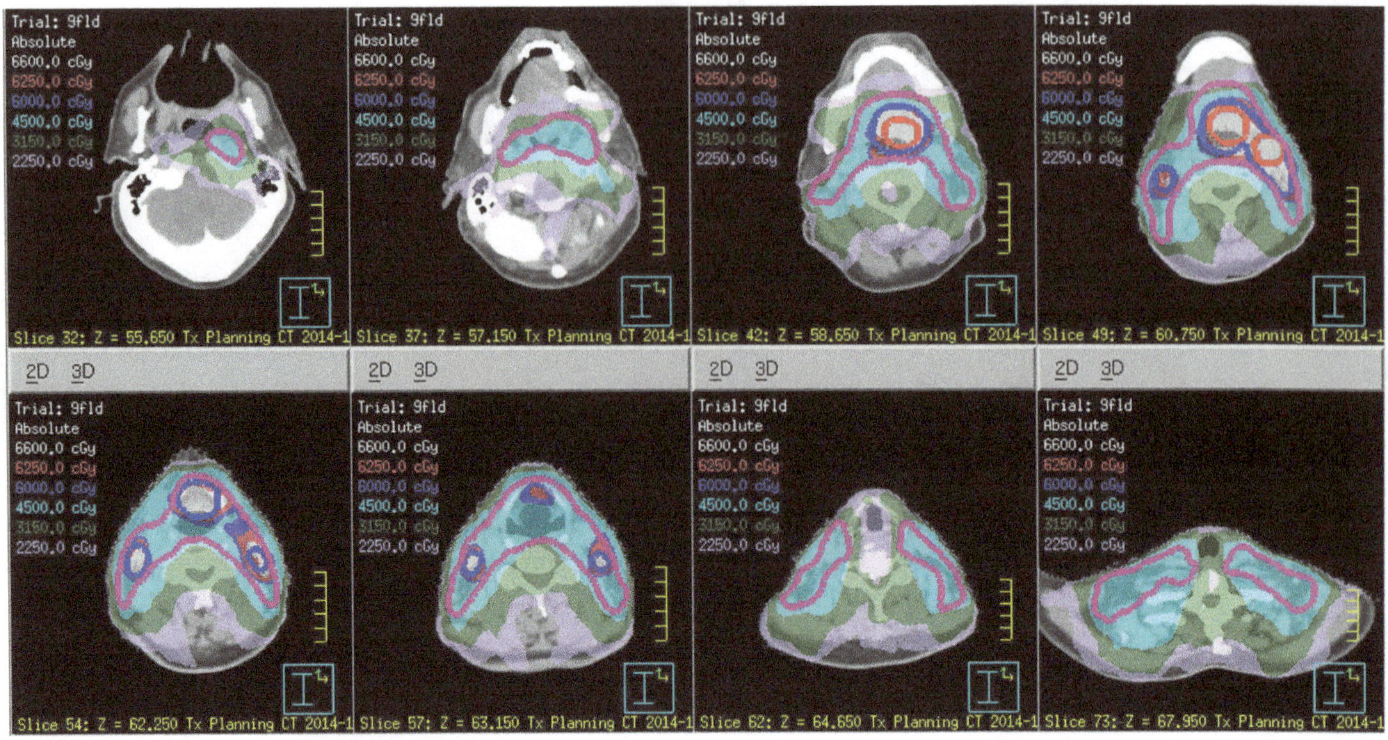

Figure 7.62 IMRT final dose distribution.

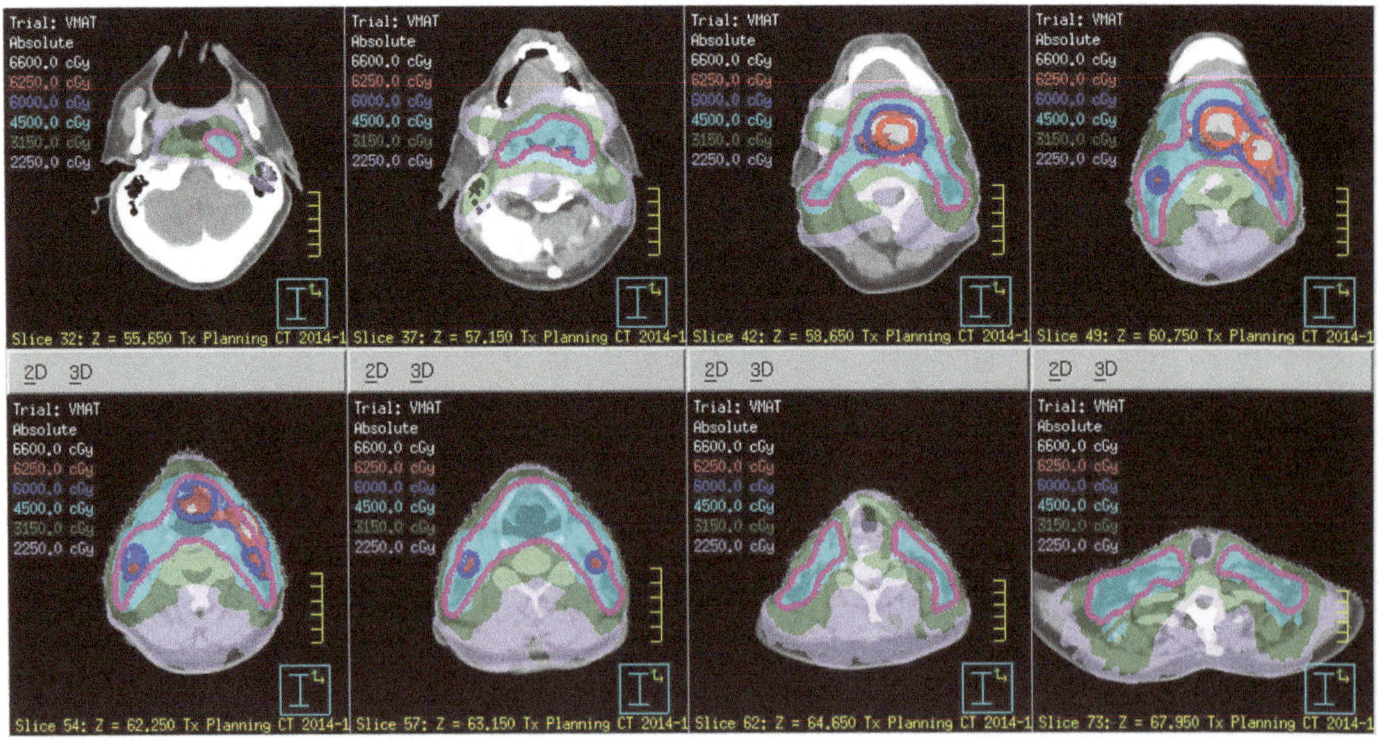

Figure 7.63 VMAT final dose distribution.

Figure 7.62 shows the final dose distribution for an IMRT plan, and Figure 7.63 shows the final dose distribution for a VMAT plan. Depending on the complexity of the target, the dose distribution might not be completely conformal to the shape of the target or completely homogeneous. It is important to look at the dose distribution along with the

DVH (Figures 7.64 and 7.65). Coverage might vary from institution to institution, but in this case the targets must get prescription dose at least at 95% of the target volume. The isodose lines must not spill out to cover organs at risk or form streaks of full dose prescription through their paths. The dose to organs at risk must be evaluated with the values from the DVH compared to the tolerance dose limits for organs at risk. Hotspots must not be in areas of low dose and, if possible, they must be within the GTV.

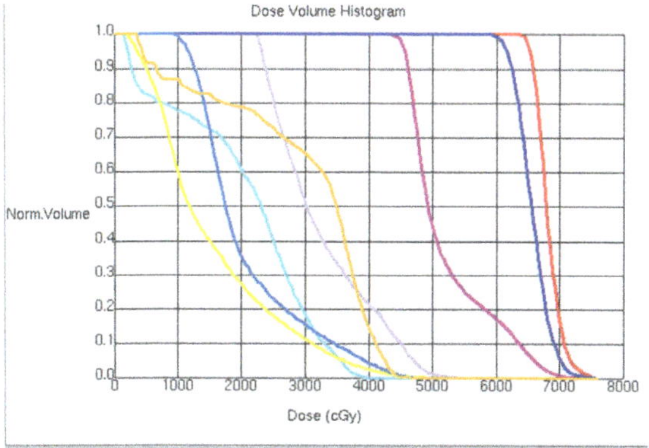

ROI Statistics

Line Type	ROI	Trial or Record	Min.	Max.	Mean	Std. Dev.
	Cord	VMAT	136.3	3945.9	2038.4	1065.9
	GTV6250	VMAT	6214.4	7569.4	6813.6	201.1
	Larynx	VMAT	2072.7	5361.5	3241.8	778.9
	Lt parotid	VMAT	162.6	4892.2	1522.9	1001.4
	PTV4500	VMAT	3550.2	7569.4	5195.9	663.3
	PTV6000	VMAT	5670.5	7569.4	6566.6	264.0
	Rt parotid	VMAT	638.5	4673.4	2043.8	635.9
	esophagus	VMAT	362.4	4636.9	2948.1	1223.5

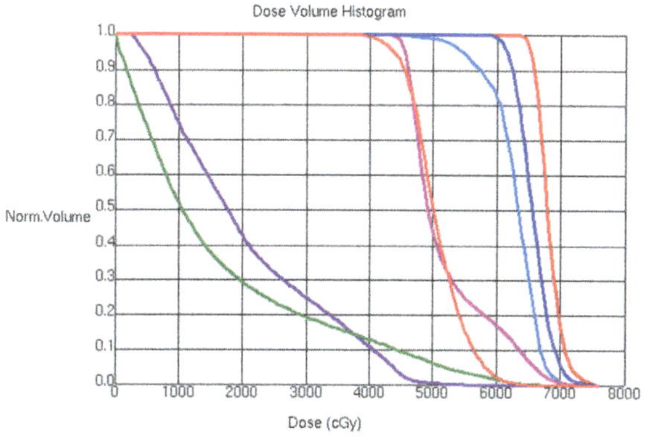

ROI Statistics

Line Type	ROI	Trial or Record	Min.	Max.	Mean	Std. Dev.
	GTV6250	VMAT	6214.4	7569.4	6813.6	201.1
	Lt submandibular	VMAT	2837.2	7241.0	6272.5	419.5
	Mandible	VMAT	227.3	5506.3	2039.4	1255.0
	Oral cavity	VMAT	0.0	6675.0	1655.9	1604.3
	PTV4500	VMAT	3550.2	7569.4	5195.9	663.3
	PTV6000	VMAT	5670.5	7569.4	6566.6	264.0
	Rt submandibular	VMAT	3864.0	6492.2	5049.6	436.5

Figure 7.64 Final DVH for case pHN02 (VMAT).

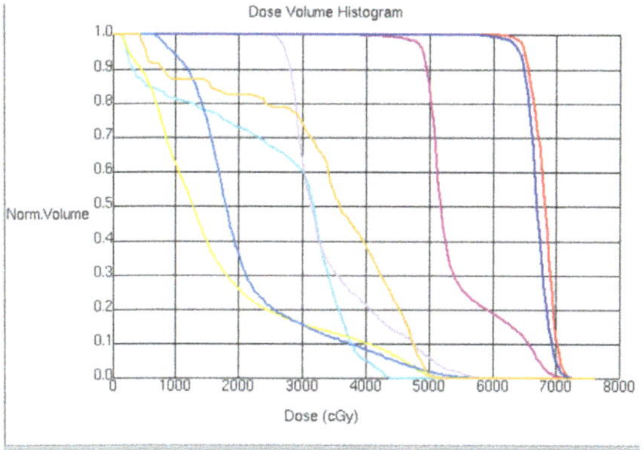

ROI Statistics

Line Type	ROI	Trial or Record	Min.	Max.	Mean	Std. Dev.
	Cord	9fld	149.1	4378.0	2614.3	1248.2
	GTV6250	9fld	5840.9	7253.4	6800.4	182.4
	Larynx	9fld	2499.9	5803.8	3439.1	741.9
	Lt parotid	9fld	88.5	5111.7	1652.2	1264.7
	PTV4500	9fld	2530.5	7253.4	5412.6	607.3
	PTV6000	9fld	5257.4	7253.4	6697.4	192.2
	Rt parotid	9fld	622.5	5644.8	2084.2	993.6
	esophagus	9fld	437.3	5168.0	3346.8	1327.0

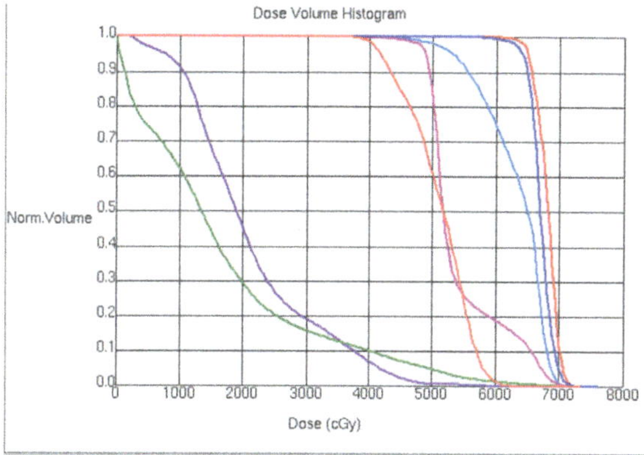

ROI Statistics

Line Type	ROI	Trial or Record	Min.	Max.	Mean	Std. Dev.
	GTV6250	9fld	5840.9	7253.4	6800.4	182.4
	Lt submandibular	9fld	2224.0	7102.9	6307.1	535.6
	Mandible	9fld	211.7	5856.2	2100.0	1031.5
	Oral cavity	9fld	0.0	7241.1	1644.8	1469.2
	PTV4500	9fld	2530.5	7253.4	5412.6	607.3
	PTV6000	9fld	5257.4	7253.4	6697.4	192.2
	Rt submandibular	9fld	3682.9	6223.1	5096.5	508.7

Figure 7.65 Final DVH for case pHN02 (IMRT).

Case Number pHN03: Head and Neck Radiotherapy Treatment

General Information

Brief Clinical History

63-year-old female patient with a locally extensive SCC in the left oropharynx. She refuses both surgery and chemotherapy.

Prescription

Left neck:
PTV 6250 = 250 cGy × 25 fractions
PTV 4500 = 180 cGy × 25 fractions

Simulation/Comments

The simulation was done with a thermoplastic mask and a moldable pillow. Fusion of images was done with a recent PET/CT for evaluation of the extent of the lesion and targeting by the radiation oncologist.

Target

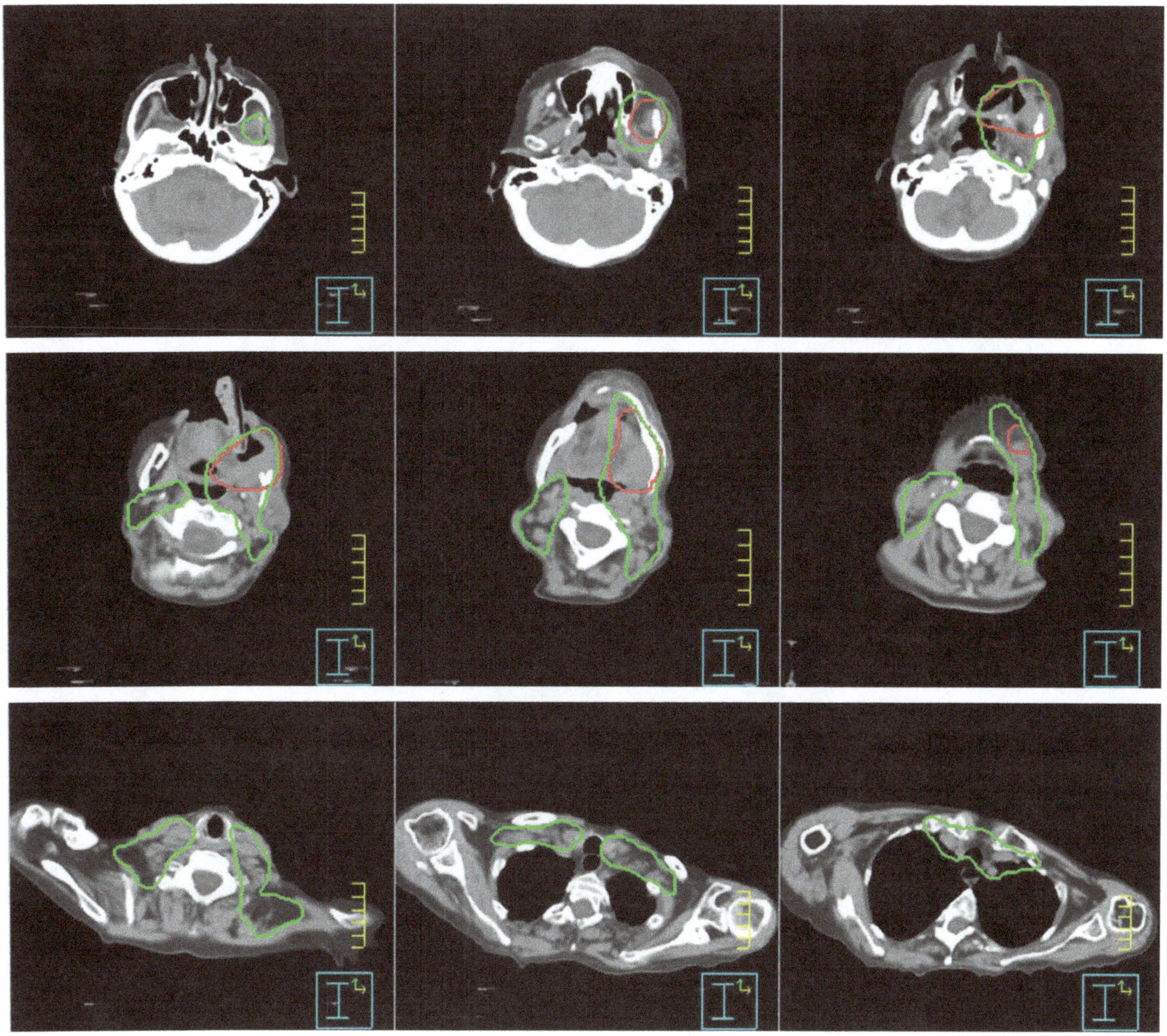

Figure 7.66 PTV contours for case pHN03: PTV 6250 (red), PTV 4500 (green).

PTV 6250 and PTV 4500 can be reviewed in Figure 7.66. Both PTV contours overlap with normal structures, such as the mandible, larynx, and the oral cavity. In this case, it is recommended that overlapping structures be accurately represented. In contrast to 3D techniques, the use of inverse planning can further minimize the dose to normal tissues.

Beam Arrangement

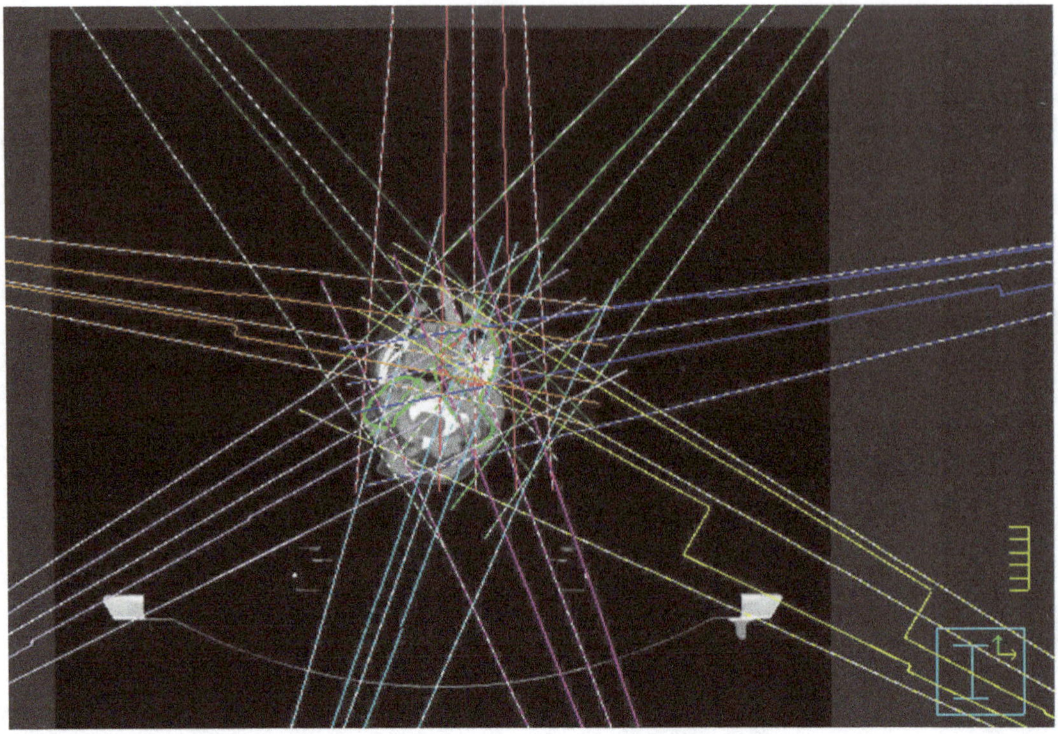

Figure 7.67 IMRT beam arrangement for case pHN03.

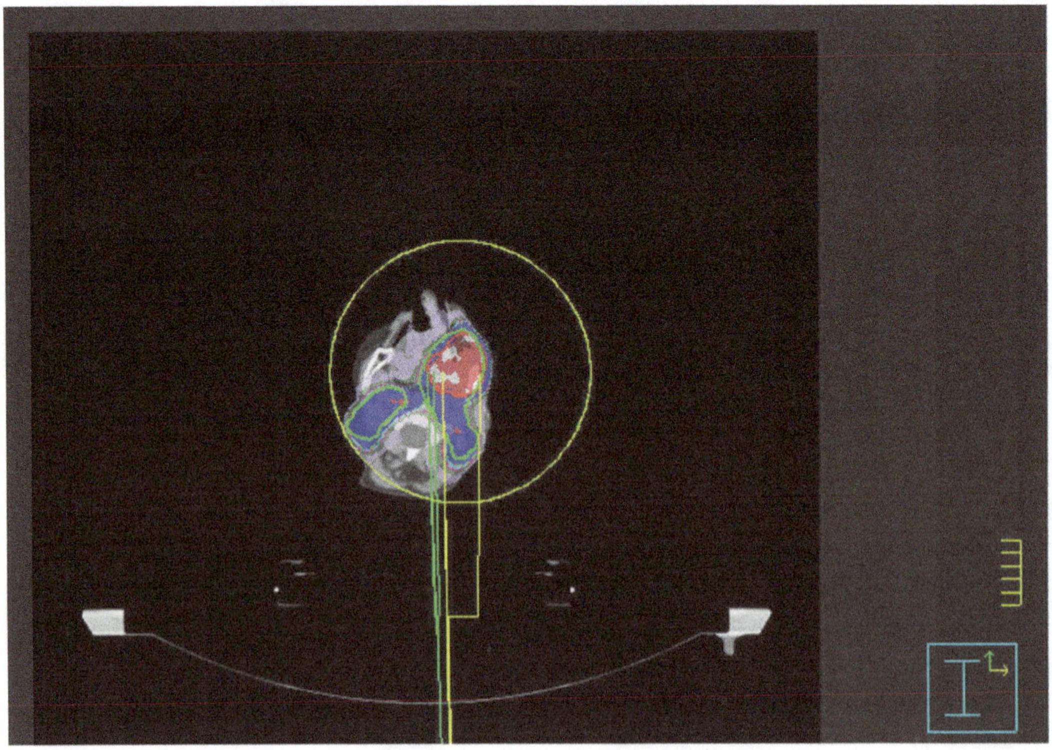

Figure 7.68 VMAT beam arrangement for case pHN03.

The IMRT plan followed a nine-field arrangement as seen in Figures 7.67 and 7.68. The beam arrangement results in appropriate dose coverage to the target. The user can place the beam isocenter in the middle of the target or in the high-dose area to achieve better dose conformity in the dose gradient region (in case of different-width MLCs). The following is a summary of the IMRT beam arrangement used in case pHN03. Notice that the order of fields was modified to follow the rotation of the gantry and optimize delivery.

Gantry Angle	Couch	Collimator	Energy (MV)
160	0	0	6
120	0	0	6
80	0	0	6
40	0	0	6
0	0	0	6
320	0	0	6
280	0	0	6
240	0	0	6
200	0	0	6

The beam arrangement for the VMAT plan had two arcs covering 360 degrees around the patient. Two arcs provide enough total beamlets to achieve a good dose distribution.

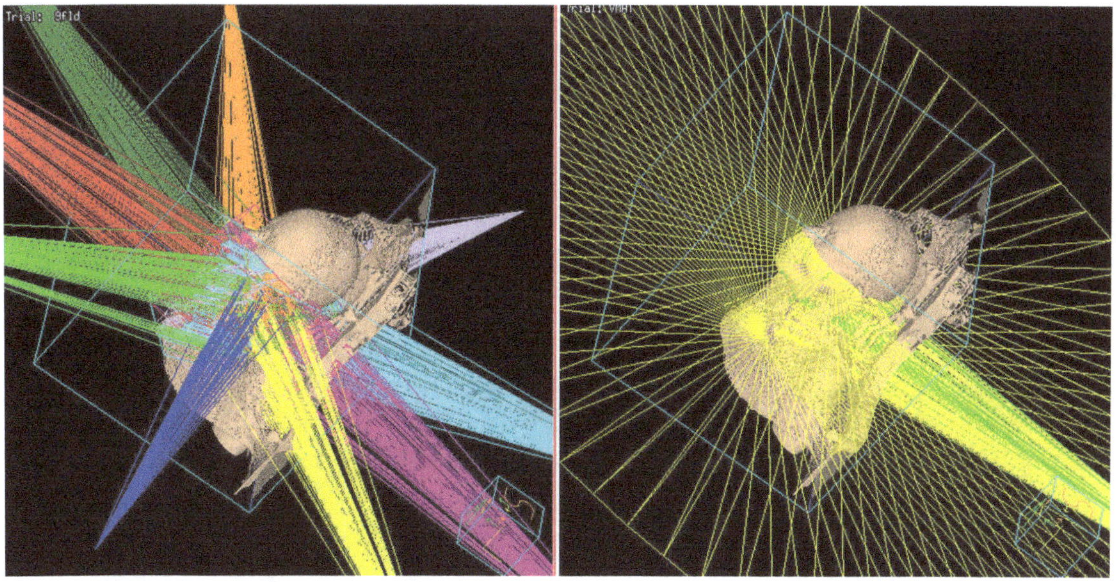

Figure 7.69 IMRT and VMAT beam arrangement for case pHN03 on a surface rendering.

Optimization

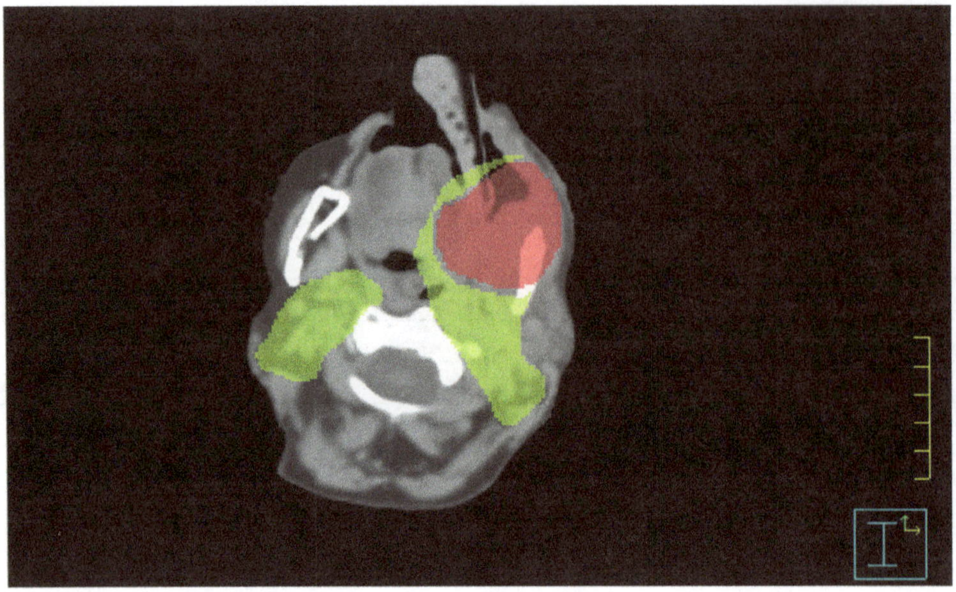

Figure 7.70 Optimization contours for different PTV levels: PTV 6250 (red colorwash), PTV 4500 (green colorwash).

Dose levels within target: There are two dose levels in this case: PTV 6250 and PTV 4500. PTV 4500 contains PTV 6250, which means that these contours need to be separated from each other for the optimization process. Figure 7.70 shows optimization contours for PTV 6250 and PTV 4500. Notice the margin around PTV 6250, which can be used instead of a PTV ring to achieve the dose gradient needed in this area.

Optimization contours: Figure 7.71 contains most of the optimization contours used in this case. The PTV ring (blue colorwash) is around the PTV 4500 contour and will help with dose conformity to the target. The body PTV (pink colorwash) encompasses most of the normal structures, while the contour with the back/neck muscles (orange colorwash) can help minimize dose to these muscles.

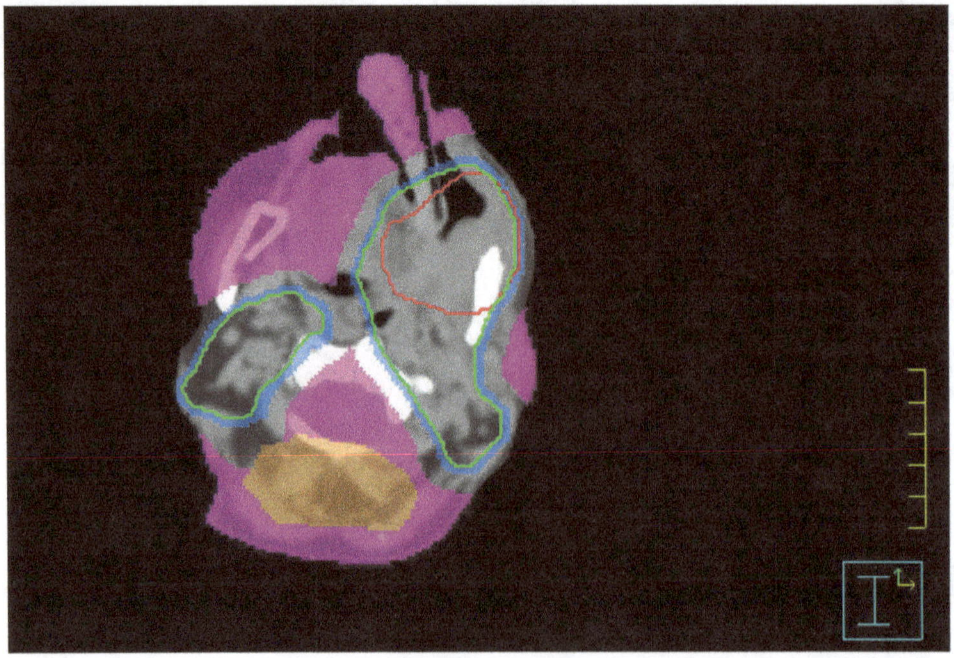

Figure 7.71 Optimization contours: PTV ring (blue colorwash), body ring (pink), and back muscles (orange).

Optimizer: Figure 7.72 shows the IMRT optimizer with a few parameters. The planner can start with the following list of objectives:

PTV 6250	→ Minimum dose	= 6250 cGy
PTV 6250	→ Maximum dose	= 6750 cGy (108% dose)
PTV 4500–PTV 6250	→ Minimum dose	= 4500 cGy
PTV 4500–PTV 6250	→ Maximum dose	= 6250 cGy
PTV rings	→ Maximum dose	= dose point in the middle of the gradients
Body ring	→ Maximum dose	= 3800 cGy
Back muscle	→ Maximum dose	= 3000 cGy

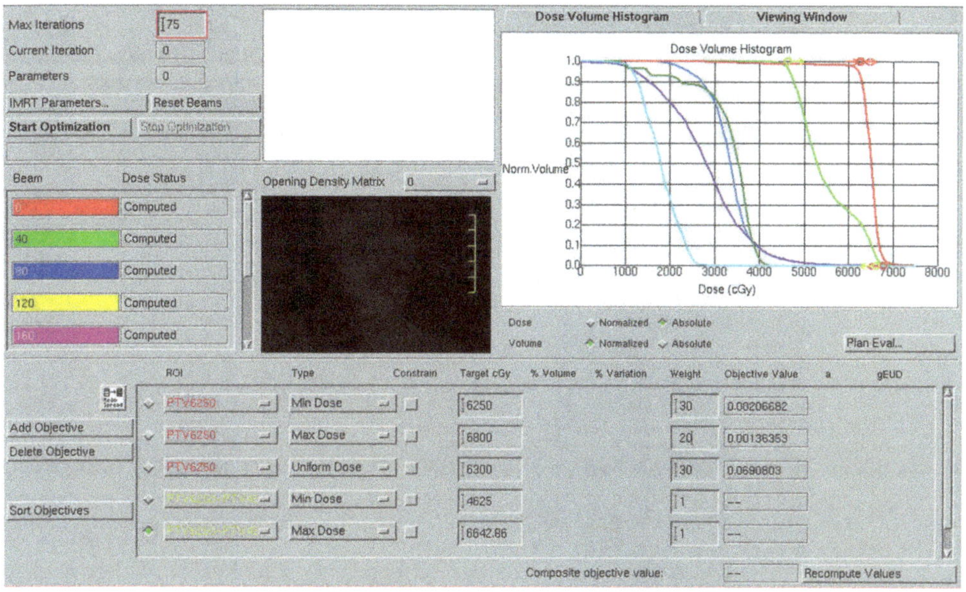

Figure 7.72 IMRT Optimizer window for case pHN03.

After the first optimization, the planner can add objectives for organs at risk until the desired dose distribution is achieved. The same parameters can be used when planning VMAT.

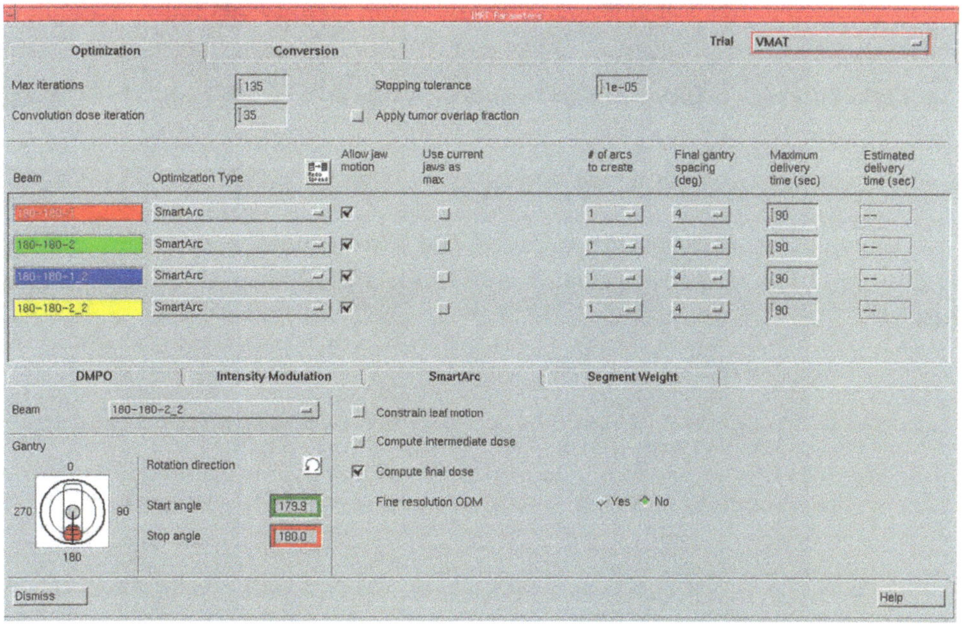

Figure 7.73 Standard VMAT parameters window for case pHN03.

Dose Distribution

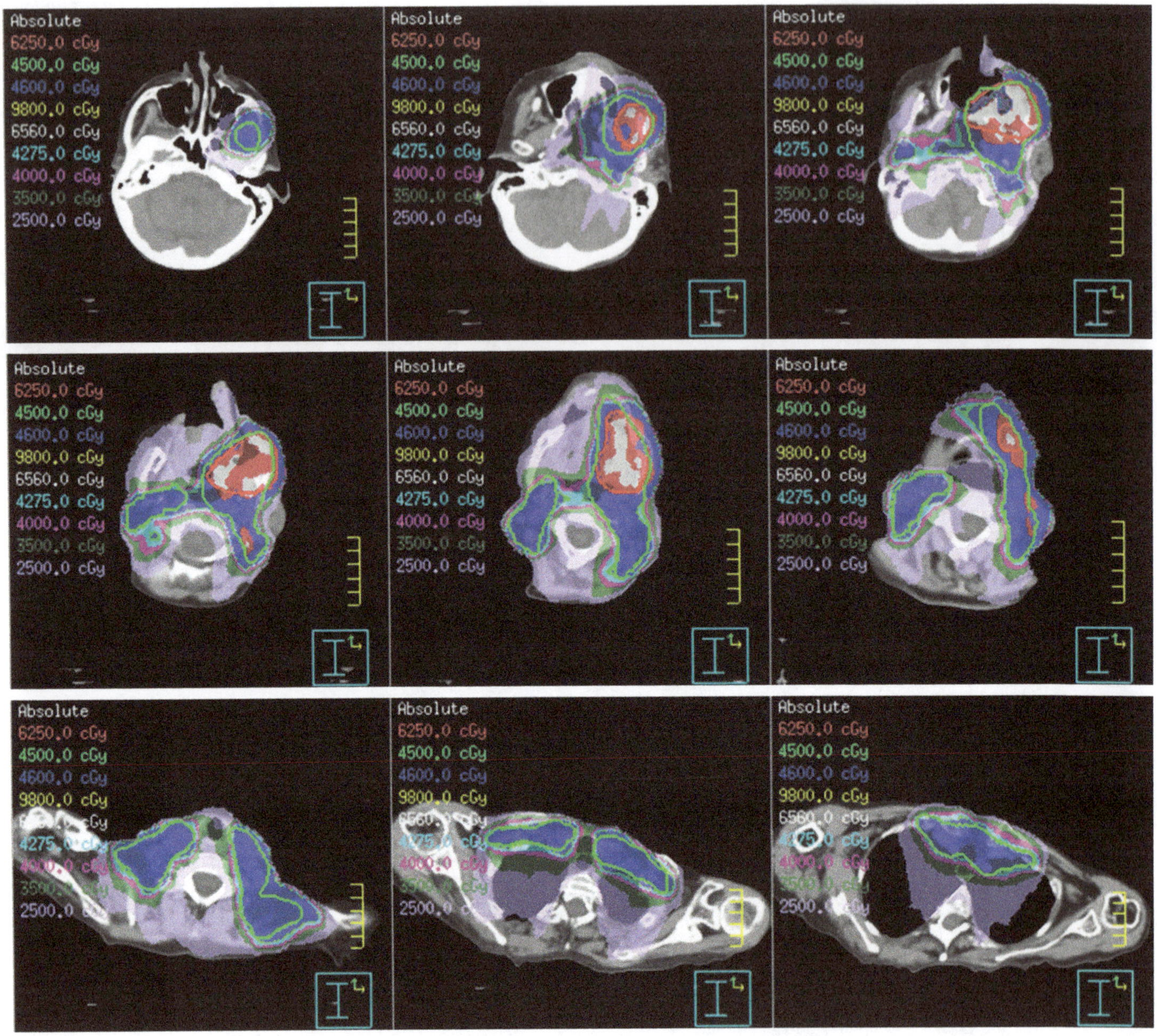

Figure 7.74 IMRT final dose distribution.

Figures 7.74 and 7.75 show the dose distribution for IMRT and VMAT plans, respectively. Hotspots must be within the PTV/GTV, and dose to OARs must be evaluated with guidelines of normal tissue tolerance according to the selected fractionation.

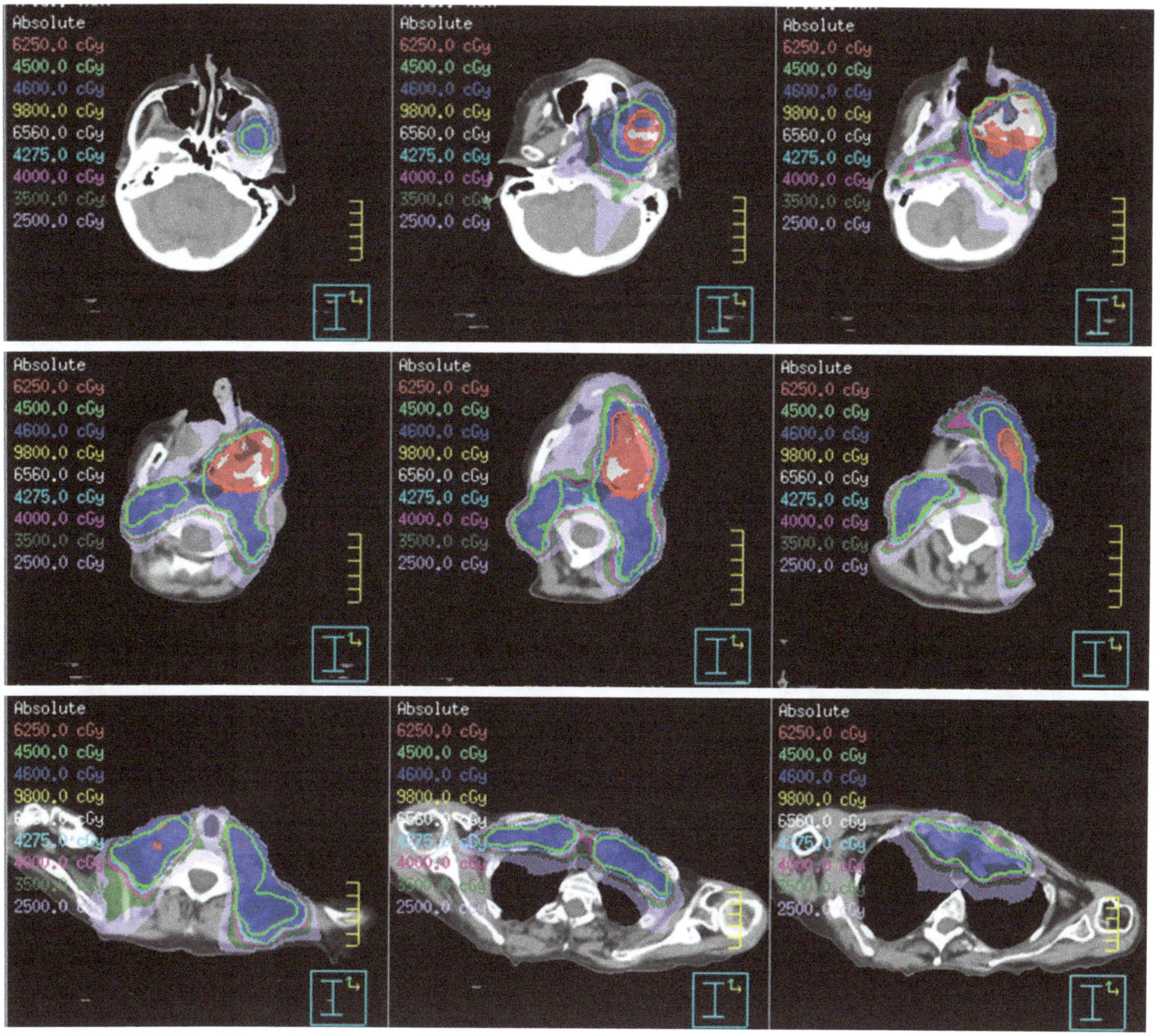

Figure 7.75 VMAT final dose distribution.

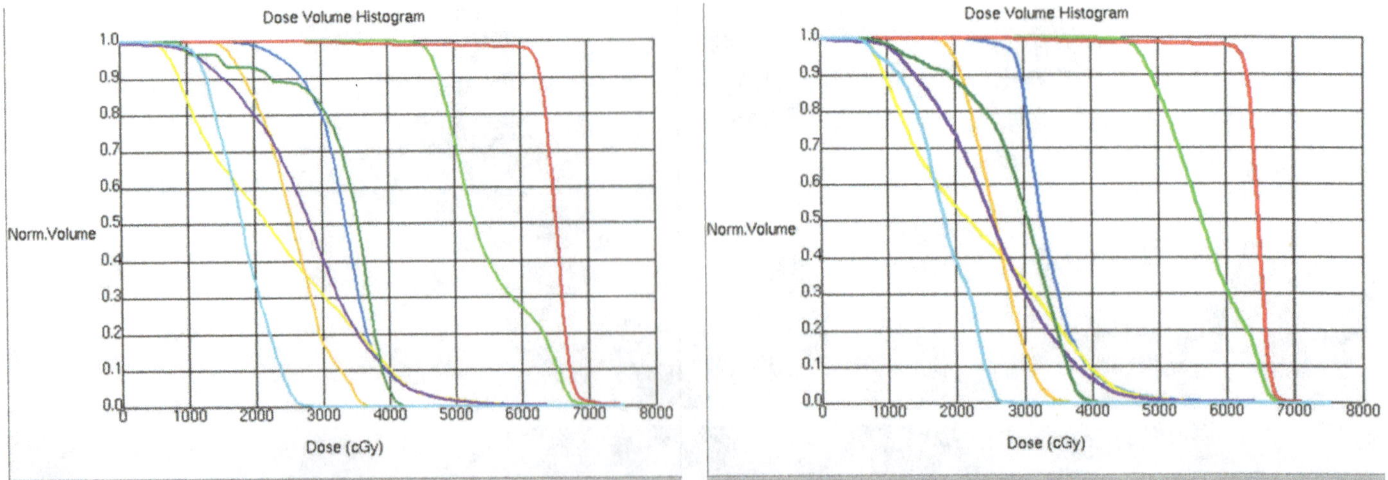

ROI Statistics

Line Type	ROI	Trial or Record	Min.	Max.	Mean	Std. Dev.
	Larynx	9fld	1483.9	6036.1	3349.6	549.1
	Left parotid	9fld	232.1	6174.4	2323.8	1200.2
	Lt ear cannal	9fld	1396.0	3676.3	2554.8	497.7
	Oral cavity	9fld	1.5	6381.5	2781.7	934.2
	PTV4500	9fld	2331.8	7474.6	5490.8	666.5
	PTV6250	9fld	2338.7	7474.6	6469.0	413.0
	esophagus	9fld	927.9	4230.9	3313.7	705.0
	spinal canal	9fld	383.0	2847.4	1804.2	414.7

ROI Statistics

Line Type	ROI	Trial or Record	Min.	Max.	Mean	Std. Dev.
	Larynx	VMAT	1937.8	6154.4	3366.1	458.0
	Left parotid	VMAT	374.0	5974.6	2345.1	1155.4
	Lt ear cannal	VMAT	1677.7	3645.5	2577.1	378.6
	Oral cavity	VMAT	4.3	6402.4	2562.8	900.0
	PTV4500	VMAT	2313.9	7099.3	5658.0	590.5
	PTV6250	VMAT	2467.3	7099.3	6429.7	392.4
	esophagus	VMAT	729.6	4052.6	2695.8	692.6
	spinal canal	VMAT	506.0	2804.9	1820.4	480.7

Figure 7.76 Final DVH for IMRT (left) and VMAT (right) plans.

Case Number pHN04: Head and Neck Radiotherapy Treatment

General Information

Brief Clinical History

75-year-old female diagnosed with pT1N0 invasive squamous cell carcinoma, moderately differentiated of the left tongue after hemiglossectomy. Radiation therapy takes place after the surgery and extends to treat the bilateral neck.

Prescription

Left tongue and bilateral neck:
PTV 6250 = 250 cGy × 25 fractions
PTV 4500 = 180 cGy × 25 fractions

Simulation/Comments

The simulation was done with a thermoplastic mask and a moldable pillow. Fusion of images was done with a recent PET/CT for evaluation of the extent of the lesion and targeting by the radiation oncologist.

Target

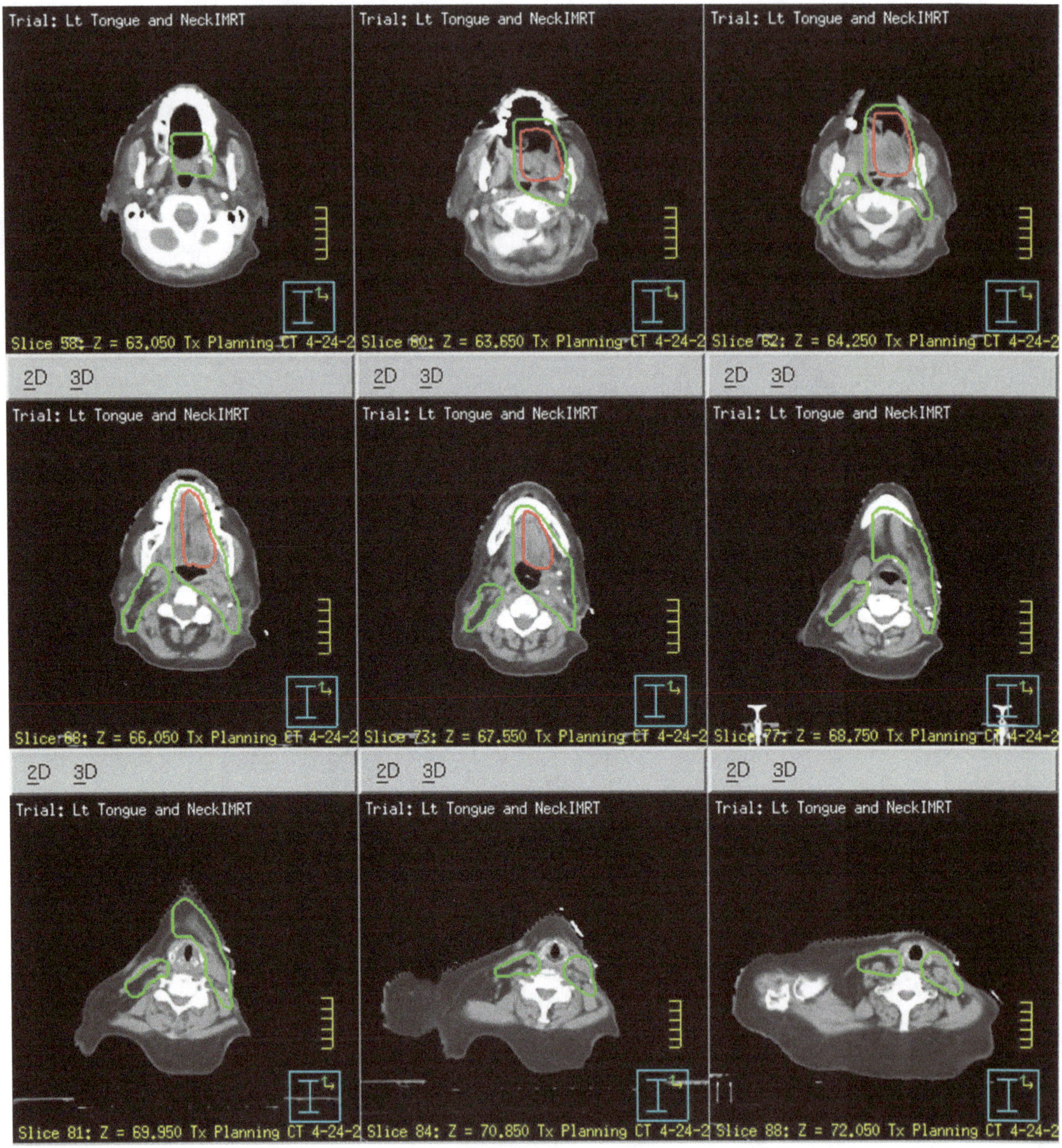

Figure 7.77 PTV contours for case pHN04: PTV 6250 (red), PTV 4500 (green).

Figure 7.77 shows PTV contours for this case. These contours encompass disease in the oral cavity that extends to neck lymph nodes. The overlapping PTV structures must be subtracted before the optimization process, and normal structures must be appropriately contoured. One can subtract the PTV from the normal structure and use it in the optimization process to avoid the conflict of objectives in the optimizer.

Beam Arrangement

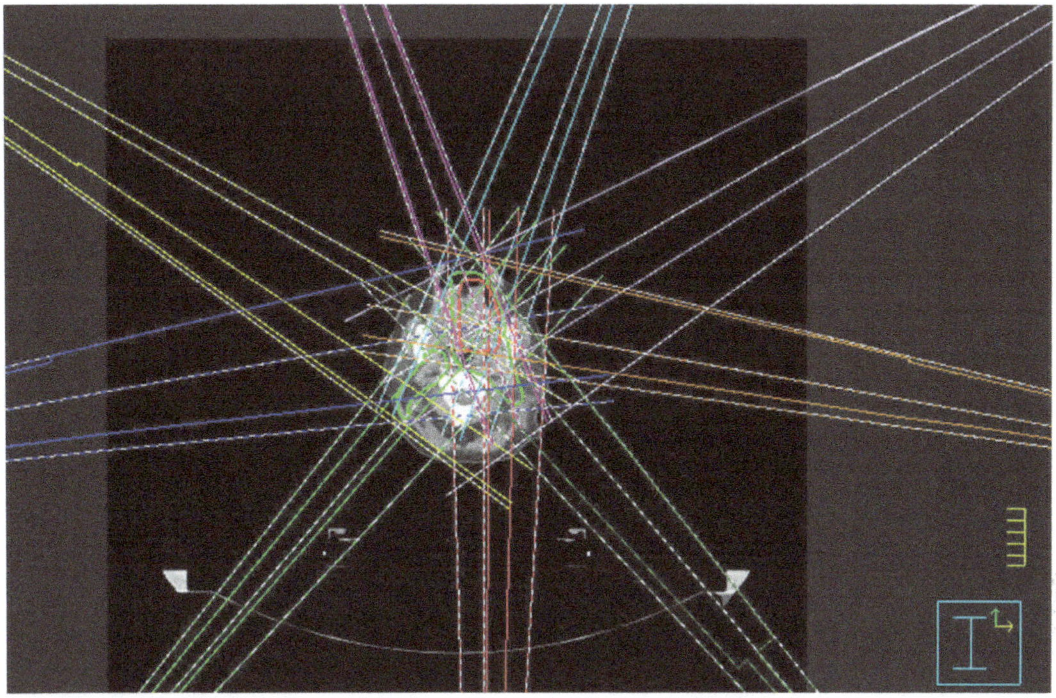

Figure 7.78 IMRT beam arrangement for case pHN04.

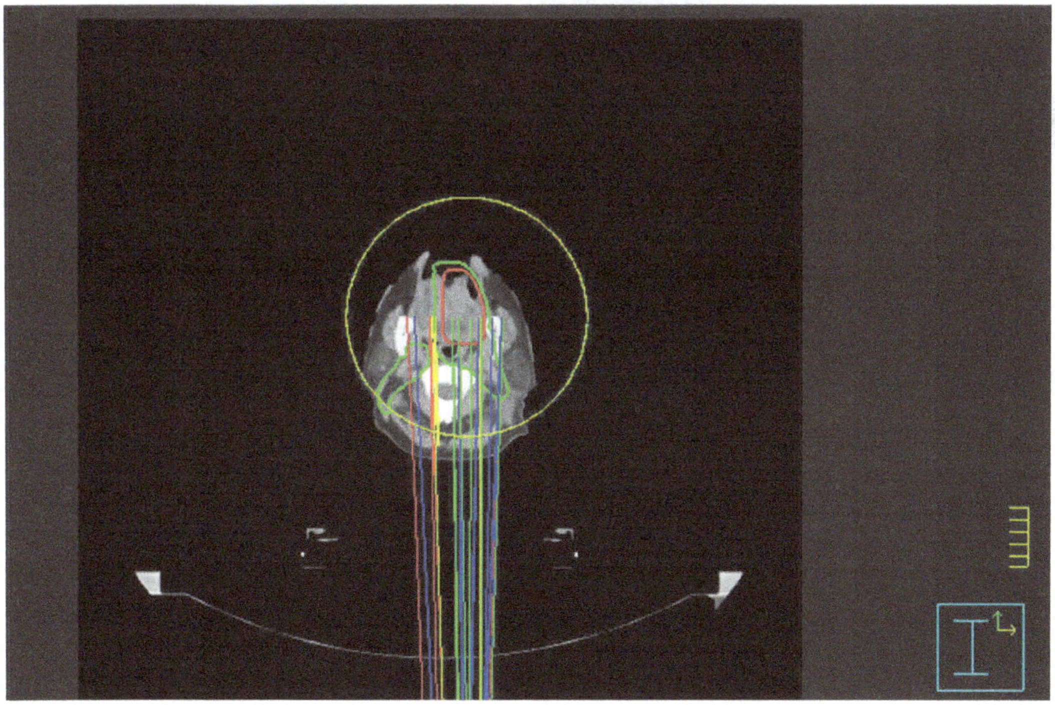

Figure 7.79 VMAT beam arrangement for case pHN04.

Figure 7.78 depicts the IMRT beam arrangement for this plan. Nine beams were placed around the patient to cover the target. Notice how none of the CAX of the beams goes through the spinal canal, i.e., while the leaves will be sparing the spinal canal on the edges of the field, they will not be blocking the central axis. Figure 7.79 shows the VMAT field arrangement. Three arcs were set to run 360 degrees around the patient, with control points every 2 degrees.

Gantry Angle	Couch	Collimator	Energy (MV)
20	0	0	6
45	0	0	6
100	0	0	6
150	0	0	6
180	0	0	6
225	0	0	6
245	0	0	6
320	0	0	6
345	0	0	6

Above is a summary of the IMRT beam arrangement used in case pHN04. Notice that the order of fields was modified to follow the rotation of the gantry and optimize delivery.

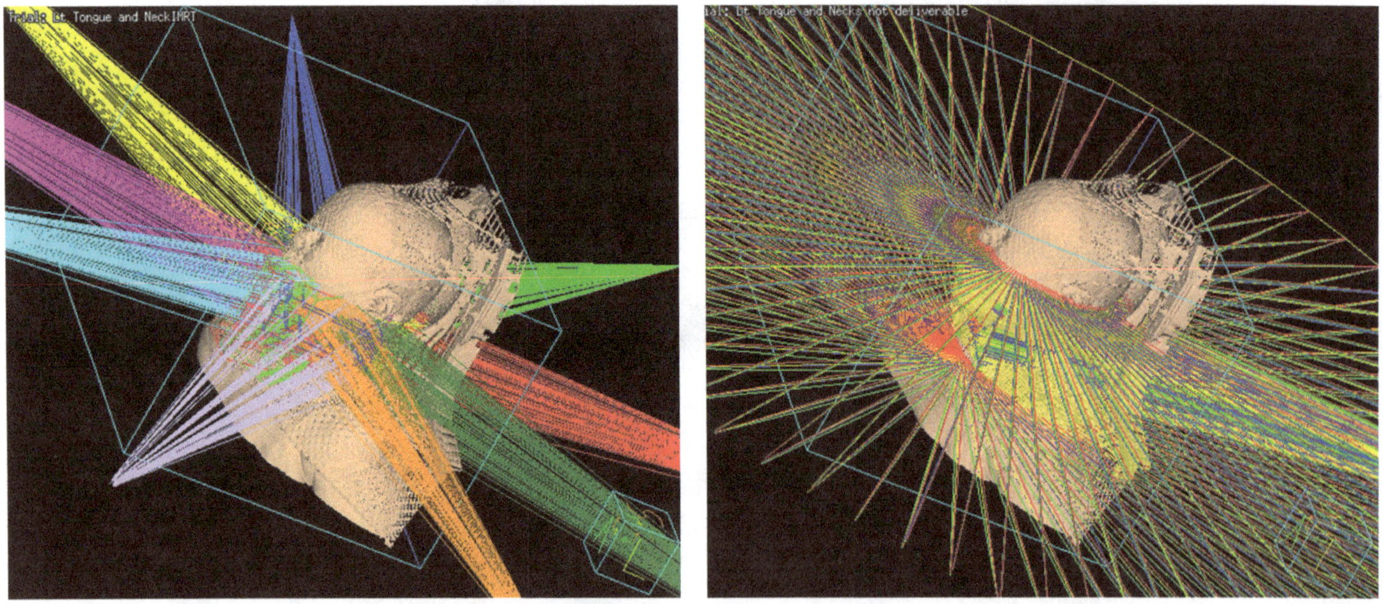

Figure 7.80 IMRT and VMAT beam arrangements for case pHN04 on a surface rendering.

Optimization

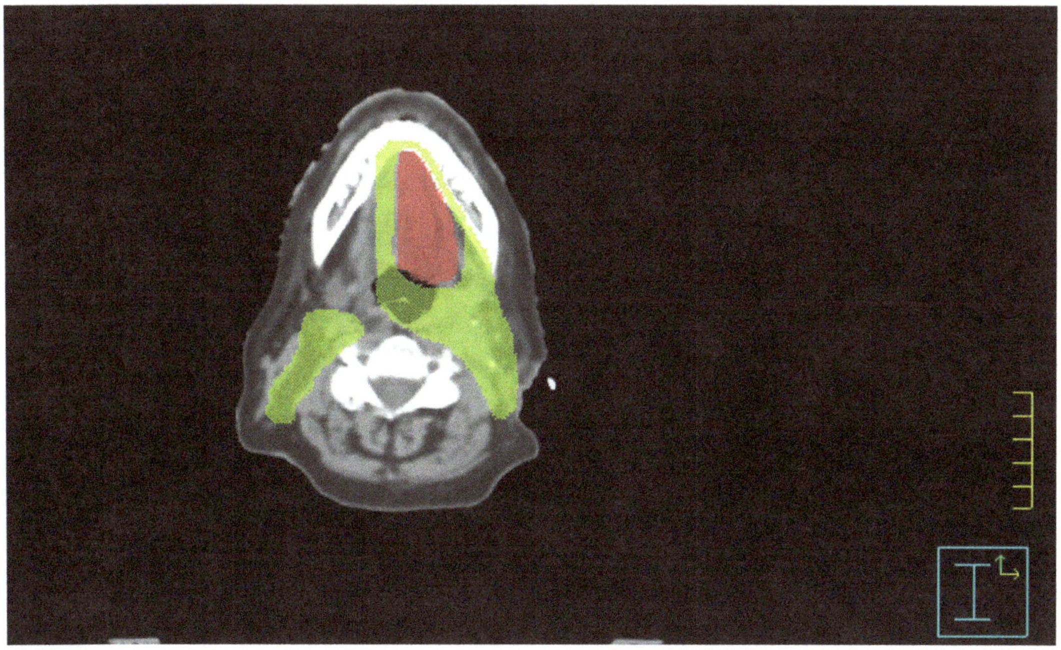

Figure 7.81 Optimization contours for different level PTV.

Dose levels within target: Figure 7.81 shows optimization contours for PTVs in colorwash display. Notice the 2-mm margin around PTV 6250, which achieves the dose gradient required in this area.

Optimization contours: Figure 7.82 displays key optimization contours for this case. There are two PTV rings, having the outer ring act as a "body ring." Other structures include a 5-millimeter expansion of the spinal canal and a triangle contour on back muscles. The cord margin helps spare the spinal canal during the optimization process, and the contour of the back muscles avoids significant dose that might cause muscle damage as a side effect.

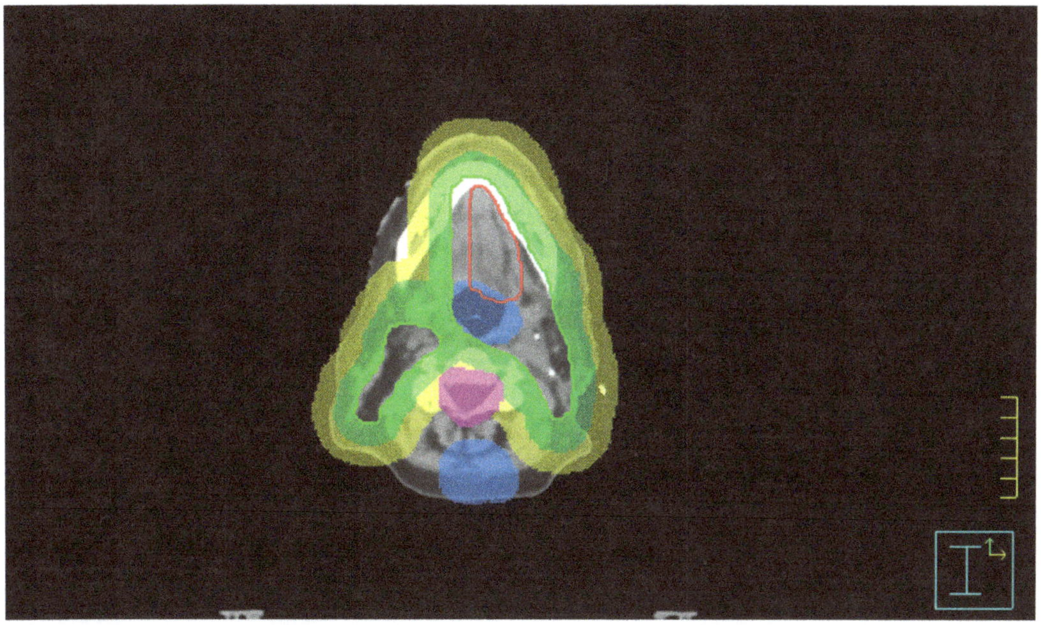

Figure 7.82 Optimization contours: PTV ring (green colorwash), outer PTV ring (yellow colorwash), cord margin (pink colorwash), back muscles (blue colorwash).

Optimizer: The planner can start with the following list of objectives:

PTV 6250	→ Minimum dose = 6250 cGy
PTV 6250	→ Maximum dose = 6750 cGy (108% dose)
PTV 4500–PTV 6250	→ Minimum dose = 4500 cGy
PTV 4500–PTV 6250	→ Maximum dose = 6250 cGy
PTV ring	→ Maximum dose = dose point in the middle of the gradients
PTV outer ring	→ Maximum dose = 3800 cGy
Back muscle	→ Maximum dose = 3000 cGy and under

After the first optimization, the planner can check the DVH and the dose distribution and add organs at risk and other structures to help shape the dose distribution to what is clinically desirable. The same parameters can be used when planning IMRT and VMAT.

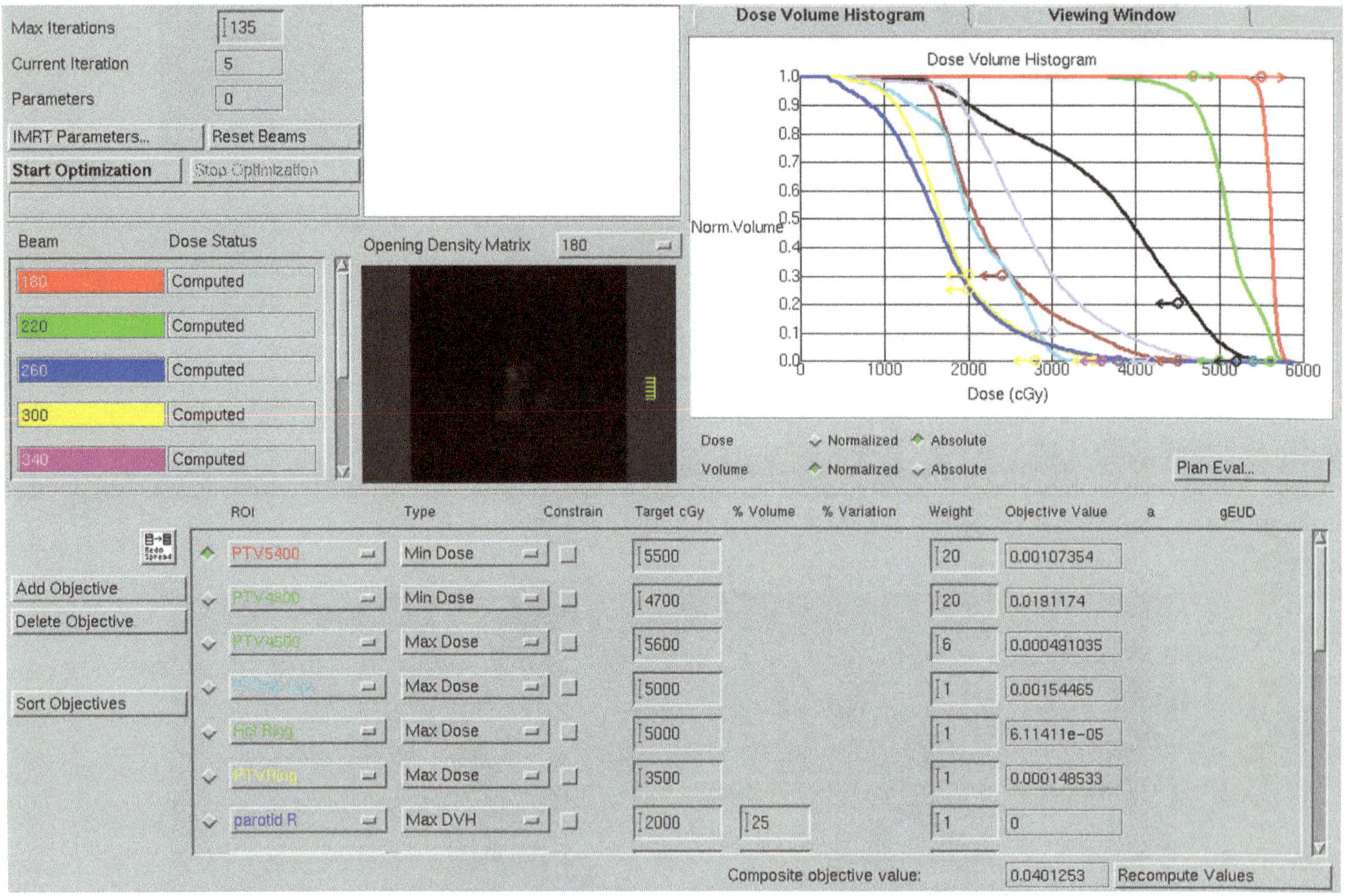

Figure 7.83 IMRT Optimizer window for case pHN04.

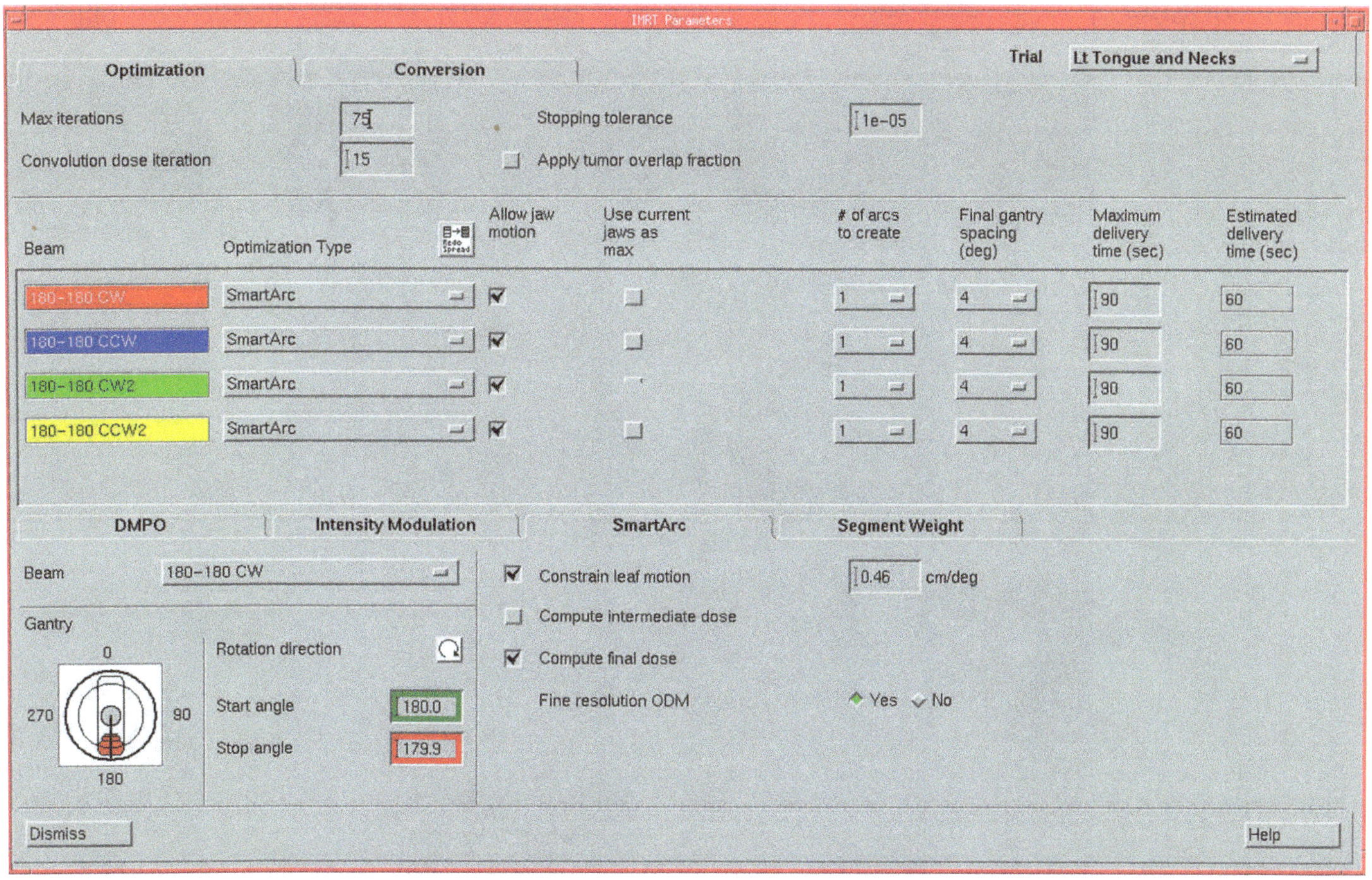

Figure 7.84 Standard VMAT parameters window for case pHN04.

Dose Distribution

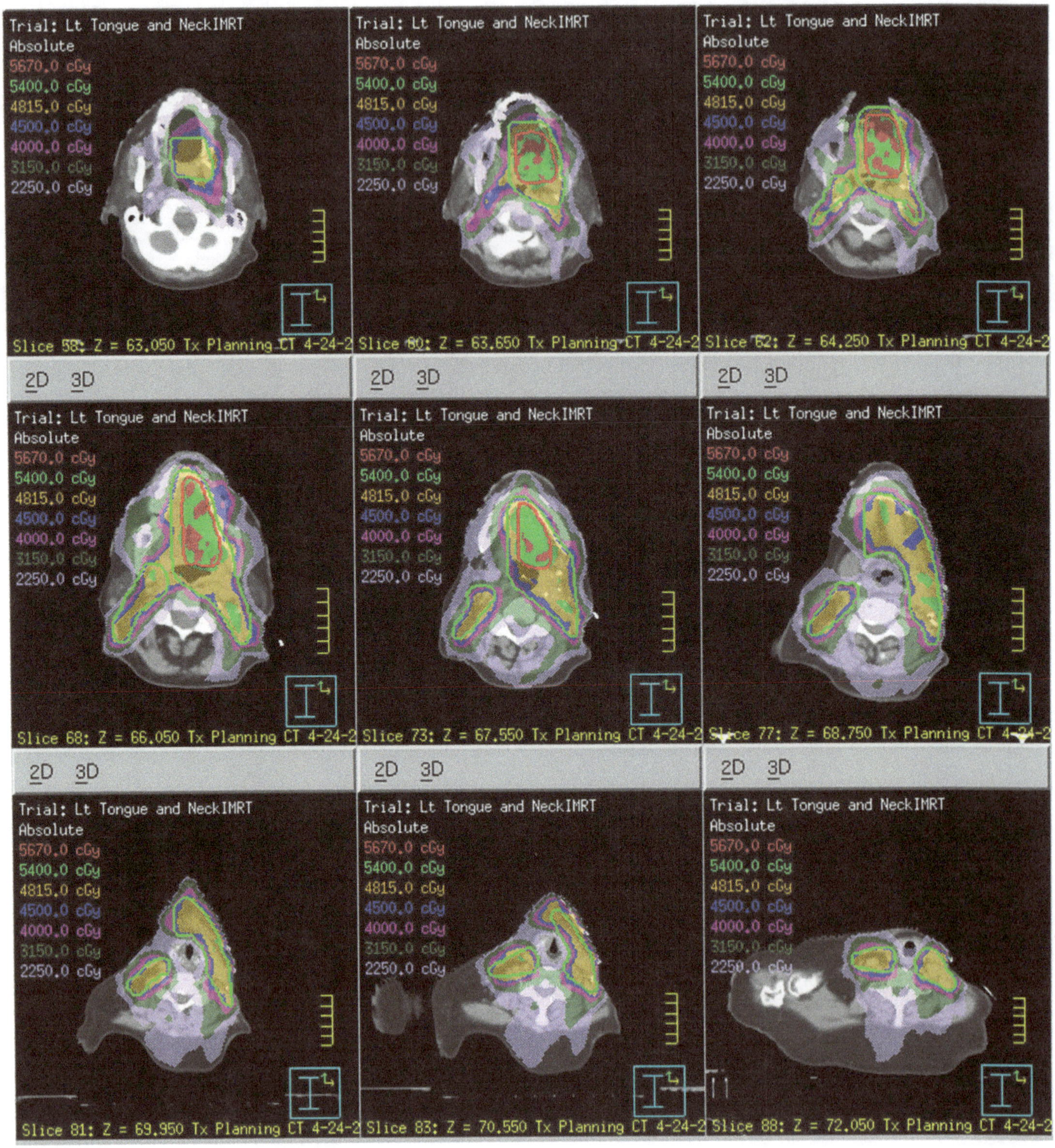

Figure 7.85 IMRT final dose distribution.

Figures 7.85 and 7.86 display the final dose distribution for IMRT and VMAT treatment plans. Maximum dose points should be inside PTV/CTV areas. Check for dose conformity and normal tissue tolerance according to the selected fractionation.

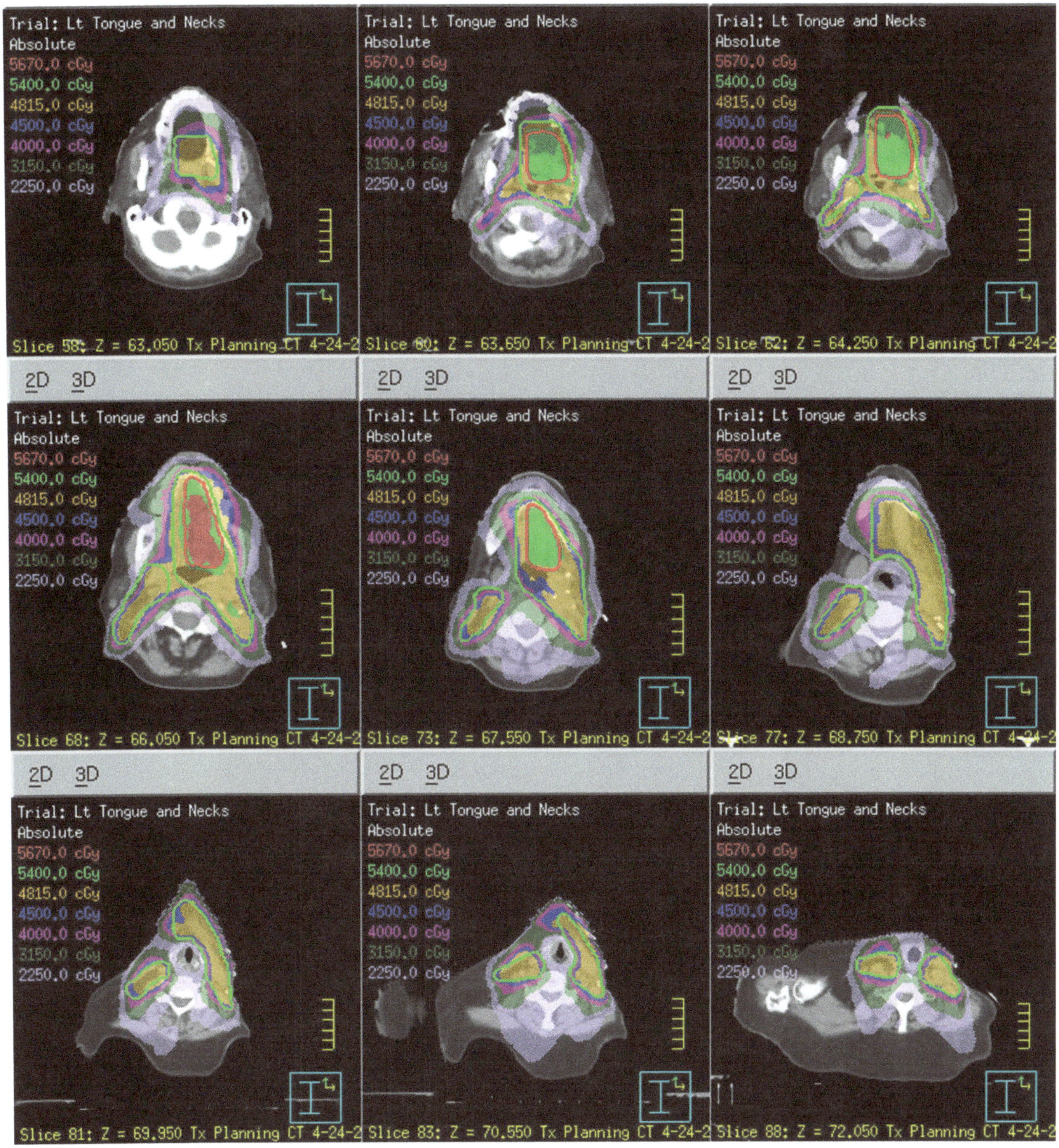

Figure 7.86 VMAT final dose distribution.

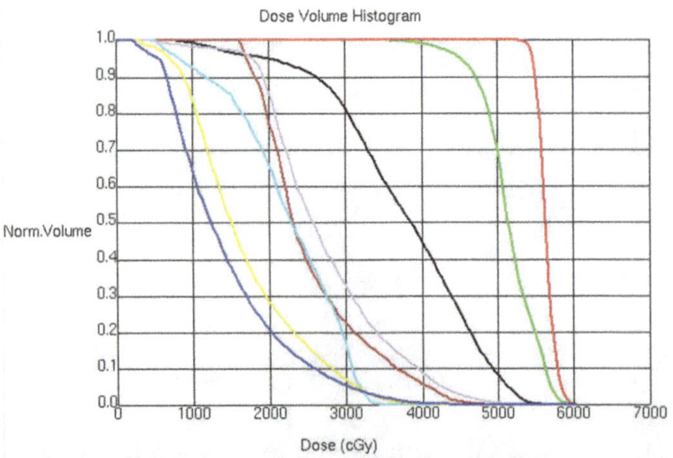

ROI Statistics

Line Type	ROI	Trial or Record	Min.	Max.	Mean	Std. Dev.
	Mandible	Lt Tongue and Neck	731.3	5571.9	3759.3	957.1
	PTV4500	Lt Tongue and Neck	976.6	5994.9	5141.2	381.1
	PTV5400	Lt Tongue and Neck	4457.2	5994.9	5639.2	109.8
	R submandibular	Lt Tongue and Neck	1601.5	4806.8	2535.6	700.3
	Spinal Canal	Lt Tongue and Neck	491.8	3441.9	2230.7	708.2
	larynx-esophagus	Lt Tongue and Neck	411.7	5322.1	2719.1	808.9
	parotid L	Lt Tongue and Neck	282.7	4161.6	1667.4	734.4
	parotid R	Lt Tongue and Neck	192.9	4352.4	1429.7	767.3

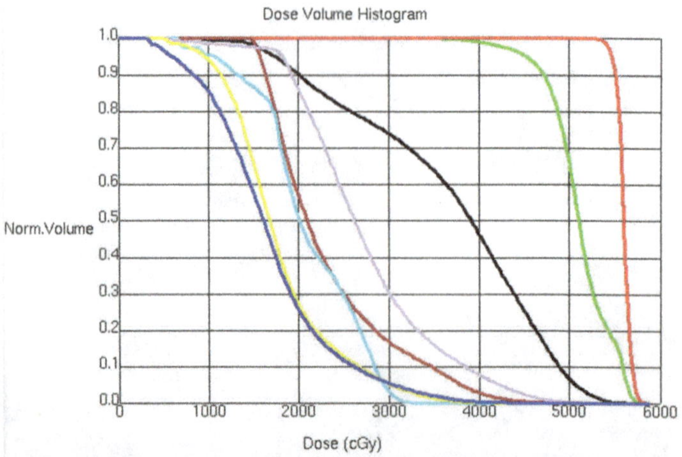

ROI Statistics

Line Type	ROI	Trial or Record	Min.	Max.	Mean	Std. Dev.
	Mandible	Lt Tongue and Neck	826.6	5631.6	3666.4	1057.2
	PTV4500	Lt Tongue and Neck	1176.4	5825.4	5104.2	360.7
	PTV5400	Lt Tongue and Neck	4498.2	5825.4	5596.6	76.6
	R submandibular	Lt Tongue and Neck	1478.8	4717.9	2315.2	685.9
	Spinal Canal	Lt Tongue and Neck	652.3	3238.0	2089.1	569.6
	larynx-esophagus	Lt Tongue and Neck	504.5	5116.7	2727.9	757.6
	parotid L	Lt Tongue and Neck	456.7	4138.3	1781.1	601.5
	parotid R	Lt Tongue and Neck	287.1	4302.2	1670.9	693.4

Figure 7.87 Final DVH for case pHN04 for IMRT (left) and VMAT (right).

Case Number pHN05: Head and Neck Radiotherapy Treatment

General Information

Brief Clinical History

78-year-old woman with a stage IVA, T4N0M0 SCC in the right lower gingiva invading into the right lower mandible, status post mandibulectomy with reconstruction.

Prescription

Left neck:
PTV 6000 = 240 cGy × 25 fractions
PTV 5000 = 200 cGy × 25 fractions

Simulation/Comments

The simulation was done with a thermoplastic mask and a moldable pillow. Fusion of images was done with a recent PET/CT for evaluation of the extent of the lesion and targeting by the radiation oncologist.

Target

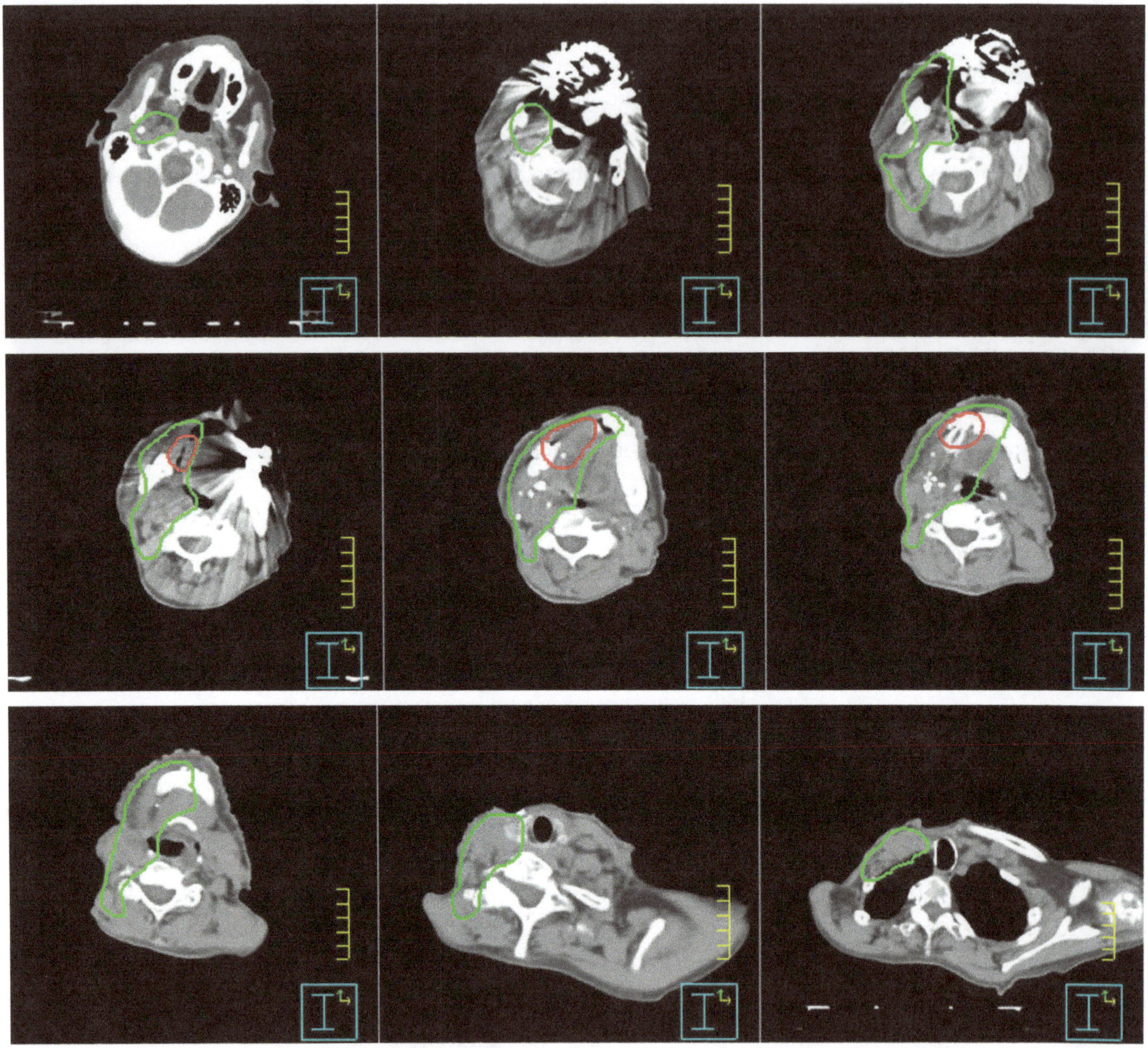

Figure 7.88 PTV contours for case pHN05: PTV 6000 (red), PTV 5000 (green).

Figure 7.88 displays the PTV contours drawn by the radiation oncologist. PTV 6000 and PTV 5000 encompass head and neck areas on one side of the body. It is recommended that both PTV contours are subtracted before the optimization process.

Beam Arrangement

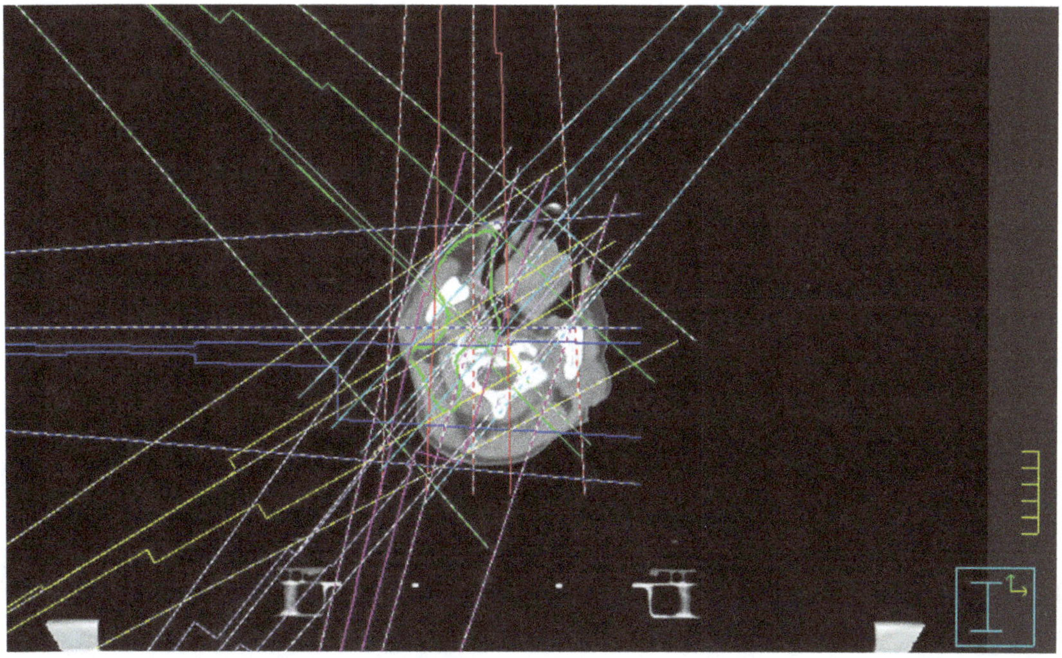

Figure 7.89 IMRT beam arrangement for case pHN05.

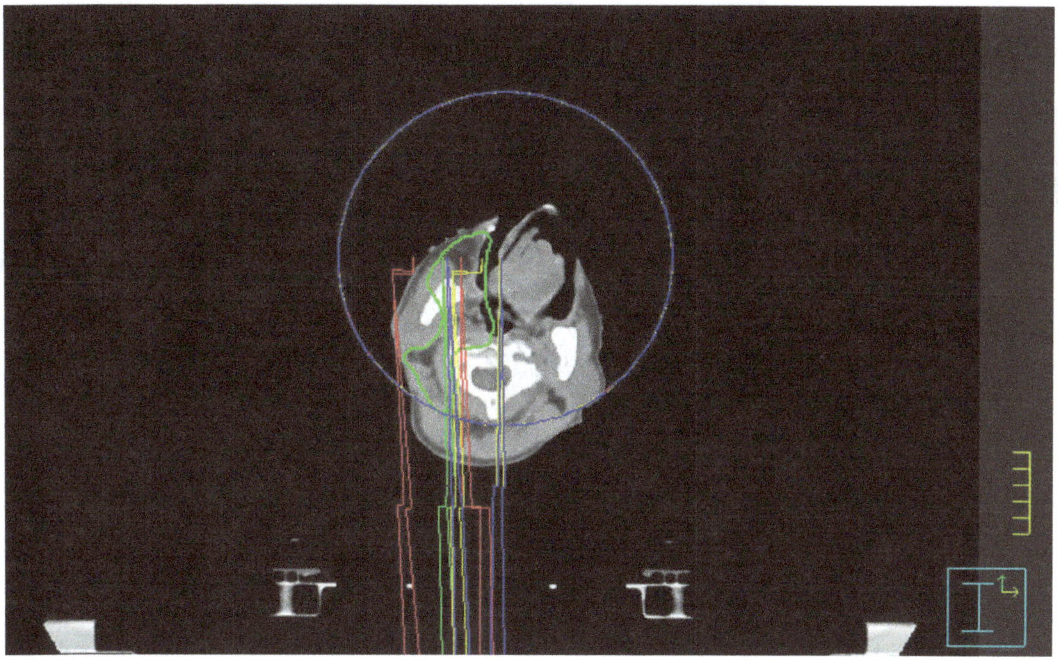

Figure 7.90 VMAT beam arrangement for case pHN05.

Figures 7.89 and 7.90 display the beam arrangements for IMRT and VMAT plans for the same case. The IMRT plan has 7 beams that cover the tumor from different angles. Notice the one-sided arrangement of beams to avoid irradiating normal tissue contralateral to the lesion side.

Gantry Angle	Couch	Collimator	Energy (MV)
35	0	0	6
0	0	0	6
325	0	0	6
270	0	0	6
240	0	0	6
220	0	0	6
195	0	0	6

The beam arrangement for the VMAT plan has 2 arcs going around the patient for complete rotations. Another option for the planner might be to set partial arcs with start and stop points close to the range covered by the IMRT plan.

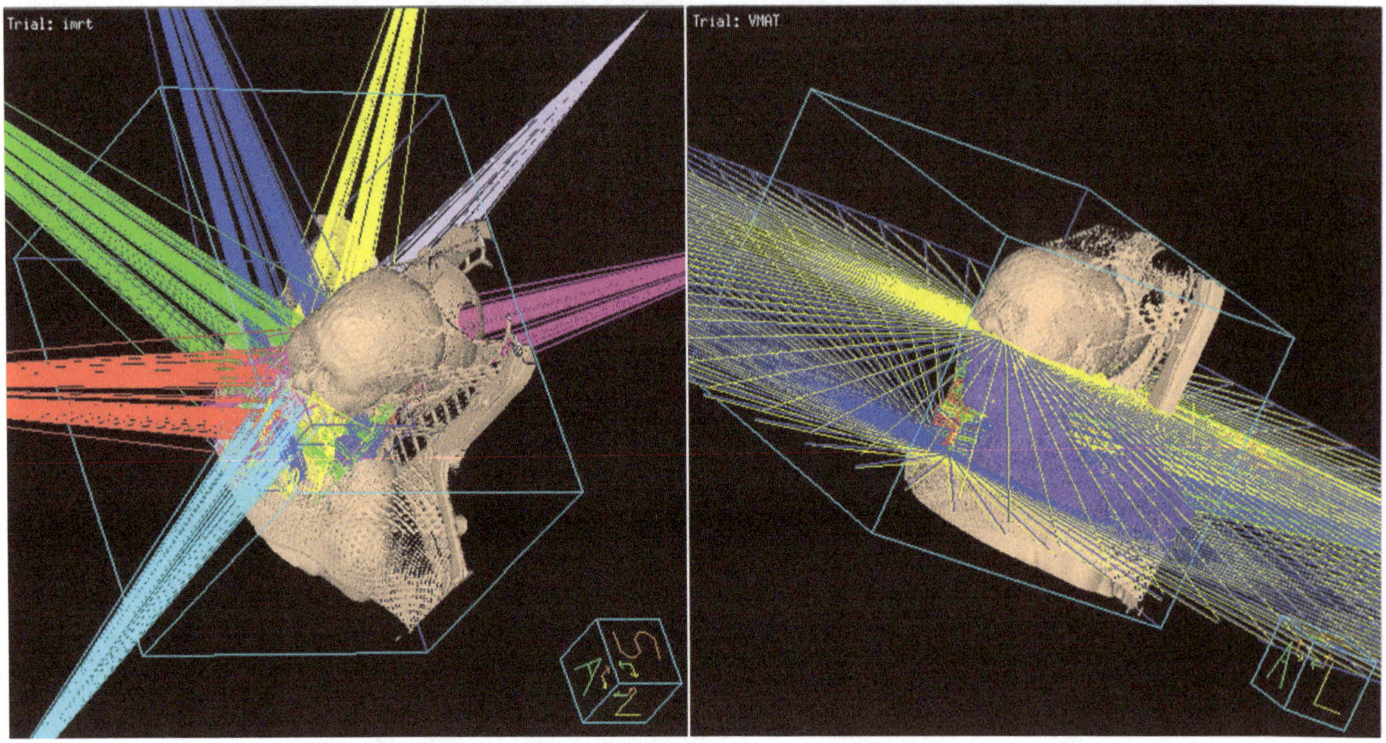

Figure 7.91 IMRT and VMAT beam arrangement for case pHN05 on a surface rendering.

Optimization

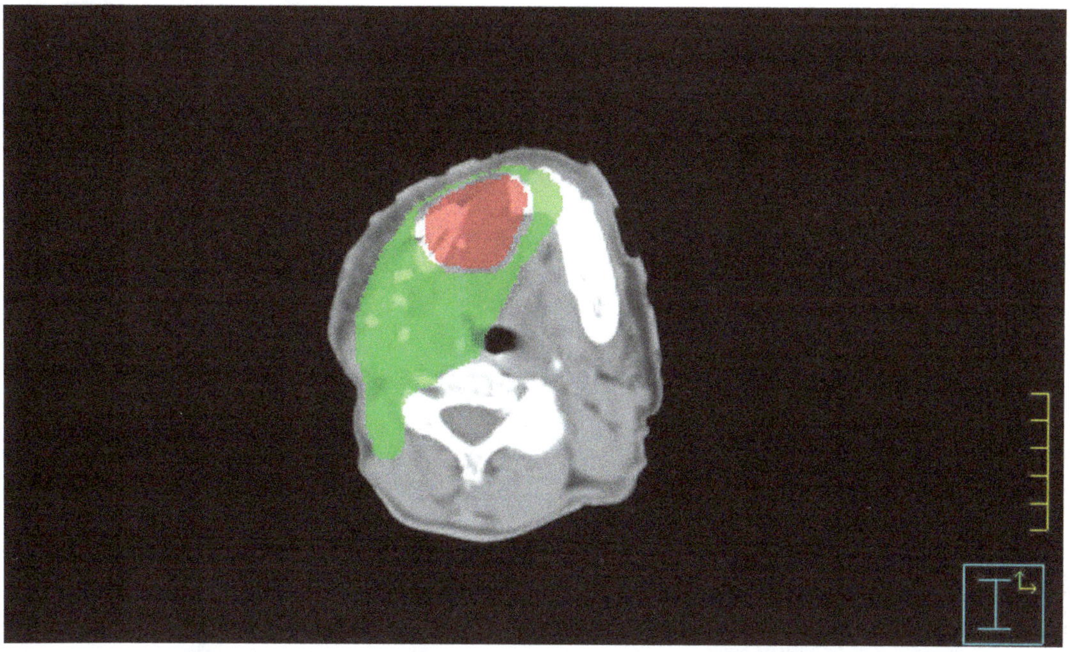

Figure 7.92 Optimization contours for different PTV levels: PTV 6000 (red colorwash), PTV 5000 (green colorwash).

Dose levels within target: There are two dose levels in this case: PTV 6000 and PTV 5000. Before running the optimization, it is important to avoid objective conflicts by separating both contours. A PTV ring can be used around PTV 6000, but a similar effect can be achieved by leaving a gap between the two dose levels.

Optimization contours: Figure 7.93 contains two of the optimization contours used in this case. A PTV ring is used to keep the dose distribution conformal to the target. In this case, the body ring has encompassed all other normal structures. It is important to mention that the existence and use of these two structures does not mean that the planner cannot place objectives on other contours of normal structures as well.

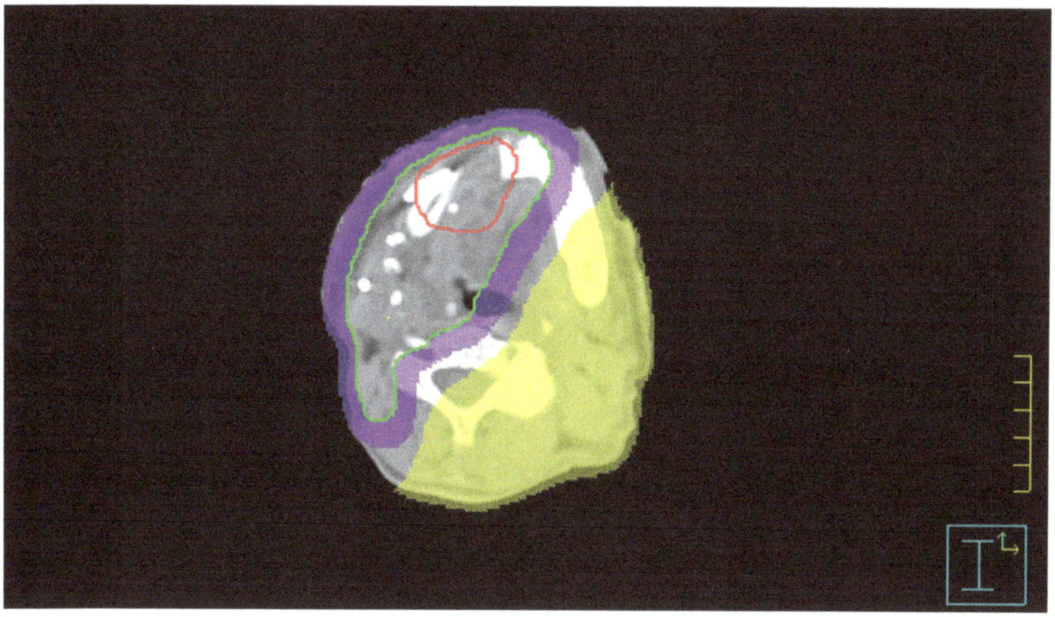

Figure 7.93 Optimization contours: PTV ring (purple colorwash), body ring (yellow).

Optimizer: Figure 7.94 shows the IMRT Optimizer with a few parameters. The planner can start with the following list of objectives:

PTV 6000	→ Minimum dose = 6000 cGy
PTV 6000	→ Maximum dose = 6480 cGy (108% dose)
PTV 5000–PTV 6000	→ Minimum dose = 5000 cGy
PTV 5000–PTV 6000	→ Maximum dose = 4250cGy
PTV ring	→ Maximum dose = dose point in the middle of the gradients
Body ring	→ Maximum dose = 3000 cGy

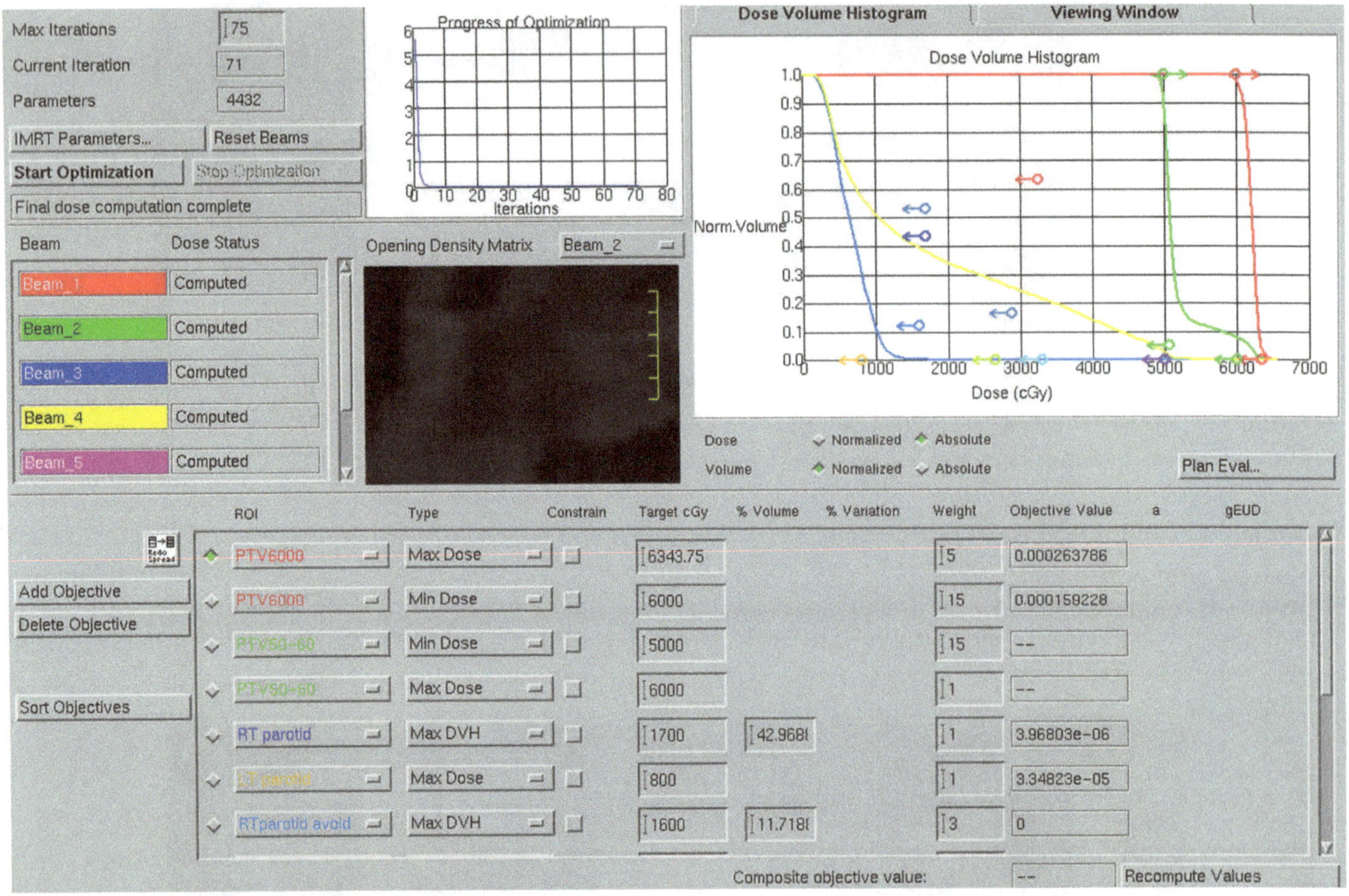

Figure 7.94 IMRT Optimizer window for case pHN05.

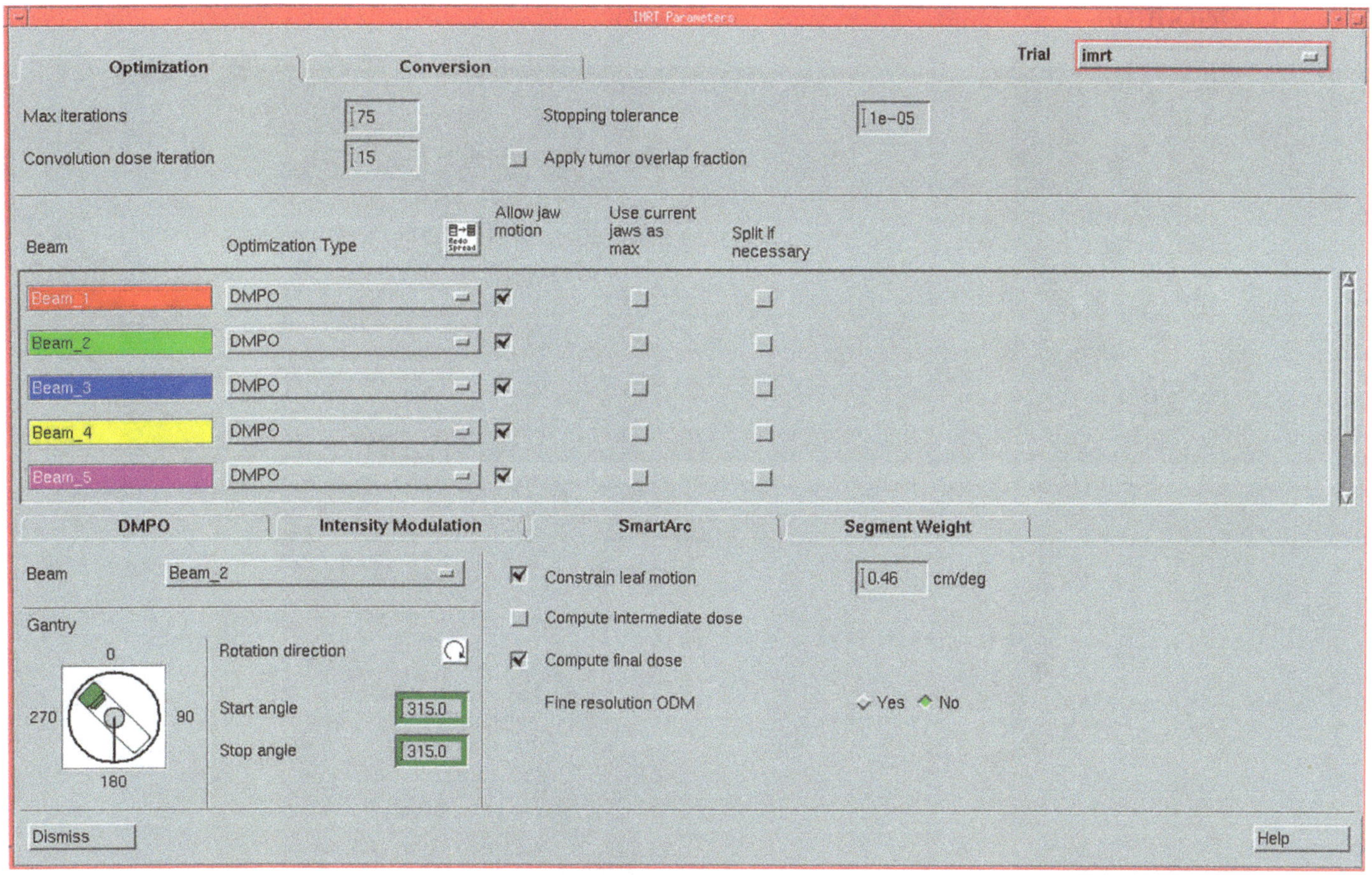

Figure 7.95 Standard VMAT parameters window for case pHN05.

Dose Distribution

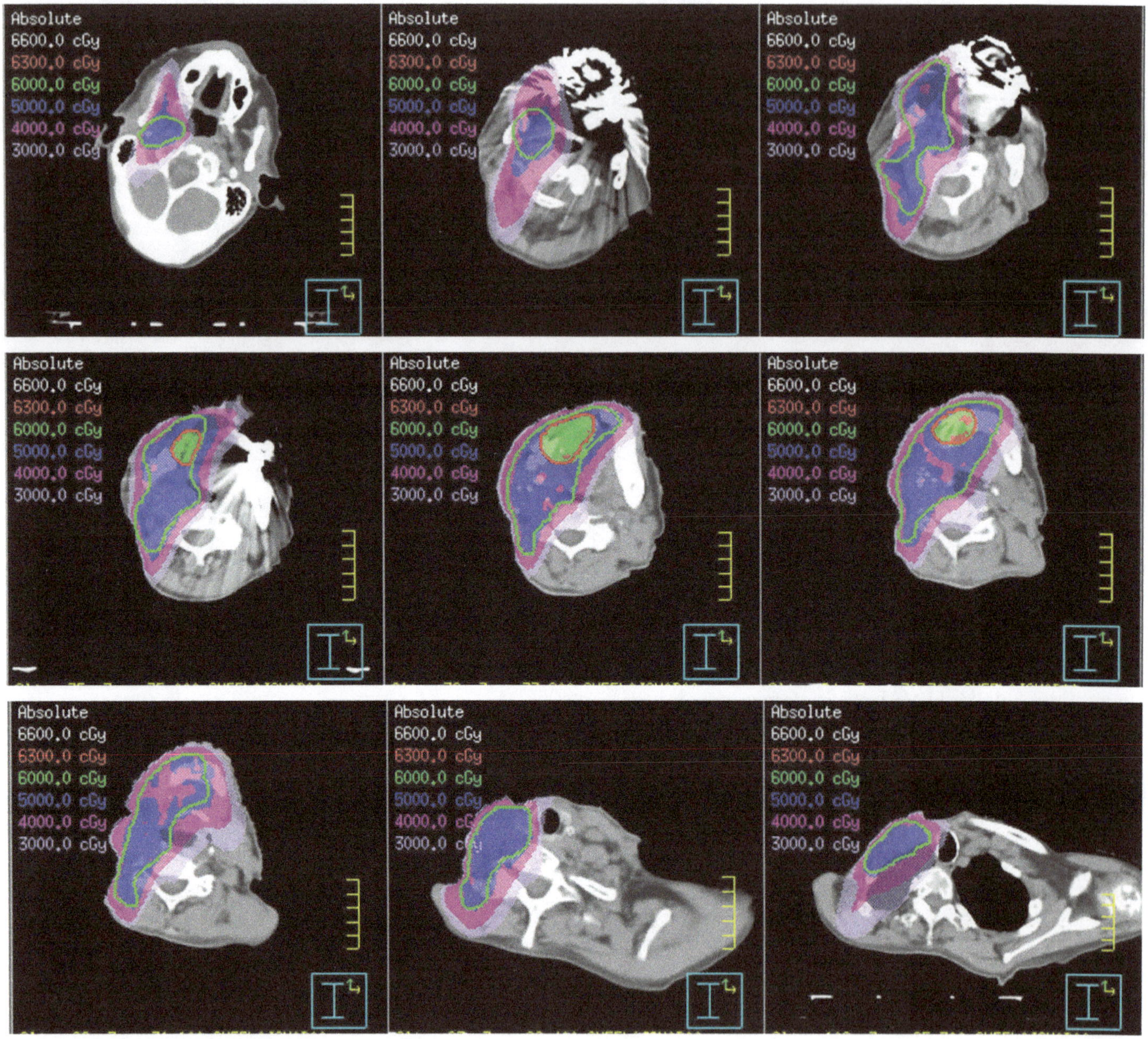

Figure 7.96 IMRT final dose distribution.

Figures 7.96 and 7.97 show the dose distribution for IMRT and VMAT plans, respectively. Hot spots must be within the PTV/GTV, and dose to OARs must be evaluated with guidelines of normal tissue tolerance according to the selected fractionation.

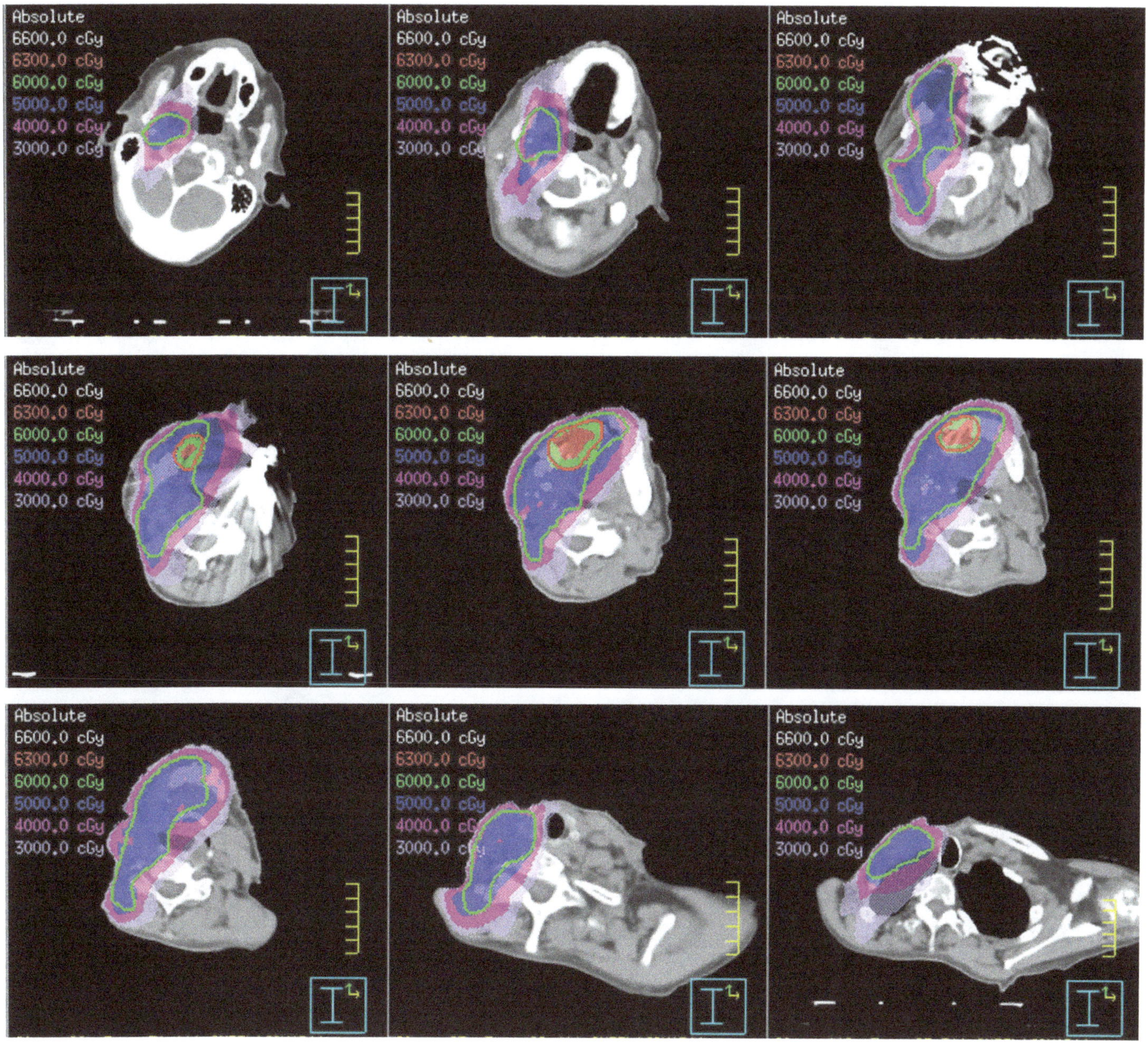

Figure 7.97 VMAT final dose distribution.

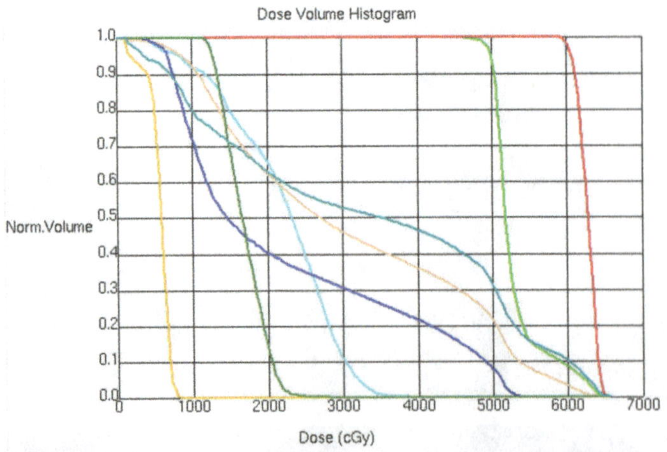

ROI Statistics

Line Type	ROI	Trial or Record	Min.	Max.	Mean	Std. Dev.
	Cord	imrt	426.3	3884.9	1875.2	1010.6
	LT parotid	imrt	105.7	835.1	565.4	136.7
	Oral cavity	imrt	--	6429.3	3072.8	1713.0
	PTV5000	imrt	1455.4	6543.4	5293.7	350.0
	PTV6000	imrt	5839.8	6543.4	6267.3	116.2
	RT parotid	imrt	190.7	5379.7	2225.5	1560.1
	esophagus	imrt	1129.9	2831.4	1678.2	269.3
	mandible	imrt	120.7	6543.4	3192.8	2095.8

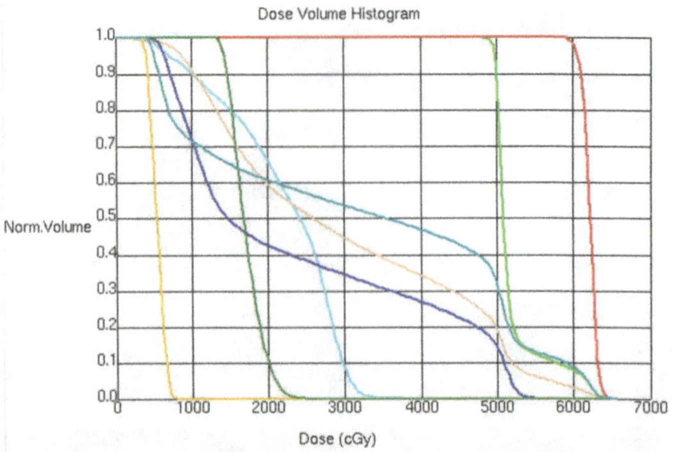

ROI Statistics

Line Type	ROI	Trial or Record	Min.	Max.	Mean	Std. Dev.
	Cord	VMAT	284.9	3741.2	1882.4	1027.5
	LT parotid	VMAT	255.5	793.8	539.7	89.1
	Oral cavity	VMAT	0.0	6476.9	2981.5	1686.0
	PTV5000	VMAT	1616.7	6477.3	5200.9	336.2
	PTV6000	VMAT	5837.4	6477.3	6211.0	86.6
	RT parotid	VMAT	316.4	5496.3	2382.8	1705.0
	esophagus	VMAT	1303.2	2481.4	1708.3	207.1
	mandible	VMAT	369.9	6477.3	3089.1	2136.8

Figure 7.98 Final DVH for IMRT (left) and VMAT (right) plans.

Figure 7.98 shows the final DVHs for this plan.

Case Number pP01: Prostate + Lymph Nodes

General Information

Brief Clinical History

81-year-old man with a diagnosis of prostate cancer. He was diagnosed 5 years prior to this therapy. The current PSA score is 33, and the Gleason score is 7. The plan for this radiotherapy treatment is to treat the prostate and lymph nodes, sparing the bowel, bladder, and rectum as much as possible.

Prescription

Prostate + lymph nodes:
PTV 7000 = 250 cGy × 28 fractions
PTV 5040 = 180cGy x 28 fractions

Simulation

The patient was simulated on a CT scanner in the supine position, with a vacuum cushion device for immobilization. The patient was asked to have a full bladder and to take anti-gas medication to reduce the gas in the bowel.

Target

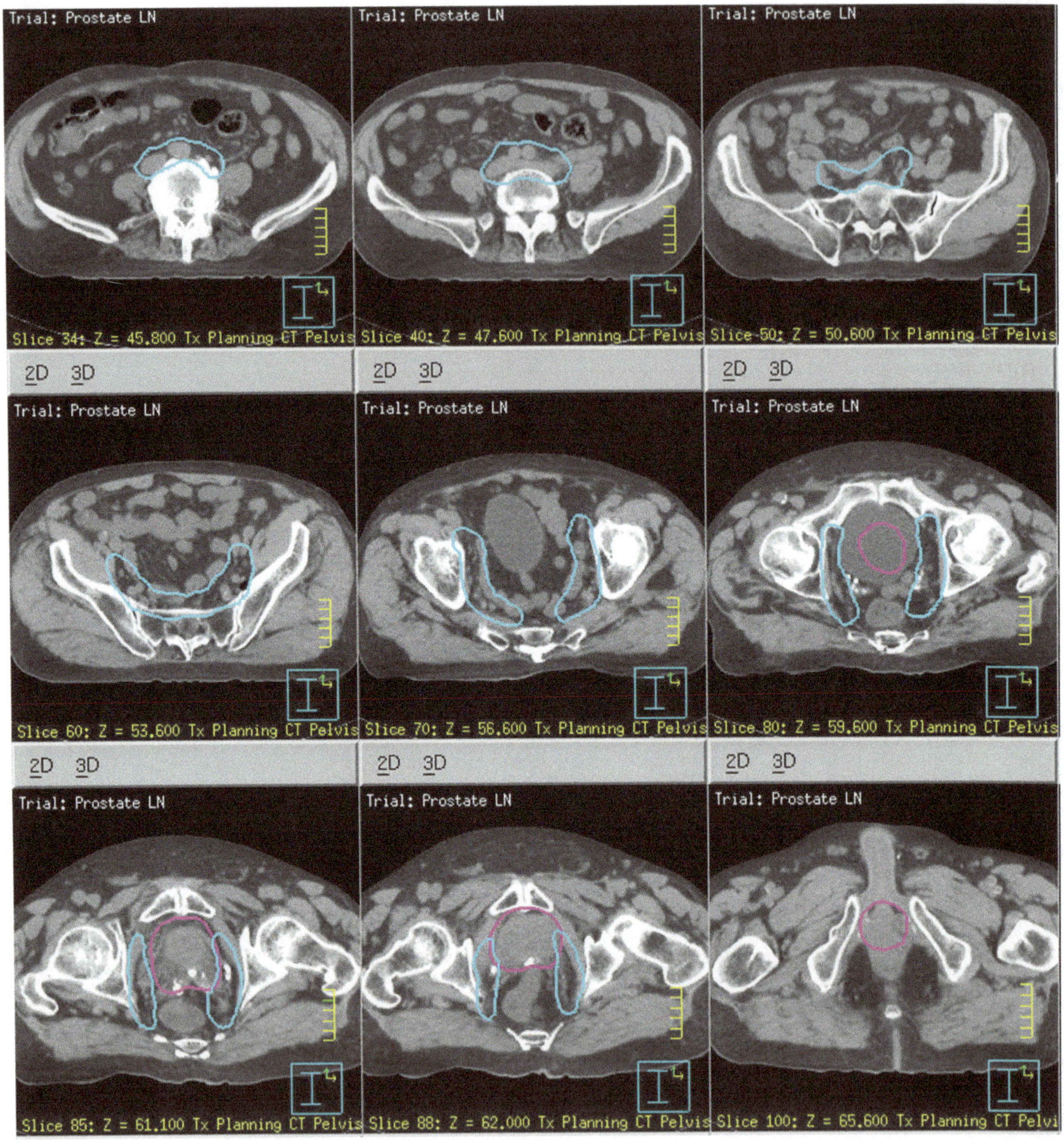

Figure 7.99 Contours of PTV 5040 (cyan) and PTV 7000 (pink).

Figure 7.99 shows axial, sagittal and coronal views of the PTV contours drawn by the radiation oncologist. These are the target contours for both PTVs: prostate and lymph nodes. The prostate PTV is pink, and the lymph nodes PTV is cyan. While the prescription dose could be delivered with a traditional technique, such as a 4-field box, inverse modulated techniques can spare OARs such as the bowel, bladder, rectum, and femoral heads.

Beam Arrangement

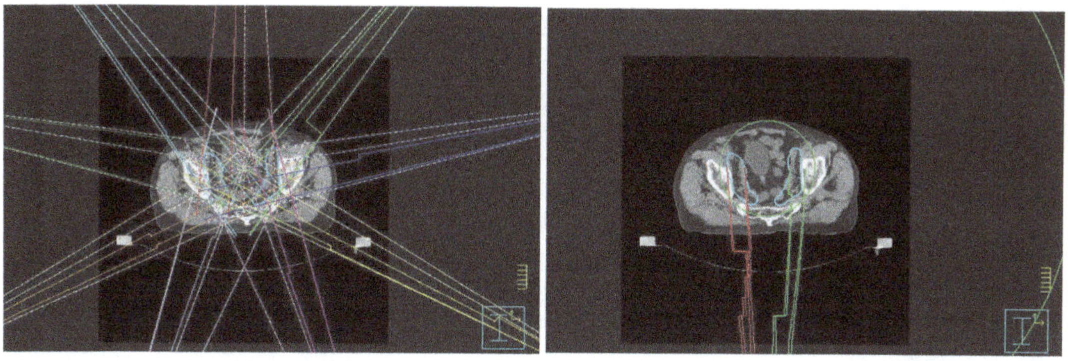

Figure 7.100 IMRT beam arrangement (left) and VMAT beam arrangement (right).

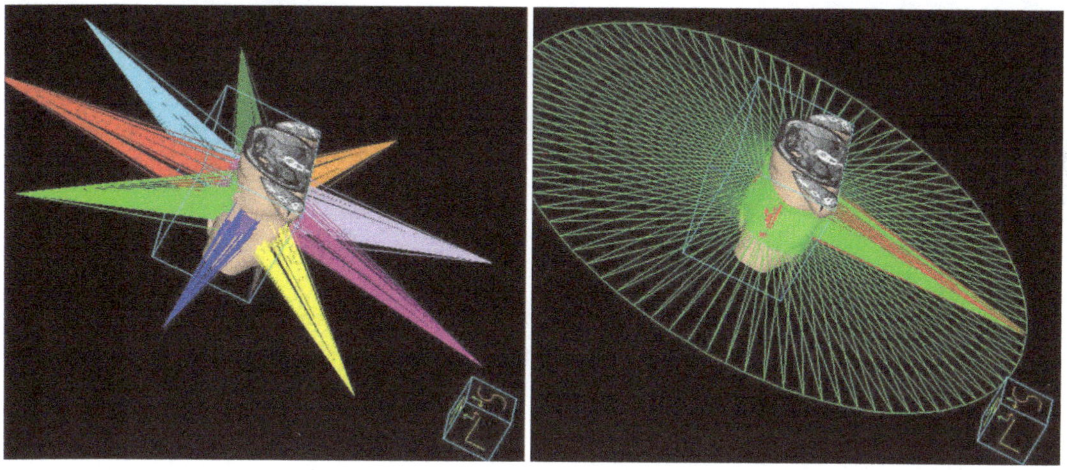

Figure 7.101 IMRT (left) and VMAT (right) beam arrangements on surface reconstruction of CT

Figures 7.100 and 7.101 show the final beam arrangement for this IMRT plan. Notice that not one of the fields opposes another field directly. The following is a summary of beam parameters for this treatment:

Gantry Angle	Couch	Collimator	Energy (MV)
200	0	0	10
240	0	0	10
280	0	0	10
320	0	0	10
0	0	0	10
40	0	0	10
80	0	0	10
120	0	0	10
160	0	0	10

The planner can start with a standard 7- or 9-beam arrangement according to the dose modulation needs. The gantry angle of each field can be modified further while taking OARs into consideration. For example, the central axis of the fields entering close to the spinal cord and the femoral heads are not directly in line with these organs. This allows an easier sparing of these organs during optimization since MLCs can cover these structures without sacrificing dose to the PTV. The beam isocenter can be chosen so that the target can be covered in the y-direction by the y-jaws. In this Varian machine, the maximum distance covered by the y-jaw is 40 cm.

Optimization

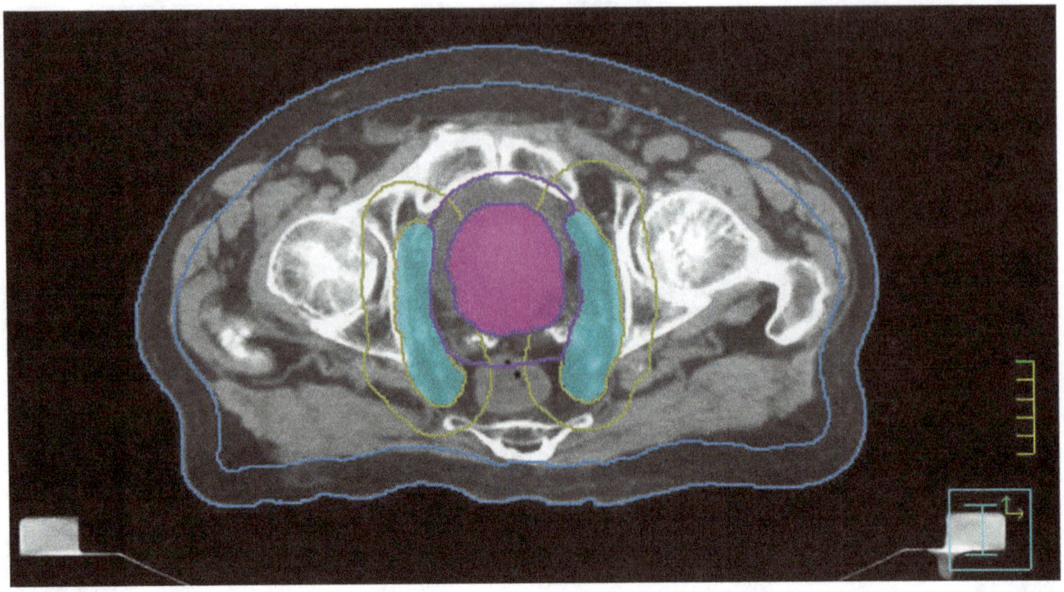

Figure 7.102 Optimization contours: body ring (blue), PTV 5040 ring (gold), and PTV 7000 ring (purple). Target contours are in colorwash display.

Dose levels within target: Prior to setting the optimizer, the planner can review the PTV contours for overlaps. In case of an overlap, the planner can create optimization contours for each dose level. This is done to avoid conflicts in the optimization objectives.

Optimization contours: Several ring structures can be used to make the dose conformal to the target and to prevent streaks of low dose.

Figure 7.102 contains some of the optimization contours that were used for this plan. The blue ring around the body is the "body PTV ring," which prevents entry dose or spillage of dose in the outer boundaries of the body contour. The gold ring and the purple ring are PTV rings. These rings control the fall-off dose from the PTV and prevent dose spillages outside the PTV. One contour not pictured in this figure is the rectum. Some planners find it useful to optimize the rectum as a contour with a margin of a few millimeters (typically 5 mm) to help spare the rectum.

Other contours can be added in between optimizations. For example, some low-dose streaks might appear when using IMRT with energies of 6 MV. In these cases, the user can decide if these defects can be corrected with additional optimization criteria, beam geometry, or number of control points (modulation).

ROI	Type	Constraint	Target (cGy)	Volume (%)	Variation (%)	Weight (%)	Objective Value
Lt Fem Head	Max Dose	No	3268	---	---	1.00	0.0000
PTV7000	Max Dose	No	7200	---	---	7.00	0.0007
PTV7000	Min Dose	No	7000	---	---	15.00	0.0004
Rt Fem Head	Max Dose	No	3268	---	---	1.00	0.0000
anterior control	Max DVH	No	3857	29.3	---	2.00	0.0000
bladder – ptv	Max DVH	No	2161	78.5	---	1.00	0.0000
bladder – ptv	Max DVH	No	4893	21.1	---	1.00	0.0000
body ring 2cm	Max DVH	No	2000	8.6	---	2.00	0.0000
body ring 2cm	Max Dose	No	2500	---	---	2.00	0.0005
bowel – ptv	Max DVH	No	1554	41.4	---	1.00	0.0001
bowel – ptv	Max DVH	No	3161	10.5	---	1.00	0.0001
ptv5040 alone	Max DVH	No	5464	4.7	---	3.00	0.0003
ptv5040 alone	Min Dose	No	5040	---	---	15.00	0.0003
rectum – ptv	Max DVH	No	3232	22.7	---	1.00	0.0001

Figure 7.103 Optimization window for case pP01.

Optimizer: Most optimizations can be started with basic objectives, covering the target and avoiding PTV rings and other optimization structures.

For example, one might start with an optimization along the following criteria:

PTV 7000	→ Minimum dose = 7000 cGy
PTV 7000	→ Maximum dose = 7350 cGy (105%)
PTV 5040 alone	→ Minimum dose = 5040 cGy
PTV 5040 alone	→ Maximum dose = 5500 cGy
PTV 5040 ring	→ Maximum dose = 5400 cGy
PTV 7000 ring	→ Maximum dose = 7000 cGy
Body ring	→ Maximum dose = 3800 cGy

Dose Distribution

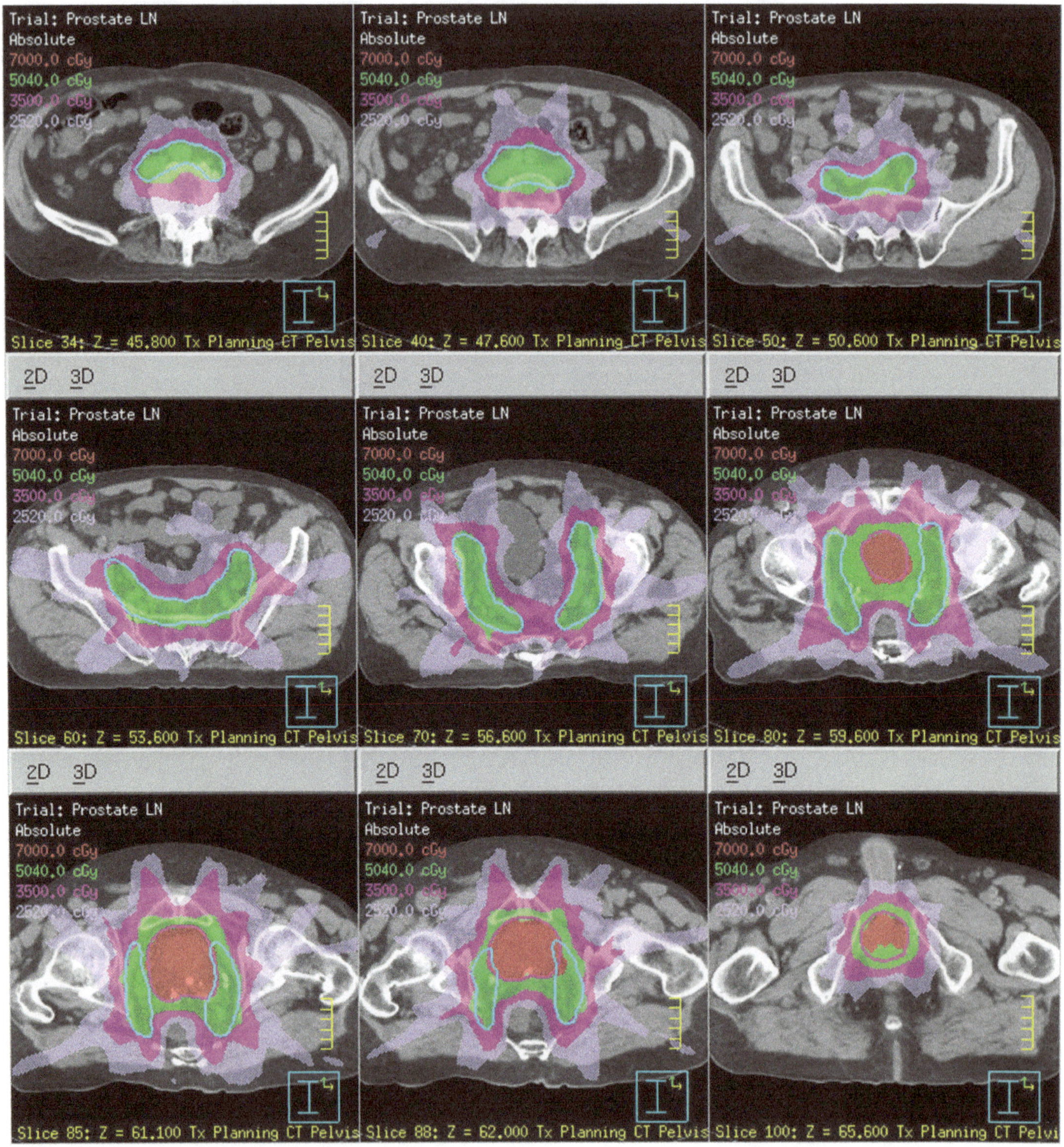

Figure 7.104 IMRT final dose distribution.

Figure 7.104 shows the final dose distribution for IMRT, and Figure 7.105 shows the final dose distribution for a VMAT plan. The planner has spared the rectum and other organs at risk during a series of optimizations. A visual inspection of this dose distribution is always necessary to evaluate the dose fall-off and distribution to different parts of the body. Low-dose streaks are acceptable (purple isodose line = 50% of 5040 cGy). Although the planner can further work on these isodose lines, reducing this low dose can also increase the undesirable hot spot in the target.

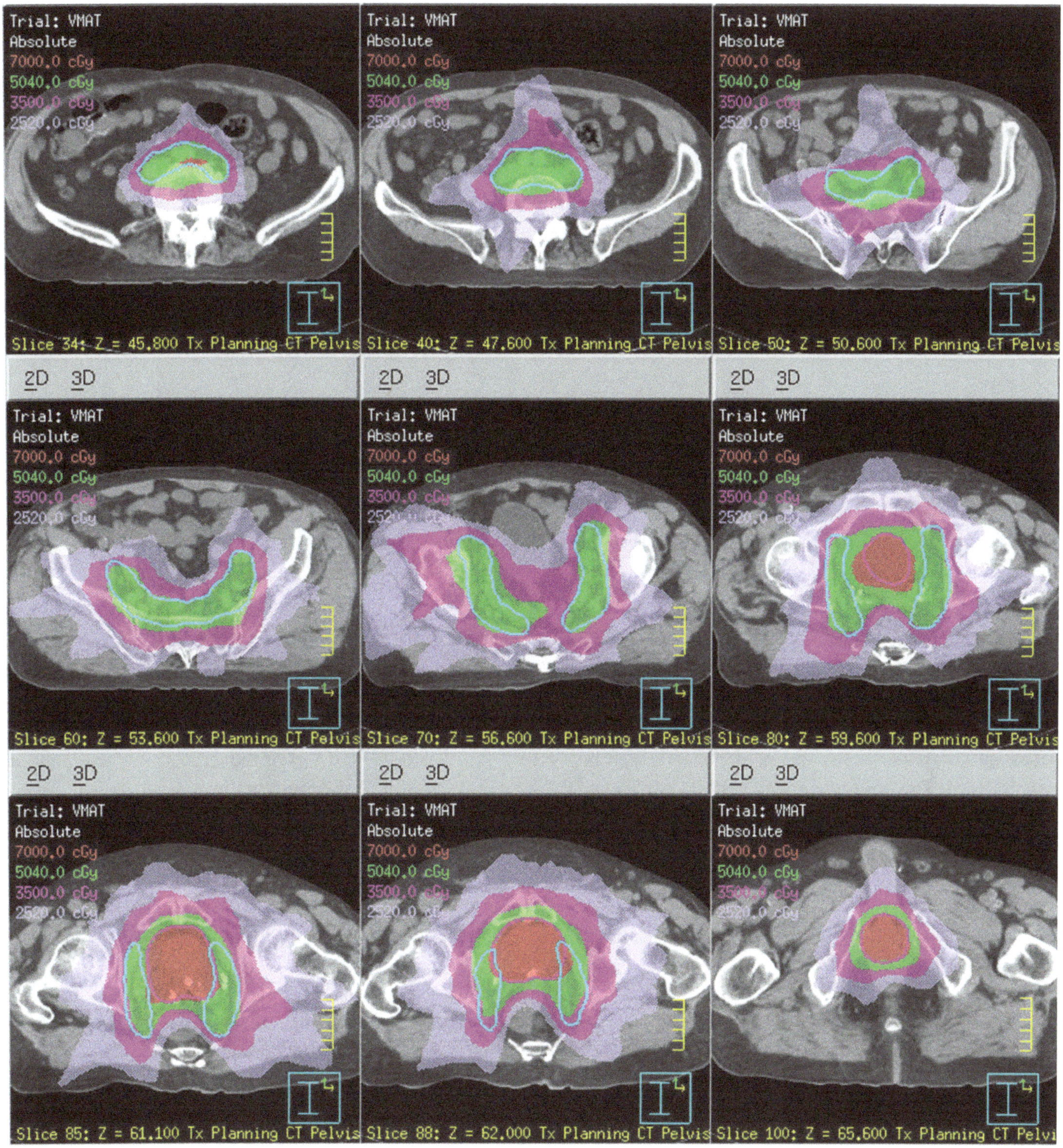

Figure 7.105 VMAT final dose distribution.

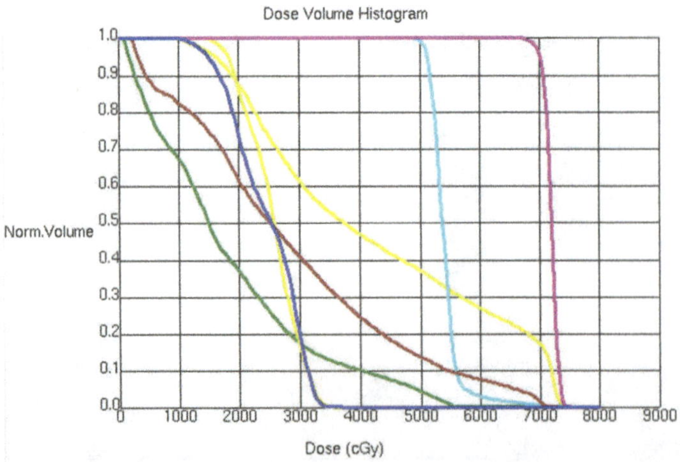

ROI Statistics

Line Type	ROI	Trial or Record	Min.	Max.	Mean	Std. Dev.
	Bladder	Prostate LN	952.2	7424.8	4215.1	2002.9
	Bowel	Prostate LN	74.0	6883.8	1248.3	1421.7
	Lt Fem Head	Prostate LN	1411.7	3575.6	2546.2	433.6
	PTV5040	Prostate LN	4562.2	7371.2	5396.8	256.3
	PTV7000	Prostate LN	6415.9	7488.4	7193.2	104.4
	Rectum	Prostate LN	192.7	7310.5	2805.3	1810.1
	Rt Fem Head	Prostate LN	880.0	3665.3	2443.8	566.9

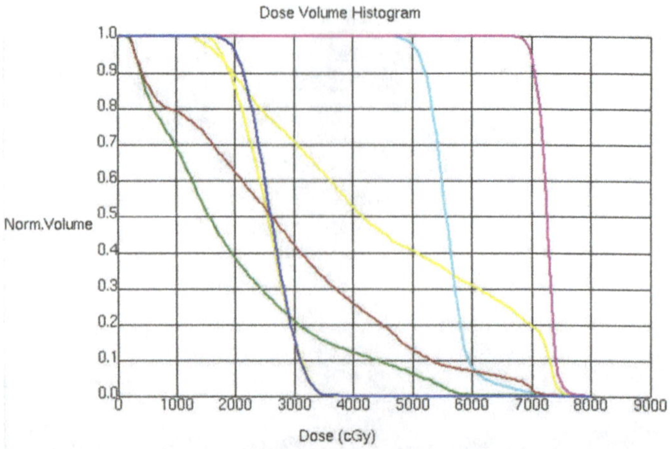

ROI Statistics

Line Type	ROI	Trial or Record	Min.	Max.	Mean	Std. Dev.
	Bladder	VMAT	1252.6	7745.7	4503.7	1989.1
	Bowel	VMAT	135.2	7539.7	1347.4	1514.7
	Lt Fem Head	VMAT	1457.5	3698.4	2527.7	442.3
	PTV5040	VMAT	3907.5	7510.4	5591.4	355.6
	PTV7000	VMAT	6318.1	7921.3	7249.1	149.2
	Rectum	VMAT	188.5	7450.4	2794.3	1845.9
	Rt Fem Head	VMAT	1625.1	3765.8	2608.6	363.6

Figure 7.106 IMRT (left) and VMAT (right) dose volume histograms.

The final DVHs for this treatment plan can be seen in Figure 7.106. The dose statistics of the targets and the organs at risk are seen. The dose maximums reported on these DVHs are not those of 0.03 cc within the contour.

Case Number pP02: Prostate Radiation Treatment

General Information

Brief Clinical History

83-year-old male with adenocarcinoma of prostate, cT3a, biopsy revealing 2/12 cores positive GS 3+3=6, MRI demonstrating PI-RADS 5, highly suspicious for malignancy.

Prescription

Prostate:
PTV 7000 = 250 cGy × 28 fractions

Simulation

The patient was simulated on a CT scanner in the supine feet-first position using a vacuum cushion for an immobilization device.

Target

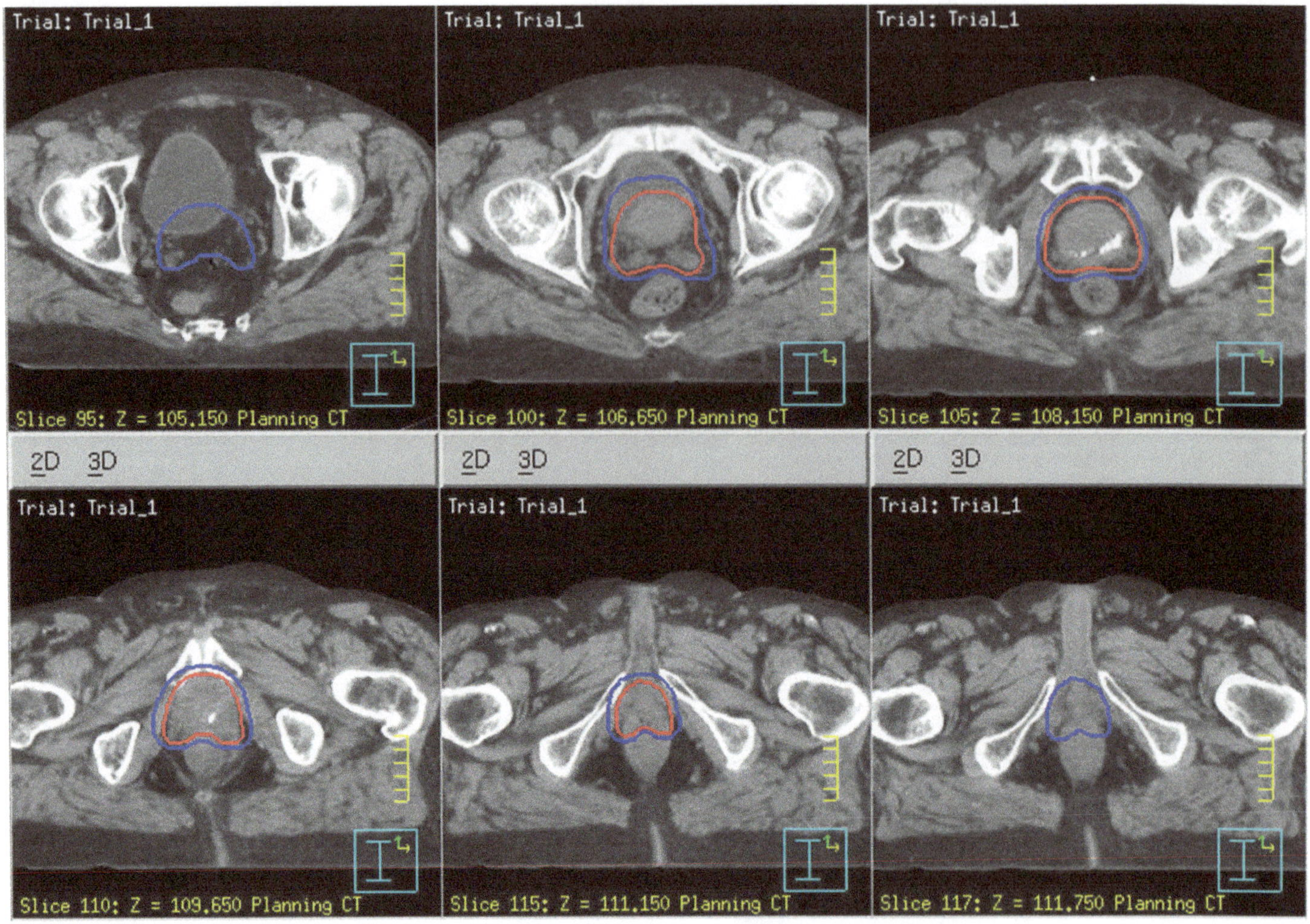

Figure 7.107 Contours of PTV 7000 (blue) and CTV (red).

Figure 7.107 shows the PTV and CTV contours drawn by the radiation oncologist. Some structures overlap other OARs, such as the rectum and the bladder. The planner must accurately represent these structures through contouring for correct interpretation of dose received by these organs on the DVH.

Beam Arrangement

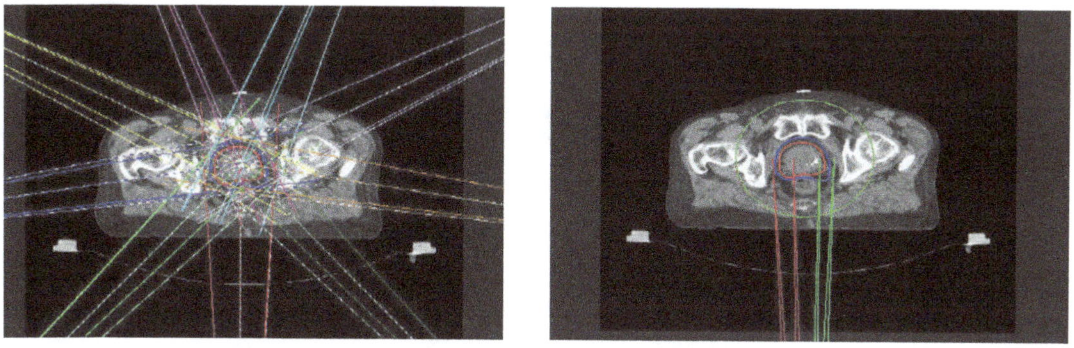

Figure 7.108 IMRT (left) and VMAT (right) beam arrangements.

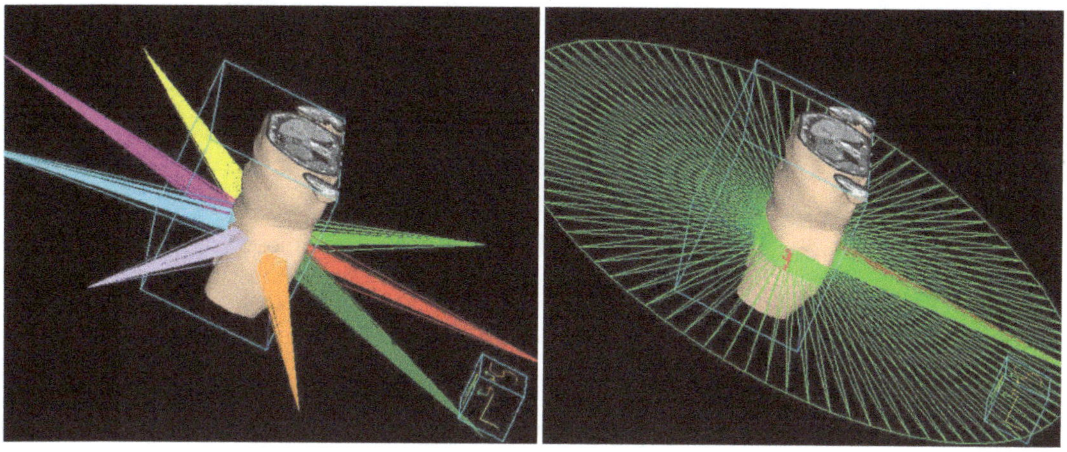

Figure 7.109 IMRT (left) and VMAT (right) beam arrangements in surface reconstruction.

Figures 7.108 and 7.109 show representations of the beam arrangement used for this case. The planner has placed 9 beams around the patient every 40 degrees. This allows for the dose modulation to avoid critical structures, such as the femoral heads. Special care must be put into not placing beams facing each other or directly opposed. Notice how the jaws of the beam are collimated in such a way that they cover the outside periphery of the target. This decreases the effects of MLC leakage. The following is a summary of beam parameters:

Gantry Angle	Couch	Collimator	Energy (MV)
180	0	0	10
220	0	0	10
260	0	0	10
300	0	0	10
340	0	0	10
20	0	0	10
60	0	0	10
100	0	0	10
140	0	0	10

The VMAT beam arrangement contains two arcs going around the patient back and forth.

Optimization

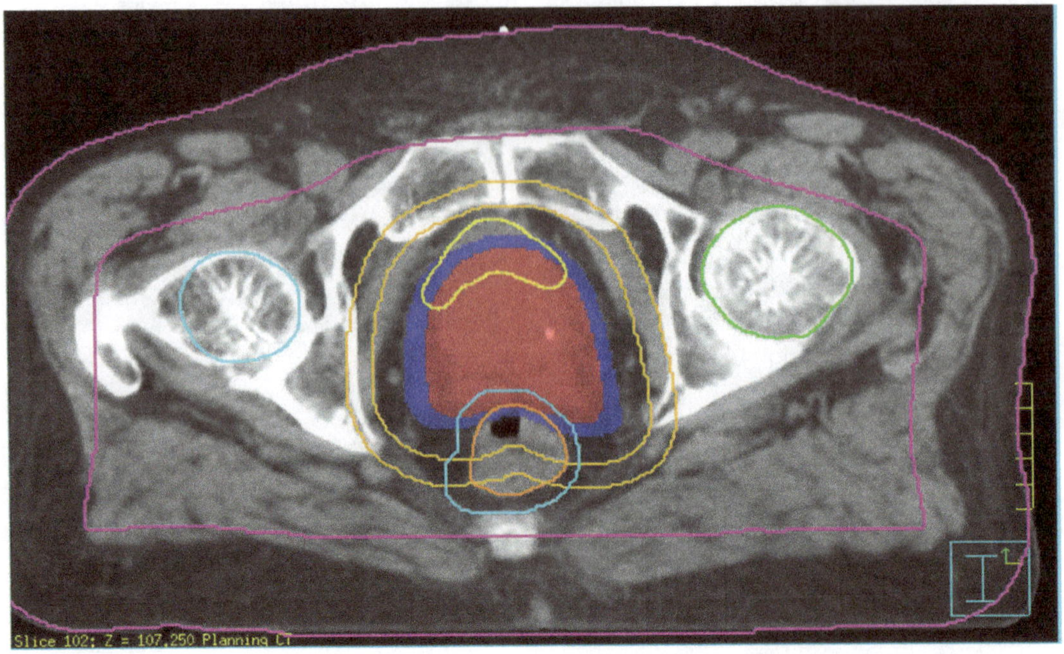

Figure 7.110 Optimization contours. The colorwash red and blue structures are the prostate CTV (red) and the PTV 7000 (blue). Other contours include the PTV ring (orange), bladder (yellow), rectum expansion (cyan), femoral heads (cyan and green), and body ring (pink).

Dose levels within target: In this case, there is only one dose prescription of 7000 cGy to the PTV, and there is no need to separate contours based on dose gradients for the optimization process. However, the planner must make sure to achieve the dose objectives for other organs at risk.

Optimization contours: Optimization contours for this plan can be seen on Figure 7.110. One of the main structures for dose conformity to the target is the PTV ring (orange). In this case, the PTV ring is about 1 cm away from the target. This was done so the planner could set the constraints to this structure to a lower isodose line. In any case, the planner could also have contoured a "hot ring" and a PTV ring to achieve the same effect. The body ring (pink) will prevent any low-dose spillage toward the surface of the patient.

Achieving dose objectives for structures such as the rectum and bladder can be a challenging task when there is an overlap from the PTV contour into these structures. In this case, there are two approaches that can help ease this task: (1) the rectum is contoured with a 5-mm margin (cyan) to improve the dosimetry of the rectum constraints and (2) the overlapping bladder can be contoured so that the maximum dose objective in the optimizer can be set to approximately prescription dose to avoid hot spots.

Optimizer: Figure 7.111 offers a screenshot of the optimizer. Not all objectives are shown. The user can start with a few basic objectives to obtain the first dose distribution for the plan. With the help of this DVH and OARs constraints, the user can add more parameters to the plan which can be changed or kept from optimization to optimization.

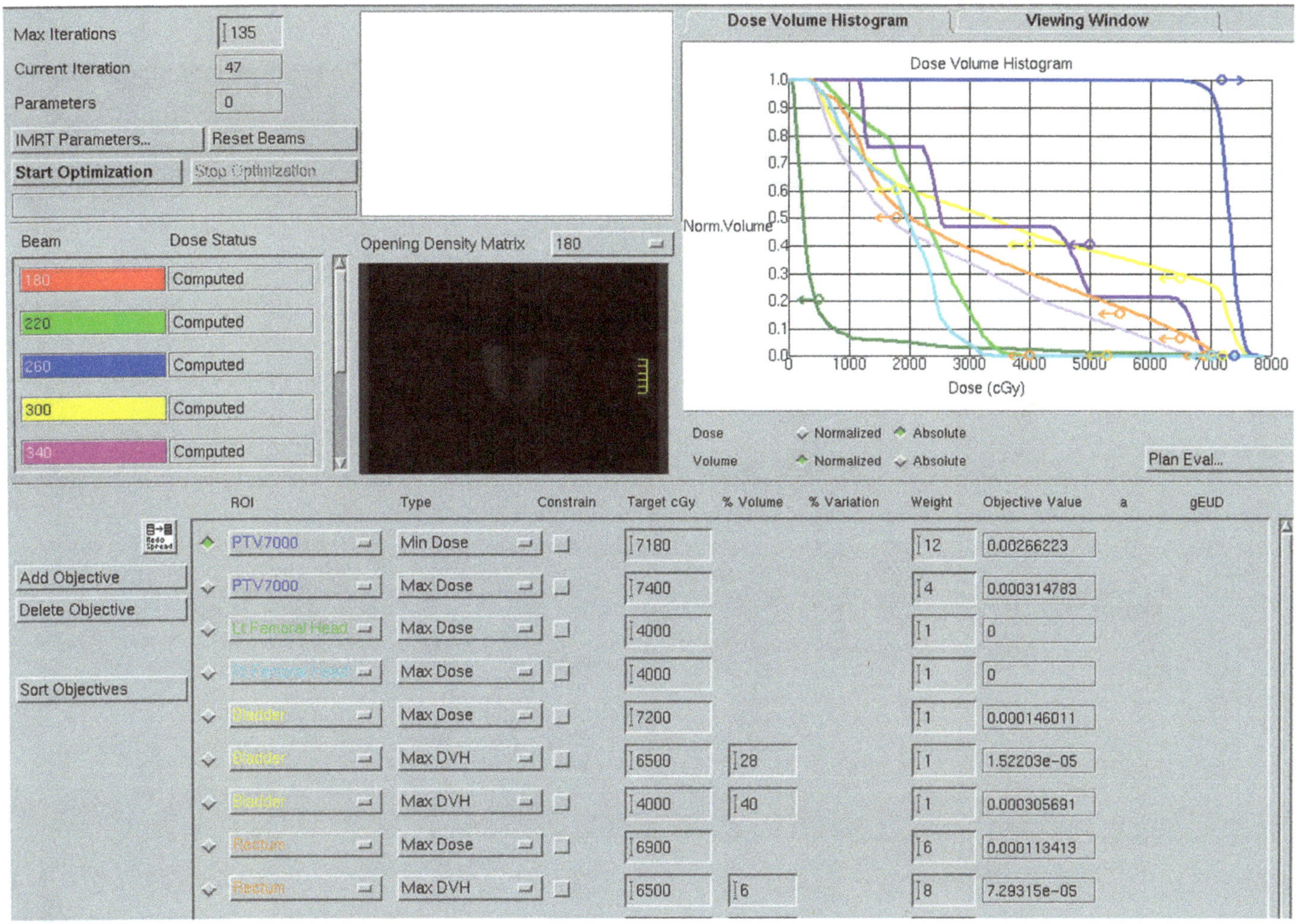

Figure 7.111 Optimizer interface.

The planner could start with basic objectives such as:

PTV 7000	→ Minimum dose = 7000 cGy
PTV 7000	→ Maximum dose = 7350 cGy (105%)
PTV 7000 ring	→ Maximum dose = 7000 cGy
Body ring	→ Maximum dose = 3800 cGy
Rectum + margin	→ Maximum dose = 7000 cGy
Bladder	→ Maximum dose = 7000 cGy

After the first optimization, the planner can add other objectives to lower doses on OARs while achieving acceptable target coverage. For example, for the rectum + margin contour, the maximum dose will be that of the PTV to avoid the loss of coverage, but the planner can add "max dvh" objectives to reduce volume irradiated at lower doses.

Dose Distribution

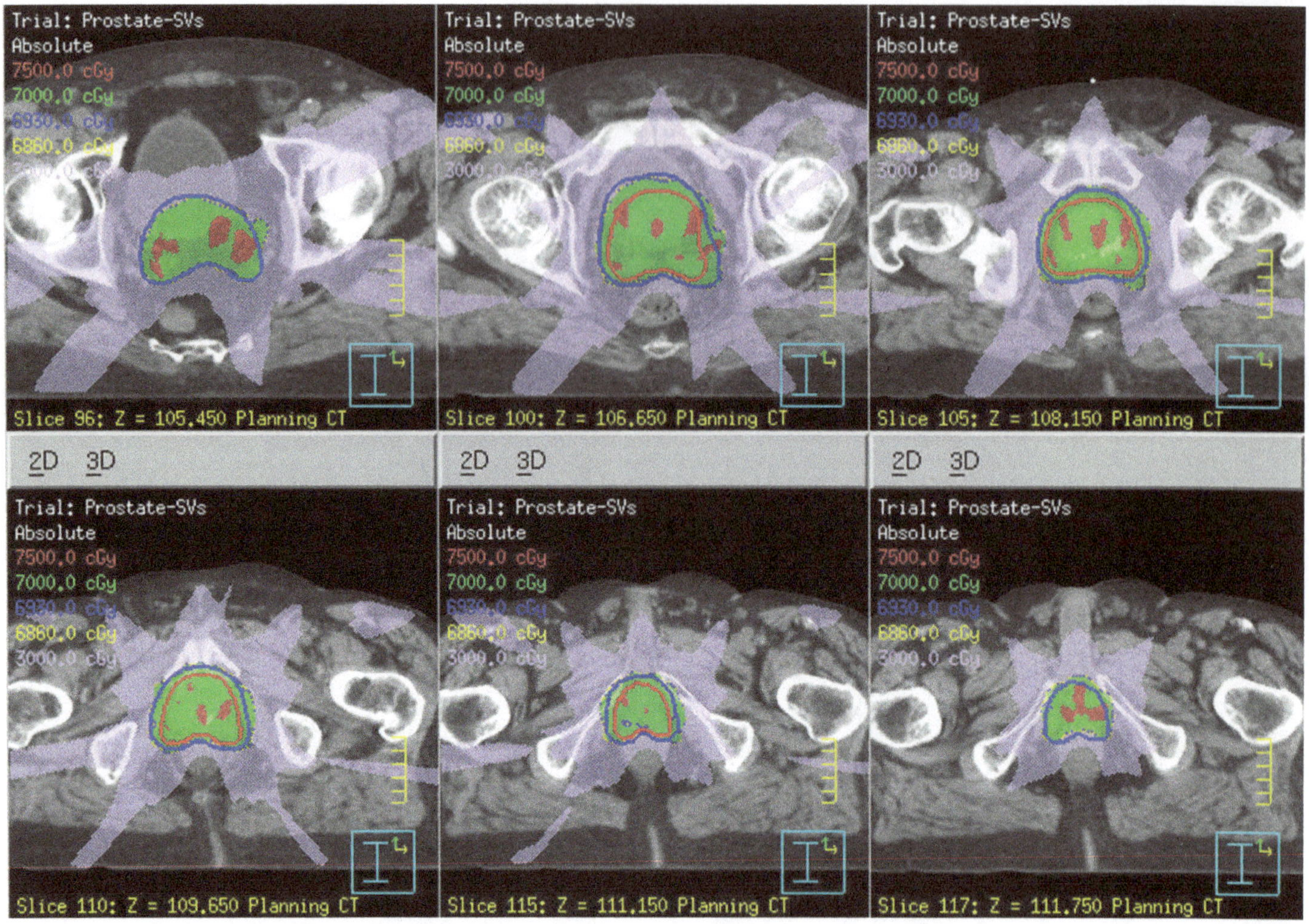

Figure 7.112 IMRT Final dose distribution.

The final dose distribution is shown in Figures 7.112 and 7.113 for IMRT and VMAT plans, respectively. The 7000 cGy (prescription) isodose line is covering 95% of the target volume. Notice how the low-isodose lines bend around the rectum, avoiding this structure.

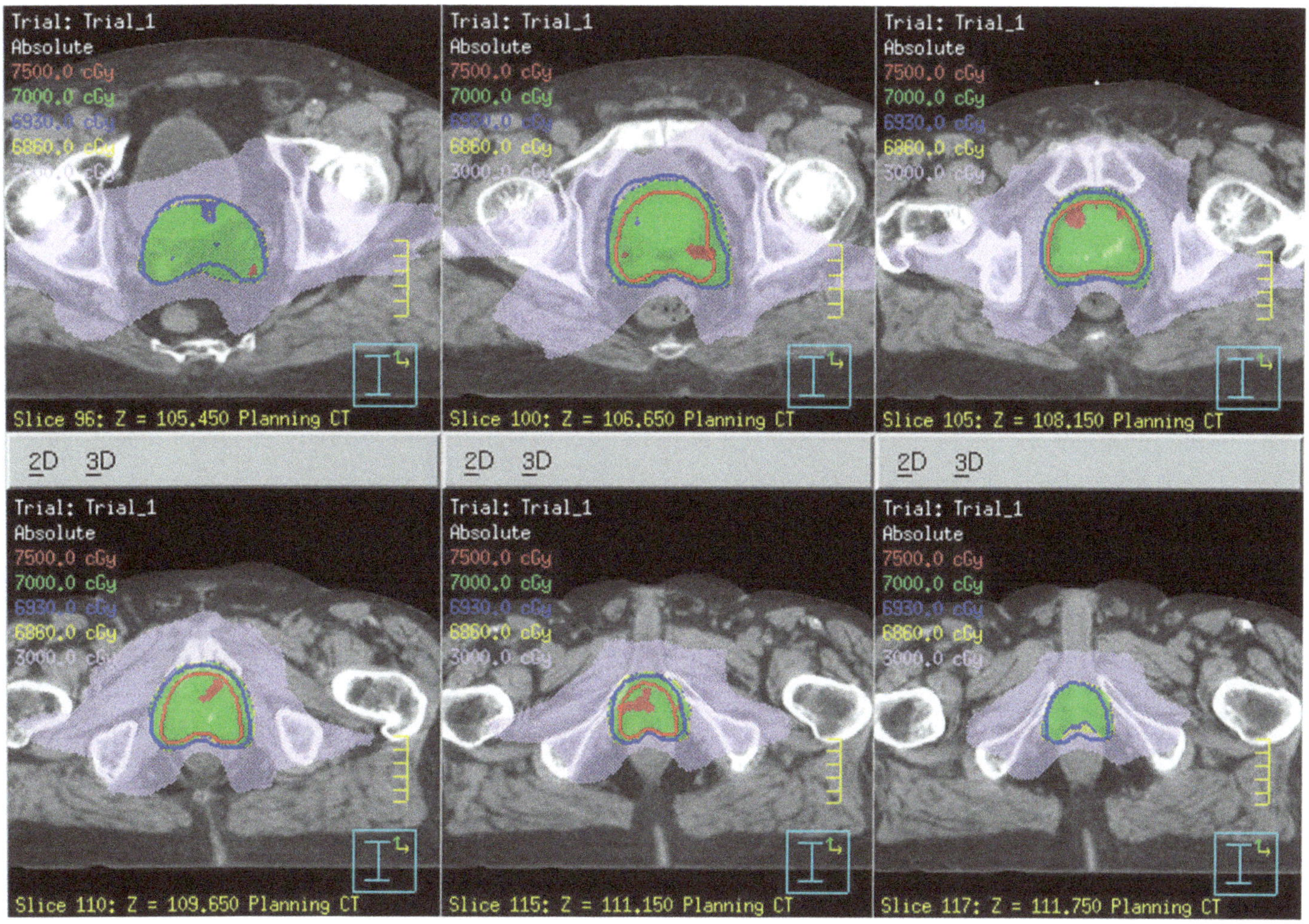

Figure 7.113 VMAT Final dose distribution.

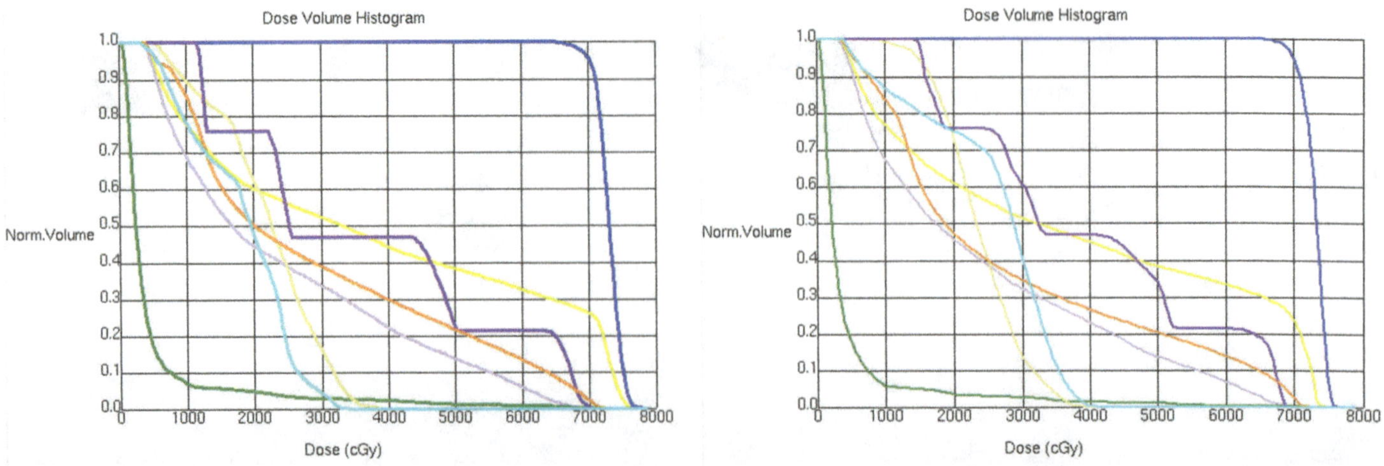

ROI Statistics

Line Type	ROI	Trial or Record	Min.	Max.	Mean	Std. Dev.
	Bladder	Prostate-SVs	316.9	7716.3	3779.9	2694.2
	Bladder-PTV	Prostate-SVs	316.9	7213.8	2401.0	1846.8
	Bowel	Prostate-SVs	57.9	7242.2	471.5	873.6
	Lt Femoral Head	Prostate-SVs	510.1	3860.3	2190.7	768.1
	PTV7000	Prostate-SVs	6025.4	7772.0	7300.1	175.5
	Penile Bulb	Prostate-SVs	1159.4	7037.9	3626.5	2038.6
	Rectum	Prostate-SVs	372.4	7232.6	2889.5	2058.6
	Rt Femoral Head	Prostate-SVs	311.8	3393.4	1804.9	749.5

ROI Statistics

Line Type	ROI	Trial or Record	Min.	Max.	Mean	Std. Dev.
	Bladder	VMAT	284.9	7477.6	3751.7	2666.8
	Bladder-PTV	VMAT	284.9	7191.2	2406.7	1880.6
	Bowel	VMAT	48.8	7167.6	422.2	748.3
	Lt Femoral Head	VMAT	832.7	3906.1	2365.9	573.3
	PTV7000	VMAT	5994.4	7638.8	7297.6	162.6
	Penile Bulb	VMAT	1413.0	6872.6	3920.2	1849.7
	Rectum	VMAT	337.5	7218.2	2780.4	2059.0
	Rt Femoral Head	VMAT	382.9	4087.8	2546.8	975.0

Figure 7.114 Final dose volume histogram for IMRT (left) and VMAT (right).

Figure 7.114 contains the final DVH for IMRT and VMAT plans. The prescription dose at the PTV 7000 covers 95% of the volume. The reason for the odd shape of the penile bulb curve on the DVH is the dose change from slice to slice of the CT data set.

Case Number pP03: Prostate Radiation Treatment

General Information

Brief Clinical History

62-year-old male with low risk prostate adenocarcinoma, presenting with PSA = 4.1 six months prior to this treatment, cT1c; Gleason score 6 (3+3), total 3 cores positive, MRI revealing diffuse hypointensity involving nearly the entire anterior to posterior right peripheral zone base to apex with a small volume of extraprostatic spread. Loss of the lateral capsule integrity and suspicion of right neurovascular bundle involvement.

Prescription

Prostate and seminal vesicles:
PTV 7000 = 250 cGy × 28 fractions

Simulation

The patient was simulated with a vacuum cushion as immobilization device. MRI and planning CT fusion were used to help target the lesion.

Target

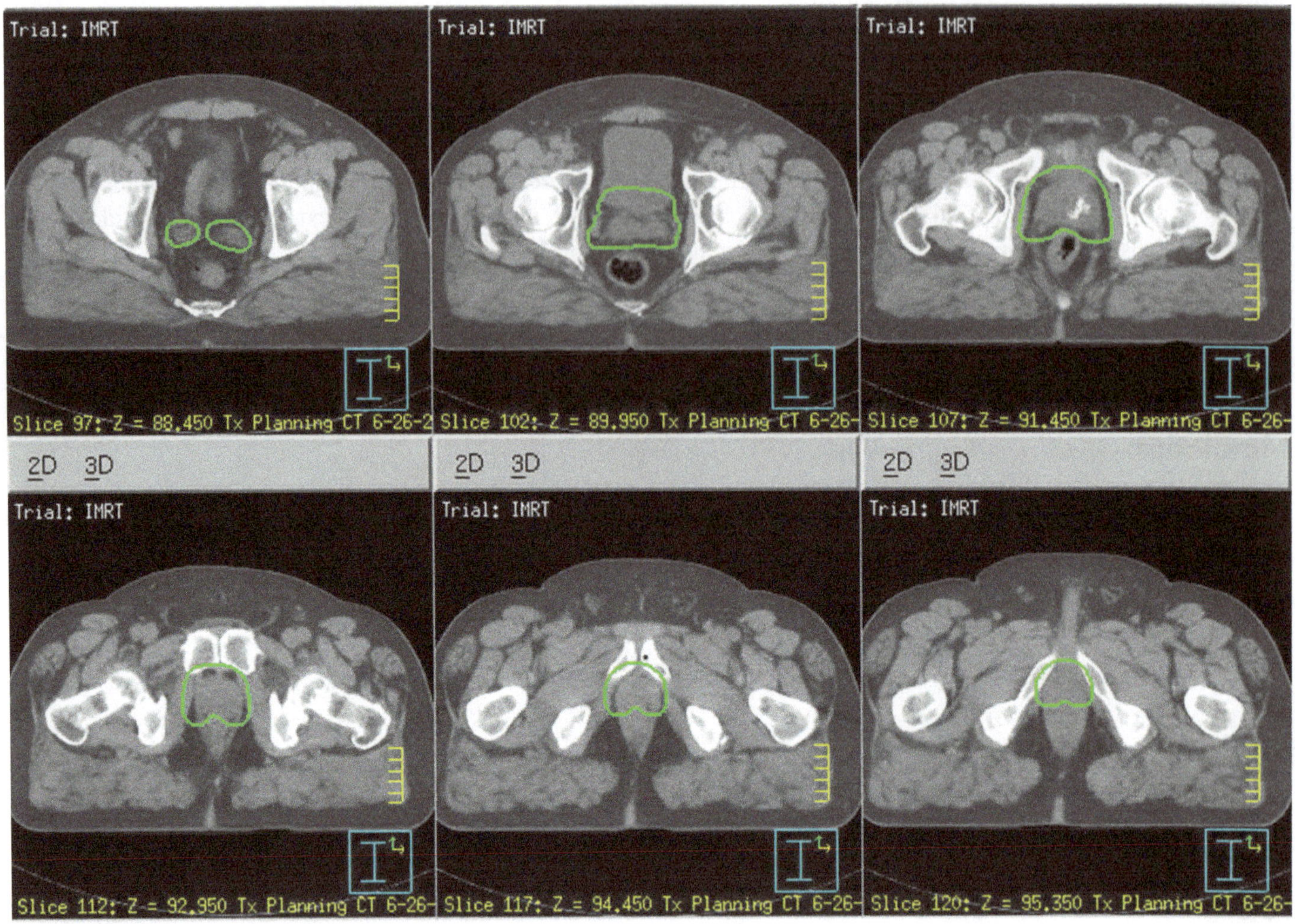

Figure 7.115 PTV 7000 contour for case pP03.

Figure 7.115 shows the contour of the target to treat for this case. During the irradiation of this PTV, the organs at risk would be the femoral heads, the bladder, the rectum, and the bowels. There is only one dose level in the target, and there is no need to make individual contours for different dose levels.

Beam Arrangement

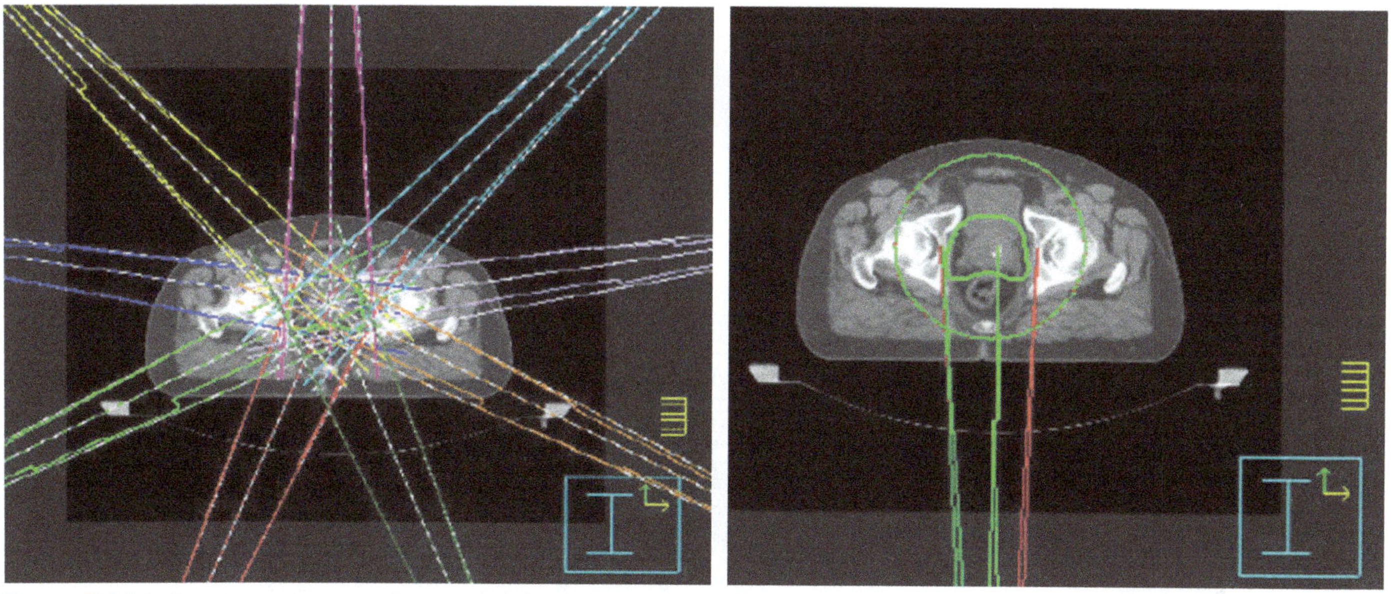

Figure 7.116 Beam arrangement for IMRT (left) and VMAT (right).

Nine beams were placed around the patient as shown in Figure 7.116. The beam's central axis does not go directly through the spinal cord or the femoral heads. This makes it feasible for the optimizer to lower the dose on these structures without sacrificing coverage at the target.

The planner can start the optimization process with a standard IMRT 7-beam arrangement and add or change the position of the beams according to the specific needs of the plan. The following is a summary of the beam parameters used for this plan.

Gantry Angle	Couch	Collimator	Energy (MV)
200	0	0	10
240	0	0	10
280	0	0	10
320	0	0	10
0	0	0	10
40	0	0	10
80	0	0	10
120	0	0	10
160	0	0	10

VMAT beam arrangement contains two arcs going around the patient back and forth.

Optimization

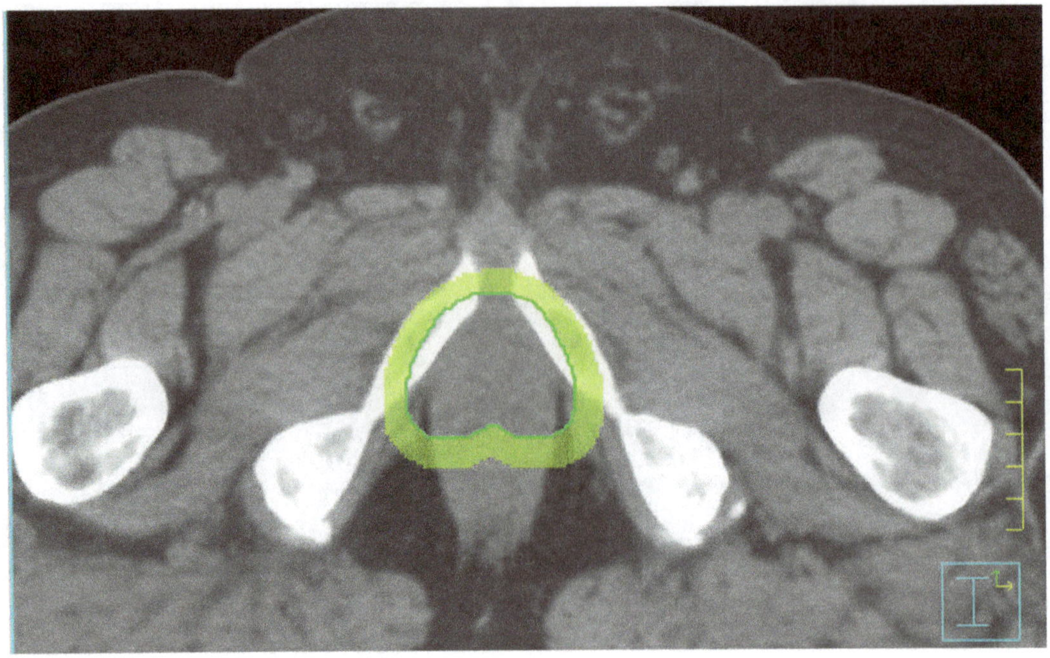

Figure 7.117 Optimization structures. PTV ring (yellow colorwash).

Before optimizing the plan, structures or contours are necessary to redirect the dose to where its distribution is optimal. Figure 7.117 shows a PTV ring which can help make the dose distribution conformal to the target if it is used in the optimization process.

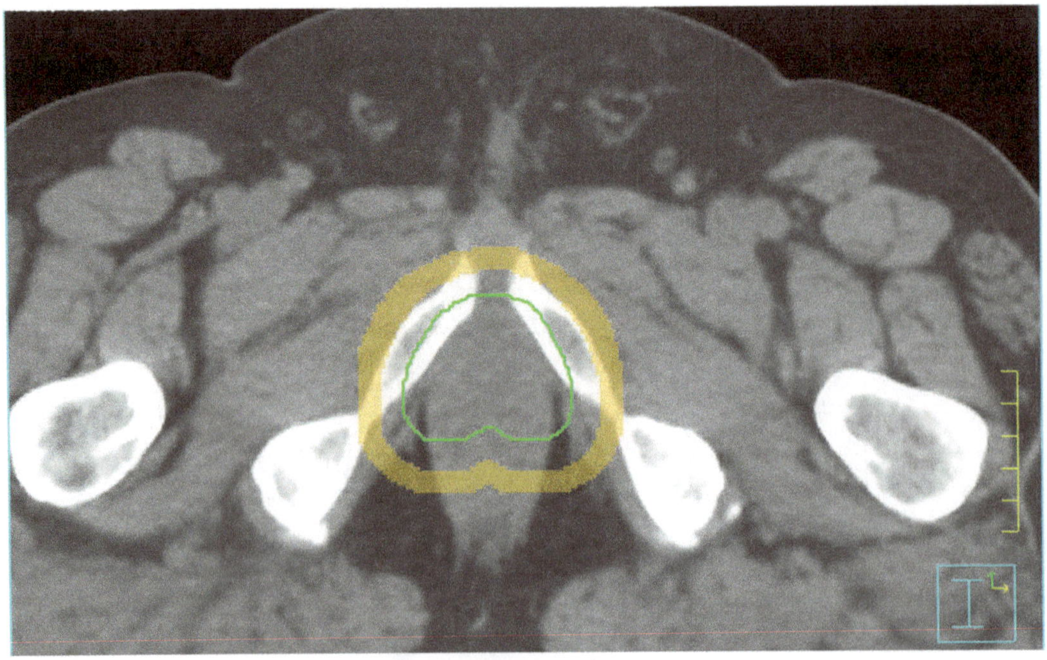

Figure 7.118 Optimization structures. Outer ring (orange colorwash).

Figure 7.118 shows an outer ring such as the orange ring or the body ring (or both at the same time) are necessary for the desired dose falloff from target to normal tissue. Constraints to PTV ring and outer ring are always lower than prescription, assuming that the dose will fall-off with these "Matryoshka" rings.

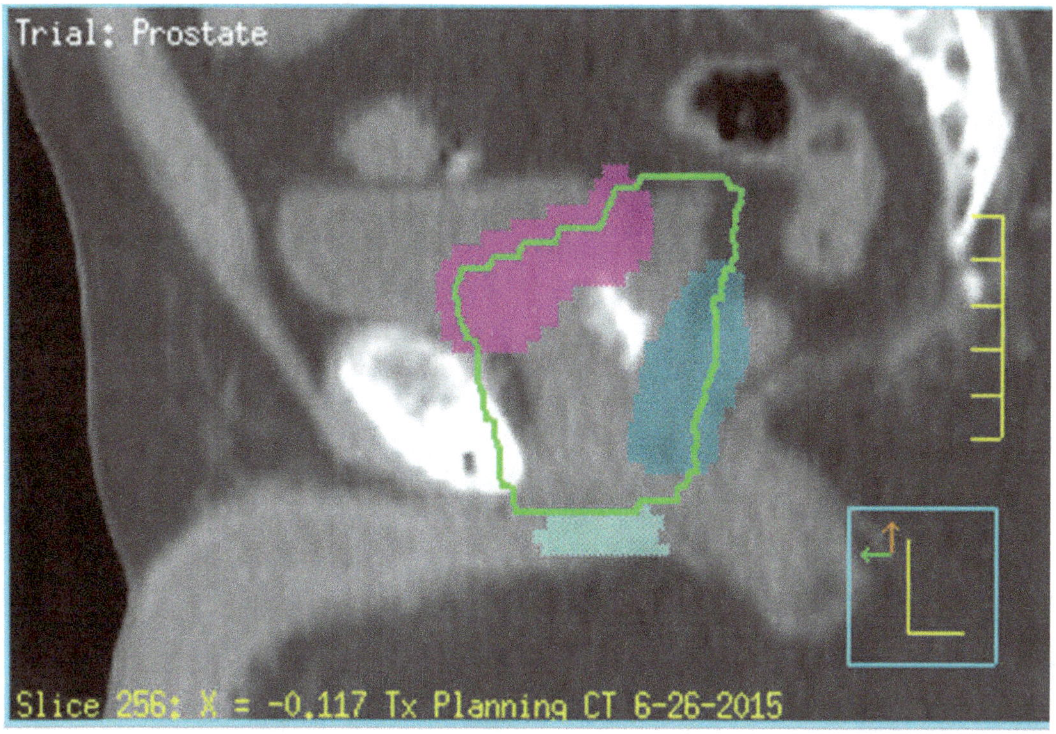

Figure 7.119 Optimization contours. Bladder overlap (pink), rectum overlap (green), and urethral area (cyan blue).

Additional contours during optimization are necessary to avoid areas of high dose where organs at risk overlap. Figure 7.119 shows pink, green, and blue structures protecting the bladder, rectum, and other structures from high-dose spilling or hot spots. Objectives for these structures must be placed on the optimizer for these contours to accomplish their purpose.

Dose Distribution

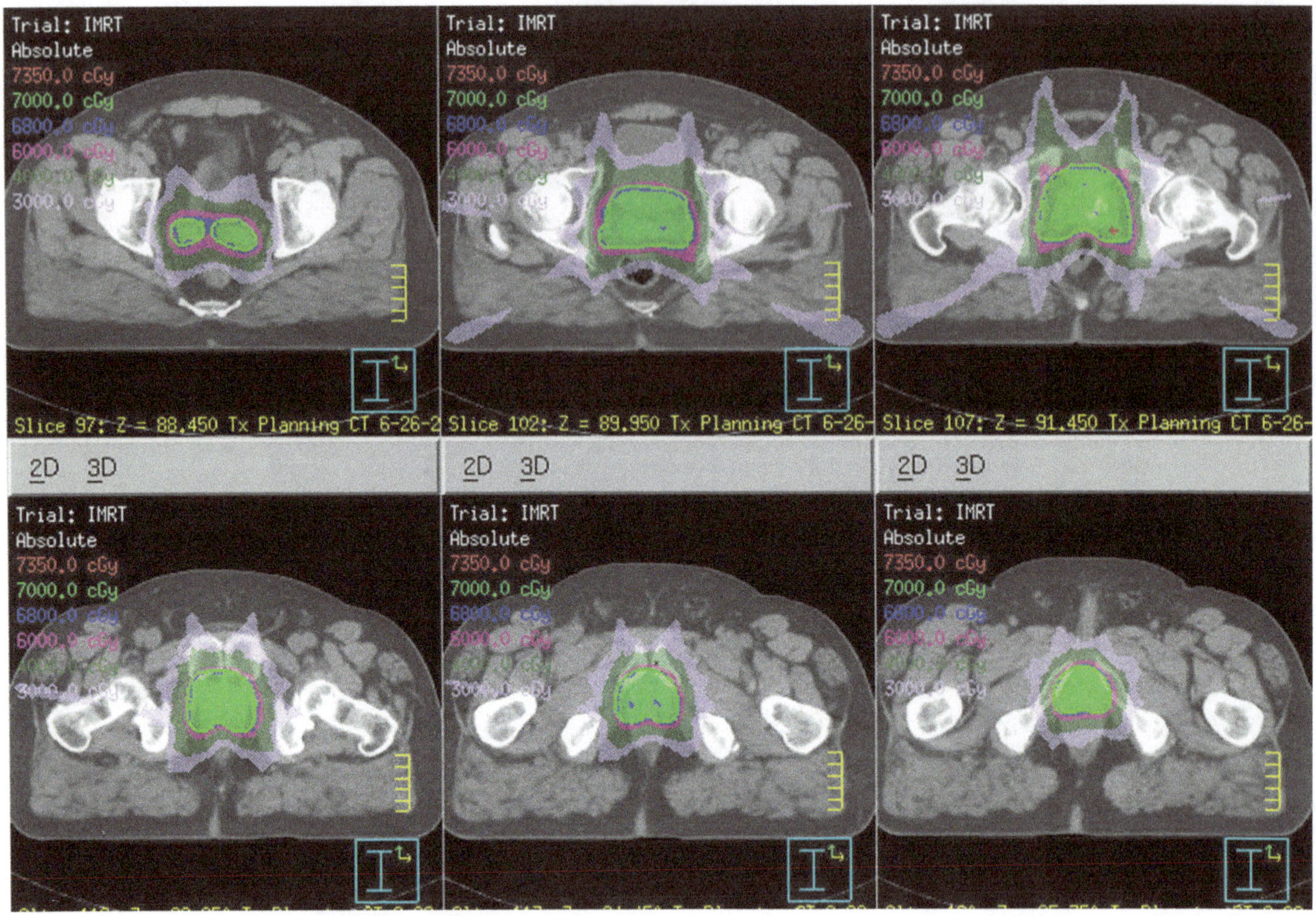

Figure 7.120 Final dose distribution for IMRT.

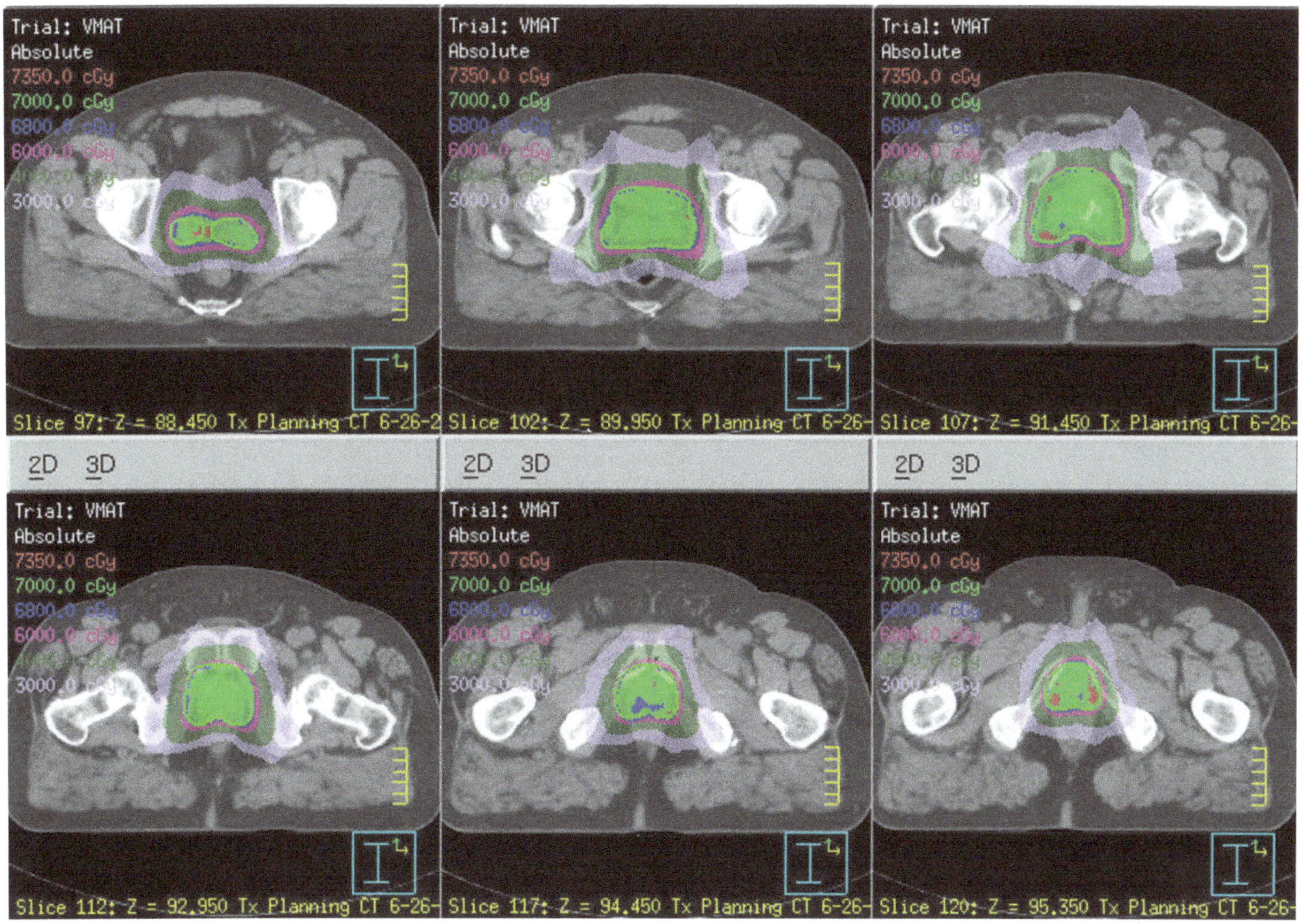

Figure 7.121 Final dose distribution for VMAT.

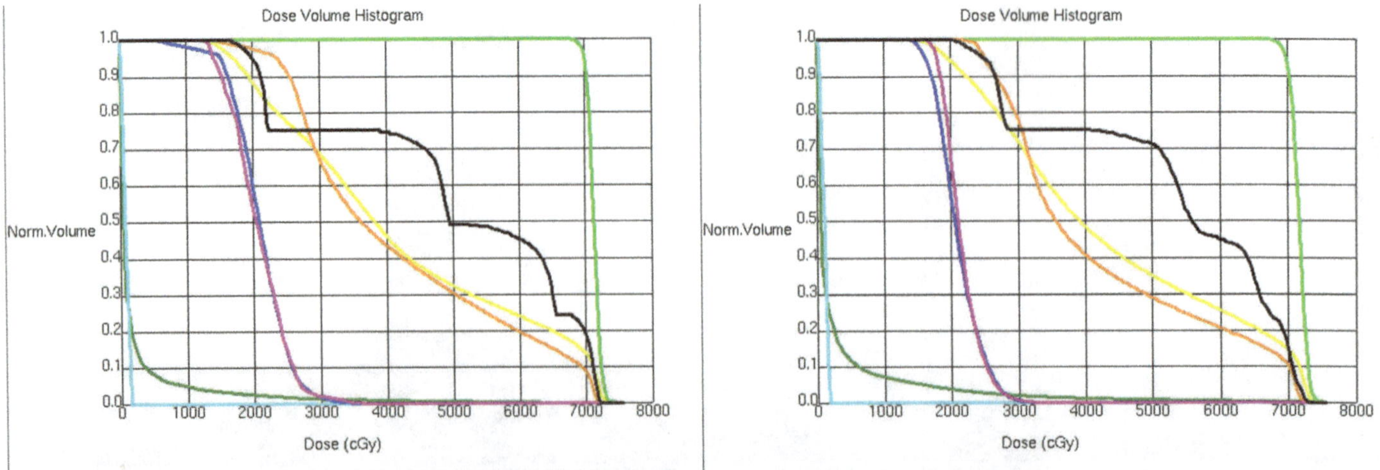

ROI Statistics

ROI	Trial or Record	Min.	Max.	Mean	Std. Dev.
Bladder	IMRT	1177.8	7339.9	4188.5	1839.0
Bowel	IMRT	5.9	6276.9	86.6	397.5
Cauda Equina	IMRT	20.4	138.3	79.6	33.2
Lt FH	IMRT	505.4	3455.0	2081.7	410.5
PTV7000	IMRT	6578.0	7436.6	7129.3	81.2
Rectum	IMRT	1378.4	7221.0	4161.8	1606.3
Rt FH	IMRT	833.8	3602.8	2062.8	407.8
penile bulb	IMRT	1639.9	7205.9	5015.8	1900.3

ROI Statistics

ROI	Trial or Record	Min.	Max.	Mean	Std. Dev.
Bladder	VMAT	1415.6	7426.7	4358.3	1763.4
Bowel	VMAT	10.7	7388.2	122.6	506.5
Cauda Equina	VMAT	32.9	170.6	105.5	38.5
Lt FH	VMAT	1342.5	3179.2	2088.3	318.2
PTV7000	VMAT	6382.3	7532.0	7157.6	108.3
Rectum	VMAT	1911.5	7303.9	4228.3	1548.4
Rt FH	VMAT	1617.0	3217.6	2145.7	262.3
penile bulb	VMAT	1978.2	7279.2	5292.1	1690.9

Figure 7.122 DVH for IMRT (left) and VMAT (right).

Figure 7.122 containing the final dose distribution is very conformal to the target. There might be some streaks of 50% or 40% isodose line present in a clinically insignificant amount. This is the effect of patient thickness and the beam energy chosen.

Case Number pP04: Prostate + Lymph Nodes

General Information

Brief Clinical History

75-year-old man with prostate cancer. Gleason score 8, 9 of 12 cores positive.

Prescription

Prostate + lymph nodes:
PTV 7000 = 250 cGy x 28 fractions
PTV 5600 = 200 cGy x 28 fractions
PTV 5040 = 180 cGy x 28 fractions

Simulation

The patient was simulated in the supine feet first position with a vacuum cushion as the immobilization device.

Target

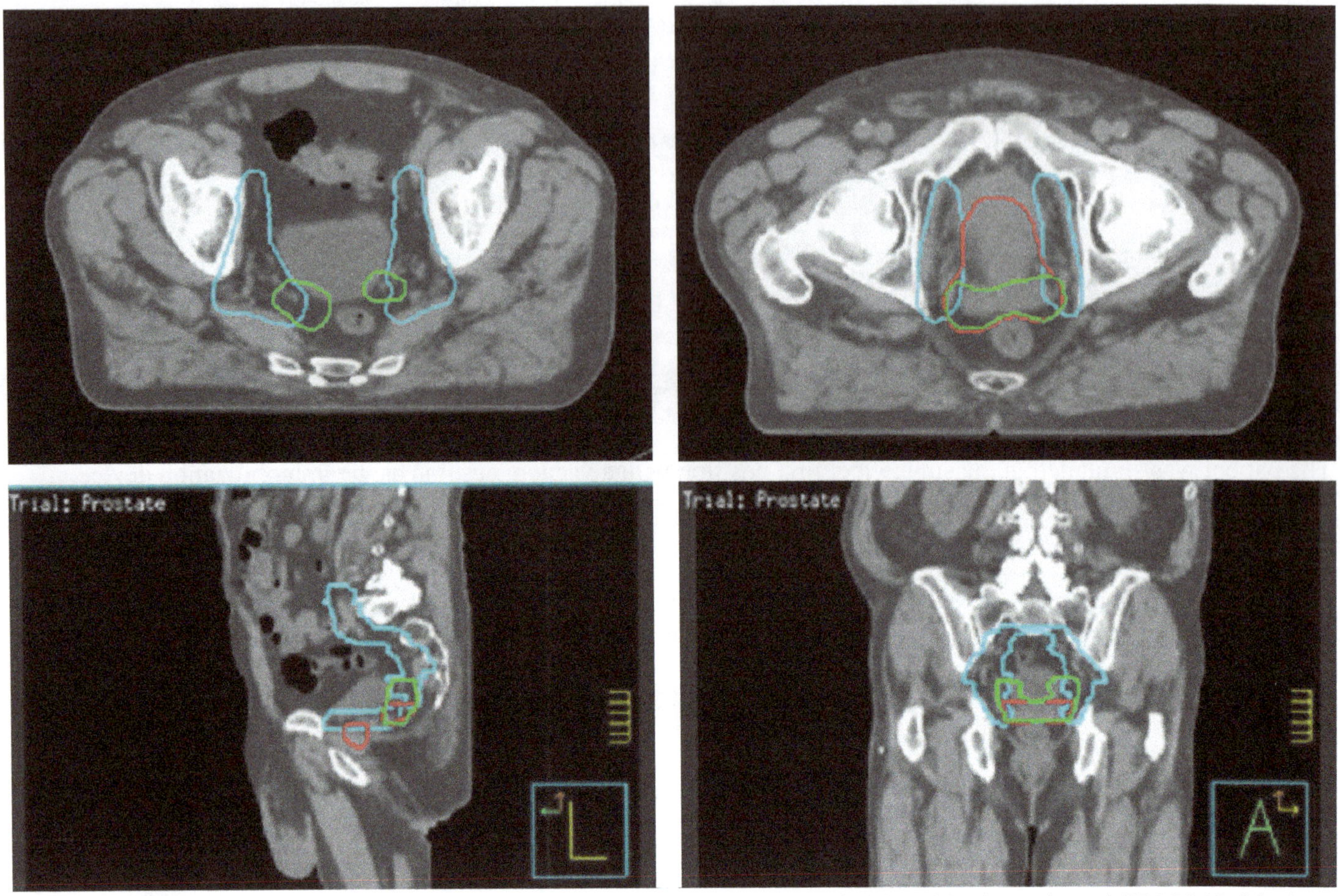

Figure 7.123 PTV contours drawn by the radiation oncologist: PTV 7000 (red), PTV 5600 (green), and PTV 5040 (blue).

The PTV drawn by the radiation oncologist can be seen in Figure 7.123. The irregular and unusual shape of the target is an indication to use IMRT or another inversely optimized technique. The organs at risk in this case are the rectum, the femoral heads, the bladder, the bowels, and the spinal cord. It is important that the planner represents these structures as accurately as possible through contouring.

Contours belonging to different dose levels should be separated for an easier optimization process and less dose conflicts on the optimization objectives. In this case, the planner should made sure that optimization contours for PTV 5040, PTV 5600, and PTV 7000 do not overlap with each other. The overlap should always prioritize the high-dose areas. For example, if PTV 7000 and PTV 5600 overlap, the 5600-cGy isodose line should only occupy the area of contour PTV 5600–PTV 7000.

Beam Arrangement

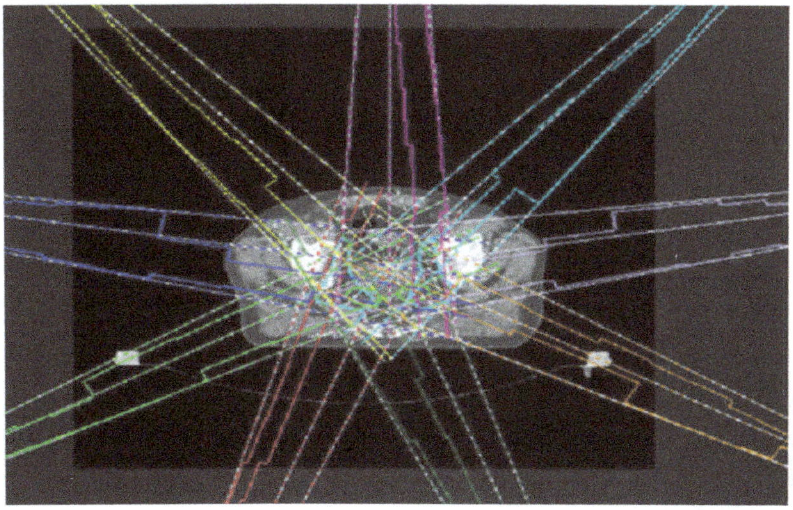

Figure 7.124 Final beam arrangement on axial slice.

Figure 7.125 Surface reconstruction of patient CT with beam positioning representation.

The beam arrangement for this case has 9 beams delivering a total of 90 control points. Figures 7.124 and 7.125 depict the positioning of the beams. Each beam delivers a set number of subfields, control points, or beamlets. Beams that go through structures at risk can still block the structure with the use of control points. When setting up the beams, the planner can start with a basic 7-field arrangement and change this arrangement as dosimetric needs become apparent in the optimization process. The following is a beam parameters summary for this treatment plan:

Gantry Angle	Couch	Collimator	Energy (MV)
200	0	0	10
240	0	0	10
280	0	0	10
320	0	0	10
0	0	0	10
40	0	0	10
80	0	0	10
120	0	0	10
160	0	0	10

Optimization

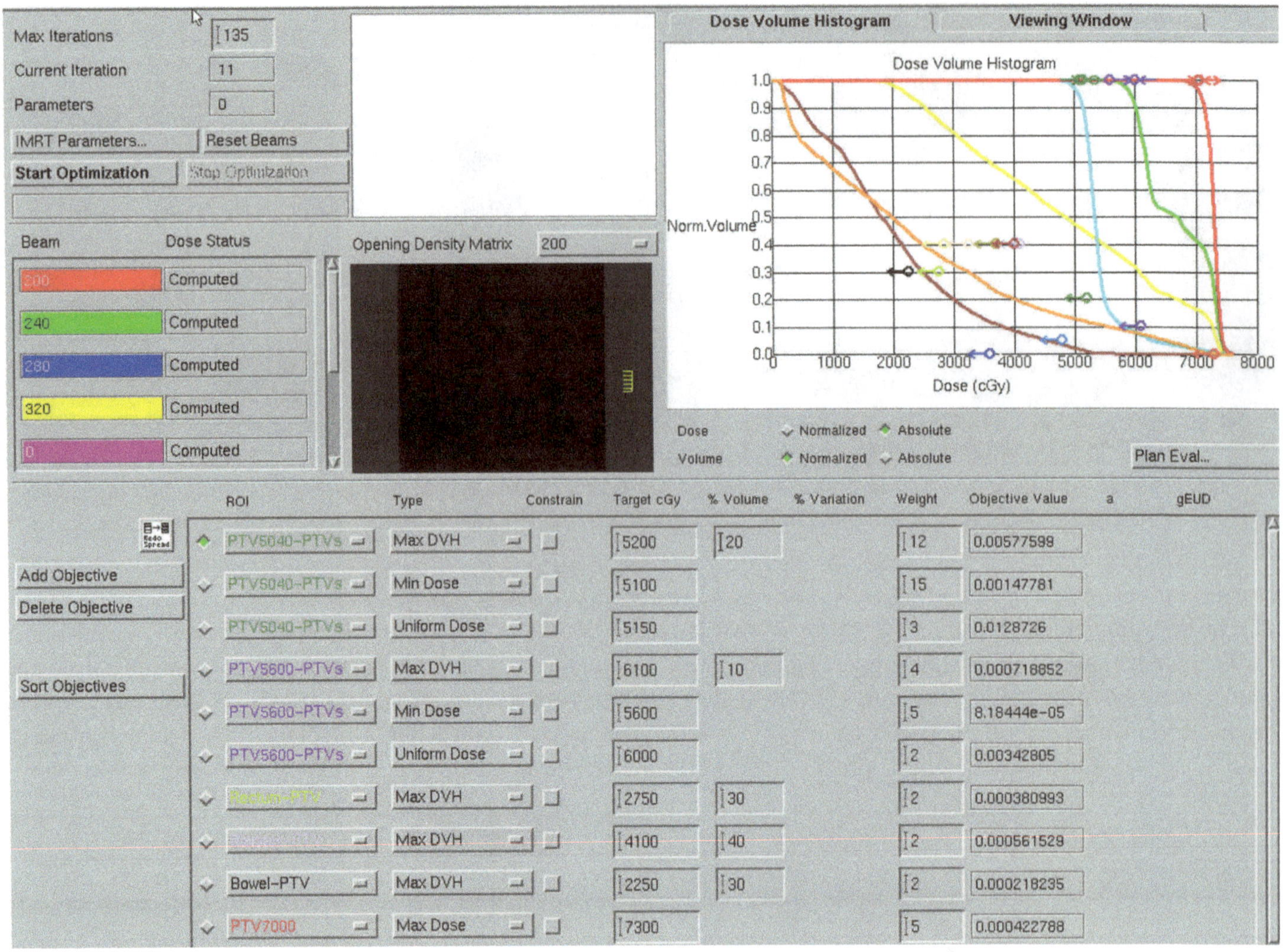

Figure 7.126 Optimizer for this treatment plan.

The optimization takes place after setting up the various contours that help "paint the dose." For example, the planner can start the optimization by setting up the following dose objectives:

PTV 7000	→ Minimum dose = 7000 cGy
PTV 7000	→ Maximum dose = 7350 cGy (105%)
PTV 7000 ring	→ Maximum dose = 7000 cGy
PTV 5600	→ Minimum dose = 5600 cGy
PTV 5600	→ Maximum dose = <7000 cGy
PTV 5600 ring	→ Maximum dose = 6300 cGy (midway between 5600 and 7000)
PTV 5040	→ Minimum dose = 5040 cGy
PTV 5040	→ Maximum dose = 5600 cGy
PTV 5040 ring	→ Maximum dose = 5320 cGy (midway between 5040 and 5600)
Body ring	→ Maximum dose = 3800 cGy

After looking at the dose volume histogram for the first optimization, the planner can add objectives for other structures, as well as modify the existing objectives.

Dose Distribution

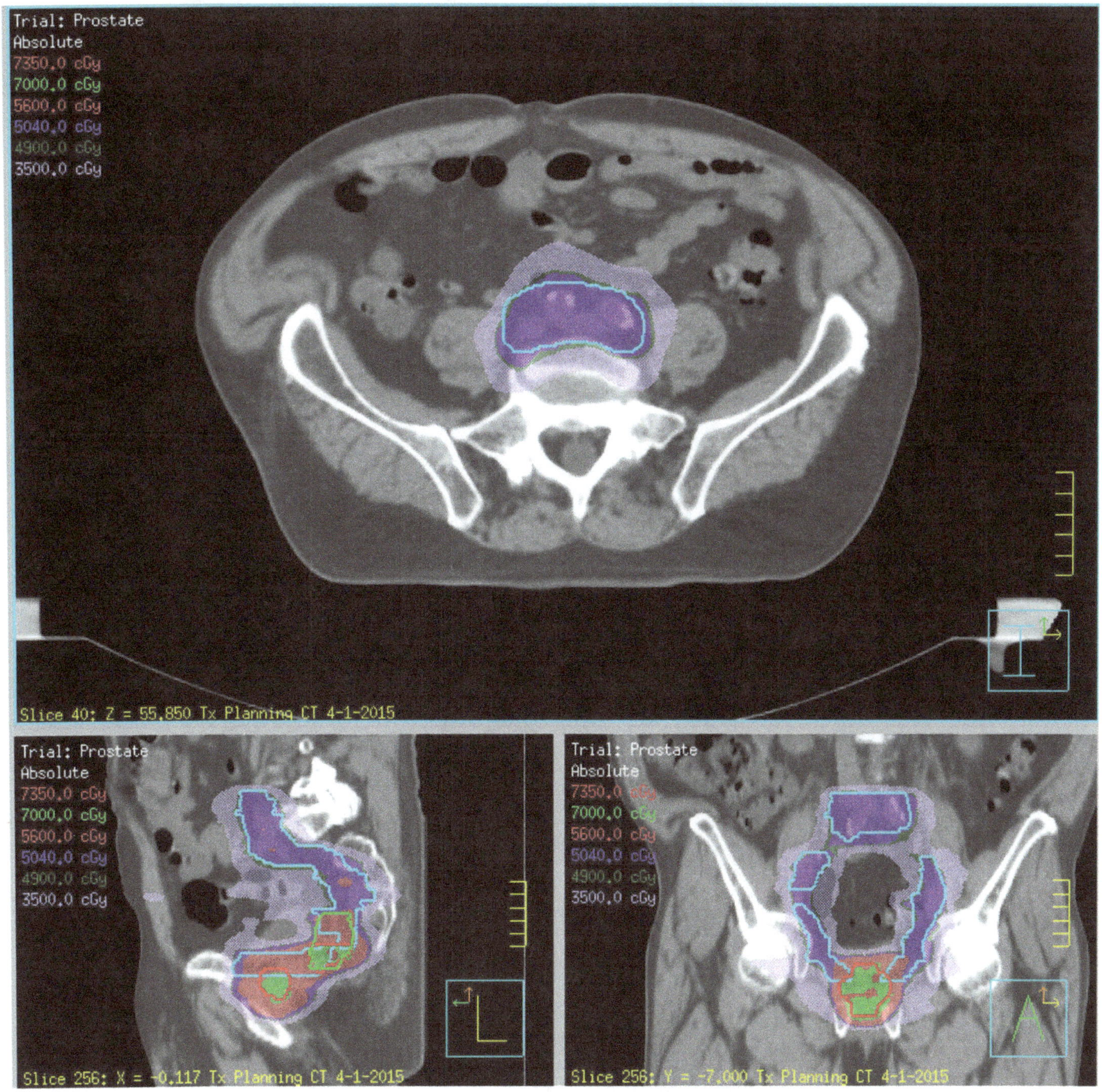

Figure 7.127 Final dose distribution on axial, sagittal, and coronal slices.

Figures 7.127 and 7.128 show the final dose distribution for case pP04. Notice the dose conformity to the target as well as the fall-off with the 50% isodose line of the highest prescription PTV. There are no significant low-dose streaks through the body, and also there are no high-dose spots in the surface of the patient. Figure 7.129 contains the final DVH for this case.

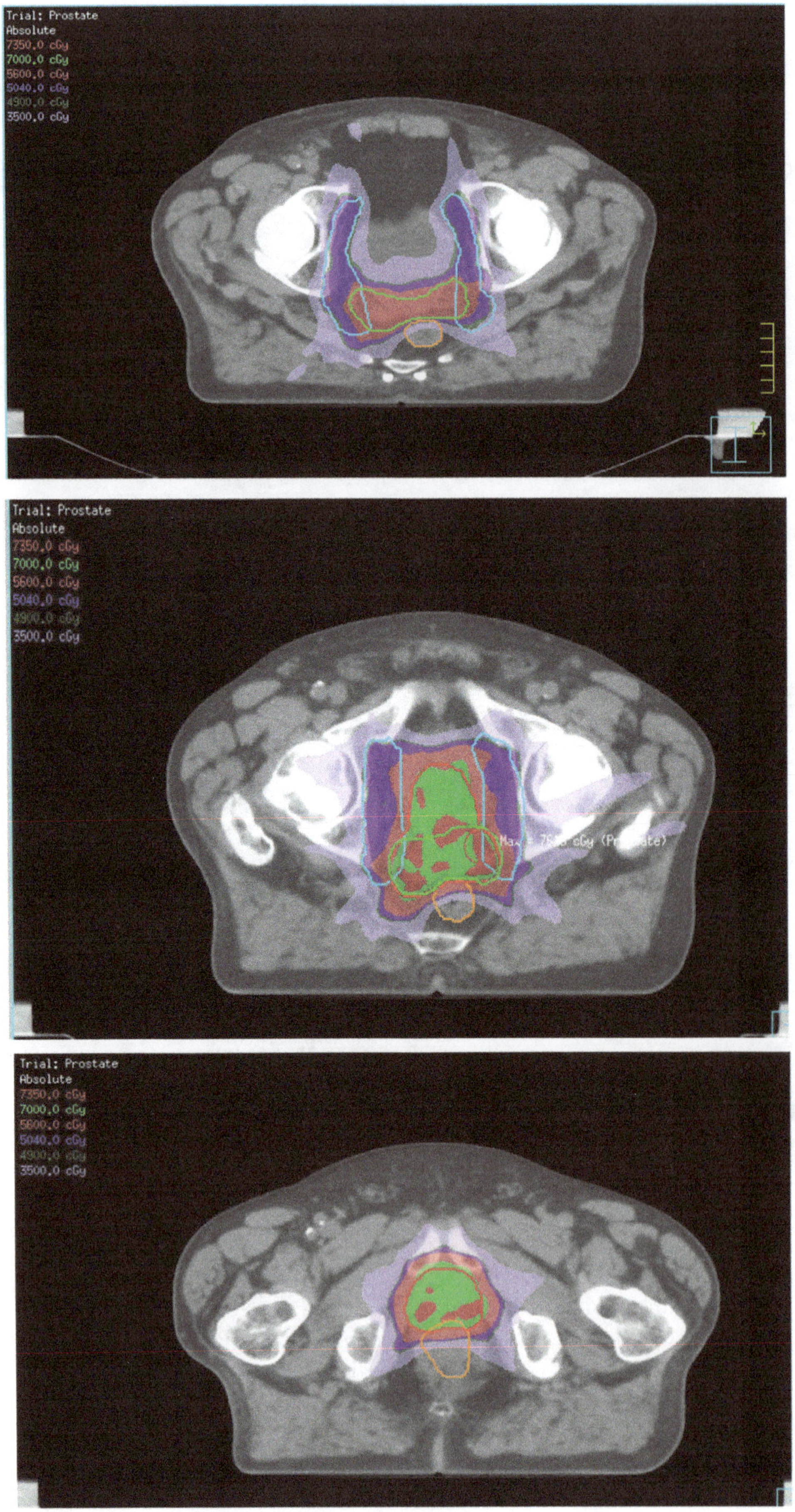

Figure 7.128 Dose distribution on different axial slices for case pP04.

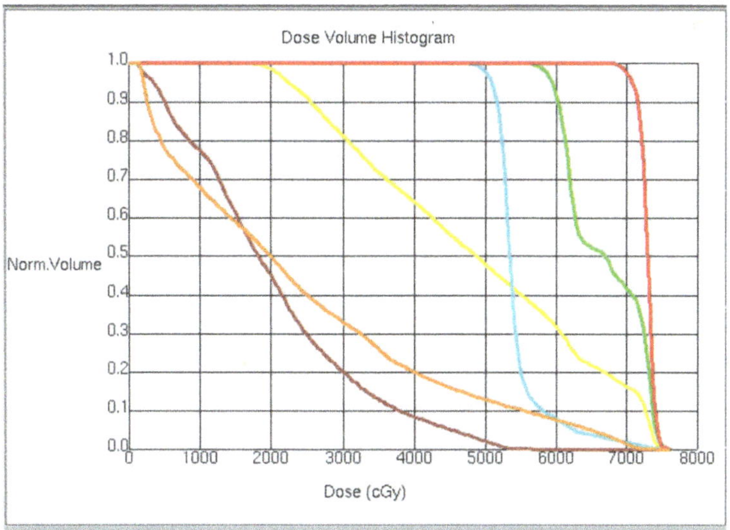

ROI Statistics

Line Type	ROI	Trial or Record	Min.	Max.	Mean	Std. Dev.
	Bladder	Prostate	1582.0	7535.3	4840.1	1700.2
	Bowel	Prostate	96.4	5645.7	862.4	1276.0
	PTV5040	Prostate	4057.5	7600.0	5427.8	390.5
	PTV5600	Prostate	5287.4	7600.0	6661.7	567.4
	PTV7000	Prostate	6530.1	7600.0	7280.9	110.6
	Rectum	Prostate	159.0	7388.4	2373.3	1942.7

Figure 7.129 Dose volume histogram for case pP04.

Case Number pP05: Prostate + Lymph Nodes

General Information

Brief Clinical History

61-year-old male with history of prostate cancer, post prostatectomy five months prior this plan, with GS 3+4=7 with tertiary 5, 0/4 LNs; 5% of gland involved by tumor, bilateral disease, no extraprostatic extension, no invasion of bladder neck, negative margins, closest margin 0.1 cm, pT2cN0. He has had a rising PSA and has met criteria for biochemical failure.

Prescription

Prostate bed:
PTV 5040 = 180 cGy × 28 fractions
PTV 5600 = 200 cGy × 28 fractions

Simulation

The patient was simulated in the supine feet first position with a vacuum cushion as immobilization device.

Target

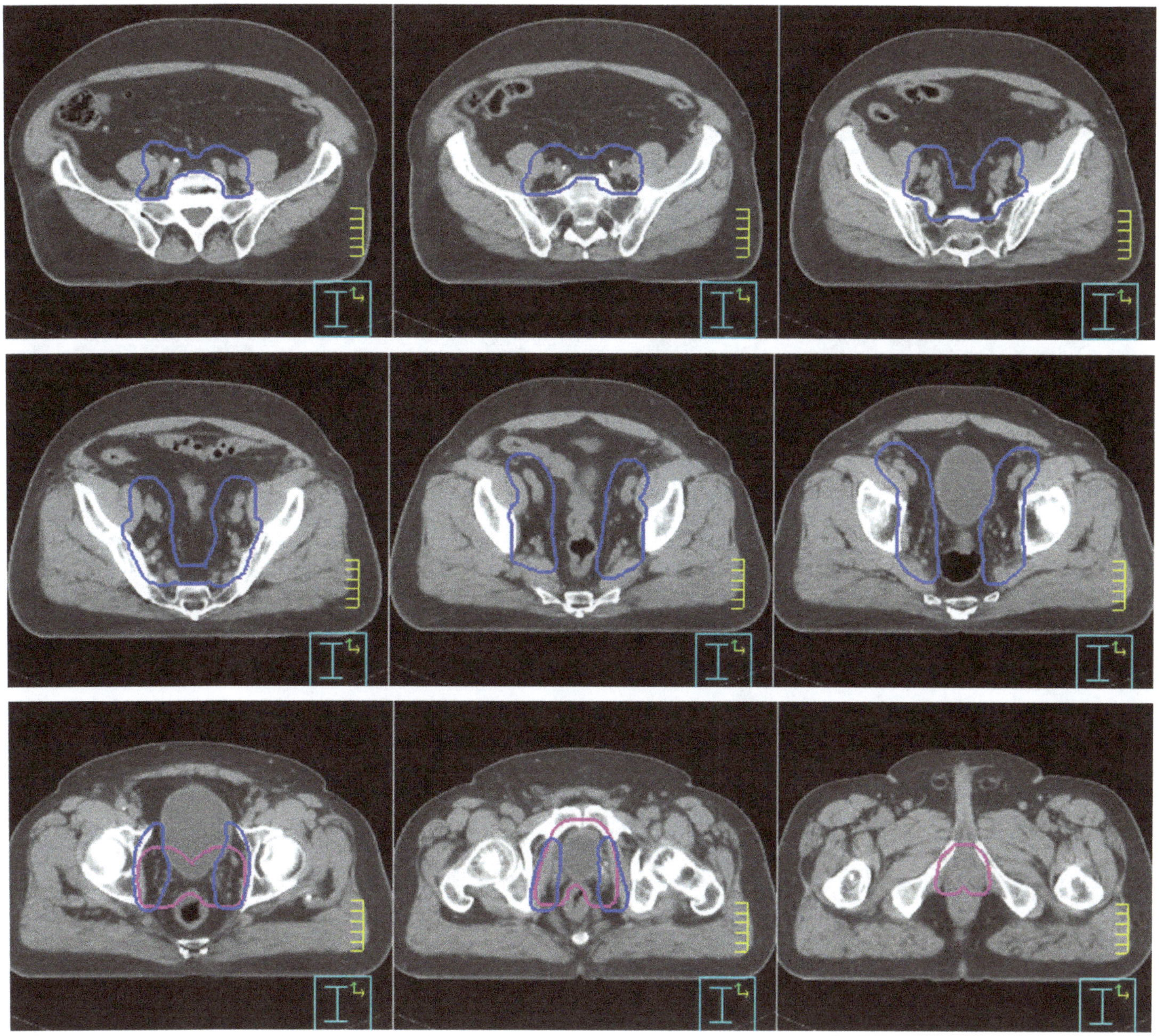

Figure 7.130 PTV contours drawn by the radiation oncologist: PTV 5040 (blue) and PTV 5600 (pink).

The PTV contours are found in Figure 7.130 at different axial cuts. Treating the PTV with an inverse modulated technique such as IMRT or VMAT will help to spare healthy tissue. The organs at risk are the bladder, the bowel, the rectum, the femoral heads, and the spinal canal.

The choice of technique is given not only on a clinical or technical basis, but also on availability. In this case, VMAT offers more control points from many angles around the patient. In fact, clinics that have linear accelerators with a low and high energy (for example, 6 MV and 15 MV) can choose to do VMAT to have the benefits of using a medium-energy linac (for example, 10 MV) while using their lowest-energy linac (6 MV).

Beam Arrangement

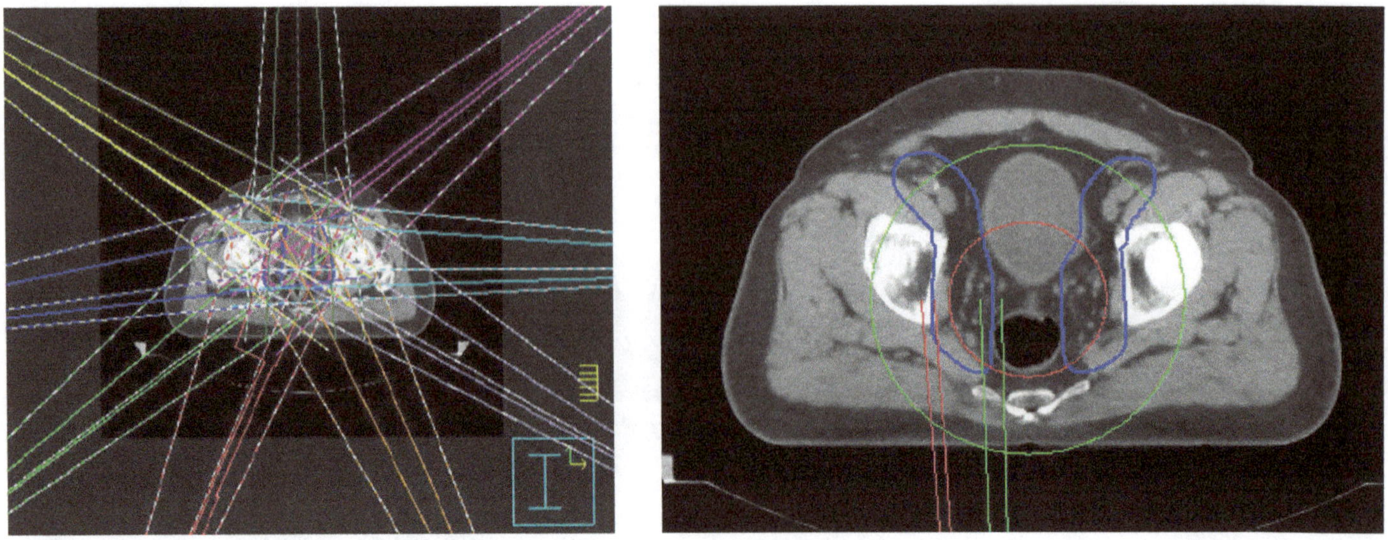

Figure 7.131 IMRT and VMAT beam arrangements.

The beam arrangement for IMRT and VMAT is depicted in Figure 7.131. The planner has used a simple two 360-degree beam arrangement pointed at the isocenter of the target. The final gantry spacing is 4 degrees for each arc (control points every 4 degrees), with the plan having a total of 182 control points. The number of control points can increase according to the capabilities of the system.

The following is a summary of beam parameters for this treatment plan.

VMAT:

Gantry Angle	Couch	Collimator	Energy (MV)
180–179.9	0	0	10
179.9-180	0	0	10

IMRT:

Gantry Angle	Couch	Collimator	Energy (MV)
0	0	0	10
40	0	0	10
80	0	0	10
120	0	0	10
160	0	0	10
200	0	0	10
230	0	0	10
250	0	0	10
280	0	0	10

Optimization

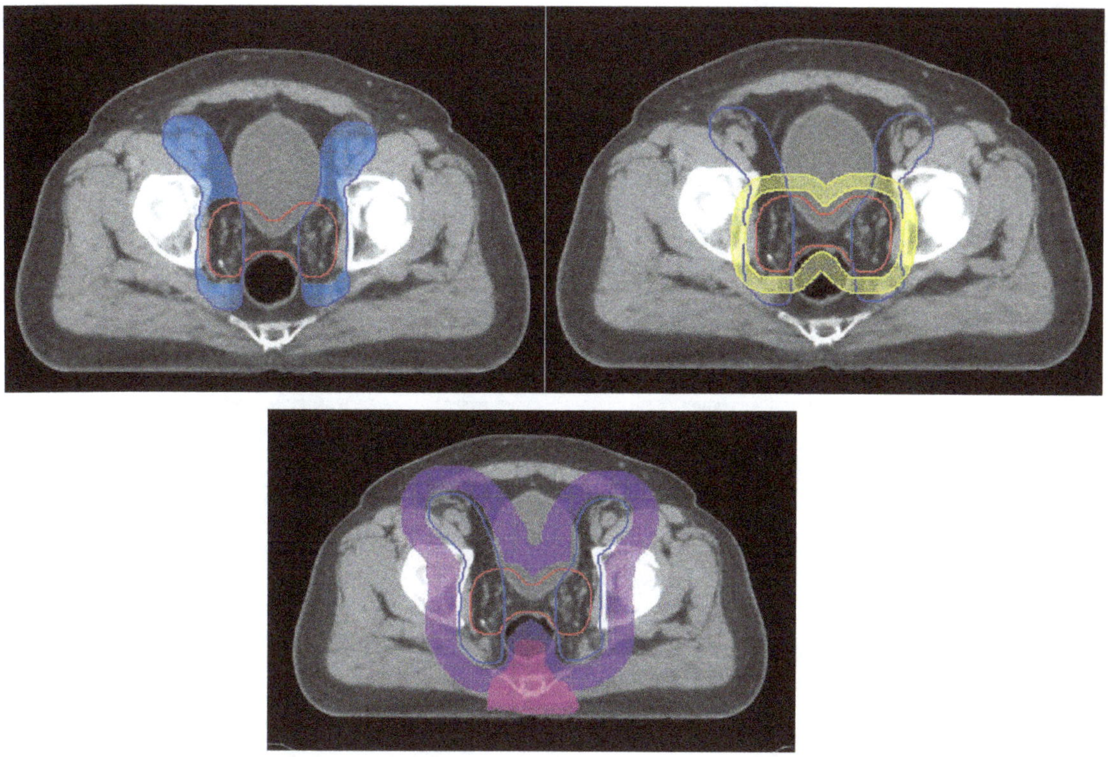

Figure 7.132 Optimization contours: PTV 5040 only (blue colorwash, top left), PTV 5600 ring (yellow colorwash, top right), and PTV 5040 ring (purple colorwash, bottom).

The basic approach for IMRT is followed for VMAT with respect to optimization contours. Figure 7.132 shows some optimization contours used on this case. The blue colorwash contour contains only the low-dose region without any overlaps. The yellow ring is a PTV 5600 ring to control the distribution in the high-dose region. It is acceptable to have overlaps to other regions as long as the maximum dose objective does not pose a conflict to the other dose levels. The purple contour is a PTV ring to avoid spills outside of the target. The pink triangle contour over the back side of the rectum is to avoid any hot spots in that area, since the dose distribution tends to bend in that region.

Dose Distribution

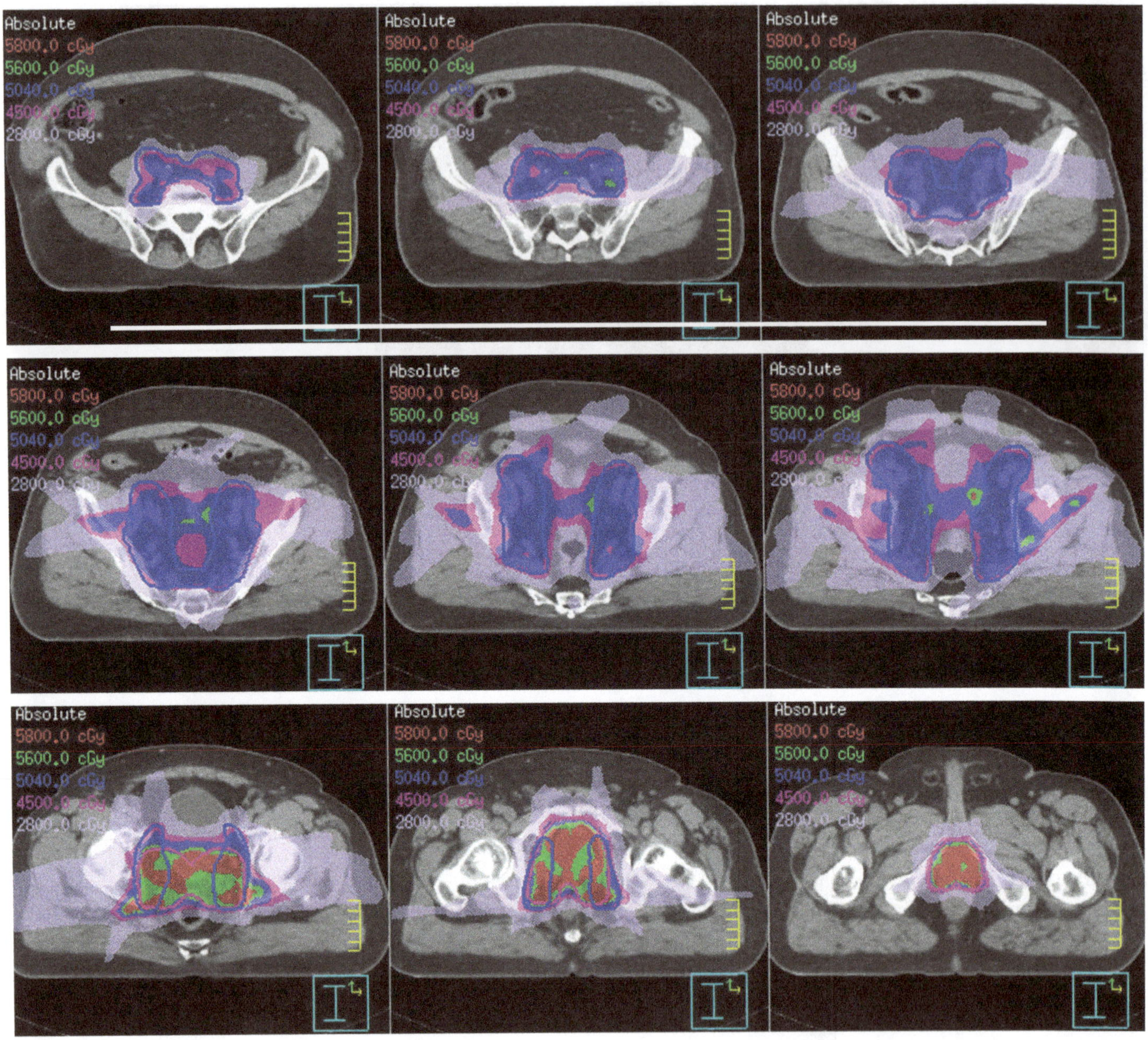

Figure 7.133 Final dose distribution for IMRT plan.

Figures 7.133 and 7.134 show the final dose distributions for this plan. Notice how the 50% isodose line is shaped to avoid structures such as the bladder, femoral heads, and rectum. Some dose falloff of 5040 cGy from the PTV 5600 contour is inevitable, but it can be minimized.

The dose distribution is conformal to the target while avoiding unnecessary irradiation to other structures. The 5600-cGy isodose line is bright green, while the 5040-cGy isodose line is dark green. The DVH for the final plan shows the degree of coverage of the targets, as well as the sparing of organs at risk. It is important to spare organs at risk as much as possible if a boost or re-treatment of the area is necessary to minimize the risk of long-term and short-term side effects. Figure 7.135 contains the DVH of the final dose distribution.

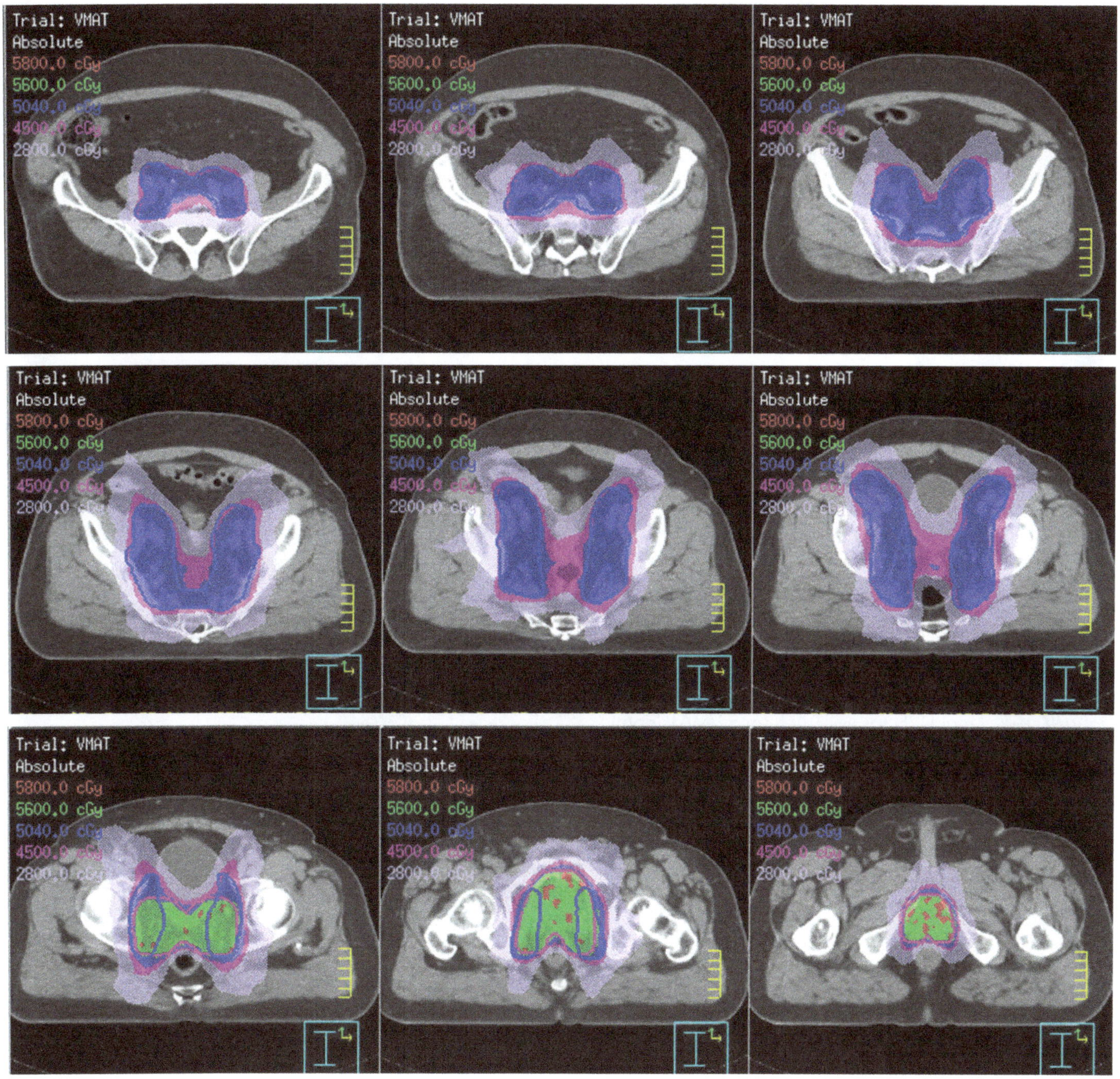

Figure 7.134 Final dose distribution for VMAT plan.

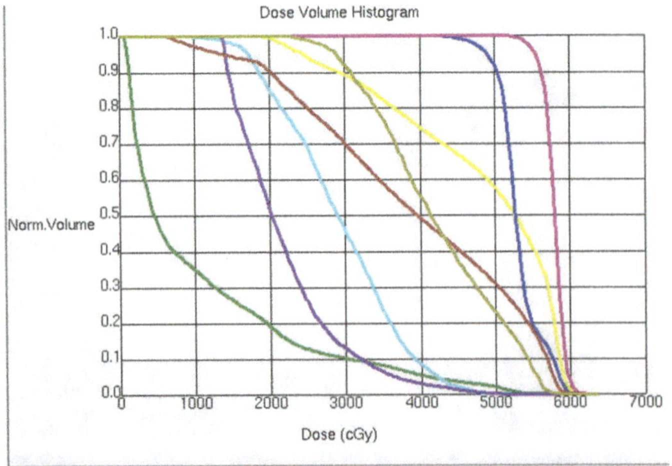

ROI Statistics

Line Type	ROI	Trial or Record	Min.	Max.	Mean	Std. Dev.
	Bladder	IMRT	1699.0	6132.0	4788.3	1148.0
	Bowel	IMRT	64.6	5720.6	1078.9	1260.7
	L FH	IMRT	666.5	5209.5	2898.3	775.8
	PTV5040	IMRT	3438.4	6151.6	5311.9	279.0
	PTV56+10prostatebed	IMRT	4639.5	6183.7	5783.1	134.1
	R FH	IMRT	991.0	5077.4	2187.2	706.3
	Rectum	IMRT	633.6	6095.0	3884.8	1426.8
	Sigmoid	IMRT	2116.8	5831.9	4194.7	855.4

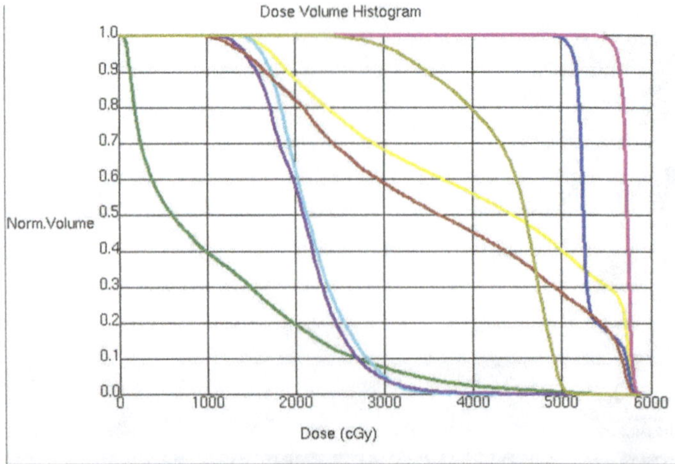

ROI Statistics

Line Type	ROI	Trial or Record	Min.	Max.	Mean	Std. Dev.
	Bladder	VMAT	1256.1	5892.0	4085.9	1500.0
	Bowel	VMAT	62.0	5316.4	1076.9	1093.6
	L FH	VMAT	1302.8	3915.5	2195.6	416.3
	PTV5040	VMAT	4346.4	5874.3	5312.9	198.1
	PTV5600	VMAT	4956.5	5892.0	5727.4	59.3
	R FH	VMAT	1046.8	4281.7	2109.9	458.5
	Rectum	VMAT	672.6	5818.0	3649.6	1516.8
	Sigmoid	VMAT	2075.5	5115.0	4385.1	563.1

Figure 7.135 Final DVH for IMRT (left) and VMAT (right).

8. Eclipse TPS Sample IMRT and VMAT Plans

Case Number eTA01: Thorax and Abdomen—Left Lung SBRT

General Information

Brief Clinical History

79-year-old female diagnosed with squamous cell carcinoma of the left lower lung.

Prescription

Left lower lobe SBRT:
PTV 6000 = 1200 cGy × 5 fractions

Simulation/Comments

The simulation was accomplished with a vacuum cushion as an immobilization device placed within a butterfly board. A headrest and moldable pillow was placed under her head for comfort and immobilization.

> **NOTE:** All sample cases in this chapter are for educational purposes only. Hypofractionated regimens, SRS and SBRT procedures must be tested and implemented clinically in a separate commissioning process from basic IMRT/VMAT implementation. All dose distributions are not suggestions for clinical treatments. Objectives in the optimizer as well as beam arrangements are for illustration/educational purposes only.

Targets

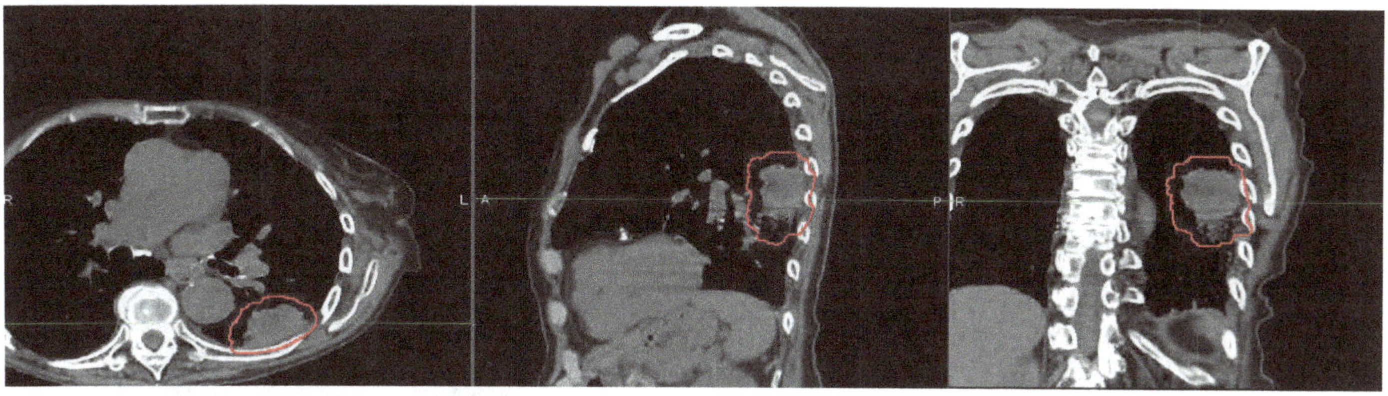

Figure 8.1 Target contour PTV 6000 (red).

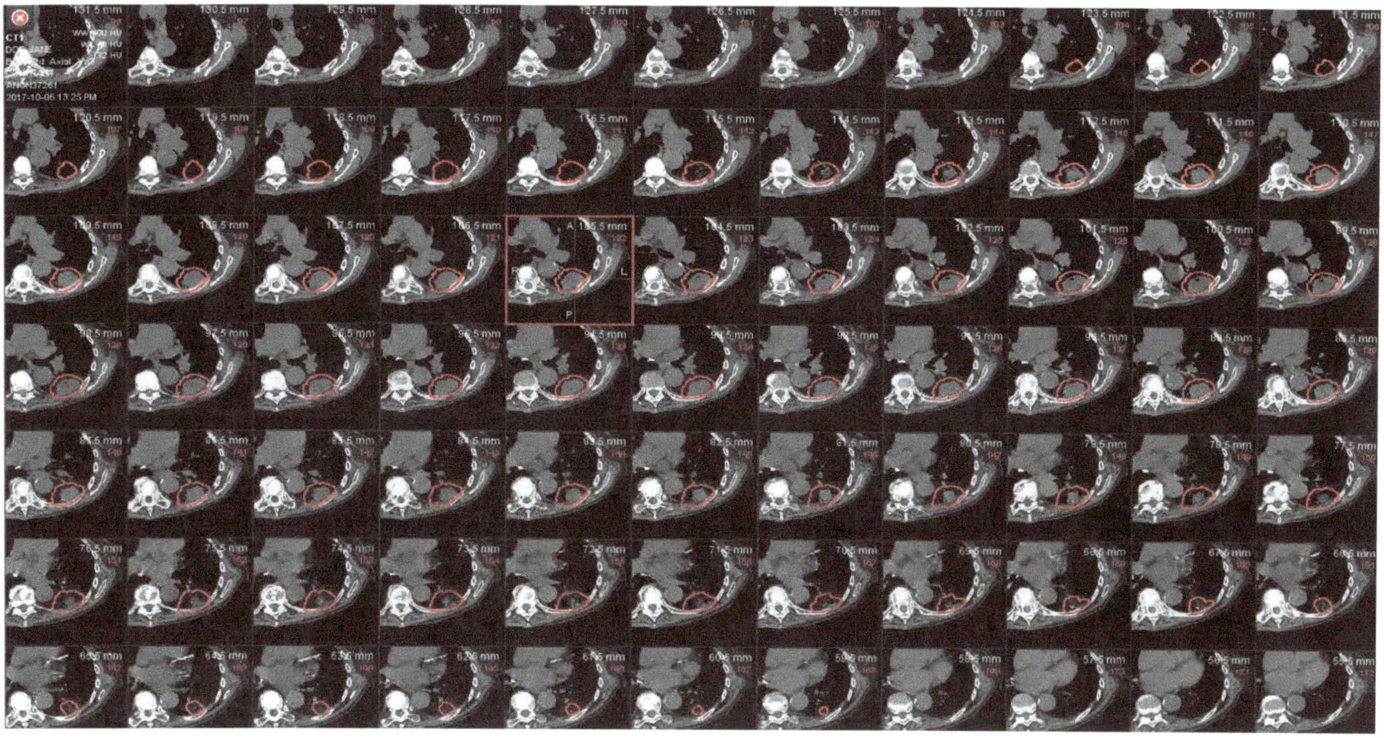

Figure 8.2 Target contour PTV 6000 (red) in axial slices.

Beam Arrangement

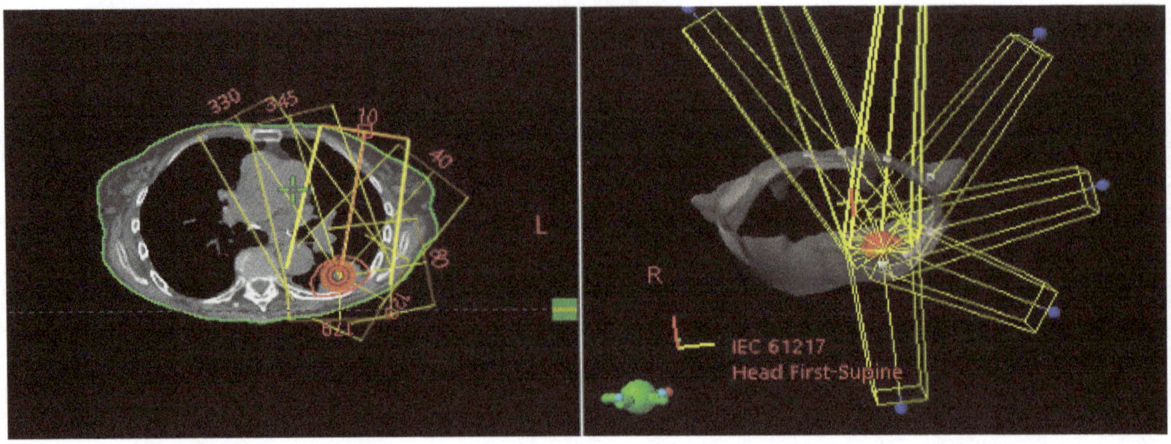

Figure 8.3 IMRT 7-field beam arrangement.

Gantry Angle	Couch	Collimator	Energy (MV)
179	0	0	10 FFF
120	0	0	10 FFF
80	0	0	10 FFF
40	0	0	10 FFF
10	0	0	10 FFF
345	0	0	10 FFF
30	0	0	10 FFF

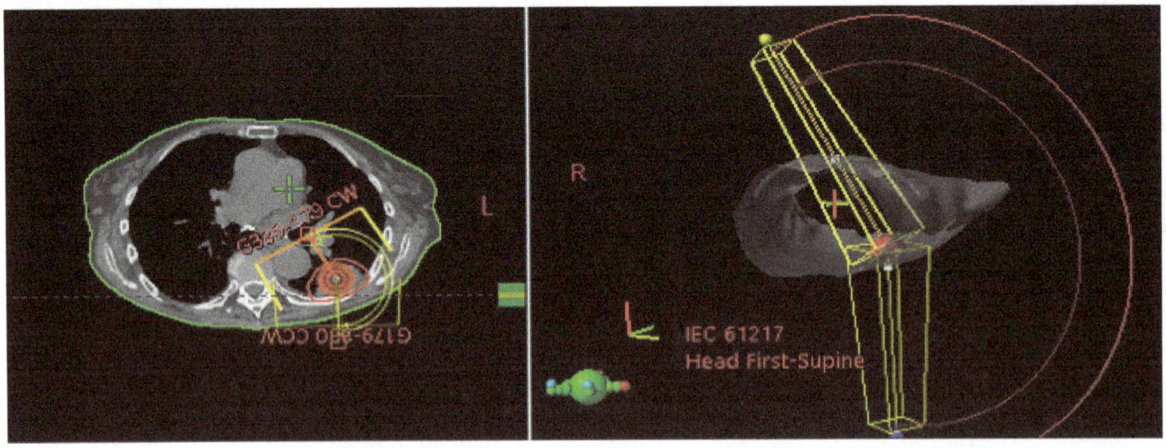

Figure 8.4 VMAT partial-arc beam arrangement.

Gantry Angle	Couch	Collimator	Energy (MV)
179–330 CCW	0	30	10 FFF
330–179 CW	0	330	10 FFF

Optimization

☑ Use Normal Tissue Objective — Priority: 100 — Define NTO Settings...

BODY		Volume [cc]:	12673	Points:	280186	Resolution [mm]:	4.50
Canal		Volume [cc]:	42	Points:	15681	Resolution [mm]:	2.65
	Upper	Volume [%]:	0.0	Dose [cGy]:	1500.0	Priority:	160
Esophagus		Volume [cc]:	31	Points:	16464	Resolution [mm]:	2.40
Heart		Volume [cc]:	725	Points:	64814	Resolution [mm]:	3.00
ITV		Volume [cc]:	52	Points:	9769	Resolution [mm]:	2.87
L Lung		**Volume [cc]:**	**1129**	**Points:**	**120933**	**Resolution [mm]:**	**3.00**
	Upper	Volume [%]:	12.0	Dose [cGy]:	2000.0	Priority:	160
L Lung-GTV		Volume [cc]:	1091	Points:	122131	Resolution [mm]:	3.00
PTV 6000		Volume [cc]:	84	Points:	12337	Resolution [mm]:	3.00
	Upper	Volume [%]:	0.0	Dose [cGy]:	7800.0	Priority:	220
	Lower	Volume [%]:	100.0	Dose [cGy]:	6000.0	Priority:	250
R Lung		Volume [cc]:	1713	Points:	159300	Resolution [mm]:	3.00
	Upper	Volume [%]:	0.0	Dose [cGy]:	1200.0	Priority:	160
Trachea		Volume [cc]:	50	Points:	14805	Resolution [mm]:	2.82

Figure 8.5 Initial optimization parameters.

The objectives above were set for the initial optimization to produce steep fall-off while achieving normal tissue dose tolerance. The allowable maximum dose within the PTV was set by the radiation oncologist to be 130% of the prescription dose. During optimization iterations, the dose distribution can be evaluated for improvement. Common problems during this stage are uneven areas of high dose within the target, missing coverage in the target, and spilled prescription dose outside the target.

☑ Use Normal Tissue Objective — Priority: 100 — Define NTO Settings...

BODY		**Volume [cc]:**	**12673**	**Points:**	**280186**	**Resolution [mm]:**	**4.50**
Canal		Volume [cc]:	42	Points:	15681	Resolution [mm]:	2.65
	Upper	Volume [%]:	0.0	Dose [cGy]:	1500.0	Priority:	160
Esophagus		Volume [cc]:	31	Points:	16464	Resolution [mm]:	2.40
Heart		Volume [cc]:	725	Points:	64814	Resolution [mm]:	3.00
ITV		Volume [cc]:	52	Points:	9769	Resolution [mm]:	2.87
L Lung		Volume [cc]:	1129	Points:	120933	Resolution [mm]:	3.00
L Lung 2cm Ring		Volume [cc]:	2154	Points:	274883	Resolution [mm]:	3.00
	Upper	Volume [%]:	2.0	Dose [cGy]:	2800.0	Priority:	200
L Lung-GTV		Volume [cc]:	1091	Points:	122131	Resolution [mm]:	3.00
	Upper	Volume [%]:	15.0	Dose [cGy]:	2000.0	Priority:	160
PTV 6000		Volume [cc]:	84	Points:	12337	Resolution [mm]:	3.00
	Upper	Volume [%]:	0.0	Dose [cGy]:	7800.0	Priority:	220
	Lower	Volume [%]:	100.0	Dose [cGy]:	6000.0	Priority:	250
R Lung		Volume [cc]:	1713	Points:	159300	Resolution [mm]:	3.00
	Upper	Volume [%]:	0.0	Dose [cGy]:	1200.0	Priority:	160
Trachea		Volume [cc]:	50	Points:	14805	Resolution [mm]:	2.82

Figure 8.6 Final optimization parameters.

Dose Distribution

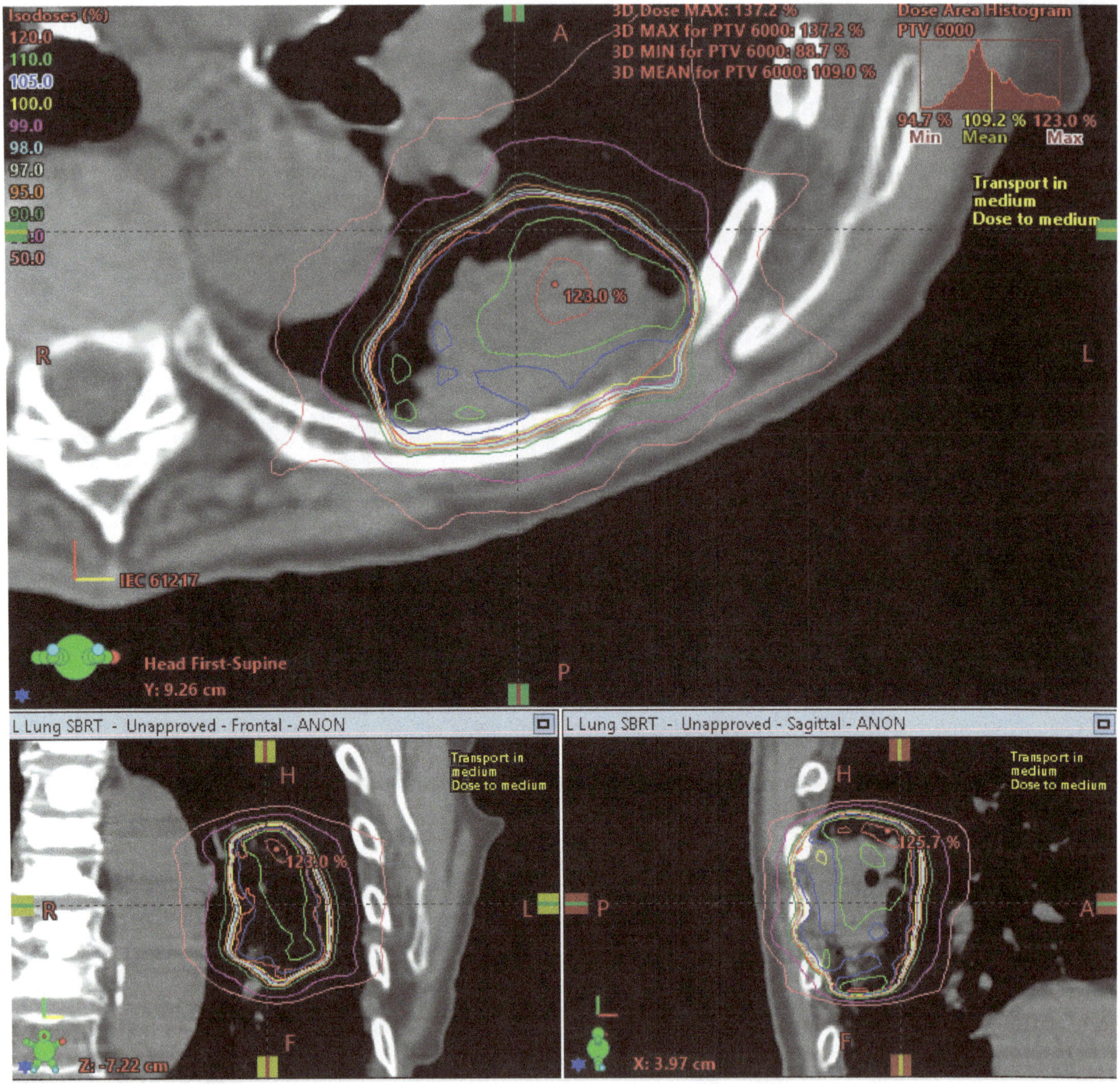

Figure 8.7 Final dose distribution.

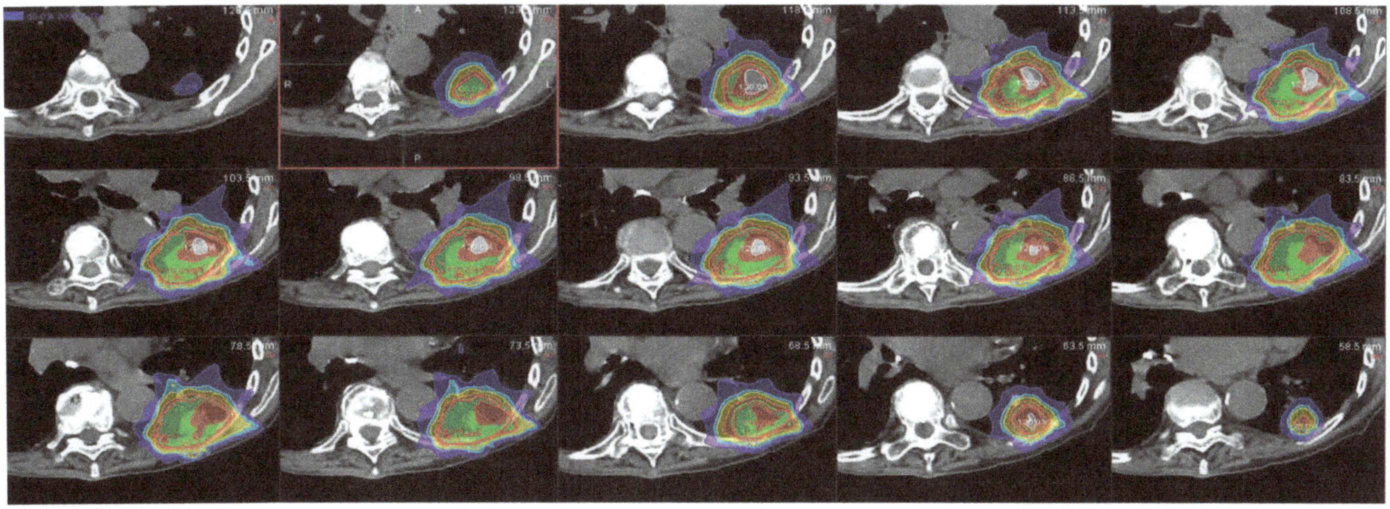

Figure 8.8 Final dose distribution on axial slices.

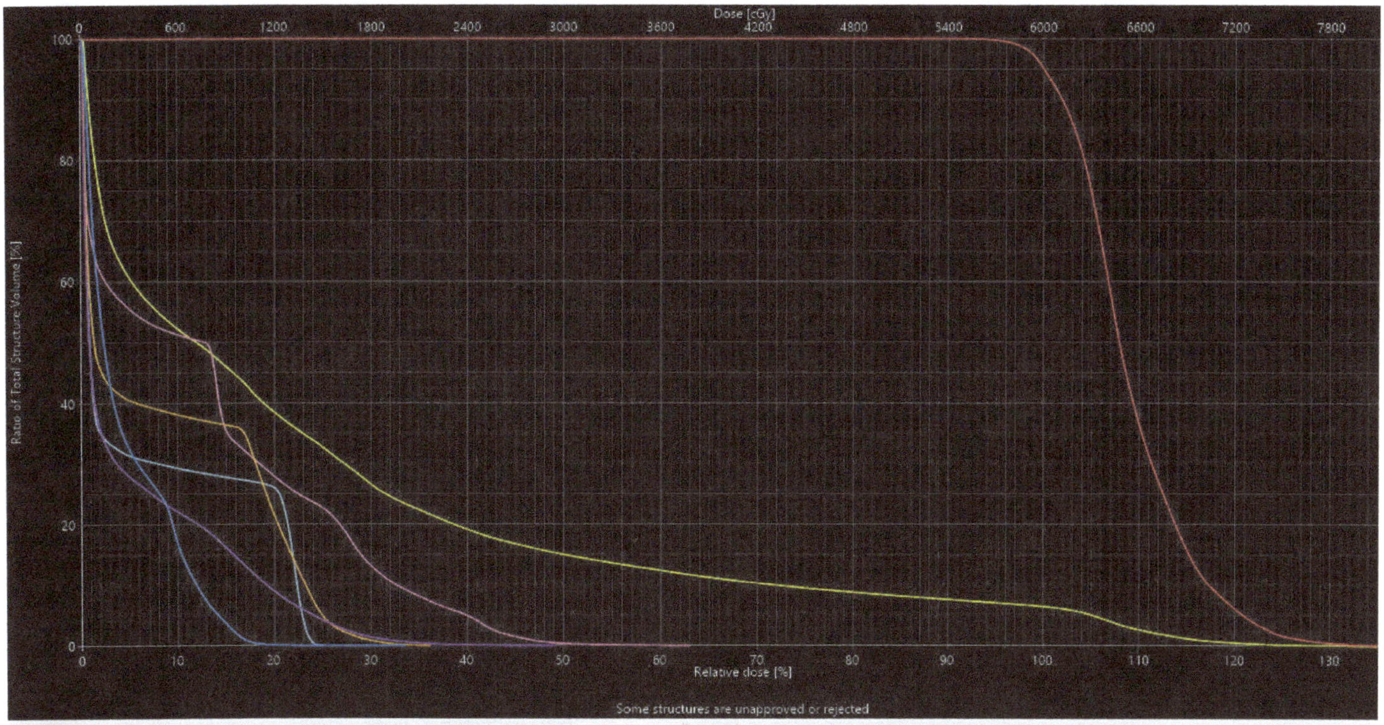

Fields | Dose | Field Alignments | Plan Objectives | Optimization Objectives | Dose Statistics | Reference Points | Calculation Models | Plan Sum

Show DVH	Structure	Approval Status	Plan	Course	Volume [cm³]	Dose Cover [%]	Sampling Cover [%]	Min Dose [%]	Max Dose [%]	Mean Dose [%]
☑	L Lung	Unapproved	L Lung SBRT	C1	1128.7	100.0	100.0	0.3	137.2	23.9
☑	Trachea	Unapproved	L Lung SBRT	C1	49.9	100.0	100.2	0.2	63.1	13.0
☑	PTV 6000	Unapproved	L Lung SBRT	C1	84.2	100.0	99.7	88.7	137.2	109.0
☑	Canal	Unapproved	L Lung SBRT	C1	41.6	100.0	99.7	0.2	24.9	6.9
☑	Esophagus	Unapproved	L Lung SBRT	C1	30.9	100.0	99.4	0.2	36.2	8.7
☑	Heart	Unapproved	L Lung SBRT	C1	724.6	100.0	99.8	0.2	49.1	5.4
☑	R Lung	Unapproved	L Lung SBRT	C1	1713.3	100.0	100.0	0.1	33.6	4.6

Figure 8.9 Final dose volume histogram.

Case Number eTA02: Thorax and Abdomen—Right Hilum and Mediastinum Radiation Treatment

General Information

Brief Clinical History

64-year-old man diagnosed with squamous cell carcinoma of the right upper lobe of the lung. Radiation therapy was prescribed with concurrent chemotherapy.

Prescription

Right hilum and mediastinum:
PTV 6000 = 200 cGy × 30 fractions

Simulation/Comments

The simulation was accomplished with a vacuum cushion placed beneath the patient's back in a butterfly board for reproducible arm positioning. A PET/CT image was fused to the planning CT to evaluate the extent of the lesion.

Targets

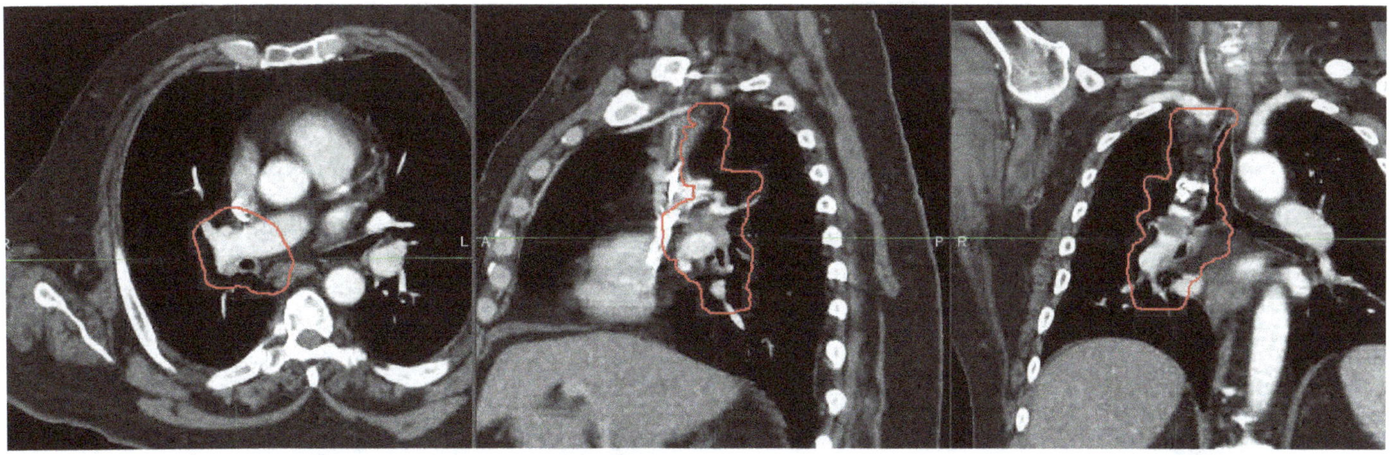

Figure 8.10 Right hilum and mediastinum PTV 6000 (red).

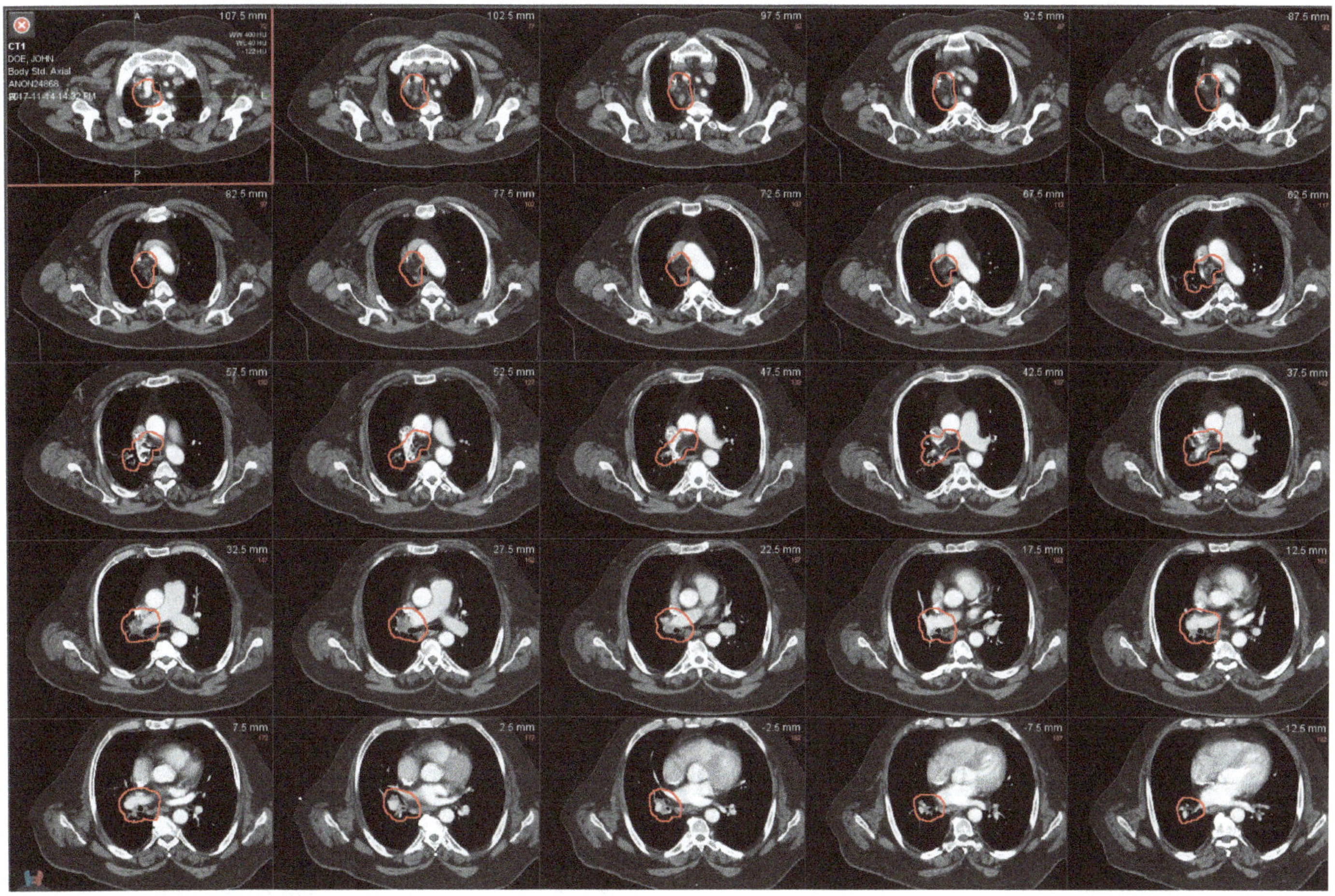

Figure 8.11 Right hilum and mediastinum PTV 6000 (red) in axial slices.

Figures 8.10 and 8.11 show the target volume in a CT with the presence of intravenous contrast. The contrast will not be administered during therapy, and departmental procedures must be followed to determine a suitable simulation dataset. One option is to contour the contrasted region and override this density to that of water in the TPS. Another option is that of setting up the patient for simulation and scan the patient twice: the first time without contrast and the second time with contrast. In this case, the first CT dataset is used for planning purposes.

Beam Arrangement

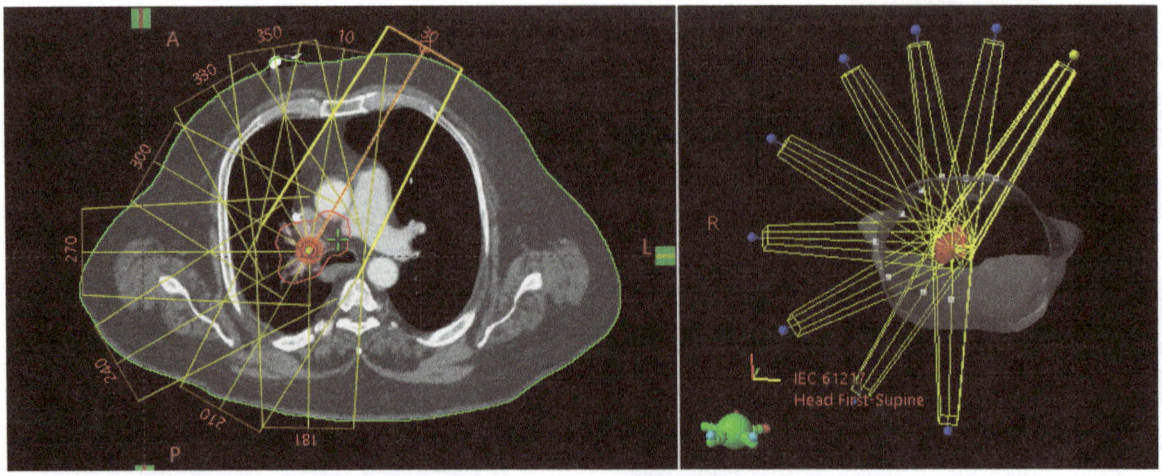

Figure 8.12 IMRT 9-field beam arrangement. Notice the sparing of the contralateral lung.

Gantry Angle	Couch	Collimator	Energy (MV)
181	0	0	6
210	0	0	6
240	0	0	6
270	0	0	6
300	0	0	6
330	0	0	6
350	0	0	6
10	0	0	6
30	0	0	6

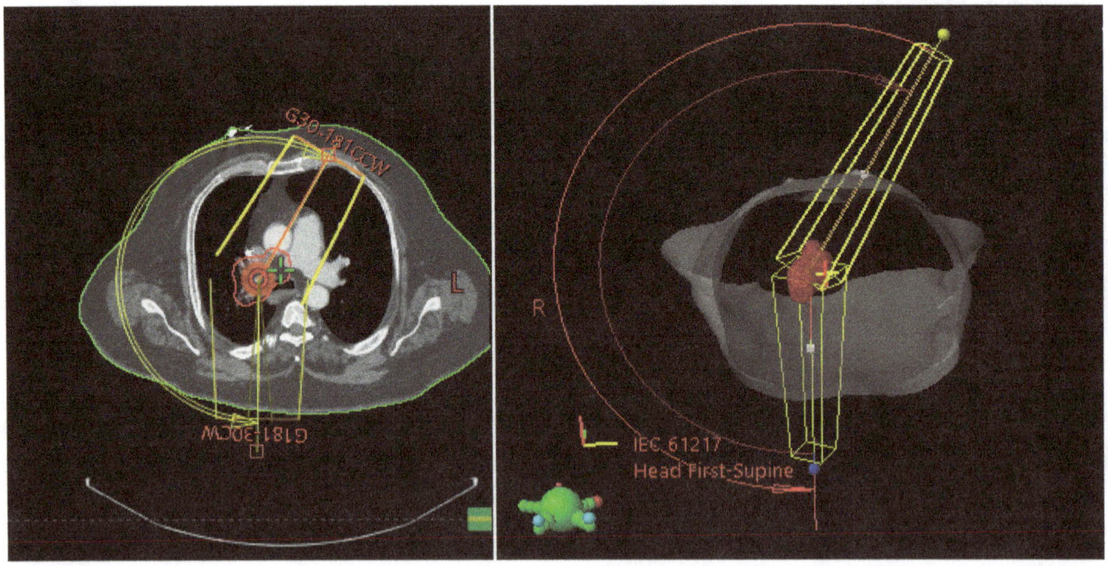

Figure 8.13 Two-arc VMAT beam arrangement. Note that depending on the location of the target, either 6-MV or 10-MV electrons may be used.

Gantry Angle	Couch	Collimator	Energy (MV)
181–30 CCW	0	0	10 FFF
30–181 CW	0	0	10 FFF

Optimization

☑ Use Normal Tissue Objective — Priority: 100 — Define NTO Settings...

BODY		**Volume [cc]:**	**38922**	**Points:**	**736543**	**Resolution [mm]:**	**4.50**
Esophagus1		Volume [cc]:	60	Points:	15660	Resolution [mm]:	2.99
Heart1		Volume [cc]:	1081	Points:	88795	Resolution [mm]:	3.00
	Upper	Volume [%]:	50.0	Dose [cGy]:	700.0	Priority:	160
	Upper		30.0		1700.0		160
	Upper		10.0		4000.0		50
L Lung1		Volume [cc]:	2206	Points:	211400	Resolution [mm]:	3.00
	Upper	Volume [%]:	3.0	Dose [cGy]:	2000.0	Priority:	160
PTV6000		Volume [cc]:	358	Points:	42970	Resolution [mm]:	3.00
	Upper	Volume [%]:	0.0	Dose [cGy]:	6400.0	Priority:	160
	Lower	Volume [%]:	100.0	Dose [cGy]:	6000.0	Priority:	250
R Lung1		Volume [cc]:	2120	Points:	195196	Resolution [mm]:	3.00
	Upper	Volume [%]:	45.0	Dose [cGy]:	2000.0	Priority:	160
Spinal Canal		Volume [cc]:	68	Points:	17239	Resolution [mm]:	3.00
	Upper	Volume [%]:	0.0	Dose [cGy]:	3800.0	Priority:	160
Trachea1		Volume [cc]:	65	Points:	14420	Resolution [mm]:	3.00

Figure 8.14 List of initial objectives.

The radiation treatment optimization focuses on sparing as much healthy tissue as possible. The target in case eTA02 is in close proximity to OARs, such as the heart and lungs. Sparing these organs at risk can yield a dose distribution that spills into the anterior/posterior aspect of the patient. During the optimization process, the planner must avoid placing large volumes of prescription dose and high dose in organs at risk.

☑ Use Normal Tissue Objective — Priority: 100 — Define NTO Settings...

100% Spill		**Volume [cc]:**	**45**	**Points:**	**33411**	**Resolution [mm]:**	**2.35**
	Upper	Volume [%]:	5.0	Dose [cGy]:	6000.0	Priority:	160
BODY		Volume [cc]:	38922	Points:	736543	Resolution [mm]:	4.50
Esophagus1		Volume [cc]:	60	Points:	15660	Resolution [mm]:	2.99
Heart1		Volume [cc]:	1081	Points:	88795	Resolution [mm]:	3.00
	Upper	Volume [%]:	50.0	Dose [cGy]:	700.0	Priority:	160
	Upper		30.0		1700.0		160
	Upper		10.0		4000.0		50
L Lung1		Volume [cc]:	2206	Points:	211400	Resolution [mm]:	3.00
	Upper	Volume [%]:	3.0	Dose [cGy]:	2000.0	Priority:	160
PTV6000		Volume [cc]:	358	Points:	42970	Resolution [mm]:	3.00
	Upper	Volume [%]:	0.0	Dose [cGy]:	6400.0	Priority:	160
	Lower	Volume [%]:	100.0	Dose [cGy]:	6000.0	Priority:	250
R Lung1		Volume [cc]:	2120	Points:	195196	Resolution [mm]:	3.00
	Upper	Volume [%]:	45.0	Dose [cGy]:	2000.0	Priority:	160
Spinal Canal		Volume [cc]:	68	Points:	17239	Resolution [mm]:	3.00
	Upper	Volume [%]:	0.0	Dose [cGy]:	3800.0	Priority:	160
Trachea1		Volume [cc]:	65	Points:	14420	Resolution [mm]:	3.00

Figure 8.15 List of final optimization objectives.

Dose Distribution

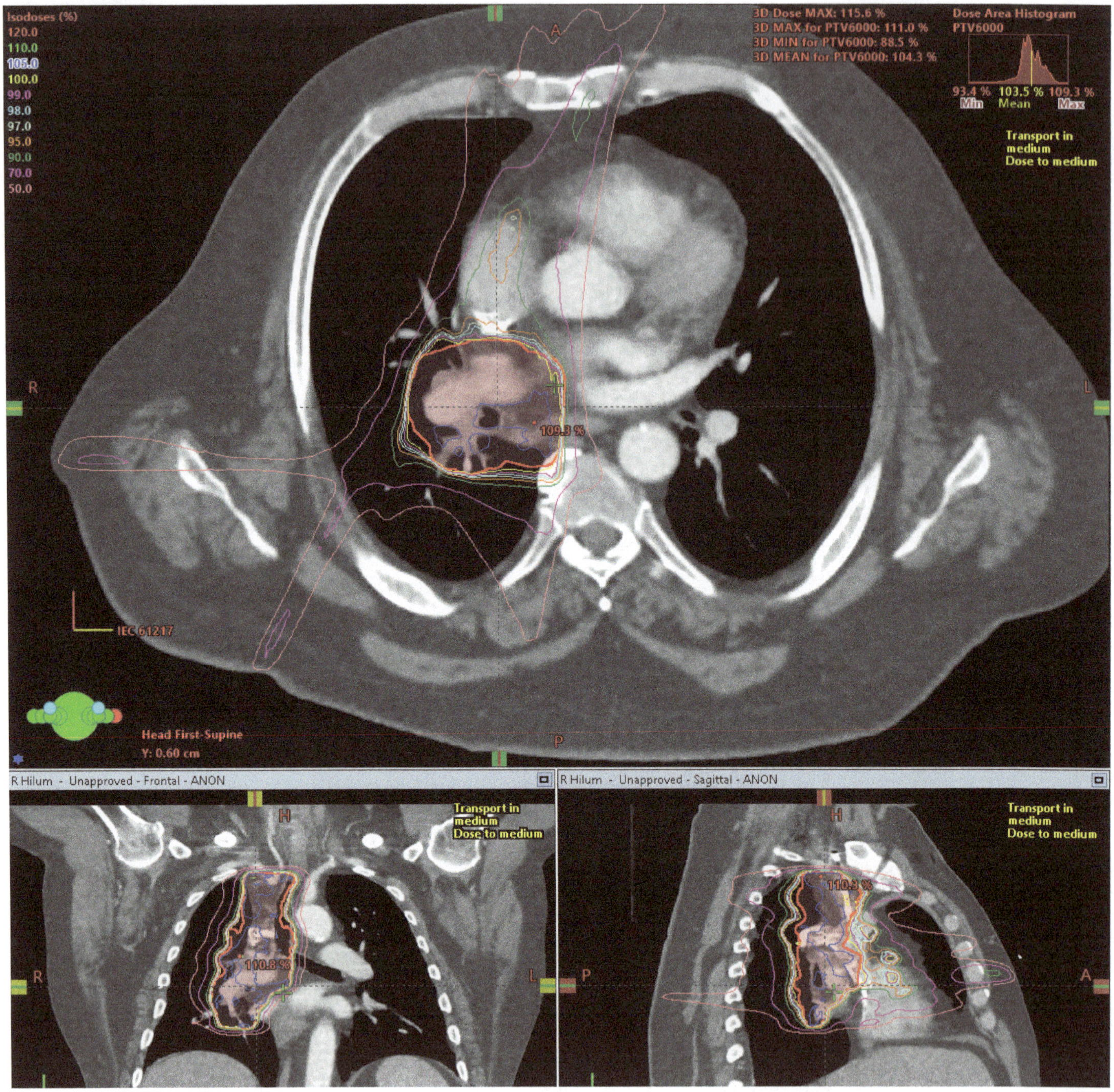

Figure 8.16 Final dose distribution.

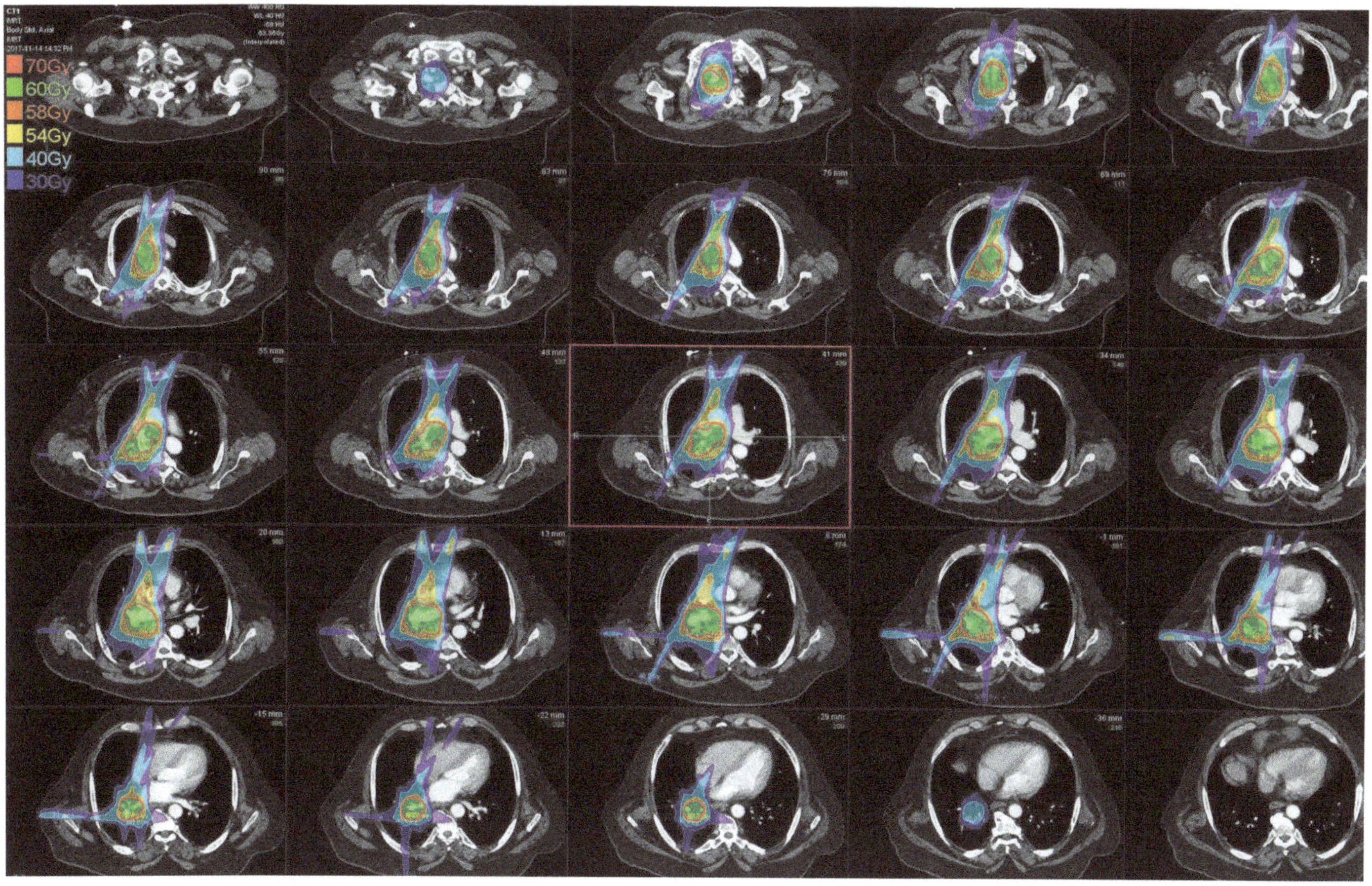

Figure 8.17 Final dose distribution on axial slices.

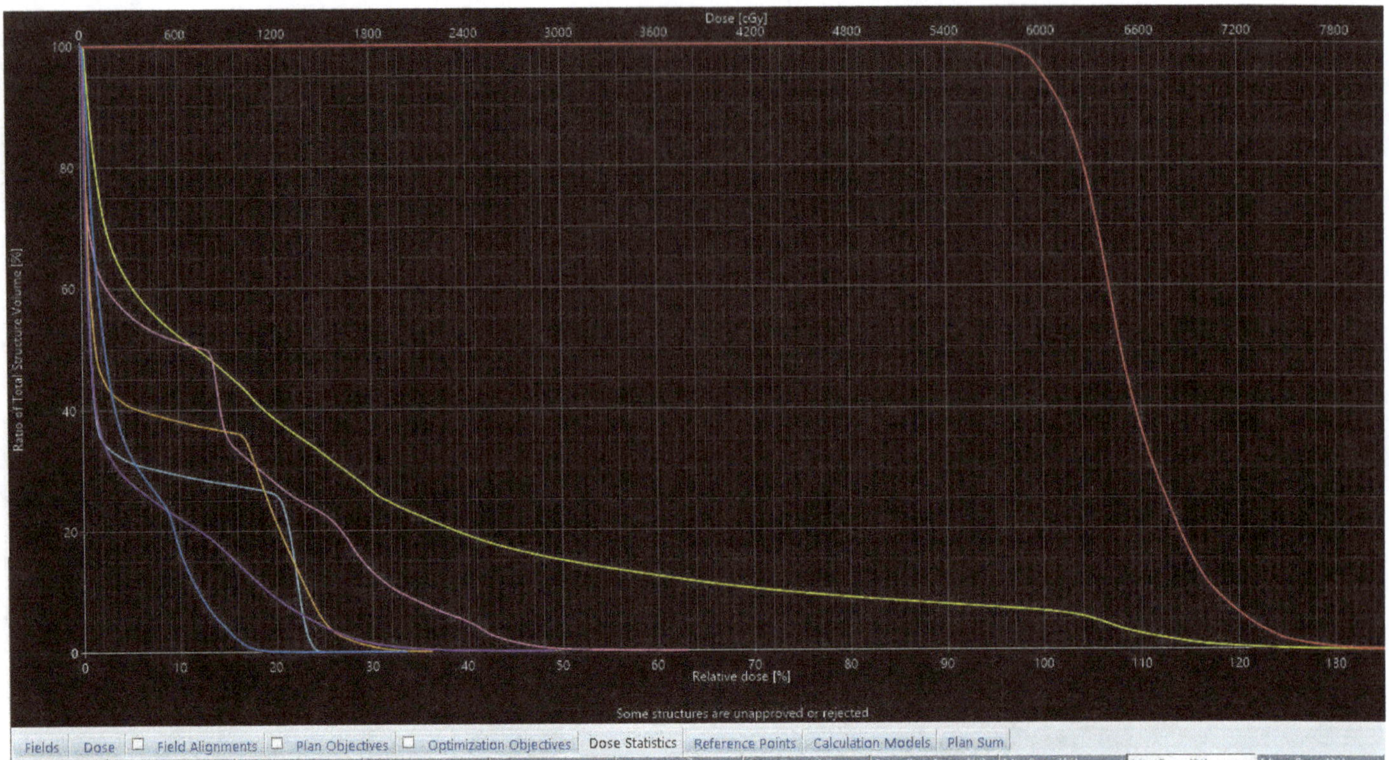

Show DVH	Structure	Approval Status	Plan	Course	Volume [cm³]	Dose Cover [%]	Sampling Cover [%]	Min Dose [%]	Max Dose [%]	Mean Dose [%]
☑	L Lung	Unapproved	L Lung SBRT	C1	1128.7	100.0	100.0	0.3	137.2	23.9
☑	Trachea	Unapproved	L Lung SBRT	C1	49.9	100.0	100.2	0.2	63.1	13.0
☑	PTV 6000	Unapproved	L Lung SBRT	C1	84.2	100.0	99.7	88.7	137.2	109.0
☑	Canal	Unapproved	L Lung SBRT	C1	41.6	100.0	99.7	0.2	24.9	6.9
☑	Esophagus	Unapproved	L Lung SBRT	C1	30.9	100.0	99.4	0.2	36.2	8.7
☑	Heart	Unapproved	L Lung SBRT	C1	724.6	100.0	99.8	0.2	49.1	5.4
☑	R Lung	Unapproved	L Lung SBRT	C1	1713.3	100.0	100.0	0.1	33.6	4.6

Figure 8.18 Final dose volume histogram.

ECLIPSE SAMPLES

Case Number eTA03: Thorax and Abdomen—Left Chest Wall Radiation Treatment

General Information

Brief Clinical History

89-year-old female diagnosed with recurrent carcinoma of the left chest wall after radical mastectomy. Inverse planning is being used to dosimetrically spare the left lung and heart.

Prescription

Left chest wall:
PTV 5040 = 180 cGy × 28 fractions

Simulation/Comments

The simulation was accomplished with a vacuum cushion placed under the patient's thorax in a wing board with the arms up.

Targets

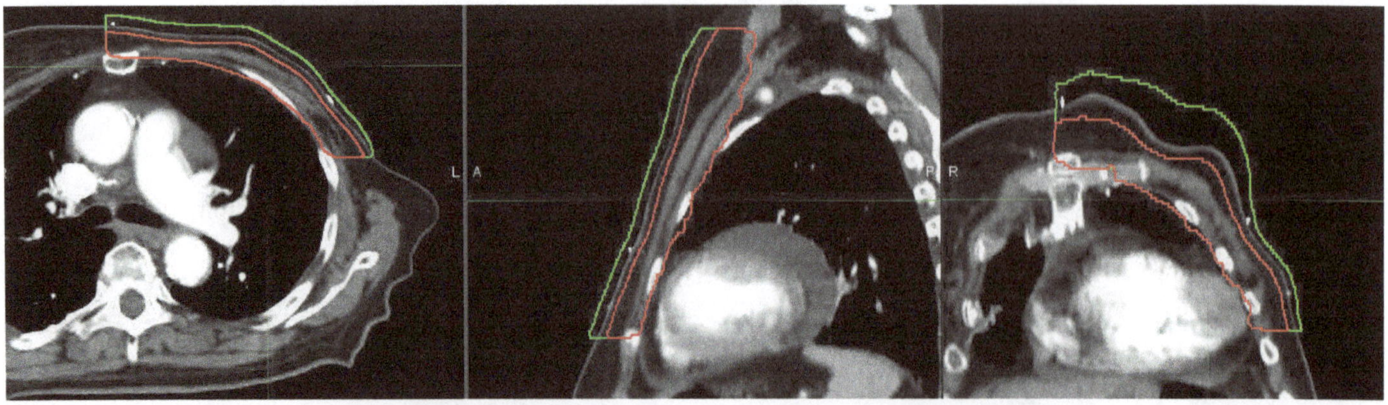

Figure 8.19 Target contours PTV 5040 (red) and optimization contour for PTV 5040 (green).

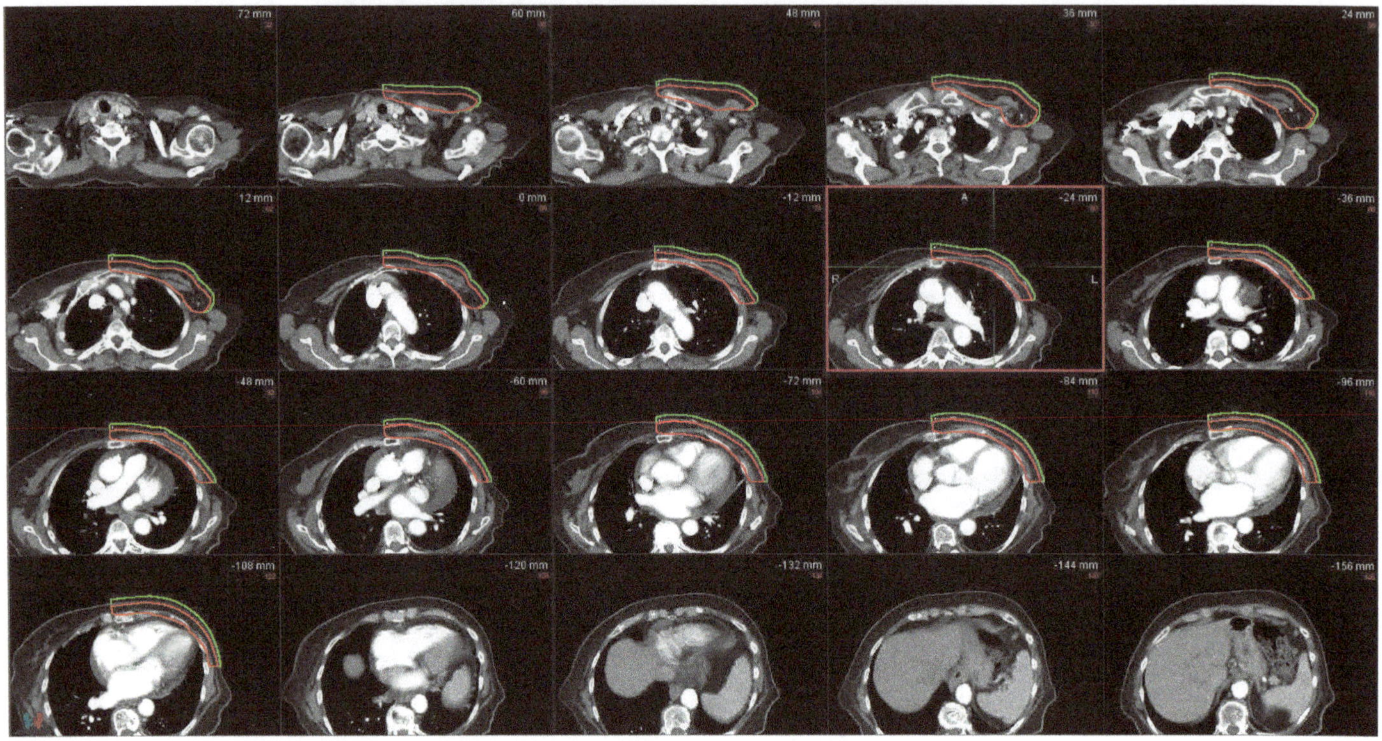

Figure 8.20 Target contours PTV 5040 (red) and optimization contour for PTV 5040 (green) in axial slices.

Figure 8.19 shows target contours PTV 5040 and its optimization contour. The green optimization contour includes a region of skin flash that expands 1 centimeter outside of the patient contour. The contrast seen in the dataset will not be administered during therapy, and departmental procedures must be followed to determine a suitable simulation dataset. One option is to contour the contrasted region and override this density to that of water in the TPS. Another option is that of setting up the patient for simulation and scan the patient twice: the first time without contrast and the second time with contrast. In this case, the first CT dataset is used for planning purposes.

Beam Arrangement

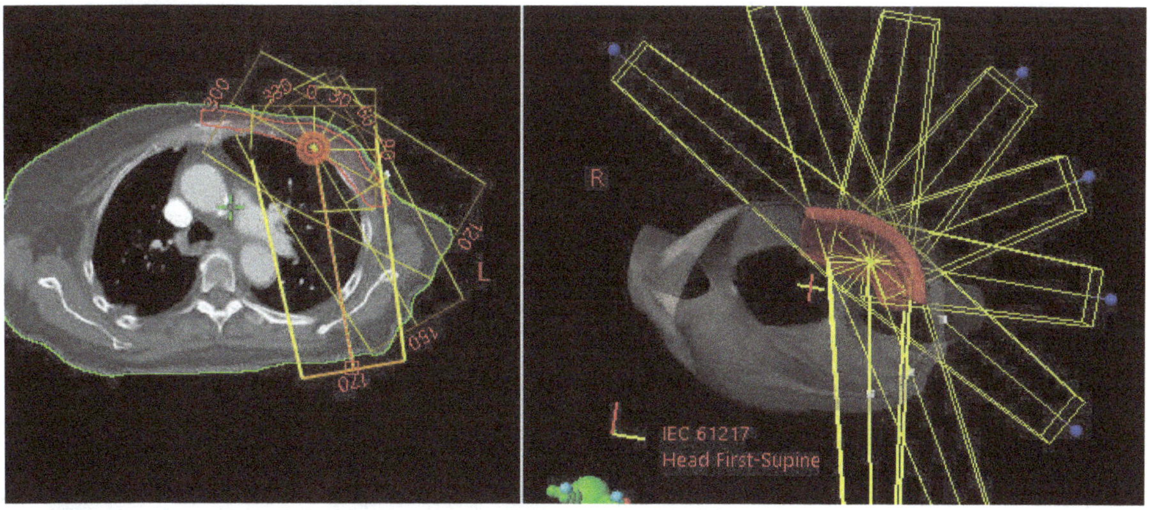

Figure 8.21 IMRT 9-field beam arrangement for case eTA03.

Gantry Angle	Couch	Collimator	Energy (MV)
300	0	0	6
330	0	0	6
0	0	0	6
30	0	0	6
60	0	0	6
90	0	0	6
120	0	0	6
150	0	0	6
170	0	0	6

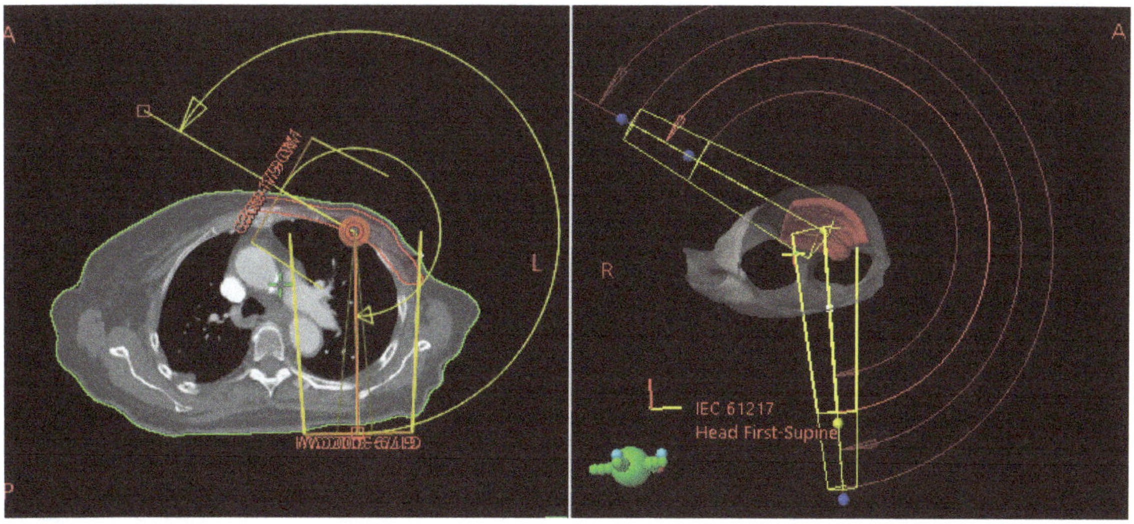

Figure 8.22 VMAT 4-arc beam arrangement.

Gantry Angle	Couch	Collimator	Energy (MV)
300–179 CW	0	30	6
179–300 CCW	0	330	6
300–179 CW 1 (split)	0	30	6
179–300 CCW 1 (split)	0	330	6

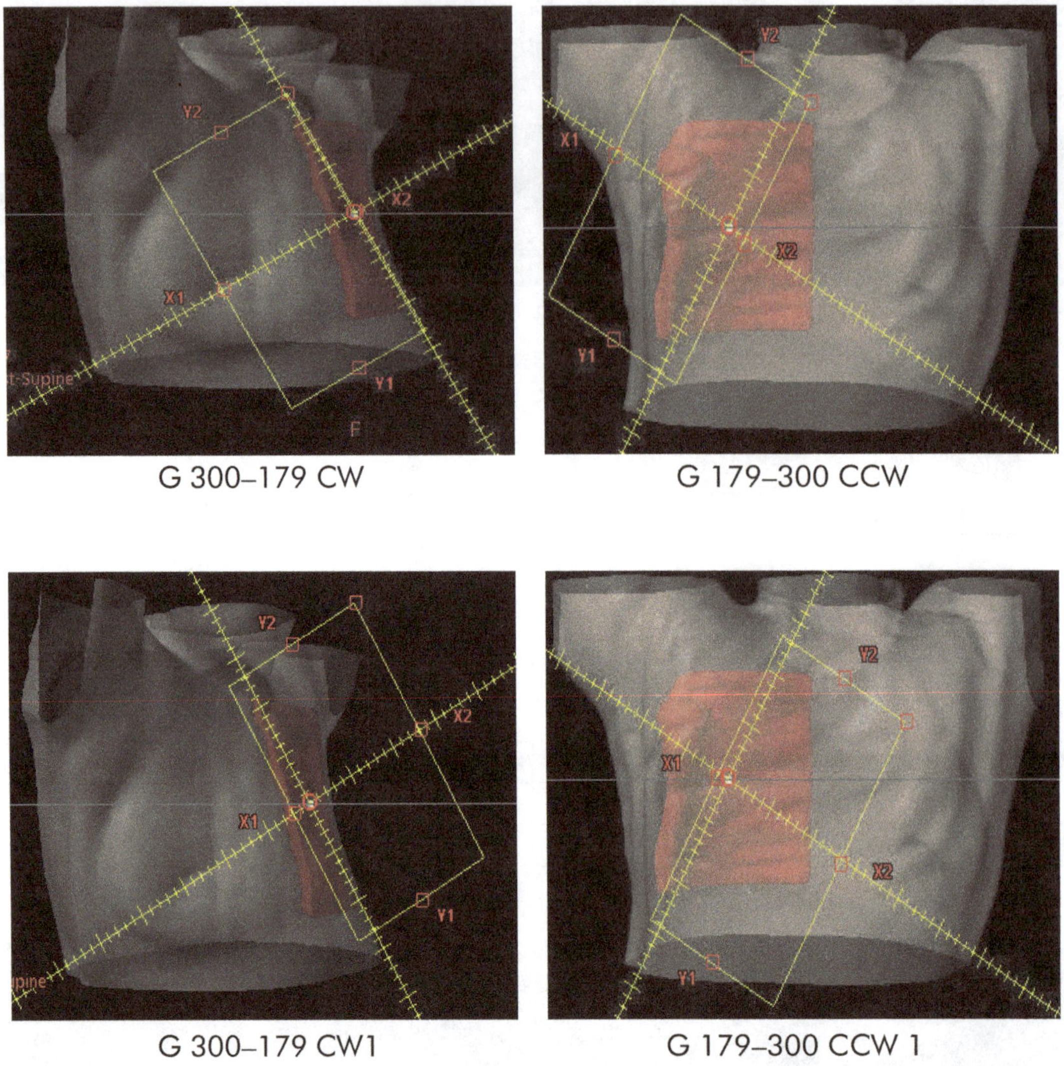

Figure 8.23 VMAT arc collimator rotations (jaw configuration not final and will be modified during optimization with tracking jaw).

Optimization

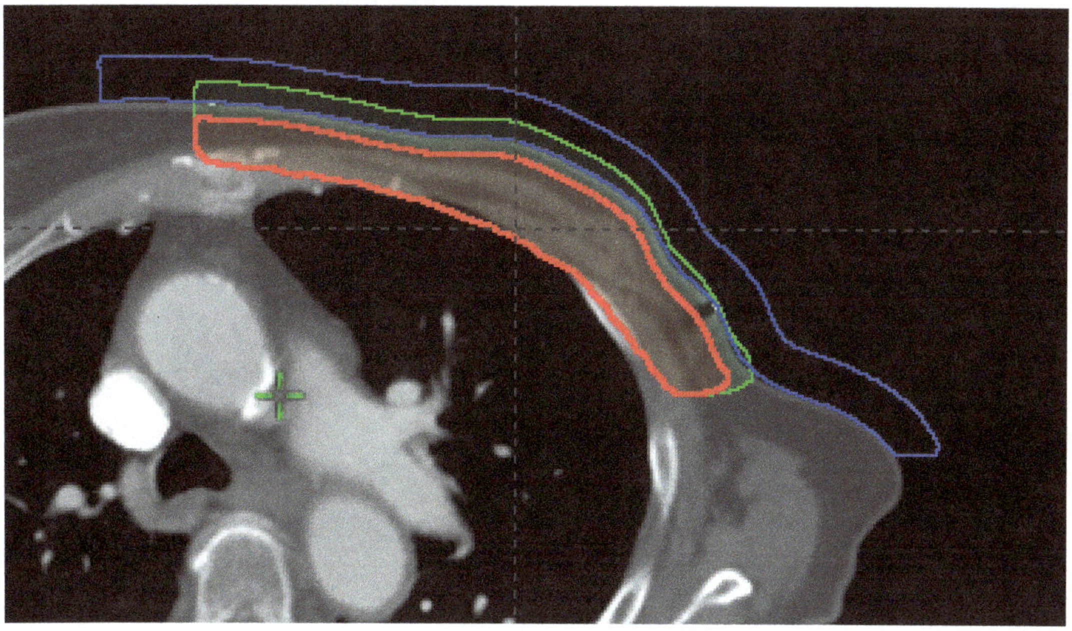

Figure 8.24 Target and optimization contours. PTV 5040 (red), Optimization PTV 5040 (green) and virtual bolus structure (blue). The bolus structure has a density of 0.4 g/cc or –600 HU in this case.

There are various approaches in the literature on the implementation of a flash region for a moving target using inverse planning techniques. The approach followed in this case can be summarized in the following steps:

(1) Add a virtual bolus structure.
(2) Optimize to the optimization PTV contour with the virtual bolus active.
(3) Override the virtual bolus density to true values (air) and recalculate the dose distribution.

If using IMRT, one can also make use of the "add flash" feature in the TPS. The authors of this workbook highly recommend that such techniques are properly researched, tested, and implemented before clinical use. The planner is encouraged to read the literature on implementing flash in breast or chest wall treatments with VMAT.

☑ Use Normal Tissue Objective Priority: 100 Define NTO Settings...

☑ BODY	Volume [cc]:	17292	Points:	429009	Resolution [mm]:	4.50
☑ Body1	Volume [cc]:	16130	Points:	309740	Resolution [mm]:	4.50
☑ Heart1	Volume [cc]:	1369	Points:	109546	Resolution [mm]:	3.00
Upper	Volume [%]:	0.0	Dose [cGy]:	4400.0	Priority:	160
Mean			Dose [cGy]:	700.0	Priority:	160
☑ L Lung2	Volume [cc]:	988	Points:	113904	Resolution [mm]:	3.00
Upper	Volume [%]:	5.0	Dose [cGy]:	5000.0	Priority:	160
Upper		20.0		2000.0		160
☑ **PTV 5040**	**Volume [cc]:**	**574**	**Points:**	**87662**	**Resolution [mm]:**	**3.00**
Upper	Volume [%]:	0.0	Dose [cGy]:	5400.0	Priority:	220
Lower	Volume [%]:	100.0	Dose [cGy]:	5040.0	Priority:	250
☑ PTV Optimization	Volume [cc]:	891	Points:	106137	Resolution [mm]:	3.00
Lower	Volume [%]:	95.0	Dose [cGy]:	5040.0	Priority:	160
☑ R Lung2	Volume [cc]:	1483	Points:	145621	Resolution [mm]:	3.00
☑ Trachea2	Volume [cc]:	29	Points:	11540	Resolution [mm]:	2.36

Figure 8.25 Initial optimization parameters.

☑ Use Normal Tissue Objective   Priority: 100   Define NTO Settings...

BODY		**Volume [cc]:**	**17292**	**Points:**	**429009**	**Resolution [mm]:**	**4.50**
Body1		Volume [cc]:	16130	Points:	309740	Resolution [mm]:	4.50
Heart1		Volume [cc]:	1369	Points:	109546	Resolution [mm]:	3.00
	Upper	Volume [%]:	0.0	Dose [cGy]:	4200.0	Priority:	160
	Mean			Dose [cGy]:	550.0	Priority:	160
L Lung2		Volume [cc]:	988	Points:	113904	Resolution [mm]:	3.00
	Upper	Volume [%]:	0.0	Dose [cGy]:	5000.0	Priority:	100
	Upper		16.0		2000.0		160
PTV 5040		Volume [cc]:	574	Points:	87662	Resolution [mm]:	3.00
	Upper	Volume [%]:	0.0	Dose [cGy]:	5300.0	Priority:	220
	Lower	Volume [%]:	100.0	Dose [cGy]:	5040.0	Priority:	250
PTV Optimization		Volume [cc]:	891	Points:	106137	Resolution [mm]:	3.00
	Upper	Volume [%]:	0.0	Dose [cGy]:	5200.0	Priority:	160
	Lower	Volume [%]:	95.0	Dose [cGy]:	5040.0	Priority:	160
R Lung2		Volume [cc]:	1483	Points:	145621	Resolution [mm]:	3.00
Trachea2		Volume [cc]:	29	Points:	11540	Resolution [mm]:	2.36

Figure 8.26 Final optimization parameters.

During the optimization process, the planner can reduce the dose to organs at risk, particularly the heart, lungs and the contralateral breast. A reduction of volume covered by high dose within the PTV can help decrease adverse skin reactions during treatment.

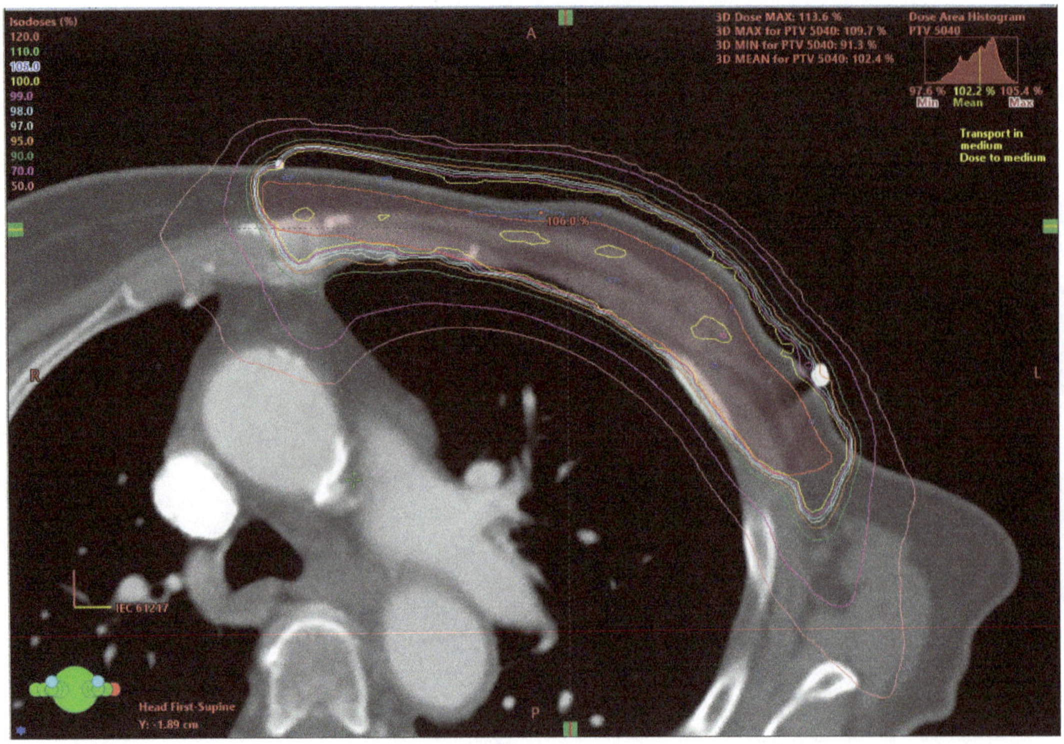

Figure 8.27 Dose distribution with virtual bolus.

Dose Distribution

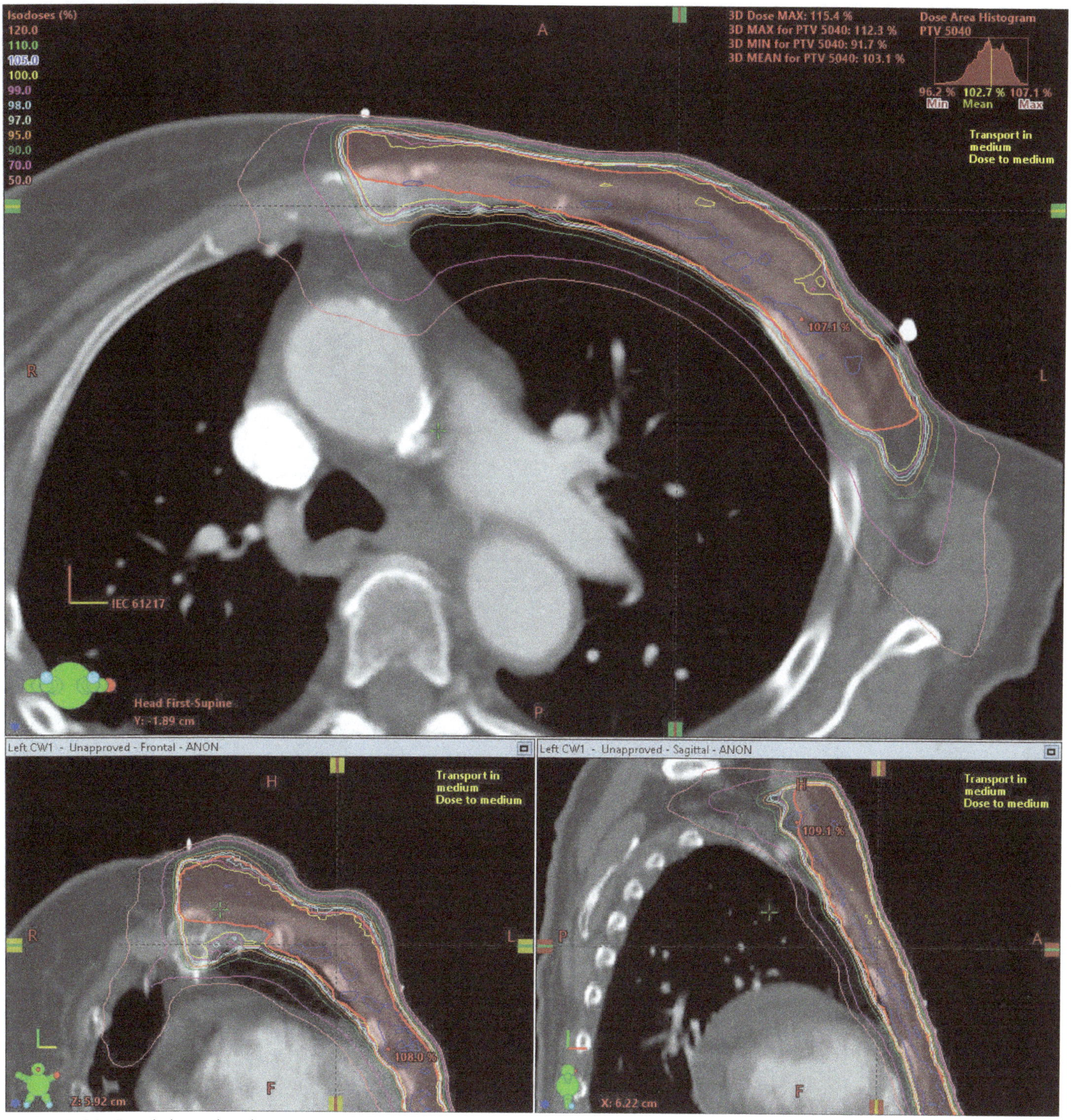

Figure 8.28 Final dose distribution.

For the final dose distribution, the planner recalculates the plan without the bolus structure but with the same beam parameters obtained from the last optimization. Re-optimizing the plan at this point is not recommended since it can delete the skin flash region.

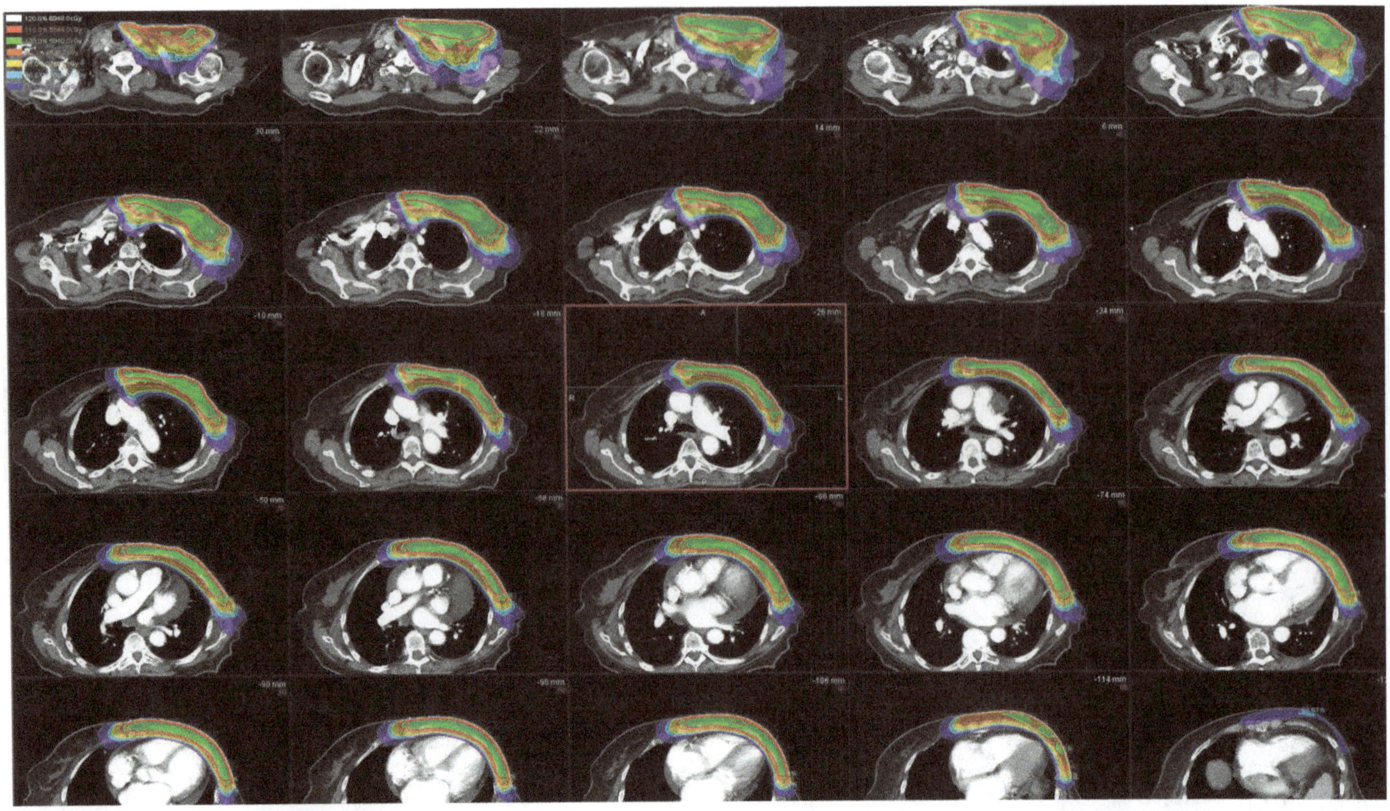

Figure 8.29 Final dose distribution in axial slices.

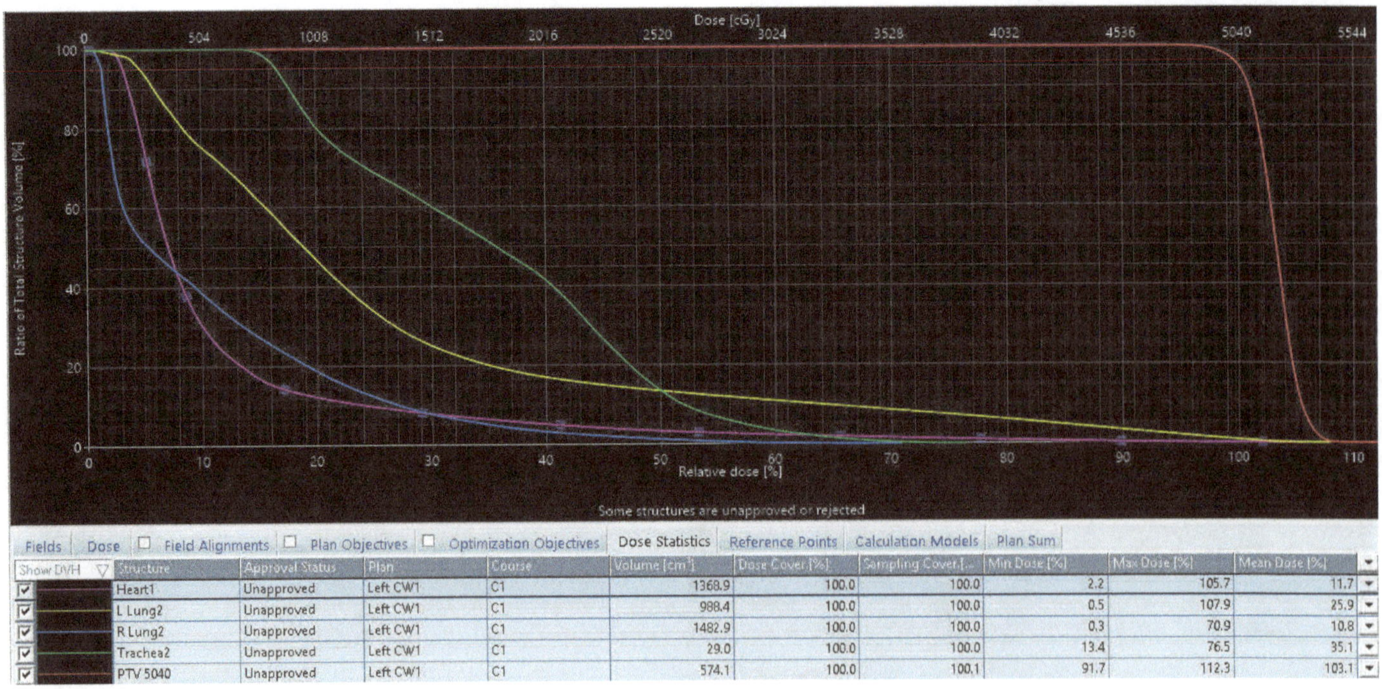

Structure	Approval Status	Plan	Course	Volume [cm³]	Dose Cover.[%]	Sampling Cover.[...	Min Dose [%]	Max Dose [%]	Mean Dose [%]
Heart1	Unapproved	Left CW1	C1	1368.9	100.0	100.0	2.2	105.7	11.7
L Lung2	Unapproved	Left CW1	C1	988.4	100.0	100.0	0.5	107.9	25.9
R Lung2	Unapproved	Left CW1	C1	1482.9	100.0	100.0	0.3	70.9	10.8
Trachea2	Unapproved	Left CW1	C1	29.0	100.0	100.0	13.4	76.5	35.1
PTV 5040	Unapproved	Left CW1	C1	574.1	100.0	100.1	91.7	112.3	103.1

Figure 8.30 Final DVH distribution.

Case Number eHN01: Head and Neck—Neck Lymphoma Radiation Treatment

General Information

Brief Clinical History

69-year-old female diagnosed with diffuse large B cell lymphoma. Patient was prescribed steroids and radiation therapy.

Prescription

Bilateral neck nodes:
PTV 4500 = 180 cGy × 25 fractions

Simulation/Comments

The simulation was accomplished with the patient on head and neck index/immobilization board, thermoplastic mask, moldable pillow, and headrest. Fusion of images was performed with a recent PET/CT for evaluation of the extent of the lesion and targeting by the radiation oncologist.

Targets

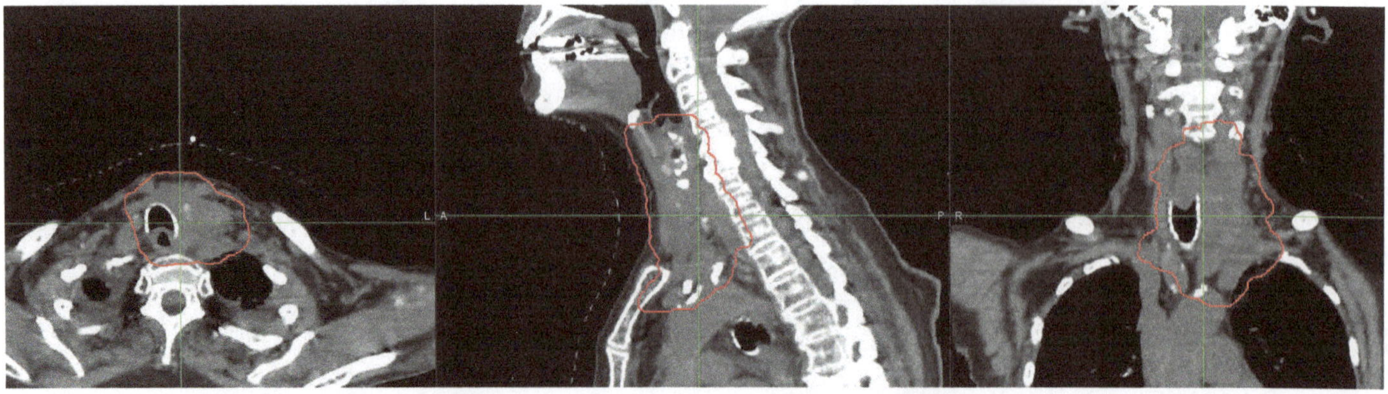

Figure 8.31 Target contour PTV 4500 (red).

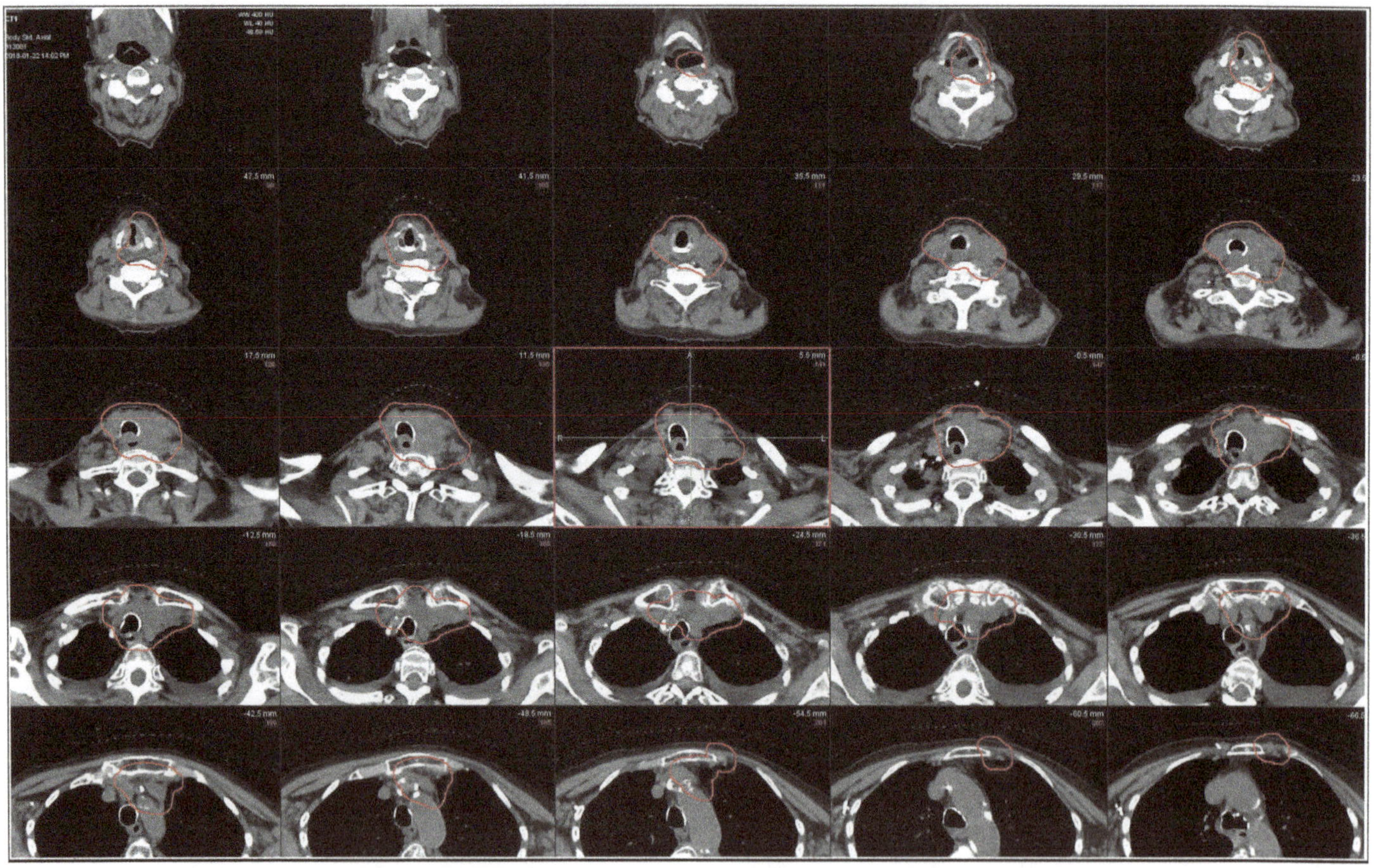

Figure 8.32 Target contour PTV 4500 (red) in axial slices.

Beam Arrangement

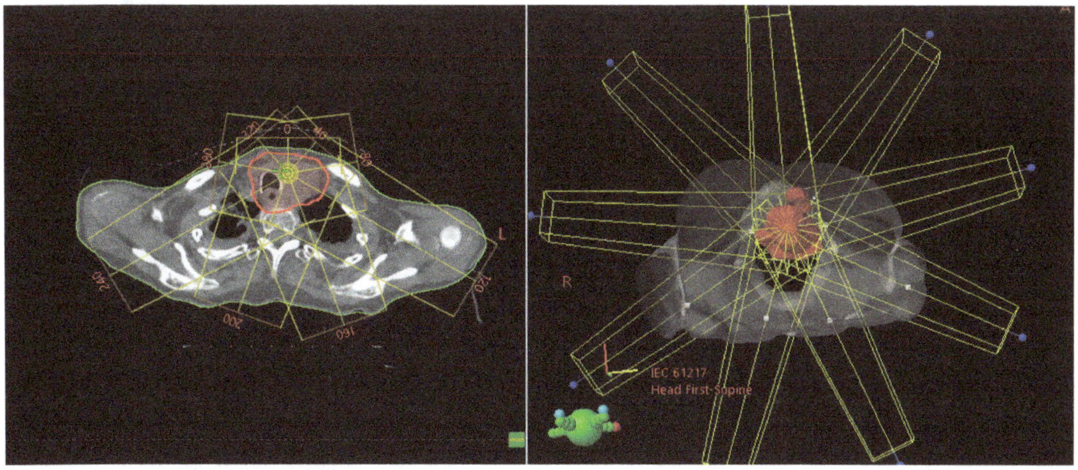

Figure 8.33 IMRT 9-field beam arrangement.

Gantry Angle	Couch	Collimator	Energy (MV)
200	0	0	6
240	0	0	6
280	0	0	6
320	0	0	6
0	0	0	6
40	0	0	6
80	0	0	6
120	0	0	6
160	0	0	6

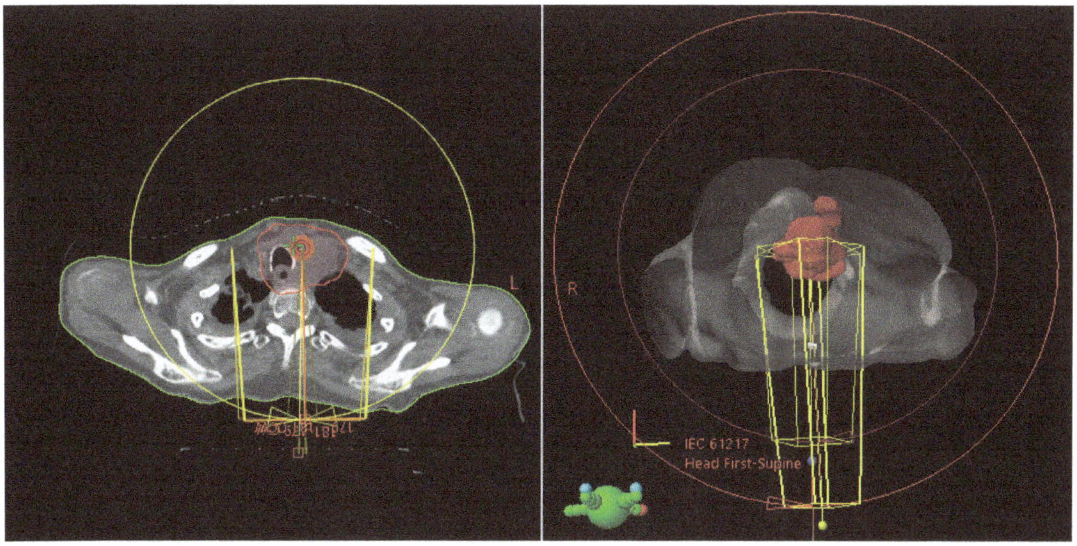

Figure 8.34 VMAT 2-arc beam arrangement.

Gantry Angle	Couch	Collimator	Energy (MV)
181–179 CCW	0	30	6
179–181 CW	0	330	6

Optimization

☑ Use Normal Tissue Objective — Priority: 100 — Define NTO Settings...

☑ **BODY**		**Volume [cc]:**	**11632**	**Points:**	**312427**	**Resolution [mm]:**	**4.50**
☑ Body-5mm		Volume [cc]:	10207	Points:	237136	Resolution [mm]:	4.50
☑ Canal Avoid		Volume [cc]:	488	Points:	57286	Resolution [mm]:	3.00
	Mean			Dose [cGy]:	1000.0	Priority:	160
☑ Canal1		Volume [cc]:	57	Points:	16059	Resolution [mm]:	2.95
	Upper	Volume [%]:	0.0	Dose [cGy]:	3000.0	Priority:	160
☑ Esophagus2		Volume [cc]:	29	Points:	14272	Resolution [mm]:	2.37
	Mean			Dose [cGy]:	1700.0	Priority:	160
☑ HighZ		Volume [cc]:	7	Points:	14973	Resolution [mm]:	1.45
☑ Lt Partial Lung		Volume [cc]:	786	Points:	80700	Resolution [mm]:	3.00
☑ PTV4500		Volume [cc]:	337	Points:	41825	Resolution [mm]:	3.00
	Upper	Volume [%]:	0.0	Dose [cGy]:	4700.0	Priority:	220
	Lower	Volume [%]:	100.0	Dose [cGy]:	4500.0	Priority:	250
☑ Revised GTV1		Volume [cc]:	191	Points:	29685	Resolution [mm]:	3.00
☑ Rt Partial Lung		Volume [cc]:	965	Points:	95045	Resolution [mm]:	3.00
☑ Trachea3		Volume [cc]:	28	Points:	15428	Resolution [mm]:	2.34

Figure 8.35 Initial objectives. Additional contours for optimization can be made (expansions on organs at risk such as the spinal canal).

During the optimization, the planner must make sure to not place regions of high dose on organs at risk.

☑ Use Normal Tissue Objective — Priority: 100 — Define NTO Settings...

☑ **BODY**		**Volume [cc]:**	**11632**	**Points:**	**312427**	**Resolution [mm]:**	**4.50**
☑ Body-5mm		Volume [cc]:	10207	Points:	237136	Resolution [mm]:	4.50
☑ Canal Avoid		Volume [cc]:	488	Points:	57286	Resolution [mm]:	3.00
	Mean			Dose [cGy]:	550.0	Priority:	160
☑ Canal1		Volume [cc]:	57	Points:	16059	Resolution [mm]:	2.95
	Upper	Volume [%]:	0.0	Dose [cGy]:	2200.0	Priority:	160
☑ Esophagus2		Volume [cc]:	29	Points:	14272	Resolution [mm]:	2.37
	Mean			Dose [cGy]:	1700.0	Priority:	160
☑ HighZ		Volume [cc]:	7	Points:	14973	Resolution [mm]:	1.45
☑ Lt Partial Lung		Volume [cc]:	786	Points:	80700	Resolution [mm]:	3.00
	Upper	Volume [%]:	50.0	Dose [cGy]:	900.0	Priority:	160
☑ PTV4500		Volume [cc]:	337	Points:	41825	Resolution [mm]:	3.00
	Upper	Volume [%]:	0.0	Dose [cGy]:	4700.0	Priority:	220
	Lower	Volume [%]:	100.0	Dose [cGy]:	4500.0	Priority:	250
☑ Revised GTV1		Volume [cc]:	191	Points:	29685	Resolution [mm]:	3.00
☑ Rt Partial Lung		Volume [cc]:	965	Points:	95045	Resolution [mm]:	3.00
	Upper	Volume [%]:	24.0	Dose [cGy]:	900.0	Priority:	160
☑ Trachea3		Volume [cc]:	28	Points:	15428	Resolution [mm]:	2.34

Figure 8.36 Final optimization objectives.

Dose Distribution

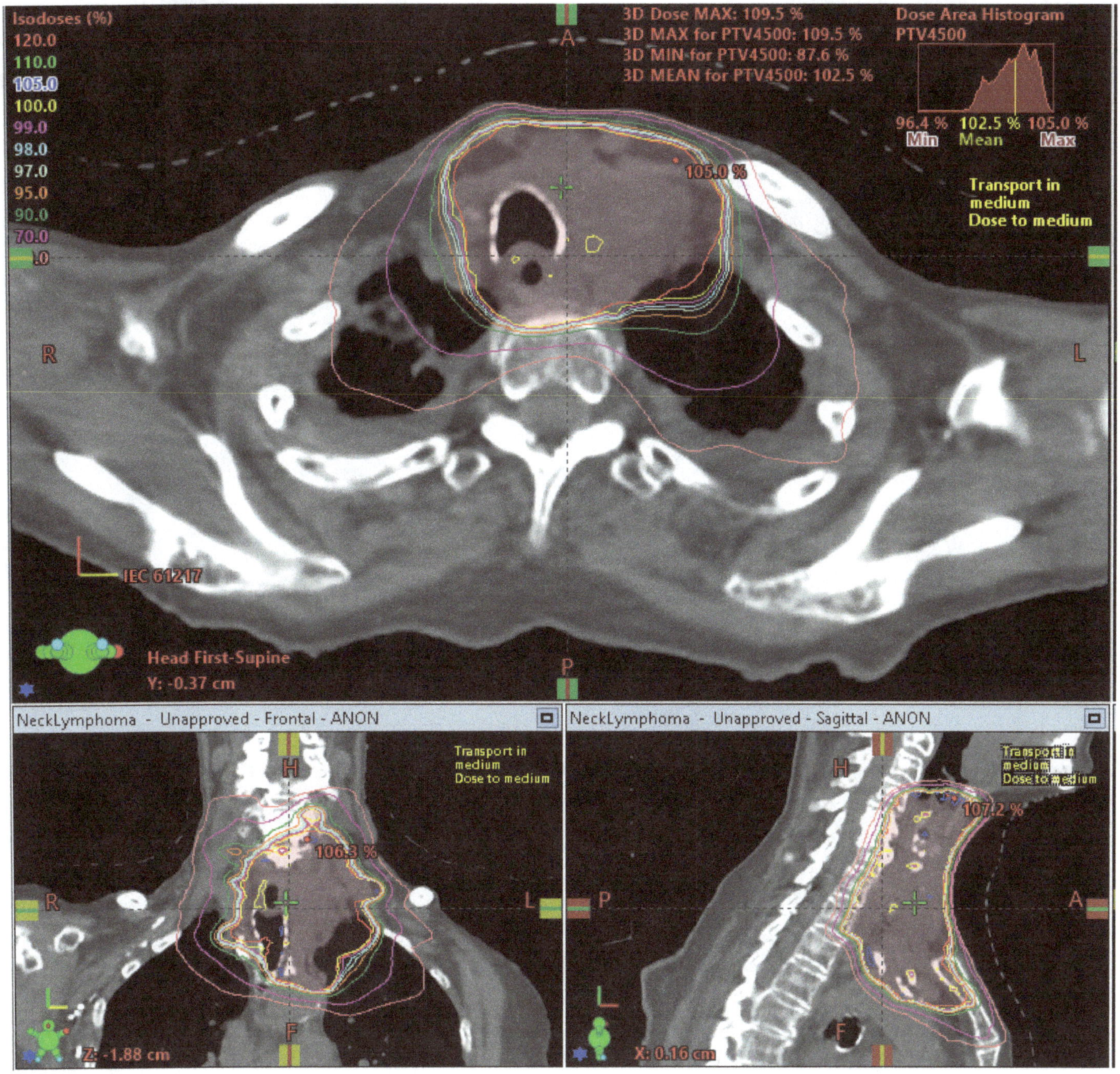

Figure 8.37 Final dose distribution.

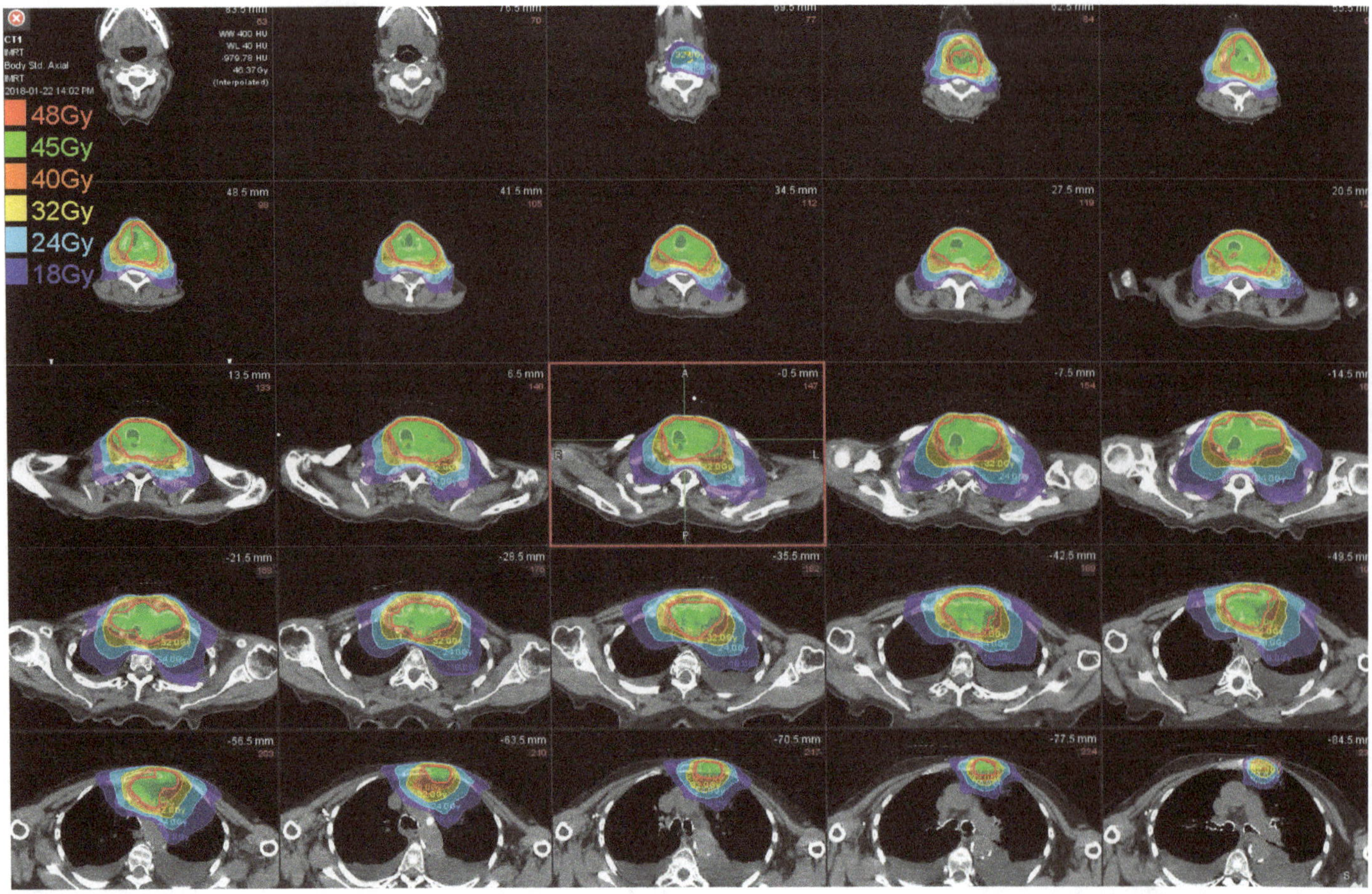

Figure 8.38 Final dose distribution on axial slices.

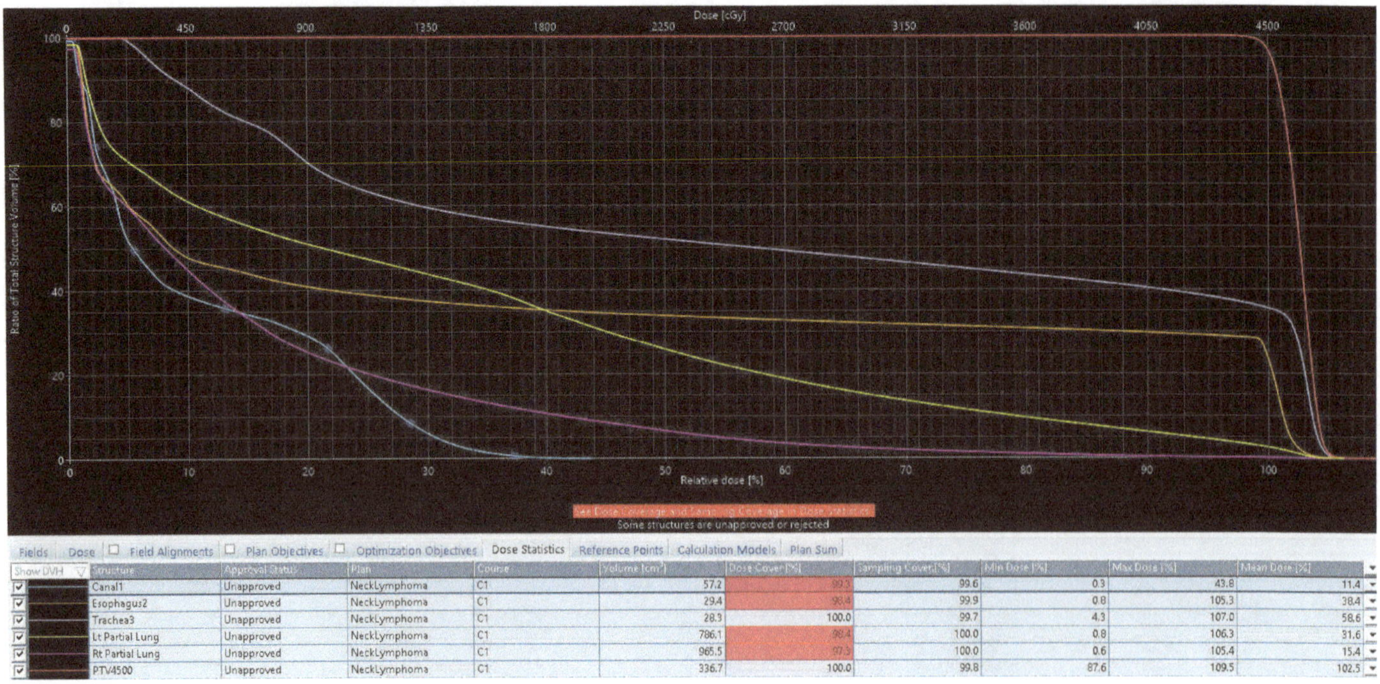

Figure 8.39 Final dose volume histogram

Case Number eHN02: Head and Neck—Bilateral Neck–Larynx Radiation Treatment

General Information

Brief Clinical History

55-year-old female diagnosed with invasive squamous cell carcinoma of the supraglottic larynx. Radiation therapy was prescribed after a total laryngectomy and bilateral neck dissection.

Prescription

Bilateral neck–larynx:
PTV 5000 = 200 cGy × 25 fractions

Simulation/Comments

The simulation was accomplished with the patient on a head-and-neck indexed/immobilization board, thermoplastic mask, moldable pillow and headrest. Fusion of images was performed with a recent PET/CT for evaluation of the extent of the lesion.

Targets

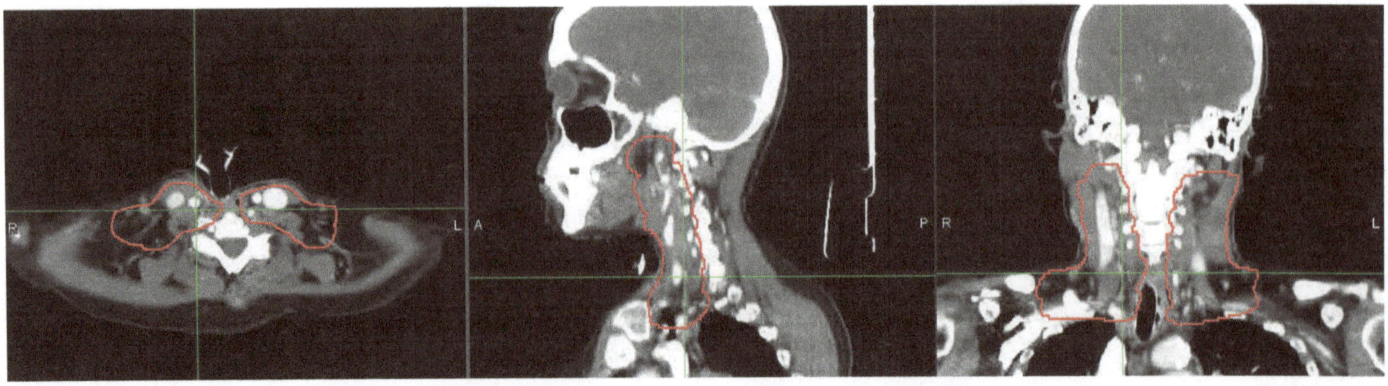

Figure 8.40 Target volume PTV 5000 (red). Notice the areas of intravenous contrast. Appropriate density correction was used for planning.

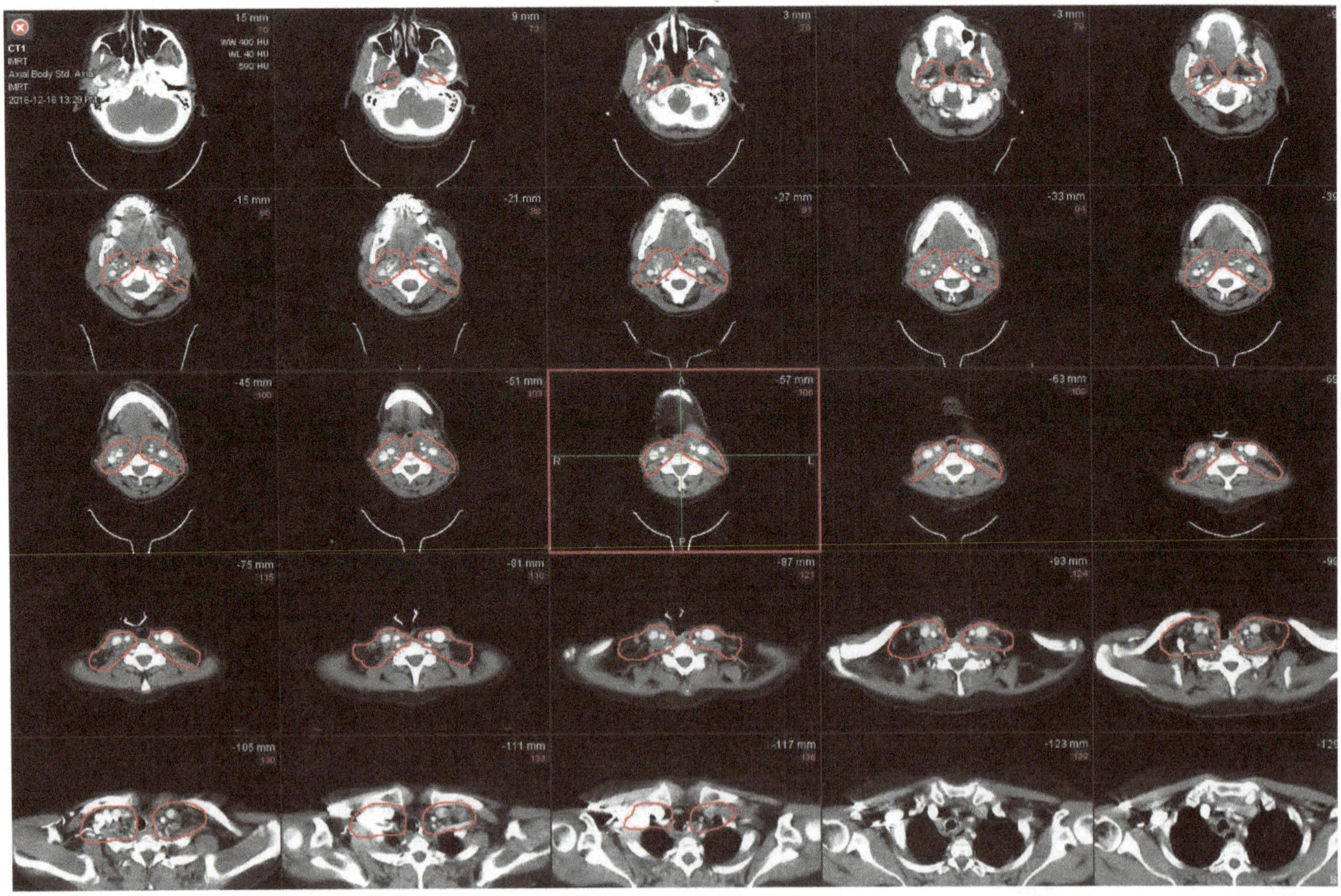

Figure 8.41 Target volume PTV 5000 (red) in axial slices. Notice the areas of intravenous contrast. Appropriate density correction was used for planning.

Beam Arrangement

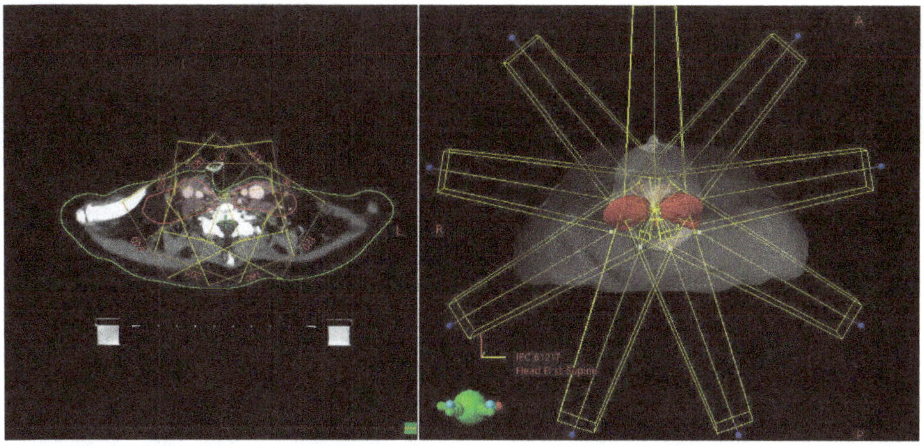

Figure 8.42 Nine-field beam arrangement.

Gantry Angle	Couch	Collimator	Energy (MV)
200	0	0	6
240	0	0	6
280	0	0	6
320	0	0	6
0	0	0	6
40	0	0	6
80	0	0	6
120	0	0	6
160	0	0	6

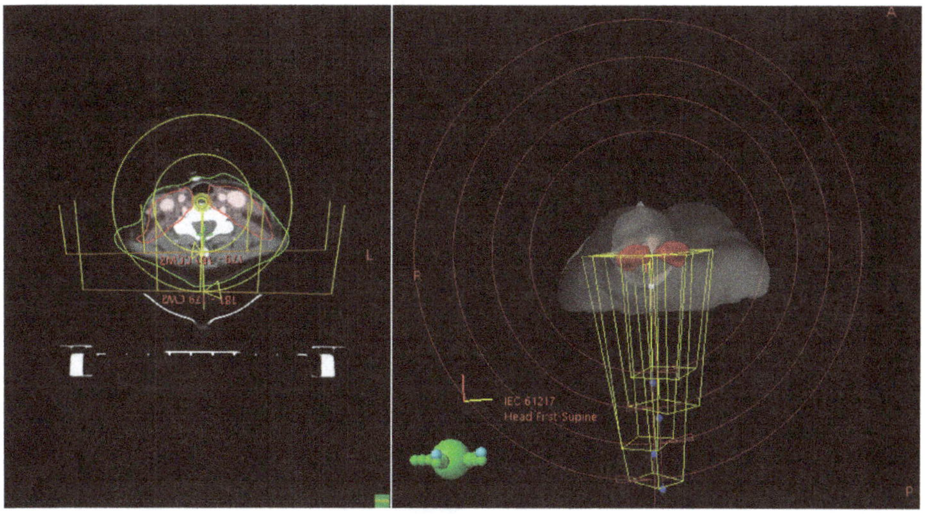

Figure 8.43 VMAT 4-arc beam arrangement. Note that in most cases a two-arc arrangement is sufficient.

Gantry Angle	Couch	Collimator	Energy (MV)
181–179 CW	0	30	6
179–181 CCW	0	330	6
181–179 CW1 (split)	0	30	6
179–181 CCW (split)	0	330	6

Optimization

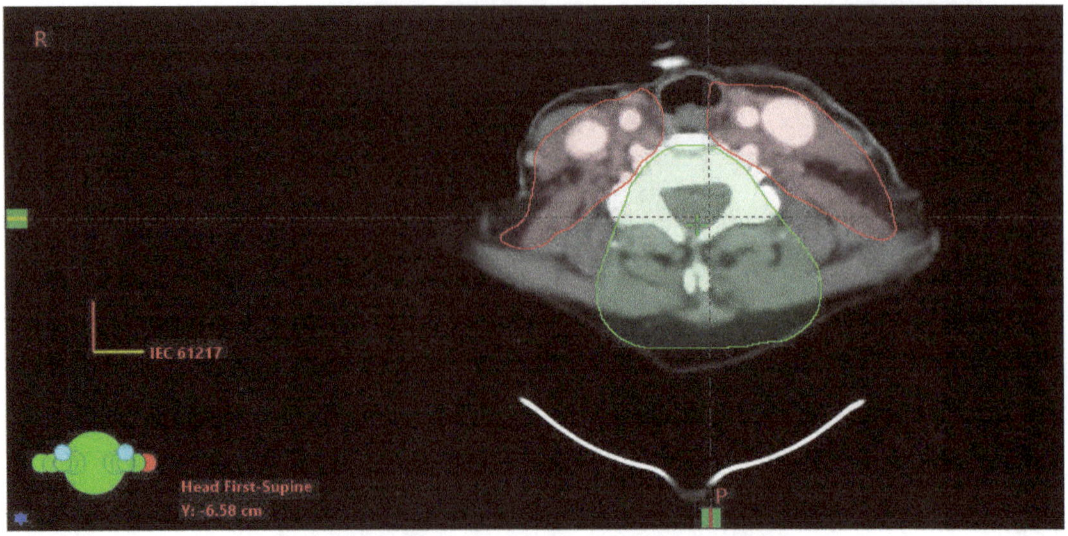

Figure 8.44 Additional optimization contours for neck (green) and target contour (red).

Figure 8.44 shows additional optimization structures. In this case, a posterior neck contour was created to avoid considerable irradiation to the area.

☑ Use Normal Tissue Objective | Priority: 100 | Define NTO Settings...

BODY		**Volume [cc]:**	**9357**	**Points:**	**307753**	**Resolution [mm]:**	**4.50**
Canal2		Volume [cc]:	68	Points:	16200	Resolution [mm]:	3.00
	Upper	Volume [%]:	0.0	Dose [cGy]:	2400.0	Priority:	160
	Mean			Dose [cGy]:	800.0	Priority:	160
Esophagus3		Volume [cc]:	14	Points:	14552	Resolution [mm]:	1.84
L Parotid		Volume [cc]:	17	Points:	10991	Resolution [mm]:	1.98
	Mean			Dose [cGy]:	2200.0	Priority:	160
L Submandibular		Volume [cc]:	9	Points:	10253	Resolution [mm]:	1.62
	Mean			Dose [cGy]:	3800.0	Priority:	160
Mandible		Volume [cc]:	71	Points:	17802	Resolution [mm]:	3.00
Oral Cavity		Volume [cc]:	99	Points:	14564	Resolution [mm]:	3.00
	Mean			Dose [cGy]:	4000.0	Priority:	160
Oropharynx		Volume [cc]:	6	Points:	9924	Resolution [mm]:	1.43
Post Avoid		Volume [cc]:	786	Points:	81071	Resolution [mm]:	3.00
	Mean			Dose [cGy]:	1000.0	Priority:	160
PTV 5000		Volume [cc]:	400	Points:	56747	Resolution [mm]:	3.00
	Upper	Volume [%]:	0.0	Dose [cGy]:	5200.0	Priority:	220
	Lower	Volume [%]:	100.0	Dose [cGy]:	5000.0	Priority:	250
R Parotid		Volume [cc]:	18	Points:	10706	Resolution [mm]:	2.03
	Mean			Dose [cGy]:	2200.0	Priority:	160
R Submandibular		Volume [cc]:	10	Points:	9918	Resolution [mm]:	1.67
	Mean			Dose [cGy]:	3800.0	Priority:	160
Trachea4		Volume [cc]:	24	Points:	12407	Resolution [mm]:	2.22

Figure 8.45 Initial optimization parameters.

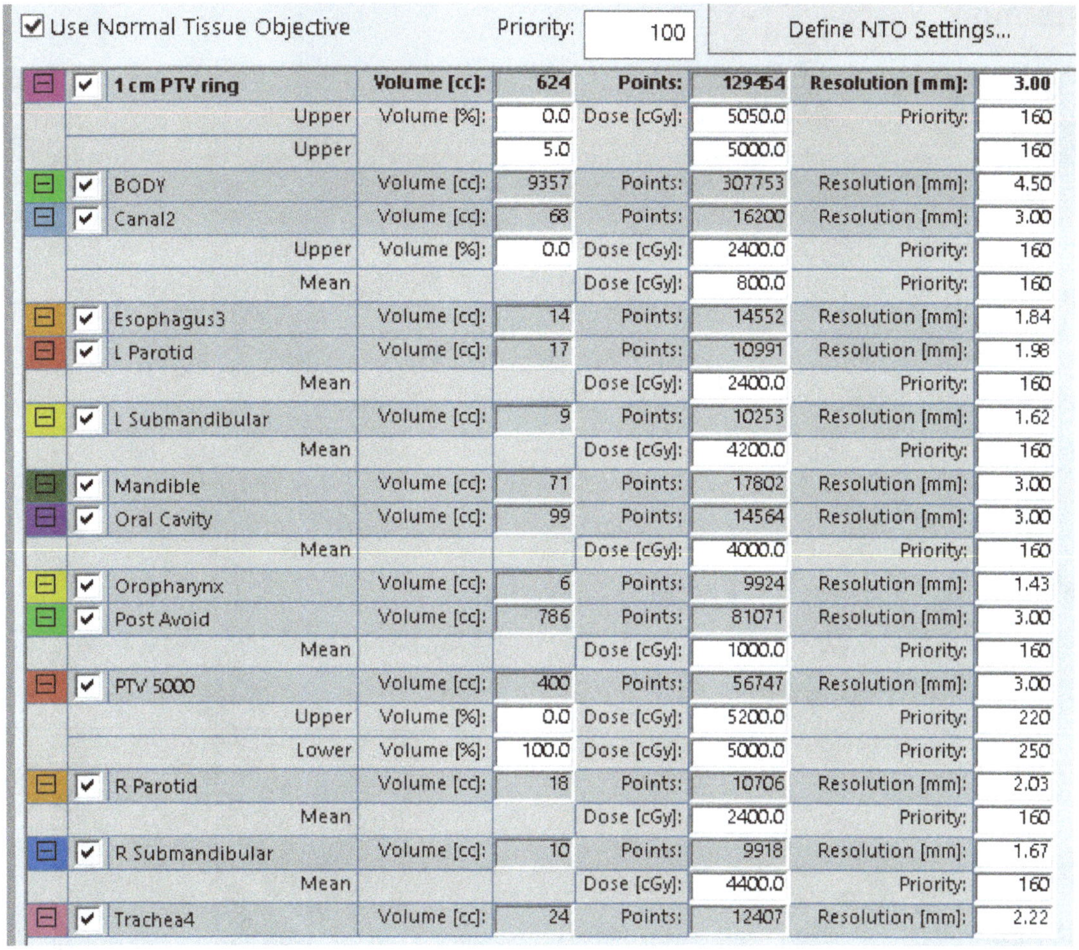

Use Normal Tissue Objective — Priority: 100 — Define NTO Settings...

1 cm PTV ring		**Volume [cc]:**	**624**	**Points:**	**129454**	**Resolution [mm]:**	**3.00**
	Upper	Volume [%]:	0.0	Dose [cGy]:	5050.0	Priority:	160
	Upper		5.0		5000.0		160
BODY		Volume [cc]:	9357	Points:	307753	Resolution [mm]:	4.50
Canal2		Volume [cc]:	68	Points:	16200	Resolution [mm]:	3.00
	Upper	Volume [%]:	0.0	Dose [cGy]:	2400.0	Priority:	160
	Mean			Dose [cGy]:	800.0	Priority:	160
Esophagus3		Volume [cc]:	14	Points:	14552	Resolution [mm]:	1.84
L Parotid		Volume [cc]:	17	Points:	10991	Resolution [mm]:	1.98
	Mean			Dose [cGy]:	2400.0	Priority:	160
L Submandibular		Volume [cc]:	9	Points:	10253	Resolution [mm]:	1.62
	Mean			Dose [cGy]:	4200.0	Priority:	160
Mandible		Volume [cc]:	71	Points:	17802	Resolution [mm]:	3.00
Oral Cavity		Volume [cc]:	99	Points:	14564	Resolution [mm]:	3.00
	Mean			Dose [cGy]:	4000.0	Priority:	160
Oropharynx		Volume [cc]:	6	Points:	9924	Resolution [mm]:	1.43
Post Avoid		Volume [cc]:	786	Points:	81071	Resolution [mm]:	3.00
	Mean			Dose [cGy]:	1000.0	Priority:	160
PTV 5000		Volume [cc]:	400	Points:	56747	Resolution [mm]:	3.00
	Upper	Volume [%]:	0.0	Dose [cGy]:	5200.0	Priority:	220
	Lower	Volume [%]:	100.0	Dose [cGy]:	5000.0	Priority:	250
R Parotid		Volume [cc]:	18	Points:	10706	Resolution [mm]:	2.03
	Mean			Dose [cGy]:	2400.0	Priority:	160
R Submandibular		Volume [cc]:	10	Points:	9918	Resolution [mm]:	1.67
	Mean			Dose [cGy]:	4400.0	Priority:	160
Trachea4		Volume [cc]:	24	Points:	12407	Resolution [mm]:	2.22

Figure 8.46 Final optimization parameters.

During optimization, the user can add optimization contours and optimization objectives and they become fit. The addition of a PTV ring can be useful to make the dose conformal, especially in areas where the target is divided to avoid an organ at risk.

Dose Distribution

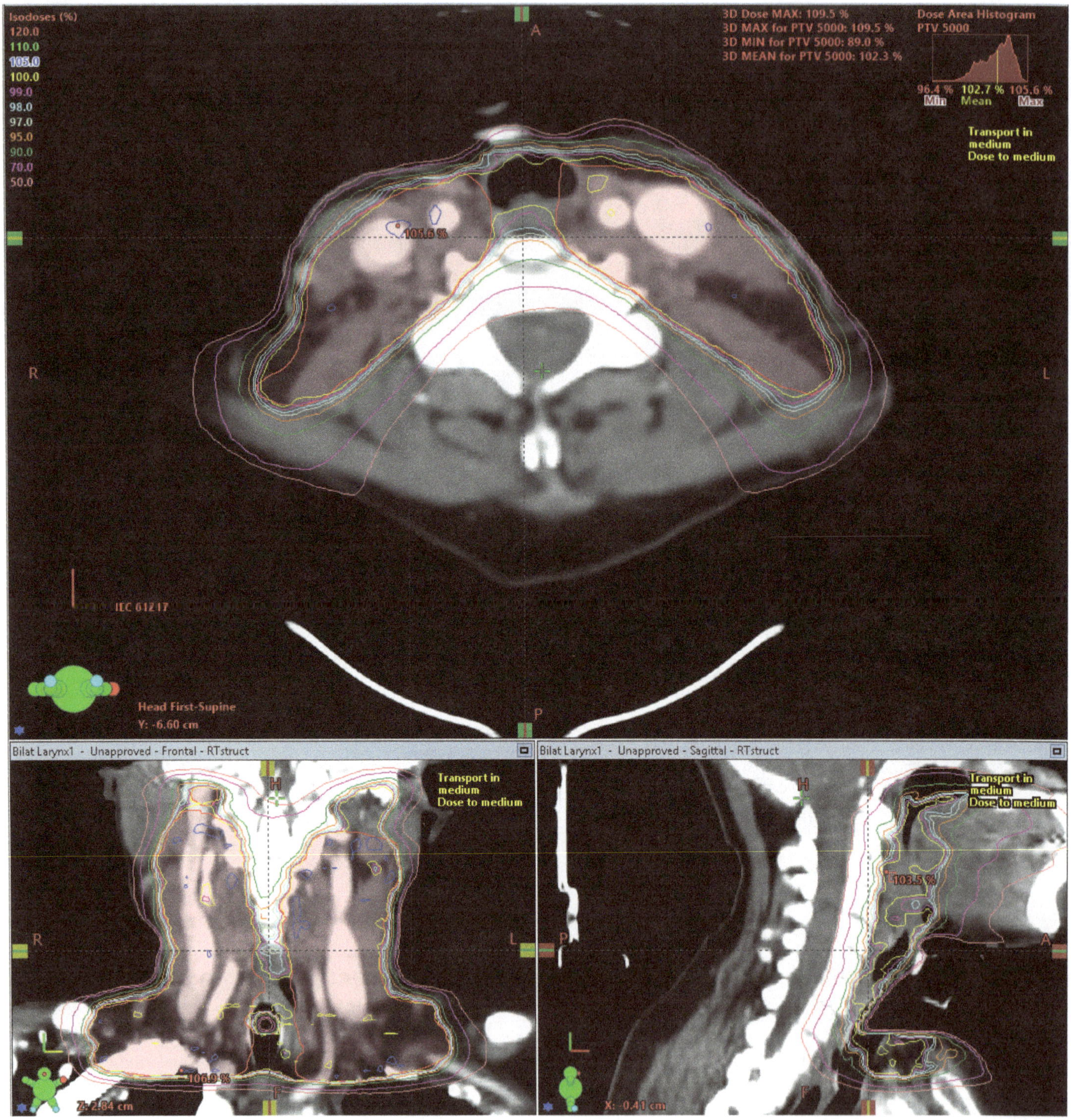

Figure 8.47 Final dose distribution.

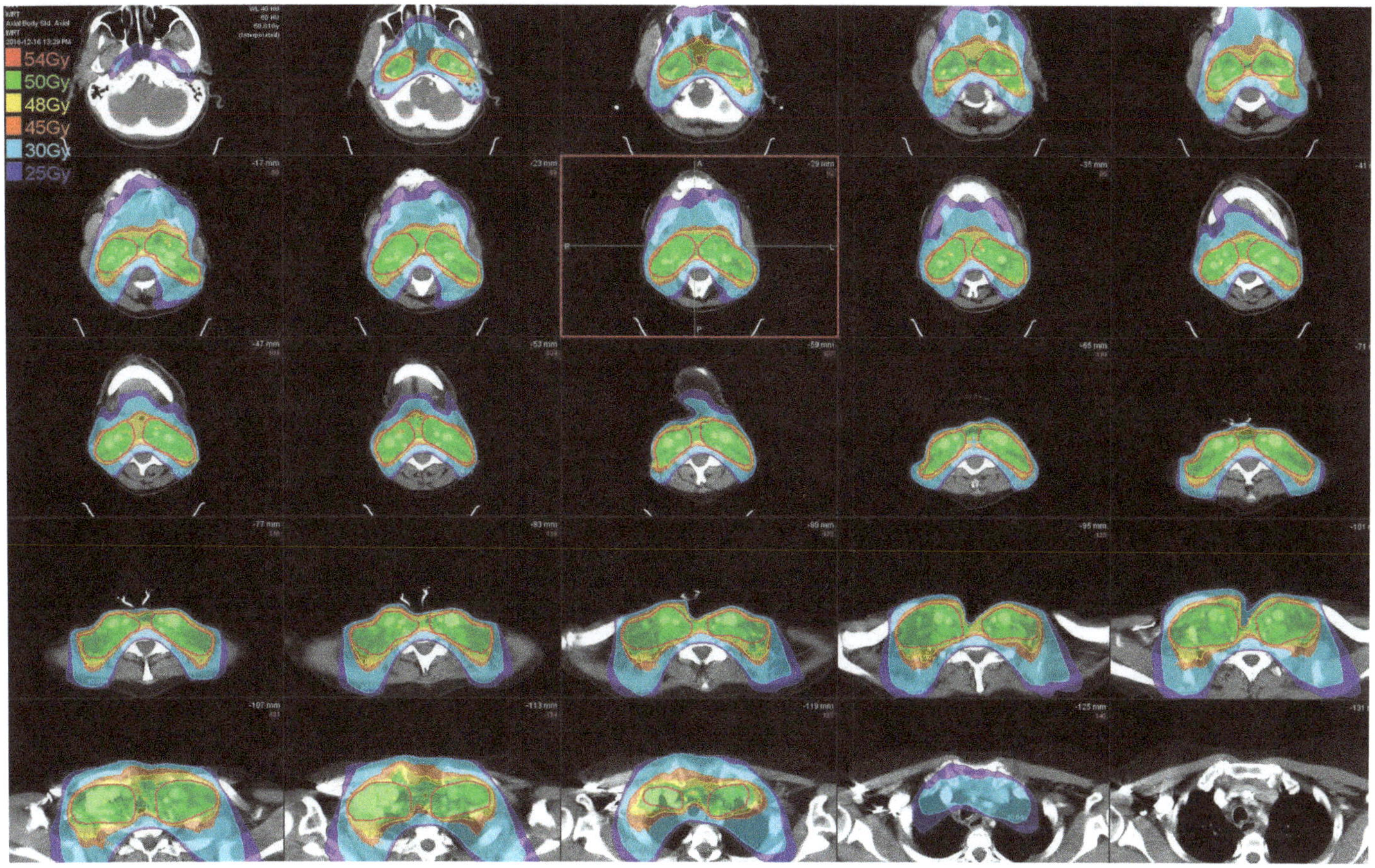

Figure 8.48 Final dose distribution in axial slices.

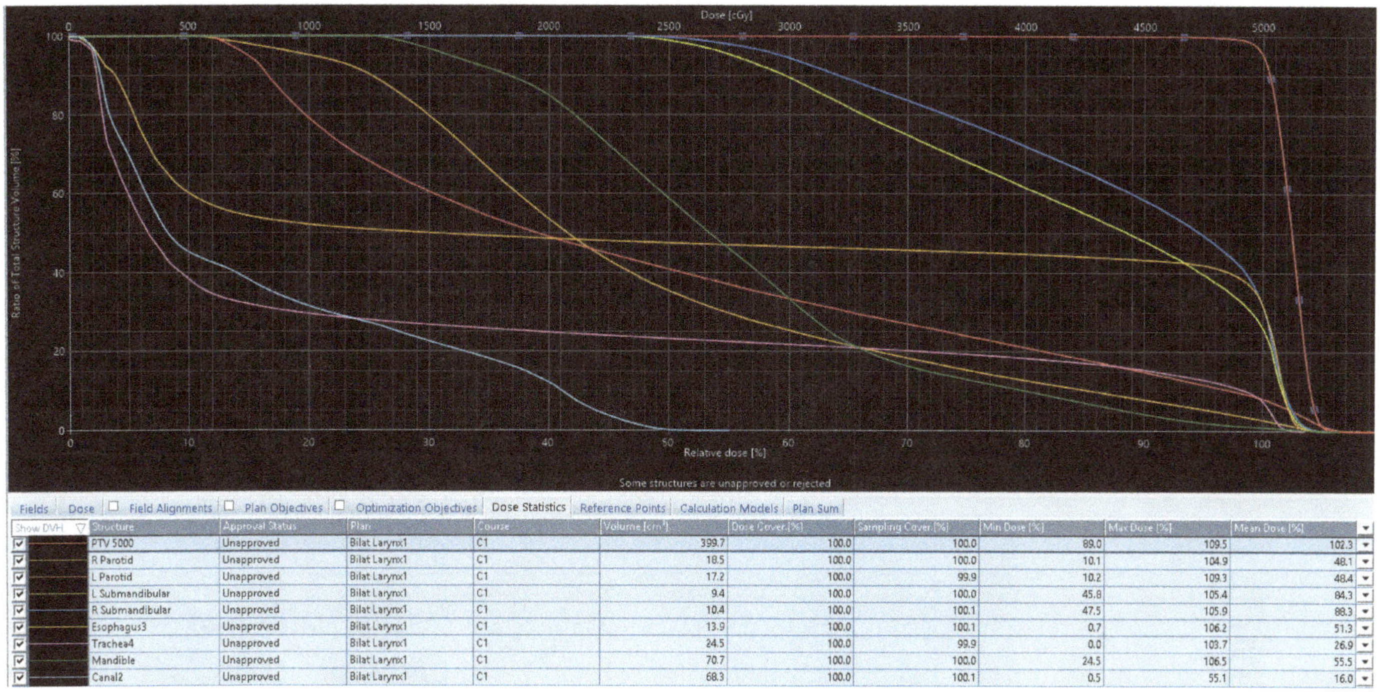

Structure	Approval Status	Plan	Course	Volume [cm³]	Dose Cover.[%]	Sampling Cover.[%]	Min Dose [%]	Max Dose [%]	Mean Dose [%]
PTV 5000	Unapproved	Bilat Larynx1	C1	399.7	100.0	100.0	89.0	109.5	102.3
R Parotid	Unapproved	Bilat Larynx1	C1	18.5	100.0	100.0	10.1	104.9	48.1
L Parotid	Unapproved	Bilat Larynx1	C1	17.2	100.0	99.9	10.2	109.3	48.4
L Submandibular	Unapproved	Bilat Larynx1	C1	9.4	100.0	100.0	45.8	105.4	84.3
R Submandibular	Unapproved	Bilat Larynx1	C1	10.4	100.0	100.1	47.5	105.9	88.3
Esophagus3	Unapproved	Bilat Larynx1	C1	13.9	100.0	100.1	0.7	106.2	51.3
Trachea4	Unapproved	Bilat Larynx1	C1	24.5	100.0	99.9	0.0	103.7	26.9
Mandible	Unapproved	Bilat Larynx1	C1	70.7	100.0	100.0	24.5	106.5	55.5
Canal2	Unapproved	Bilat Larynx1	C1	68.3	100.0	100.1	0.5	55.1	16.0

Figure 8.49 Final dose volume histogram.

Case Number eHN03: Head and Neck—Left Temporal GBM Radiation Treatment

General Information

Brief Clinical History

65-year-old male diagnosed with left temporal lobe glioblastoma.

Prescription

Left temporal GBM:
PTV 6000 = 200 cGy × 30 fractions

Simulation/Comments

The simulation was accomplished with the patient on a head-and-neck index/immobilization board, thermoplastic mask, moldable pillow, and headrest. Fusion of images was performed with a recent MRI for evaluation of the extent of the lesion.

Targets

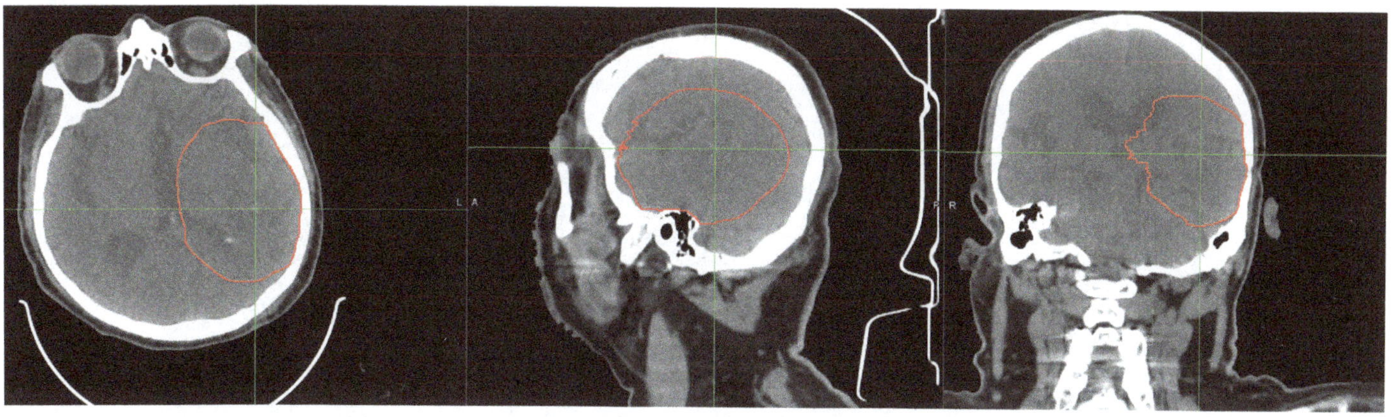

Figure 8.50 Target contour PTV 6000 (red).

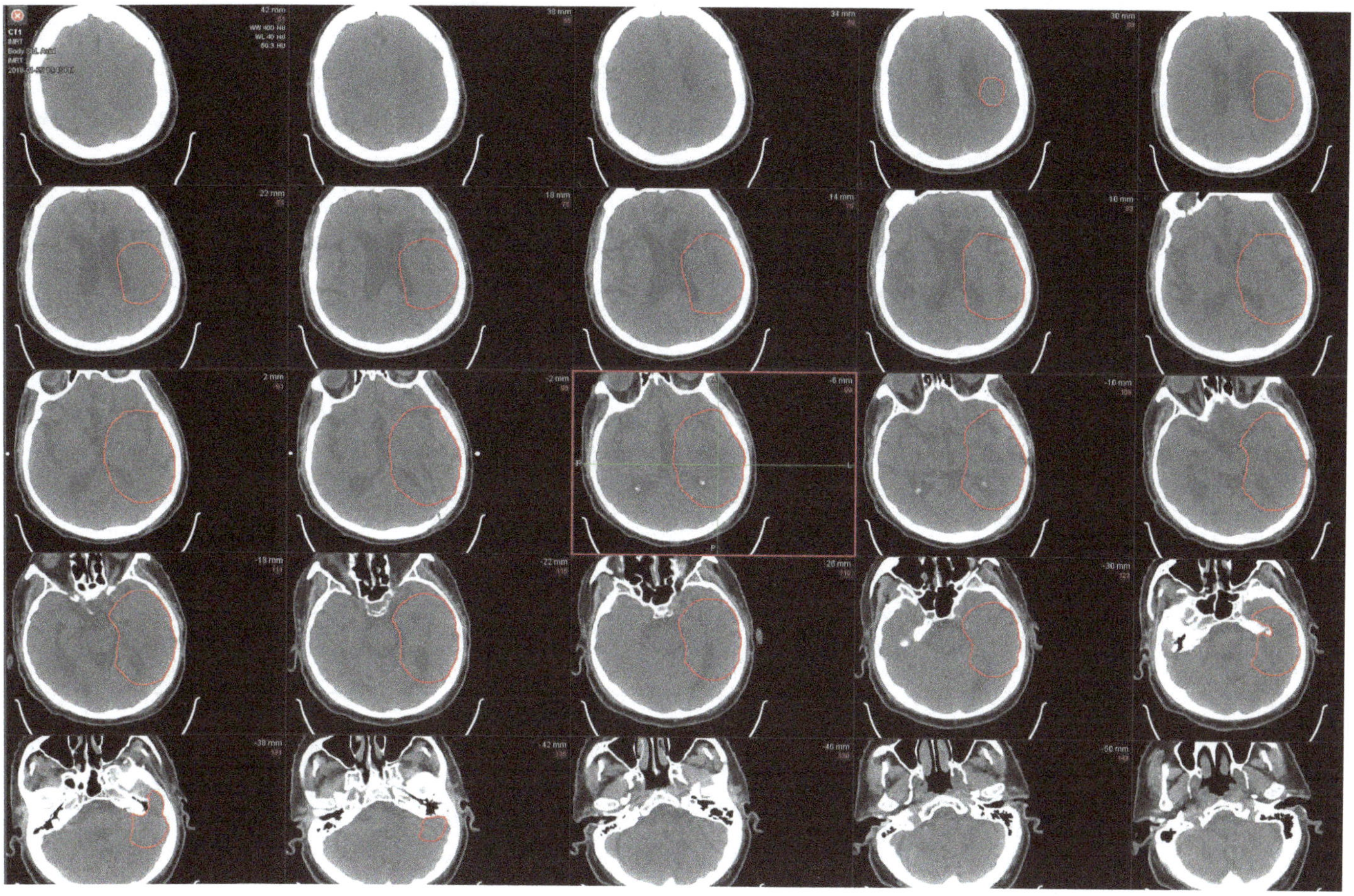

Figure 8.51 Target contour PTV 6000 (red) in axial slices.

Beam Arrangement

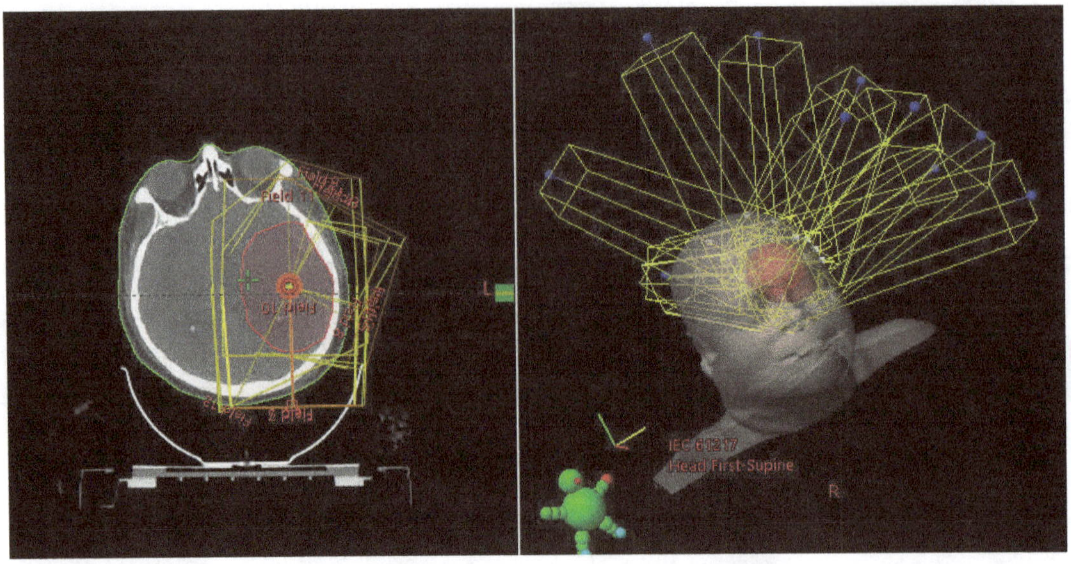

Figure 8.52 IMRT beam arrangement. Beam arrangements with fewer (5+) fields are also possible.

Gantry Angle	Couch	Collimator	Energy (MV)
179	0	0	6
105	0	0	6
30	0	0	6
30	330	0	6
105	330	0	6
179	330	0	6
179	300	0	6
105	300	0	6
30	300	0	6
100	270	0	6
30	270	0	6
225	60	0	6
225	60	0	6

It is possible to make a positive impact on dose distribution by approaching the target from different entrance angles. This method can be used to treat lesions located in the cranial region due to the freedom to place vertex-like beams. The planner must consider the entrance and exit dose sites and avoid optical structures and any other organs at risk. The planner must also be conscious of the delivery time when setting up beam arrangements and optimization options. Avoid placing vertex beams with exit doses directed at the brainstem or cord.

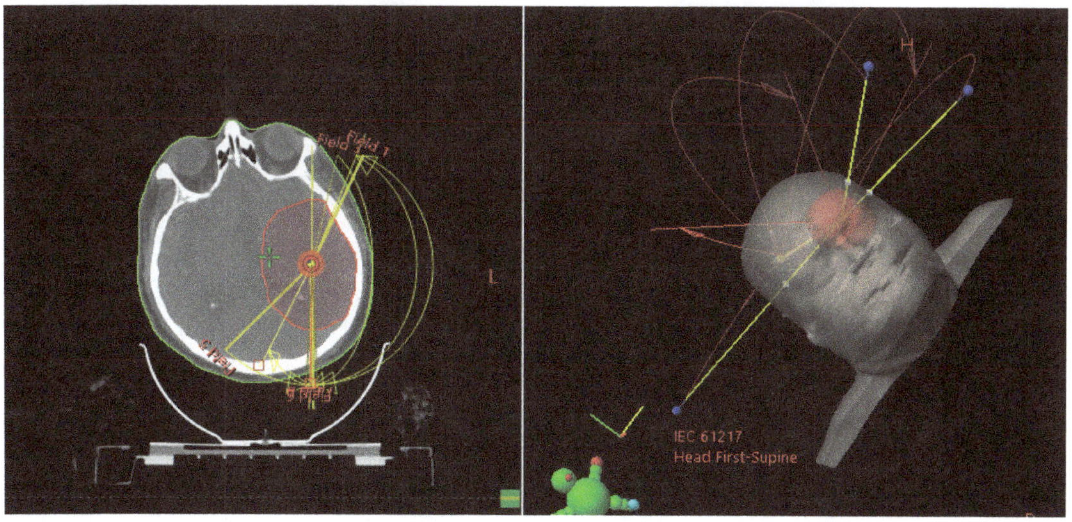

Figure 8.53 VMAT beam arrangement. Beam arrangements with fewer (4+) fields are also possible.

Gantry Angle	Couch	Collimator	Energy (MV)
25–179 CW	0	30	6
179–25 CCW	330	330	6
25–179 CW	300	30	6
179–35 CCW	270	330	6
225–181 CCW	0	30	6
181–225 CW	60	330	6

Optimization

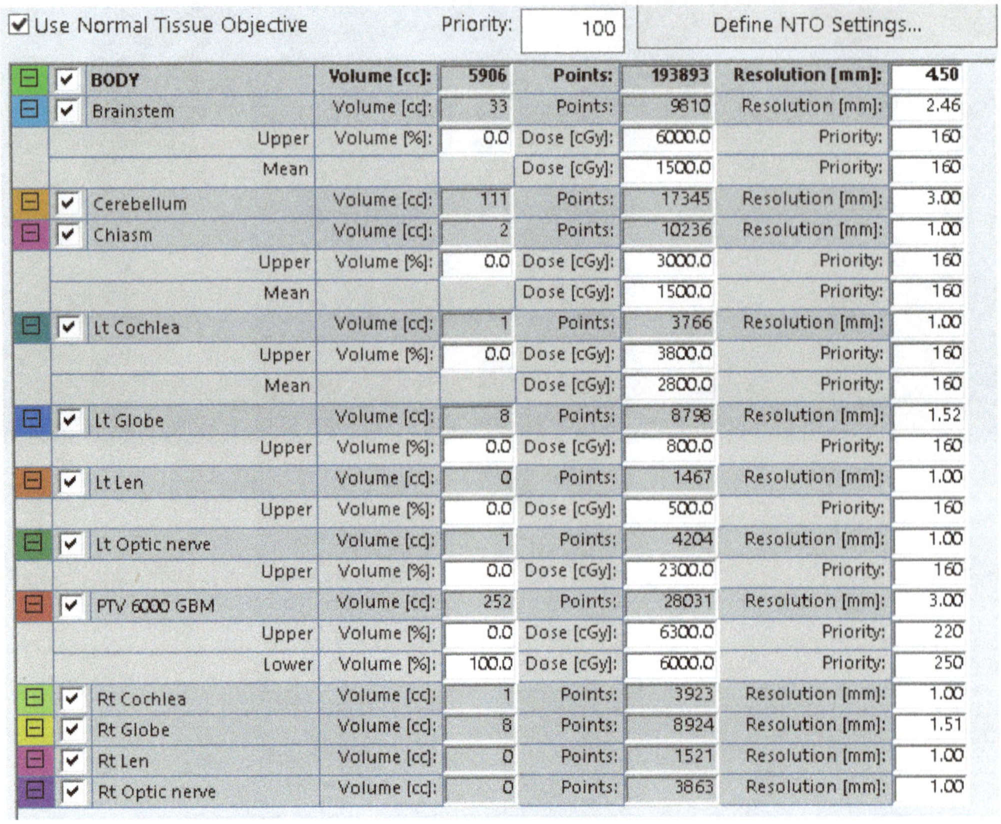
Use Normal Tissue Objective Priority: 100 Define NTO Settings...

BODY		Volume [cc]:	5906	Points:	193893	Resolution [mm]:	4.50
Brainstem		Volume [cc]:	33	Points:	9810	Resolution [mm]:	2.46
	Upper	Volume [%]:	0.0	Dose [cGy]:	6000.0	Priority:	160
	Mean			Dose [cGy]:	1500.0	Priority:	160
Cerebellum		Volume [cc]:	111	Points:	17345	Resolution [mm]:	3.00
Chiasm		Volume [cc]:	2	Points:	10236	Resolution [mm]:	1.00
	Upper	Volume [%]:	0.0	Dose [cGy]:	3000.0	Priority:	160
	Mean			Dose [cGy]:	1500.0	Priority:	160
Lt Cochlea		Volume [cc]:	1	Points:	3766	Resolution [mm]:	1.00
	Upper	Volume [%]:	0.0	Dose [cGy]:	3800.0	Priority:	160
	Mean			Dose [cGy]:	2800.0	Priority:	160
Lt Globe		Volume [cc]:	8	Points:	8798	Resolution [mm]:	1.52
	Upper	Volume [%]:	0.0	Dose [cGy]:	800.0	Priority:	160
Lt Len		Volume [cc]:	0	Points:	1467	Resolution [mm]:	1.00
	Upper	Volume [%]:	0.0	Dose [cGy]:	500.0	Priority:	160
Lt Optic nerve		Volume [cc]:	1	Points:	4204	Resolution [mm]:	1.00
	Upper	Volume [%]:	0.0	Dose [cGy]:	2300.0	Priority:	160
PTV 6000 GBM		Volume [cc]:	252	Points:	28031	Resolution [mm]:	3.00
	Upper	Volume [%]:	0.0	Dose [cGy]:	6300.0	Priority:	220
	Lower	Volume [%]:	100.0	Dose [cGy]:	6000.0	Priority:	250
Rt Cochlea		Volume [cc]:	1	Points:	3923	Resolution [mm]:	1.00
Rt Globe		Volume [cc]:	8	Points:	8924	Resolution [mm]:	1.51
Rt Len		Volume [cc]:	0	Points:	1521	Resolution [mm]:	1.00
Rt Optic nerve		Volume [cc]:	0	Points:	3863	Resolution [mm]:	1.00

Figure 8.54 Initial optimization objectives.

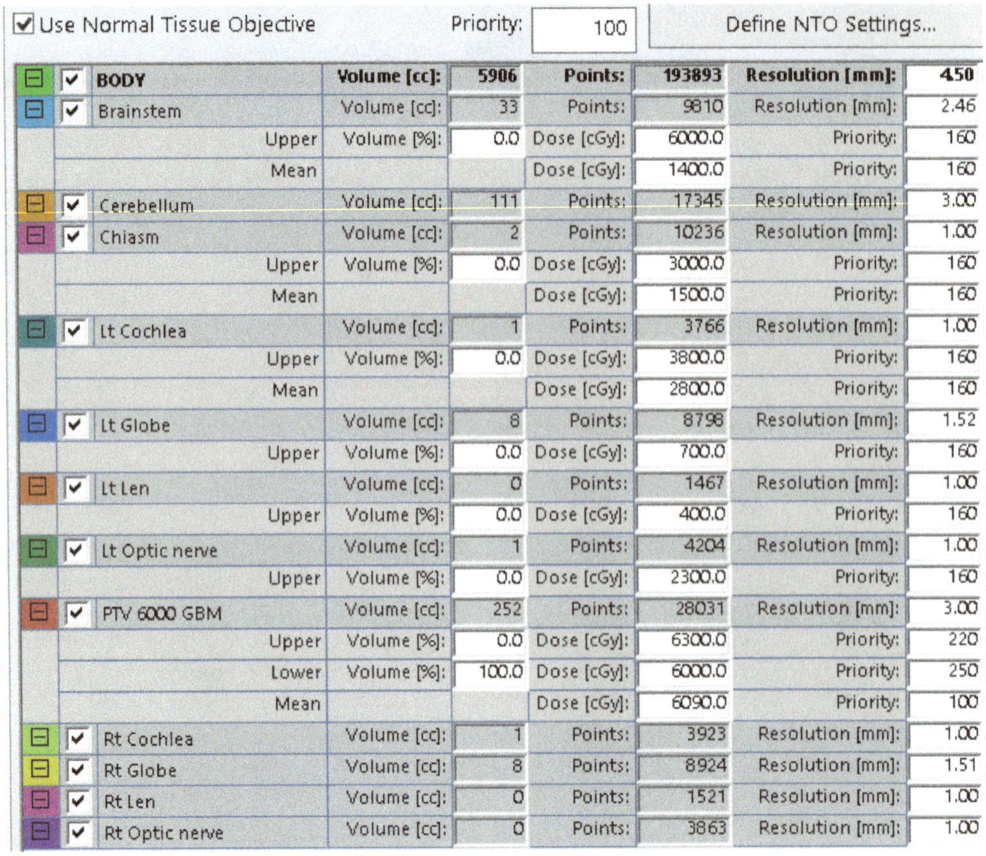
Use Normal Tissue Objective Priority: 100 Define NTO Settings...

BODY		Volume [cc]:	5906	Points:	193893	Resolution [mm]:	4.50
Brainstem		Volume [cc]:	33	Points:	9810	Resolution [mm]:	2.46
	Upper	Volume [%]:	0.0	Dose [cGy]:	6000.0	Priority:	160
	Mean			Dose [cGy]:	1400.0	Priority:	160
Cerebellum		Volume [cc]:	111	Points:	17345	Resolution [mm]:	3.00
Chiasm		Volume [cc]:	2	Points:	10236	Resolution [mm]:	1.00
	Upper	Volume [%]:	0.0	Dose [cGy]:	3000.0	Priority:	160
	Mean			Dose [cGy]:	1500.0	Priority:	160
Lt Cochlea		Volume [cc]:	1	Points:	3766	Resolution [mm]:	1.00
	Upper	Volume [%]:	0.0	Dose [cGy]:	3800.0	Priority:	160
	Mean			Dose [cGy]:	2800.0	Priority:	160
Lt Globe		Volume [cc]:	8	Points:	8798	Resolution [mm]:	1.52
	Upper	Volume [%]:	0.0	Dose [cGy]:	700.0	Priority:	160
Lt Len		Volume [cc]:	0	Points:	1467	Resolution [mm]:	1.00
	Upper	Volume [%]:	0.0	Dose [cGy]:	400.0	Priority:	160
Lt Optic nerve		Volume [cc]:	1	Points:	4204	Resolution [mm]:	1.00
	Upper	Volume [%]:	0.0	Dose [cGy]:	2300.0	Priority:	160
PTV 6000 GBM		Volume [cc]:	252	Points:	28031	Resolution [mm]:	3.00
	Upper	Volume [%]:	0.0	Dose [cGy]:	6300.0	Priority:	220
	Lower	Volume [%]:	100.0	Dose [cGy]:	6000.0	Priority:	250
	Mean			Dose [cGy]:	6090.0	Priority:	100
Rt Cochlea		Volume [cc]:	1	Points:	3923	Resolution [mm]:	1.00
Rt Globe		Volume [cc]:	8	Points:	8924	Resolution [mm]:	1.51
Rt Len		Volume [cc]:	0	Points:	1521	Resolution [mm]:	1.00
Rt Optic nerve		Volume [cc]:	0	Points:	3863	Resolution [mm]:	1.00

Figure 8.55 Final optimization objectives.

Dose Distribution

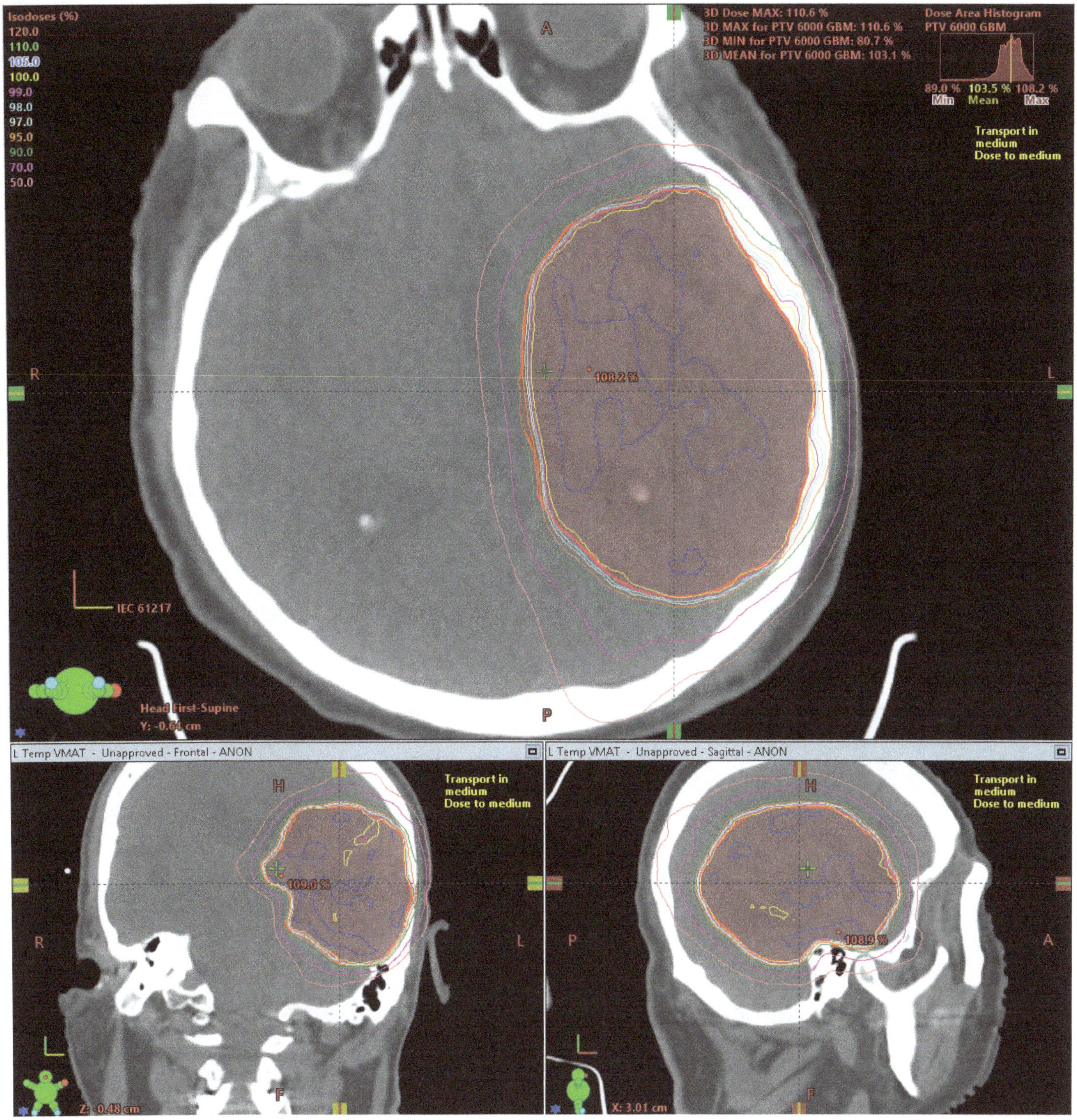

Figure 8.56 Final dose distribution.

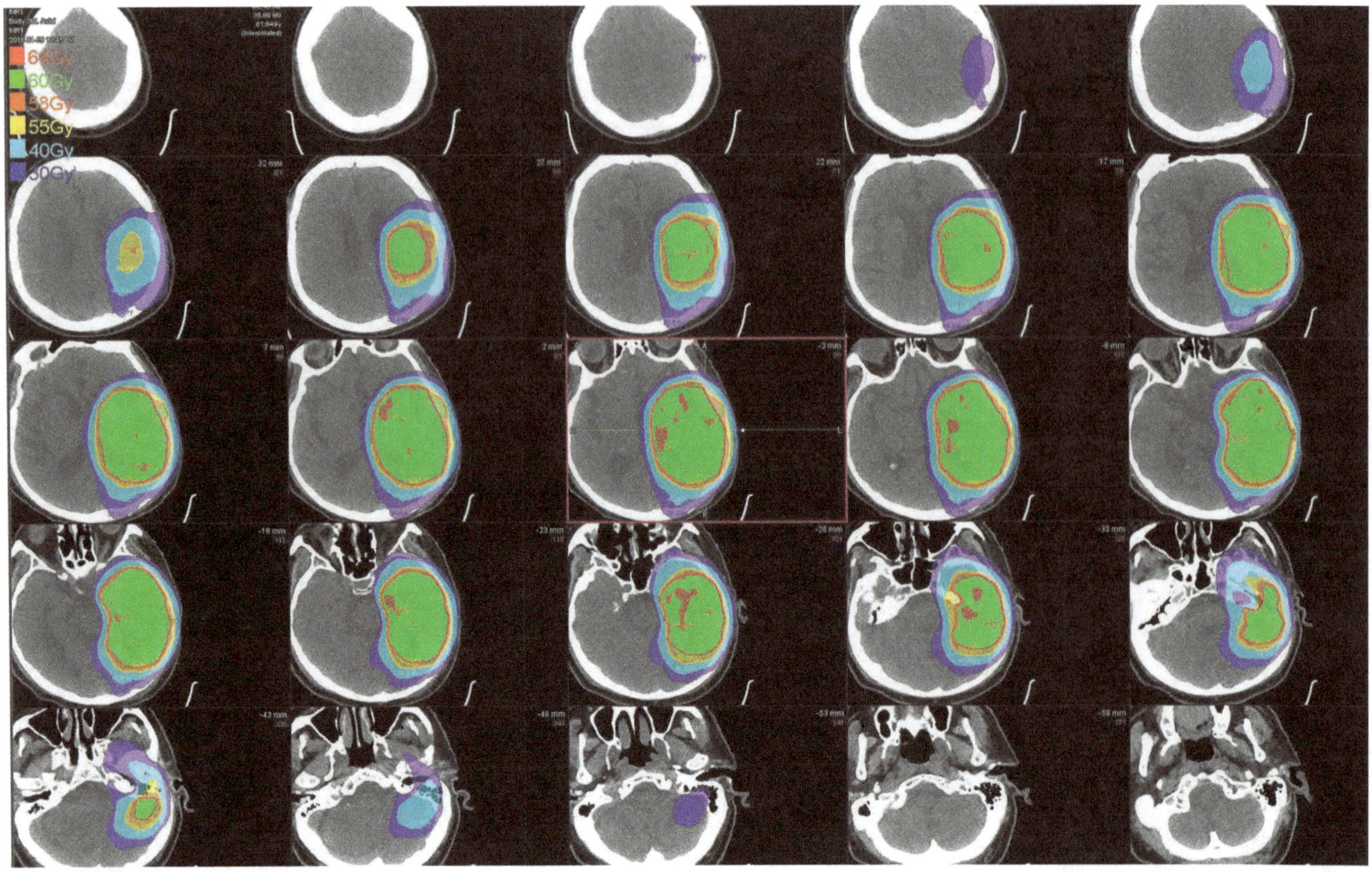

Figure 8.57 Final dose distribution in axial slices.

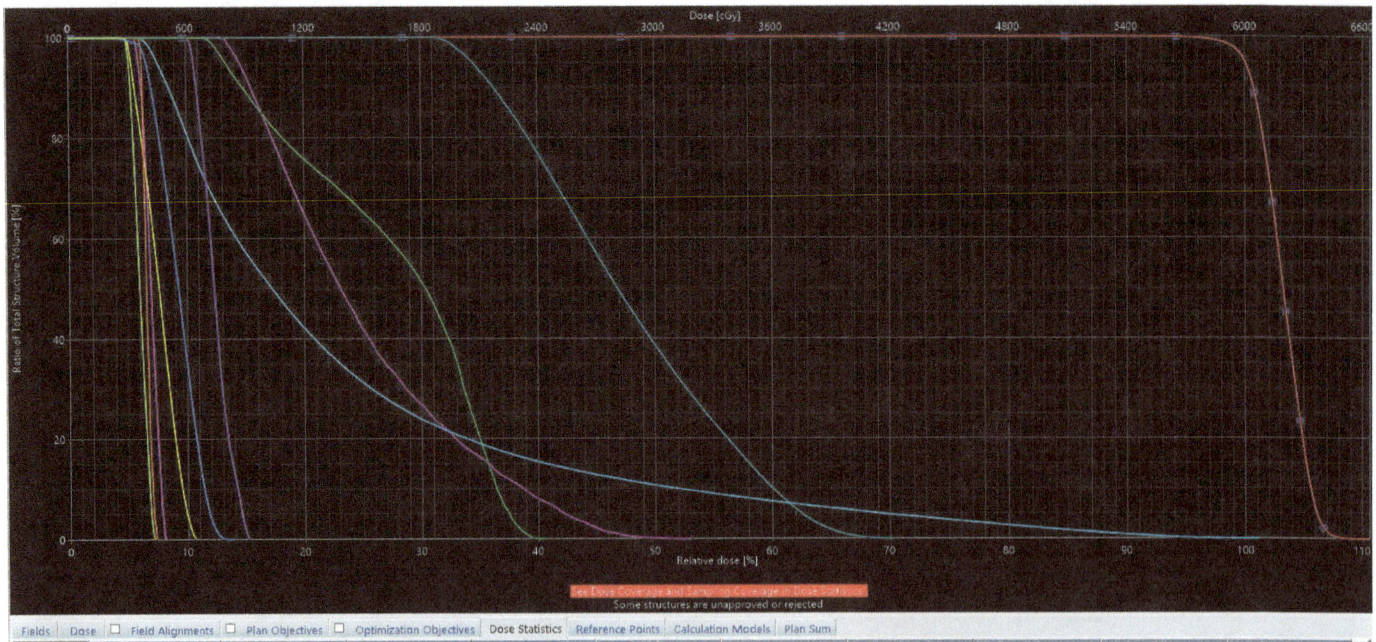

Structure	Approval Status	Plan	Course	Volume [cm³]	Dose Cover.[%]	Sampling Cover.[%]	Min Dose [%]	Max Dose [%]	Mean Dose [%]
PTV 6000 GBM	Unapproved	L Temp VMAT	C1	252.5	100.0	100.0	80.7	110.6	103.1
Chiasm	Unapproved	L Temp VMAT	C1	2.1	100.0	93.2	13.0	53.1	25.1
Lt Cochlea	Unapproved	L Temp VMAT	C1	0.6	100.0	89.7	31.2	70.2	47.6
Rt Cochlea	Unapproved	L Temp VMAT	C1	0.6	100.0	88.2	4.8	7.2	6.0
Lt Optic nerve	Unapproved	L Temp VMAT	C1	0.6	100.0	87.1	11.3	40.7	27.6
Lt Globe	Unapproved	L Temp VMAT	C1	7.8	100.0	99.1	3.9	13.9	9.2
Lt Len	Unapproved	L Temp VMAT	C1	0.1	100.0	86.9	6.0	7.4	6.6
Rt Optic nerve	Unapproved	L Temp VMAT	C1	0.5	100.0	85.9	9.8	15.3	12.4
Rt Globe	Unapproved	L Temp VMAT	C1	7.7	100.0	99.2	4.3	10.7	7.6
Rt Len	Unapproved	L Temp VMAT	C1	0.2	100.0	88.9	5.5	8.1	7.0
Brainstem	Unapproved	L Temp VMAT	C1	33.0	100.0	98.9	5.8	101.1	24.0

Figure 8.58 Final dose volume histogram.

ECLIPSE SAMPLES

Case Number eP01: Pelvis—Prostate Bed Treatment

General Information

Brief Clinical History

67-year-old man with surgically resected prostate bed.

Prescription

Prostate bed:
PTV 6800 = 200 cGy × 34 fractions

Simulation/Comments

The simulation was accomplished with a vacuum cushion as an immobilization device positioned beneath the pelvis.

Targets

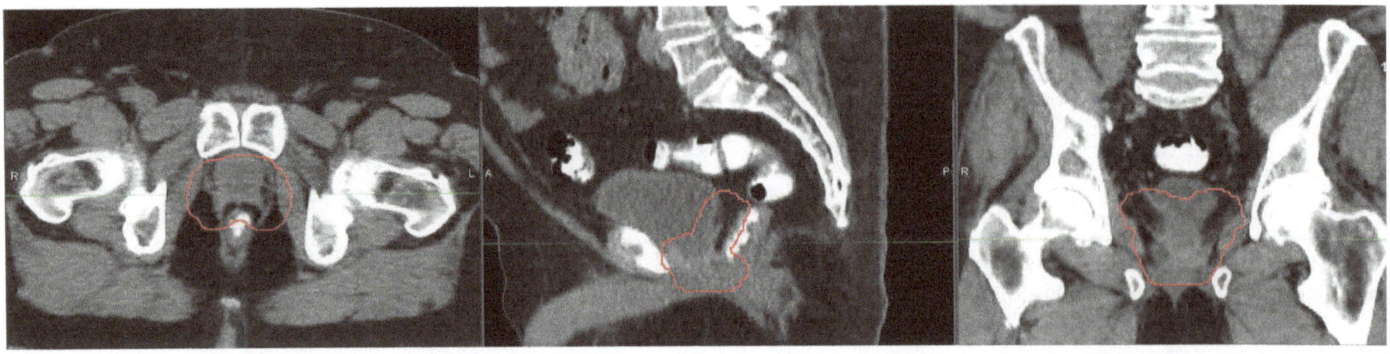

Figure 8.59 Target volume PTV 6800 (red). Notice the areas of ingested contrast. Appropriate density correction was used for planning.

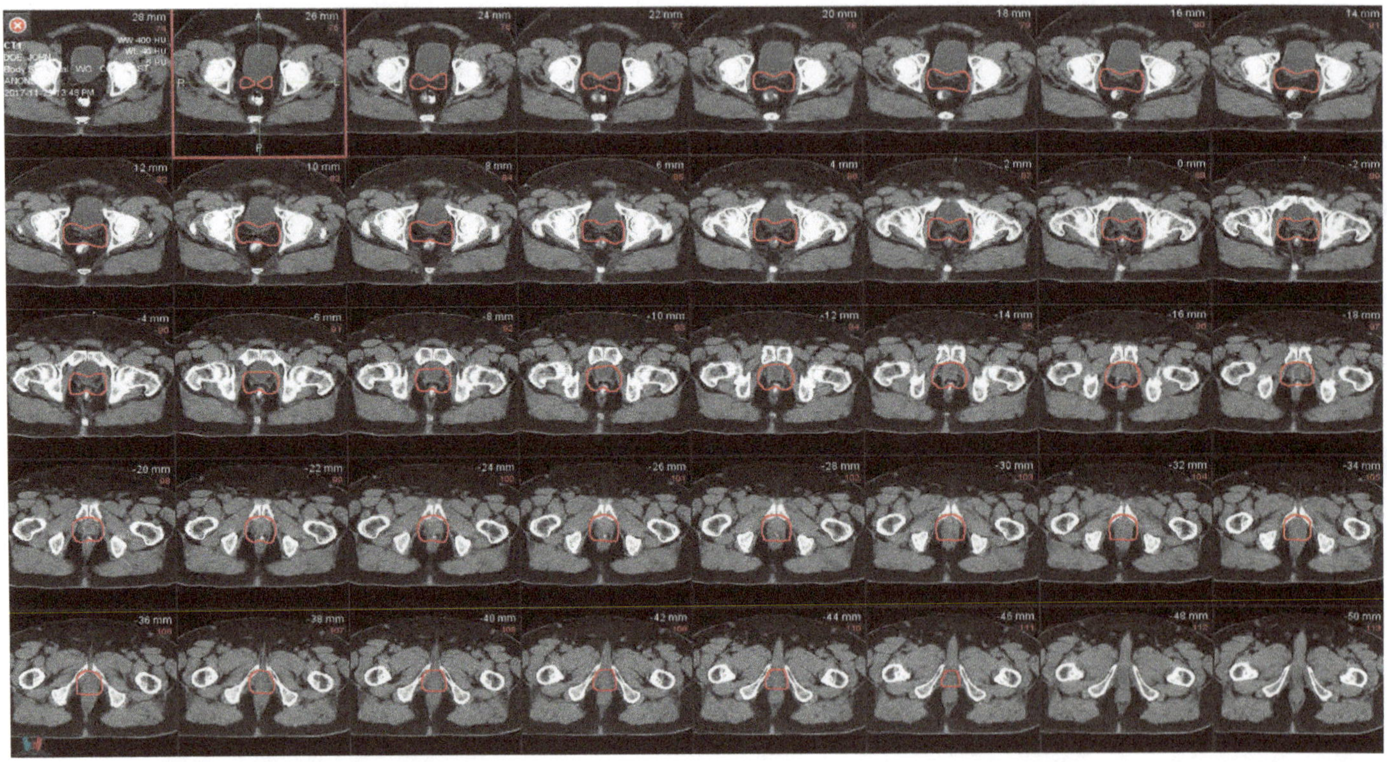

Figure 8.60 Target volume PTV 6800 (red) in axial slices. Notice the areas of ingested contrast. Appropriate density correction was used for planning.

Beam Arrangement

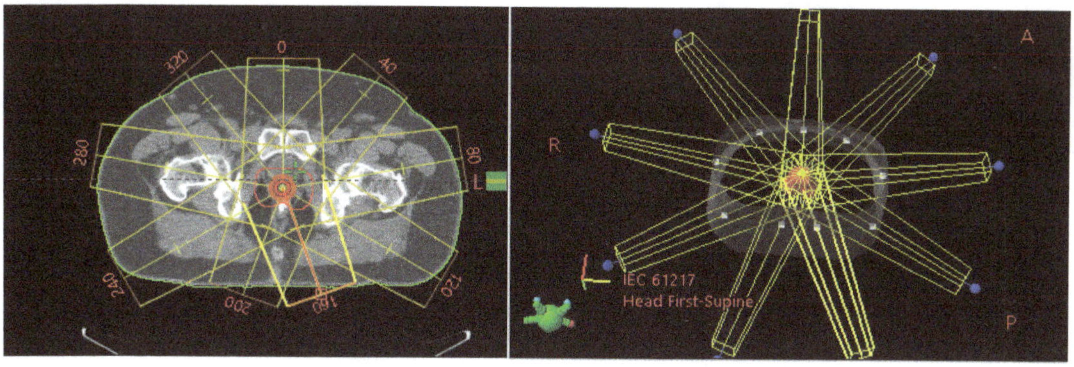

Figure 8.61 Nine-field IMRT beam arrangement.

Gantry Angle	Couch	Collimator	Energy (MV)
200	0	0	10 MV
240	0	0	10 MV
280	0	0	10 MV
320	0	0	10 MV
0	0	0	10 MV
40	0	0	10 MV
80	0	0	10 MV
120	0	0	10 MV
160	0	0	10 MV

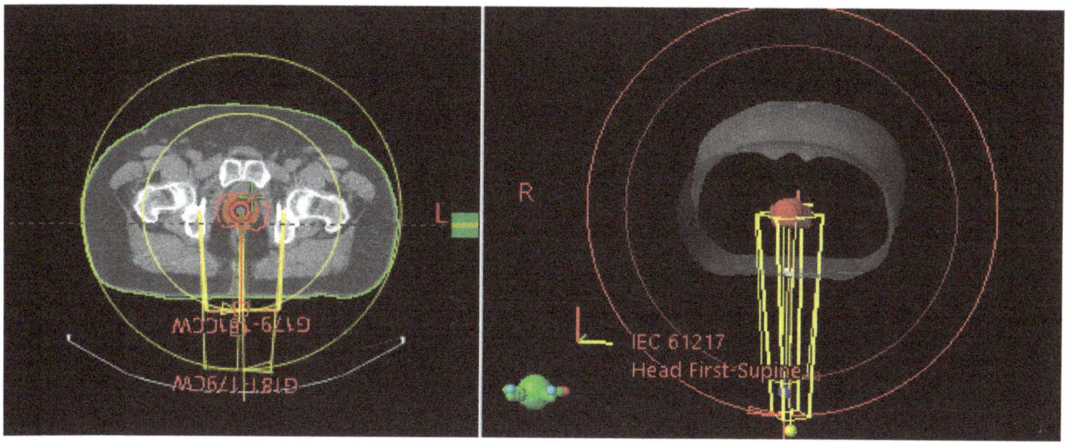

Figure 8.62 Two-arc VMAT beam arrangement.

Gantry Angle	Couch	Collimator	Energy (MV)
181–179 CW	0	30	10 FFF
179–181 CCW	0	330	10 FFF

Optimization

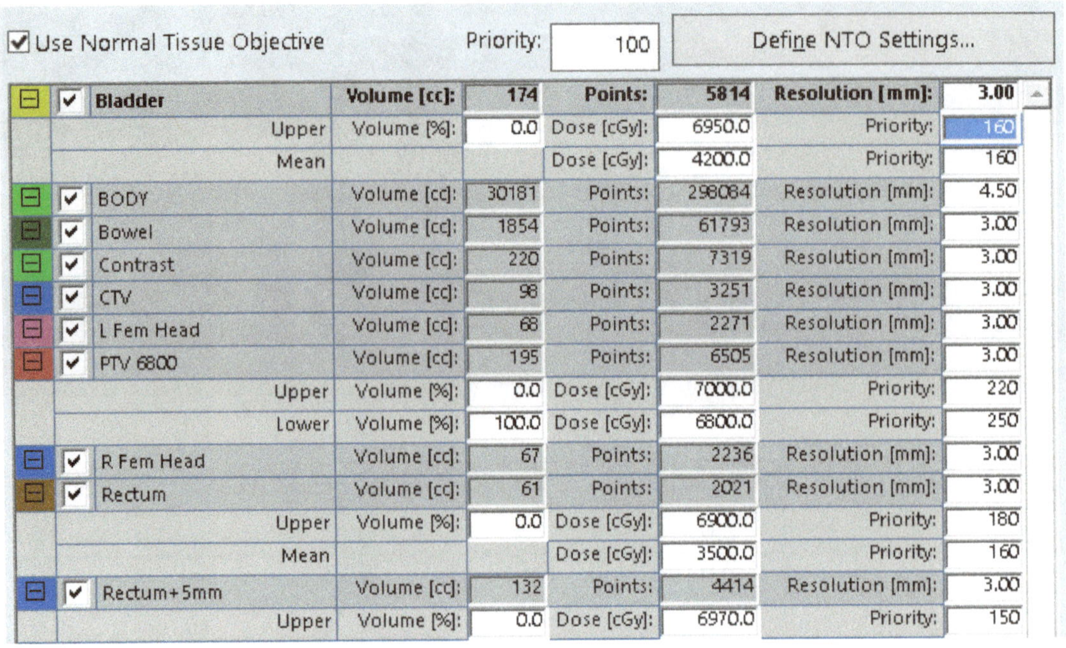

☑ Use Normal Tissue Objective — Priority: 100 — Define NTO Settings...

☑ Bladder		Volume [cc]:	174	Points:	5814	Resolution [mm]:	3.00
	Upper	Volume [%]:	0.0	Dose [cGy]:	6950.0	Priority:	160
	Mean			Dose [cGy]:	4200.0	Priority:	160
☑ BODY		Volume [cc]:	30181	Points:	298084	Resolution [mm]:	4.50
☑ Bowel		Volume [cc]:	1854	Points:	61793	Resolution [mm]:	3.00
☑ Contrast		Volume [cc]:	220	Points:	7319	Resolution [mm]:	3.00
☑ CTV		Volume [cc]:	98	Points:	3251	Resolution [mm]:	3.00
☑ L Fem Head		Volume [cc]:	68	Points:	2271	Resolution [mm]:	3.00
☑ PTV 6800		Volume [cc]:	195	Points:	6505	Resolution [mm]:	3.00
	Upper	Volume [%]:	0.0	Dose [cGy]:	7000.0	Priority:	220
	Lower	Volume [%]:	100.0	Dose [cGy]:	6800.0	Priority:	250
☑ R Fem Head		Volume [cc]:	67	Points:	2236	Resolution [mm]:	3.00
☑ Rectum		Volume [cc]:	61	Points:	2021	Resolution [mm]:	3.00
	Upper	Volume [%]:	0.0	Dose [cGy]:	6900.0	Priority:	180
	Mean			Dose [cGy]:	3500.0	Priority:	160
☑ Rectum+5mm		Volume [cc]:	132	Points:	4414	Resolution [mm]:	3.00
	Upper	Volume [%]:	0.0	Dose [cGy]:	6970.0	Priority:	150

Figure 8.63 Initial optimization objectives.

Additional optimization structures for this plan include a "rectum + 5 mm margin" contour to prevent regions of high dose in the overlap between the rectum and PTV structures. The NTO feature can be used to increase dose conformity to the target. During the optimization process, the planner must introduce new objectives and structures as they seem fit. Ideal dose distribution will reduce areas of high dose in regions of organ-at-risk overlap while avoiding large regions of low dose expanding from the target to the surface of the body.

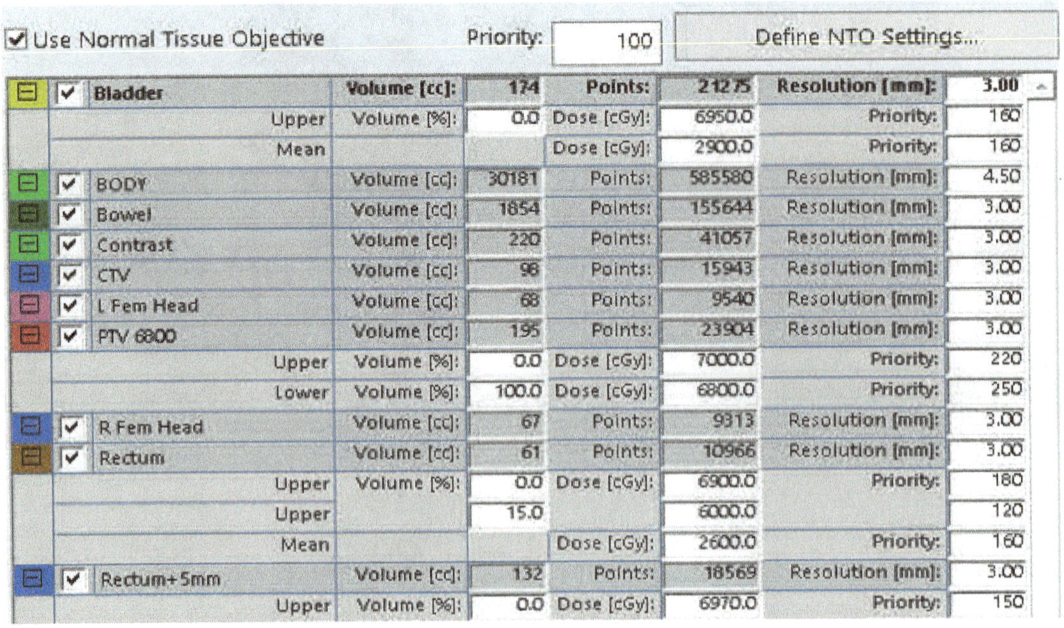

☑ Use Normal Tissue Objective — Priority: 100 — Define NTO Settings...

☑ Bladder		Volume [cc]:	174	Points:	21275	Resolution [mm]:	3.00
	Upper	Volume [%]:	0.0	Dose [cGy]:	6950.0	Priority:	160
	Mean			Dose [cGy]:	2900.0	Priority:	160
☑ BODY		Volume [cc]:	30181	Points:	585580	Resolution [mm]:	4.50
☑ Bowel		Volume [cc]:	1854	Points:	155644	Resolution [mm]:	3.00
☑ Contrast		Volume [cc]:	220	Points:	41057	Resolution [mm]:	3.00
☑ CTV		Volume [cc]:	98	Points:	15943	Resolution [mm]:	3.00
☑ L Fem Head		Volume [cc]:	68	Points:	9540	Resolution [mm]:	3.00
☑ PTV 6800		Volume [cc]:	195	Points:	23904	Resolution [mm]:	3.00
	Upper	Volume [%]:	0.0	Dose [cGy]:	7000.0	Priority:	220
	Lower	Volume [%]:	100.0	Dose [cGy]:	6800.0	Priority:	250
☑ R Fem Head		Volume [cc]:	67	Points:	9313	Resolution [mm]:	3.00
☑ Rectum		Volume [cc]:	61	Points:	10966	Resolution [mm]:	3.00
	Upper	Volume [%]:	0.0	Dose [cGy]:	6900.0	Priority:	180
	Upper		15.0		6000.0		120
	Mean			Dose [cGy]:	2600.0	Priority:	160
☑ Rectum+5mm		Volume [cc]:	132	Points:	18569	Resolution [mm]:	3.00
	Upper	Volume [%]:	0.0	Dose [cGy]:	6970.0	Priority:	150

Figure 8.64 Final optimization objectives.

Dose Distribution

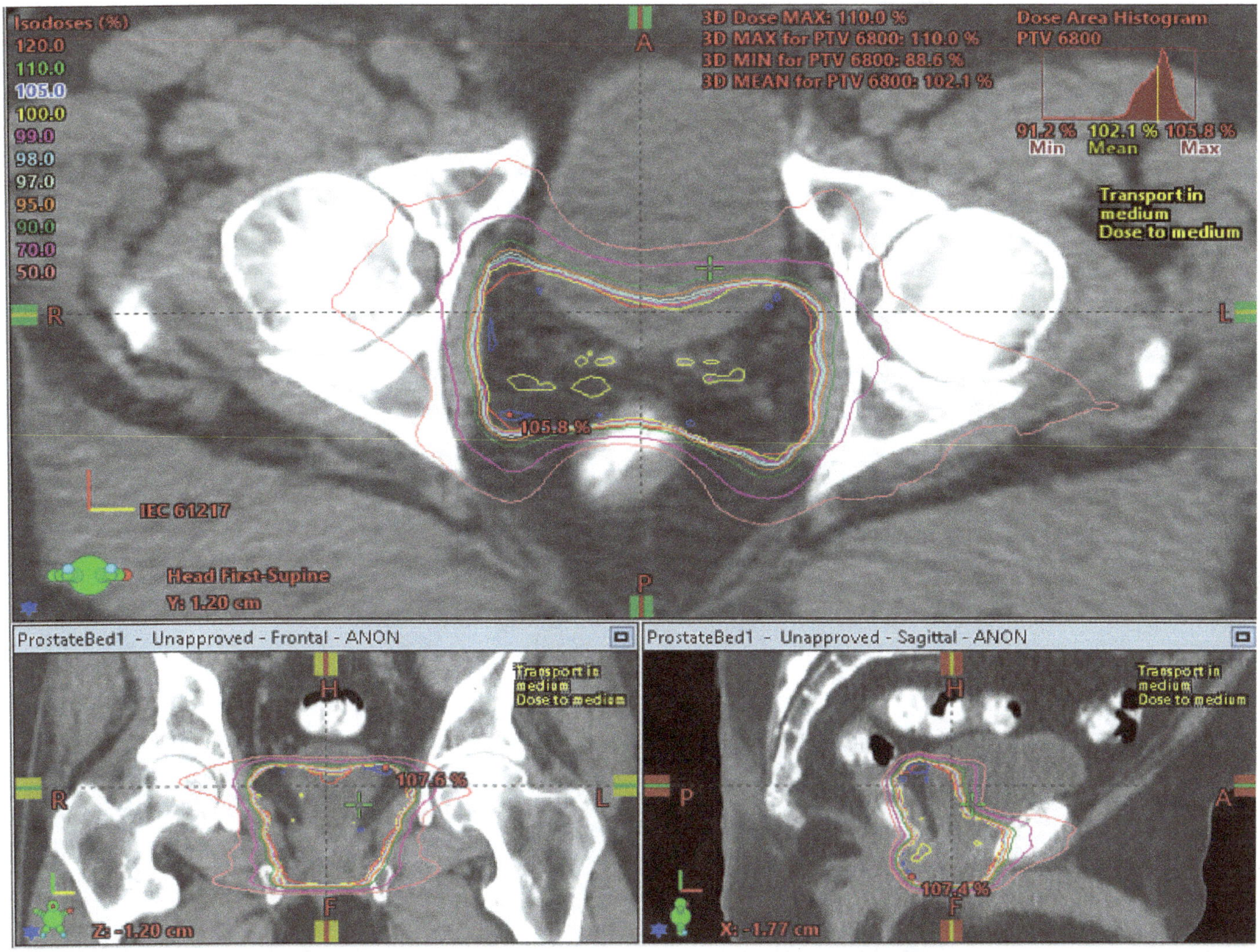

Figure 8.65 Final dose distribution.

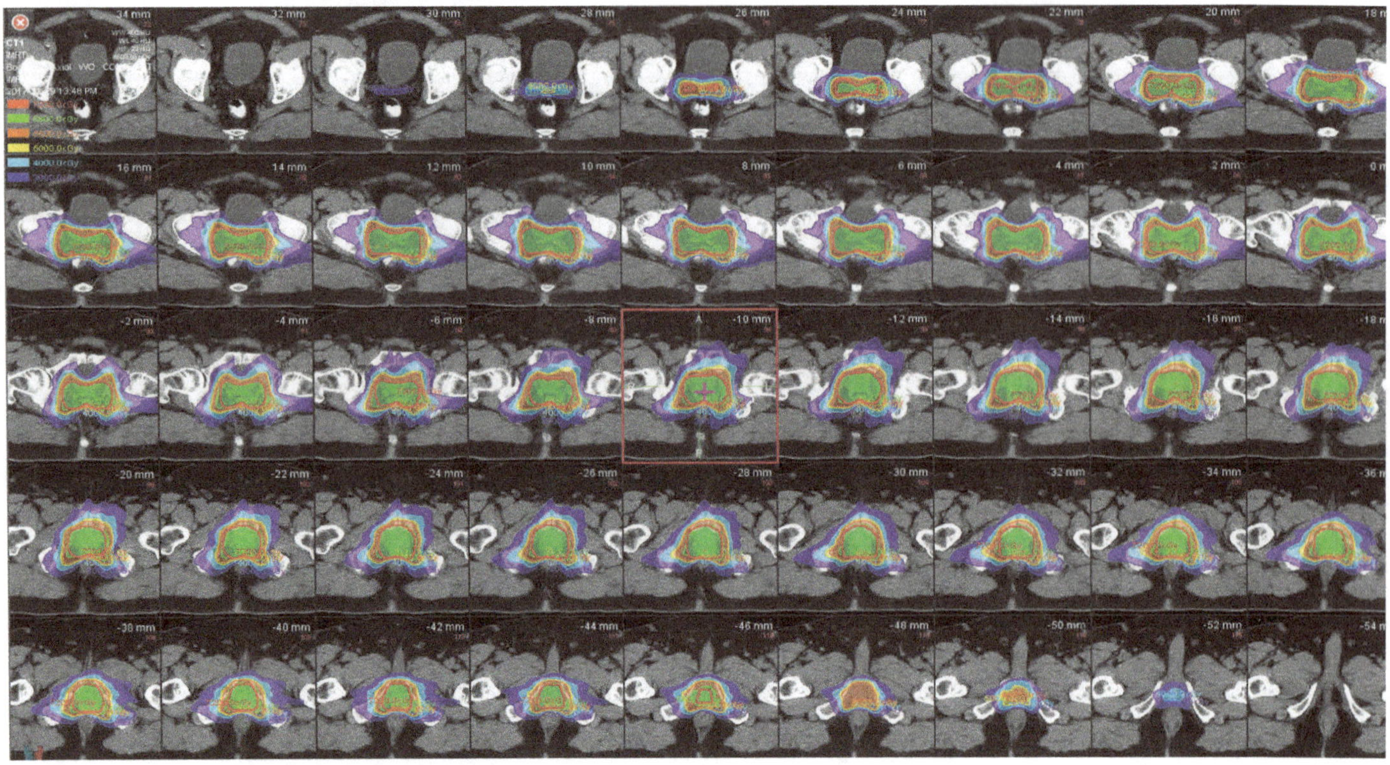

Figure 8.66 Final dose distribution on axial slices.

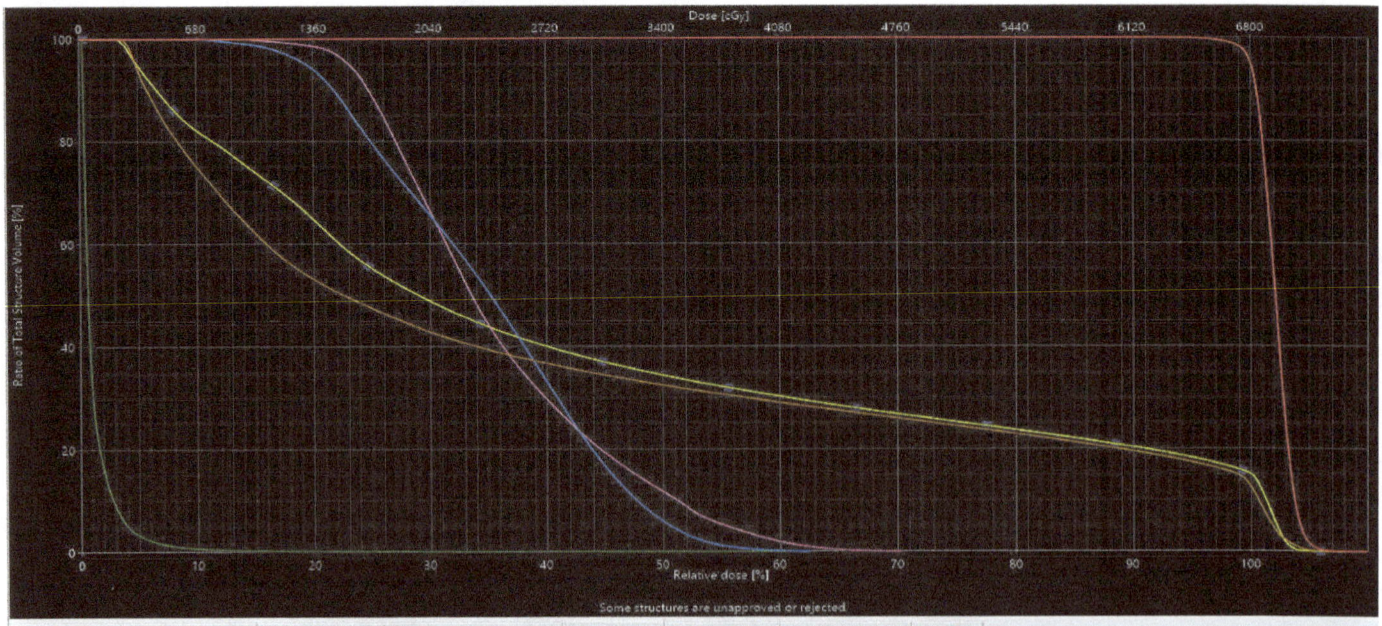

Fields | Dose | Field Alignments | Plan Objectives | Optimization Objectives | Dose Statistics | Reference Points | Calculation Models | Plan Sum

Show DVH	Structure	Approval Status	Plan	Course	Volume [cm³]	Dose Cover.[%]	Sampling Cover.[%]	Min Dose [%]	Max Dose [%]	Mean Dose [%]
☑	Bladder	Unapproved	ProstateBed1	C1	174.4	100.0	100.0	3.0	106.8	43.3
☑	Bowel	Unapproved	ProstateBed1	C1	1853.8	100.0	100.0	0.2	60.6	1.3
☑	L Fem Head	Unapproved	ProstateBed1	C1	68.2	100.0	100.0	12.2	70.3	35.8
☑	R Fem Head	Unapproved	ProstateBed1	C1	67.1	100.0	100.1	9.7	62.7	34.8
☑	Rectum	Unapproved	ProstateBed1	C1	60.7	100.0	100.1	3.5	108.3	39.9
☑	PTV 6800	Unapproved	ProstateBed1	C1	195.2	100.0	100.1	88.6	110.0	102.1

Figure 8.67 Final dose volume histogram.

ECLIPSE SAMPLES

Case Number eP02: Pelvis—Right Hip Treatment

General Information

Brief Clinical History

77-year-old male diagnosed with adenocarcinoma of the lung and multiple bone metastasis.

Prescription

Right hip:
PTV 3400 = 340 cGy × 10 fractions.

Simulation/Comments

The simulation was accomplished with the patient in the supine position with a vacuum cushion underneath the pelvis. Fusion of images was performed with a recent PET/CT to evaluate the extent of the lesion.

Targets

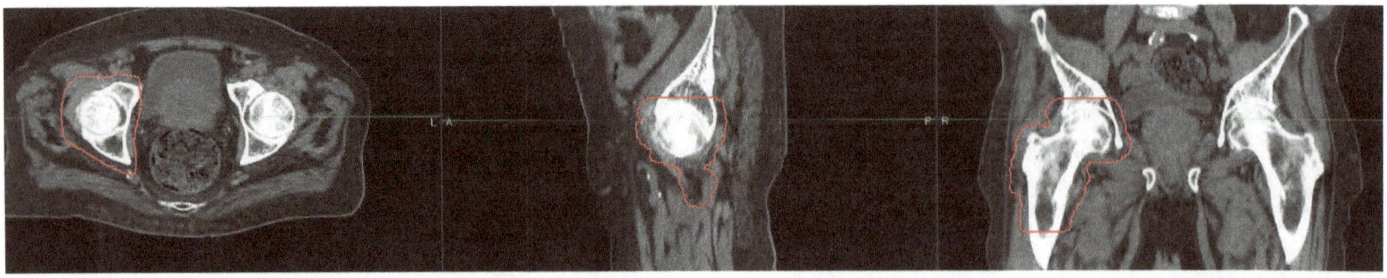

Figure 8.68 Target contour PTV 3400 (red). Notice the rectum volume which is expected to decrease during treatment.

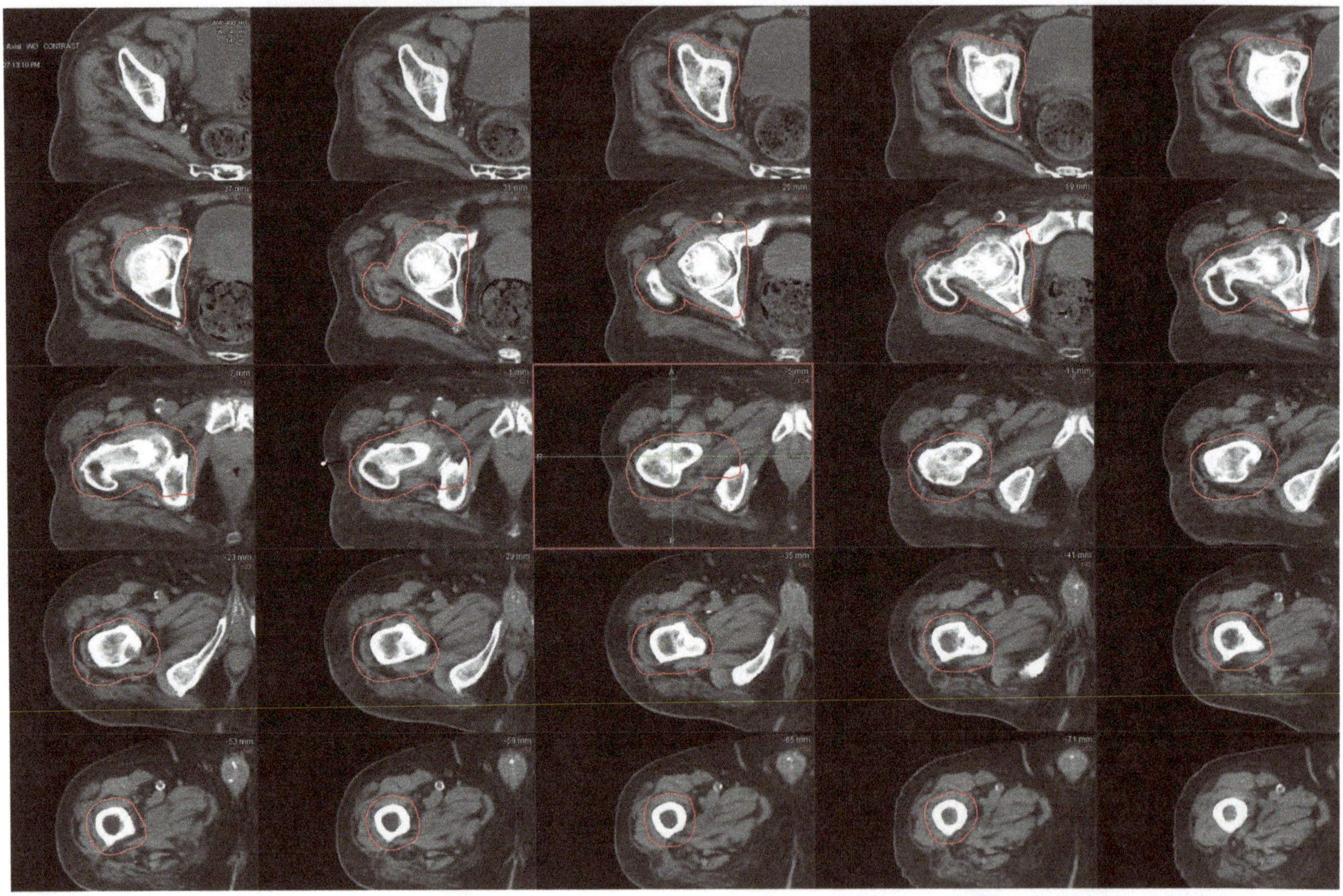

Figure 8.69 Target contour PTV 3400 (red) in axial slices. Notice the rectum volume, which is expected to decrease during treatment.

Beam Arrangement

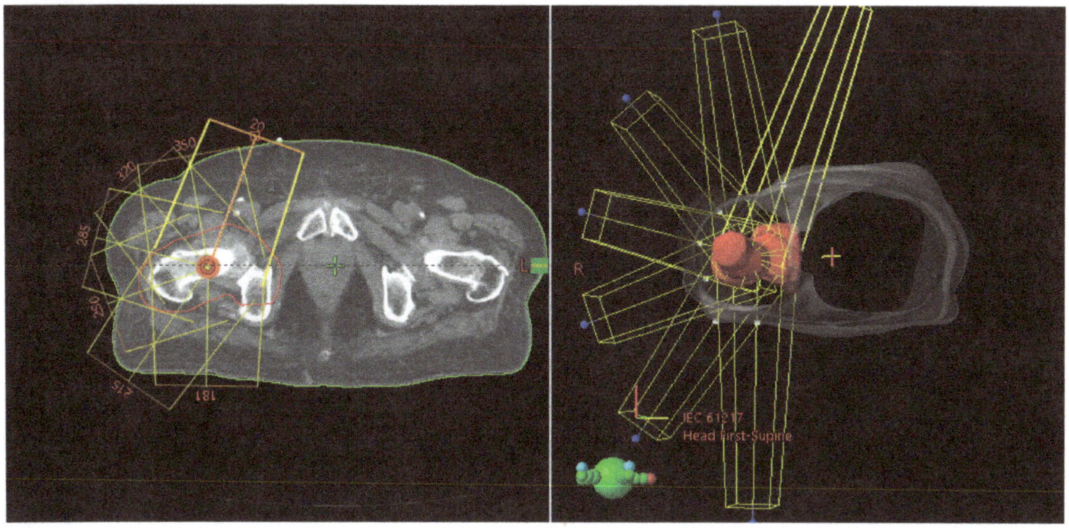

Figure 8.70 Seven-field IMRT beam arrangement.

Gantry Angle	Couch	Collimator	Energy (MV)
181	0	0	10 FFF
215	0	0	10 FFF
250	0	0	10 FFF
285	0	0	10 FFF
320	0	0	10 FFF
350	0	0	10 FFF
20	0	0	10 FFF

Figure 8.70 shows the IMRT beam arrangement around the target. Notice the avoidance of dose entrance through the rectum, bladder, and contralateral hip.

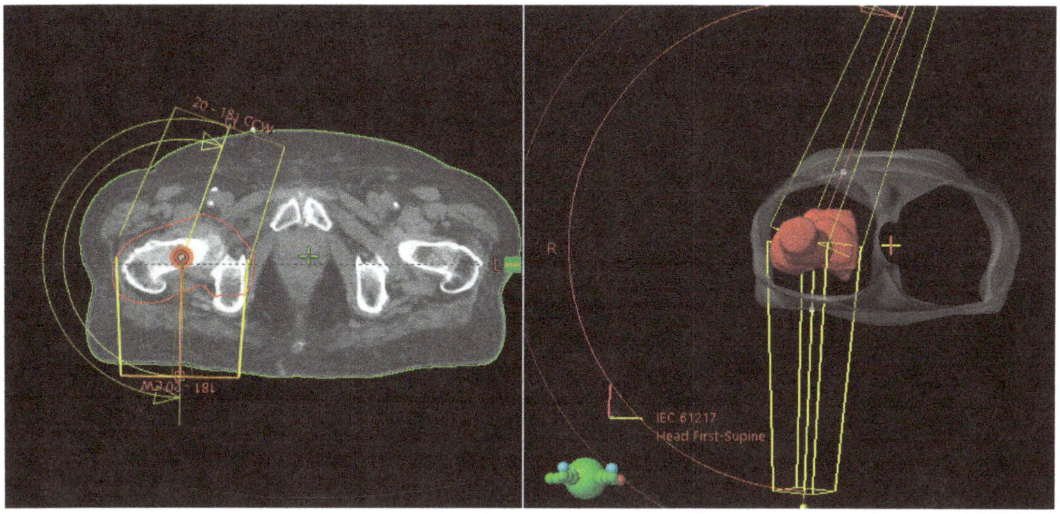

Figure 8.71 VMAT beam arrangement.

Gantry Angle	Couch	Collimator	Energy (MV)
181–30 CW	0	30	10 FFF
30–181 CCW	0	330	10 FFF

Optimization

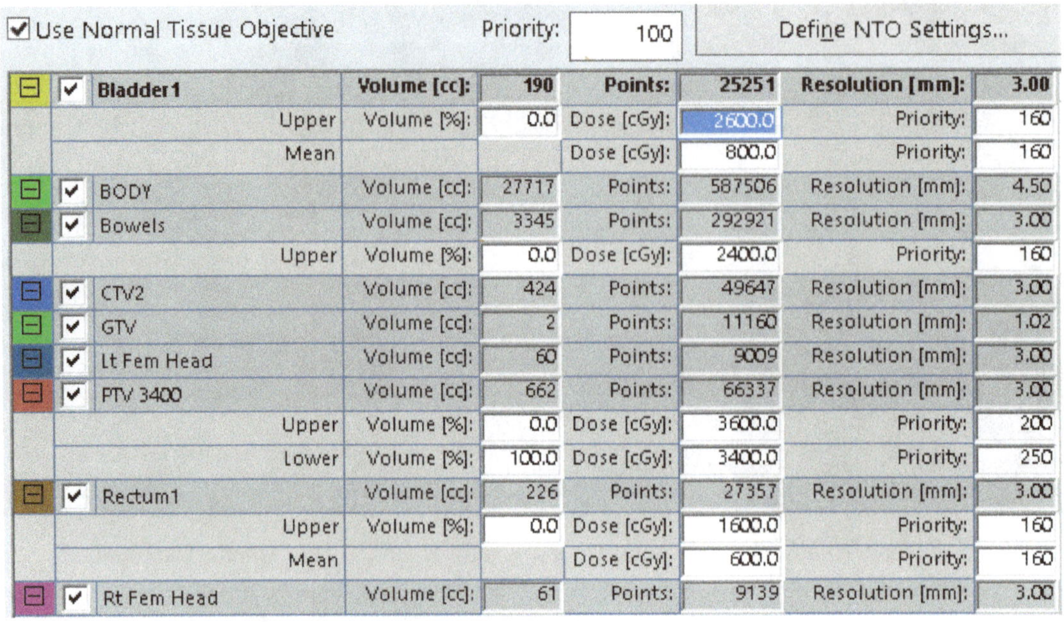

☑ Use Normal Tissue Objective — Priority: 100 — Define NTO Settings...

Bladder1		Volume [cc]:	190	Points:	25251	Resolution [mm]:	3.00
	Upper	Volume [%]:	0.0	Dose [cGy]:	2600.0	Priority:	160
	Mean			Dose [cGy]:	800.0	Priority:	160
BODY		Volume [cc]:	27717	Points:	587506	Resolution [mm]:	4.50
Bowels		Volume [cc]:	3345	Points:	292921	Resolution [mm]:	3.00
	Upper	Volume [%]:	0.0	Dose [cGy]:	2400.0	Priority:	160
CTV2		Volume [cc]:	424	Points:	49647	Resolution [mm]:	3.00
GTV		Volume [cc]:	2	Points:	11160	Resolution [mm]:	1.02
Lt Fem Head		Volume [cc]:	60	Points:	9009	Resolution [mm]:	3.00
PTV 3400		Volume [cc]:	662	Points:	66337	Resolution [mm]:	3.00
	Upper	Volume [%]:	0.0	Dose [cGy]:	3600.0	Priority:	200
	Lower	Volume [%]:	100.0	Dose [cGy]:	3400.0	Priority:	250
Rectum1		Volume [cc]:	226	Points:	27357	Resolution [mm]:	3.00
	Upper	Volume [%]:	0.0	Dose [cGy]:	1600.0	Priority:	160
	Mean			Dose [cGy]:	600.0	Priority:	160
Rt Fem Head		Volume [cc]:	61	Points:	9139	Resolution [mm]:	3.00

Figure 8.72 Initial optimization objectives.

The optimization prioritizes sparing of the organs at risk while achieving target coverage and conformity. In this case, the critical structures are physically separated from the target and no regions of overlap exist.

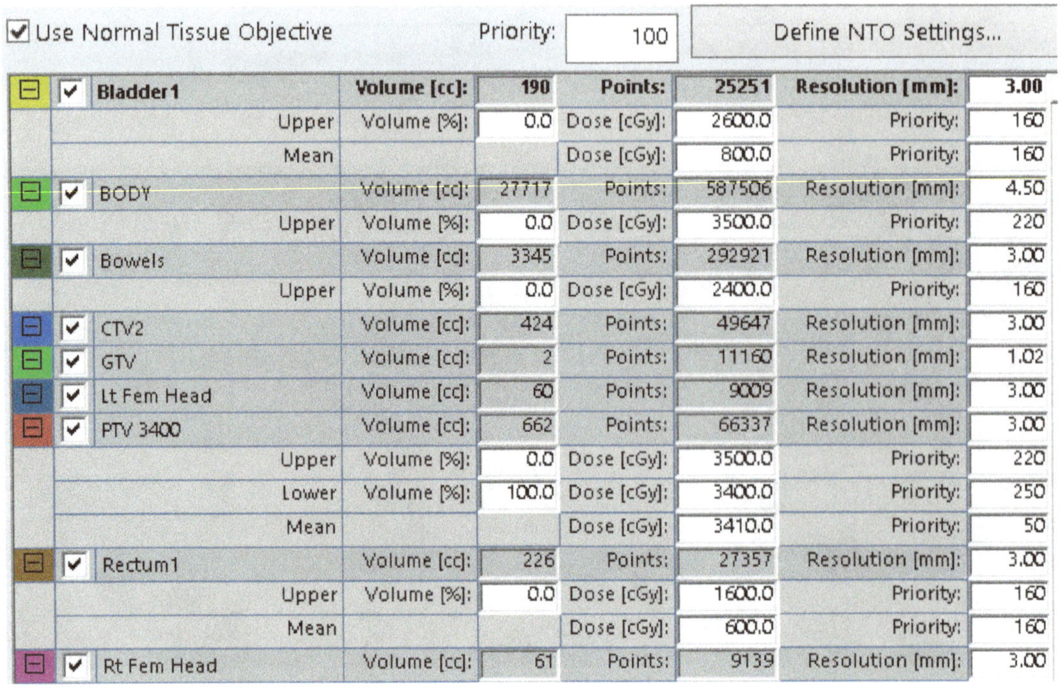

☑ Use Normal Tissue Objective — Priority: 100 — Define NTO Settings...

Bladder1		Volume [cc]:	190	Points:	25251	Resolution [mm]:	3.00
	Upper	Volume [%]:	0.0	Dose [cGy]:	2600.0	Priority:	160
	Mean			Dose [cGy]:	800.0	Priority:	160
BODY		Volume [cc]:	27717	Points:	587506	Resolution [mm]:	4.50
	Upper	Volume [%]:	0.0	Dose [cGy]:	3500.0	Priority:	220
Bowels		Volume [cc]:	3345	Points:	292921	Resolution [mm]:	3.00
	Upper	Volume [%]:	0.0	Dose [cGy]:	2400.0	Priority:	160
CTV2		Volume [cc]:	424	Points:	49647	Resolution [mm]:	3.00
GTV		Volume [cc]:	2	Points:	11160	Resolution [mm]:	1.02
Lt Fem Head		Volume [cc]:	60	Points:	9009	Resolution [mm]:	3.00
PTV 3400		Volume [cc]:	662	Points:	66337	Resolution [mm]:	3.00
	Upper	Volume [%]:	0.0	Dose [cGy]:	3500.0	Priority:	220
	Lower	Volume [%]:	100.0	Dose [cGy]:	3400.0	Priority:	250
	Mean			Dose [cGy]:	3410.0	Priority:	50
Rectum1		Volume [cc]:	226	Points:	27357	Resolution [mm]:	3.00
	Upper	Volume [%]:	0.0	Dose [cGy]:	1600.0	Priority:	160
	Mean			Dose [cGy]:	600.0	Priority:	160
Rt Fem Head		Volume [cc]:	61	Points:	9139	Resolution [mm]:	3.00

Figure 8.73 Final optimization objectives.

Dose Distribution

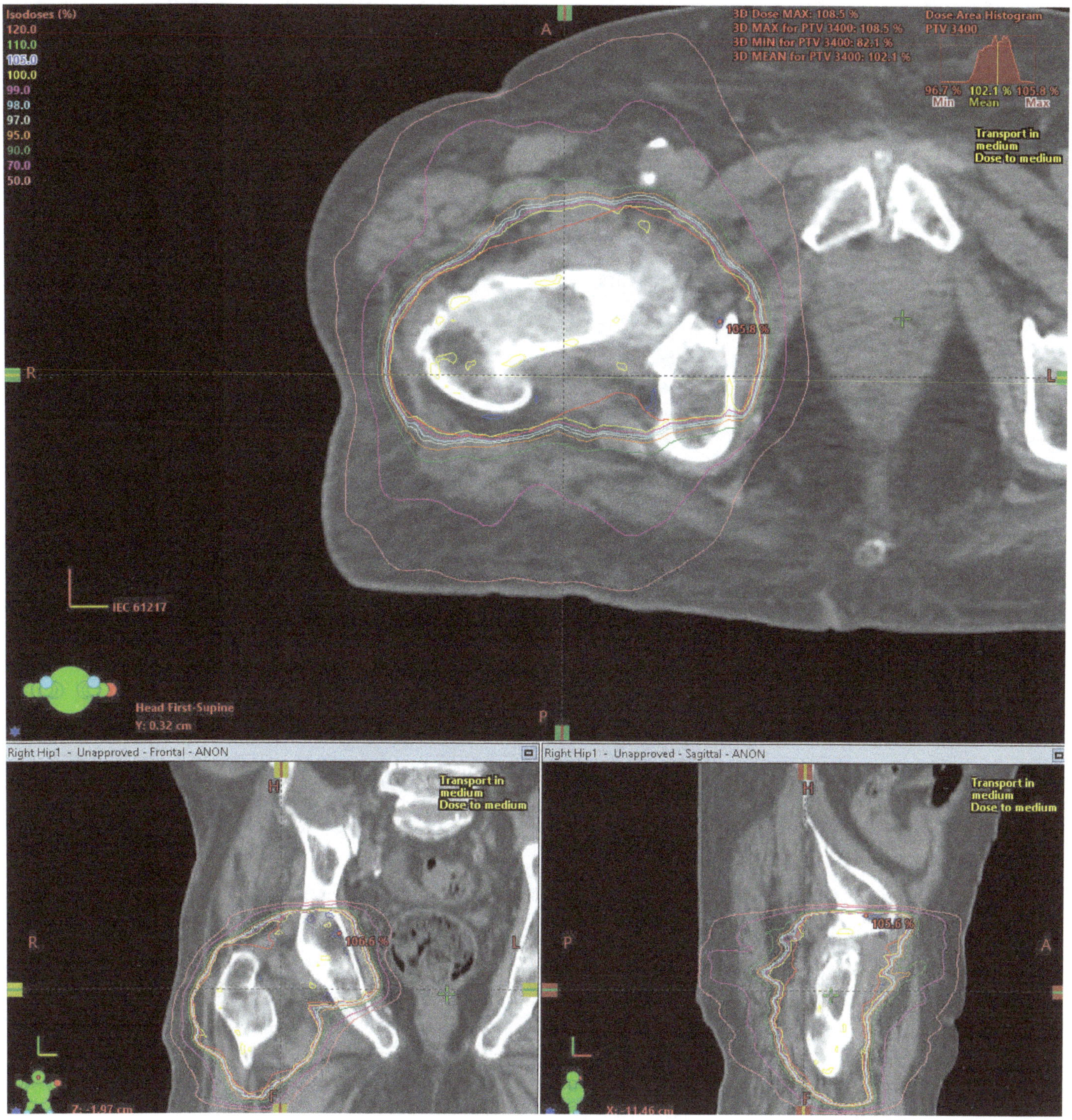

Figure 8.74 Final dose distribution.

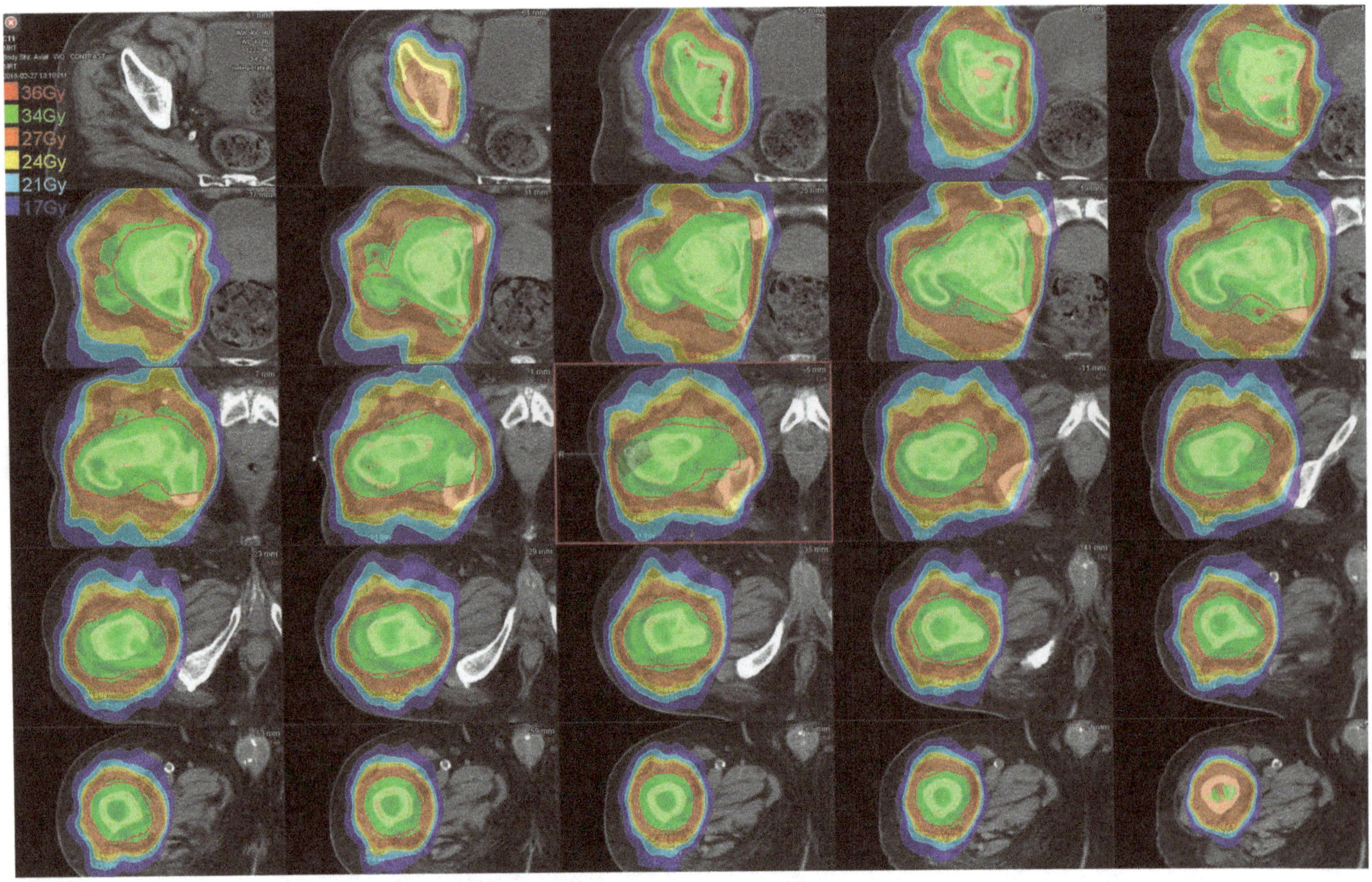

Figure 8.75 Final dose distribution in axial slices.

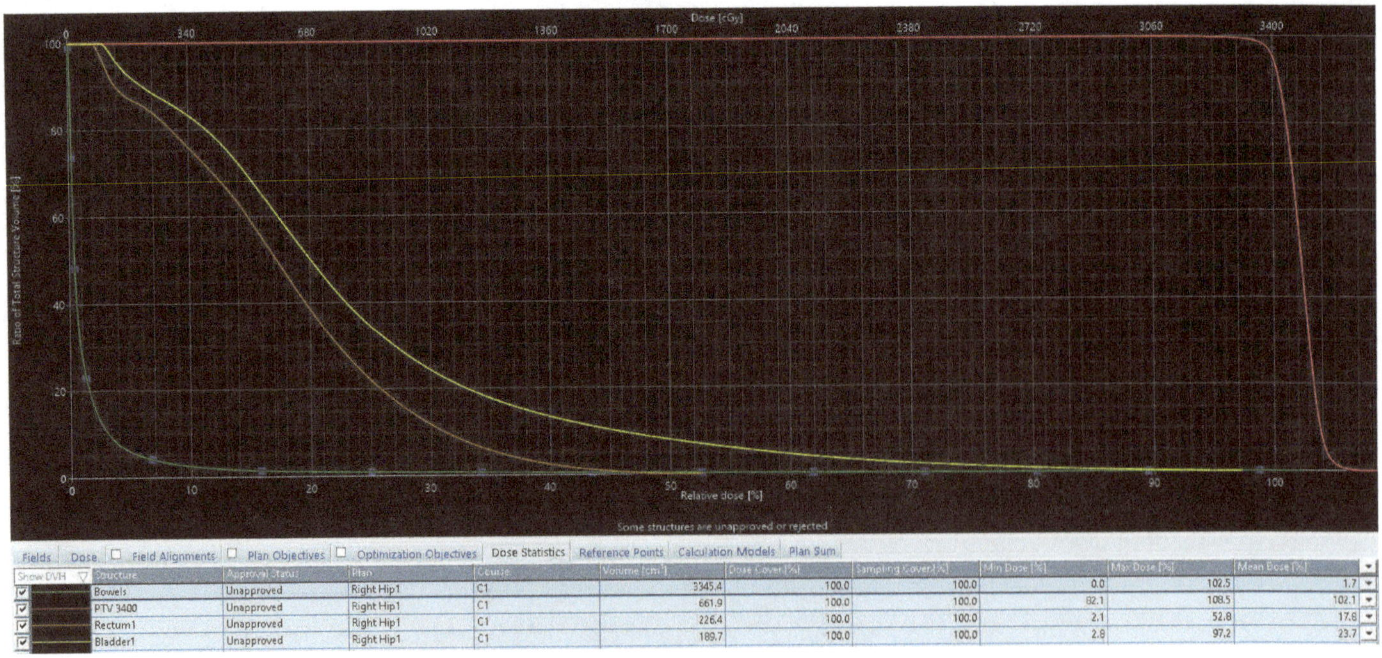

Figure 8.76 Final dose volume histogram.

Case Number eP03: Pelvis—Recurrent Rectal Cancer Treatment

General Information

Brief Clinical History

56-year-old male with a history of rectal cancer. He had previously received pre-operative radiation therapy and chemotherapy. PET/CT demonstrated recurrent disease. The patient had severe pain and urinary issues justifying palliative re-irradiation.

Prescription

Prostate and nodes:
PTV 6000 = 600 cGy × 10 fractions
PTV 5000 = 500 cGy × 10 fractions
PTV 3000 = 300 cGy × 10 fractions

Simulation/Comments

The simulation was accomplished with the patient in the supine position with a vacuum cushion underneath the pelvis. Fusion of images was performed with a recent PET/CT for targeting by the radiation oncologist.

Targets

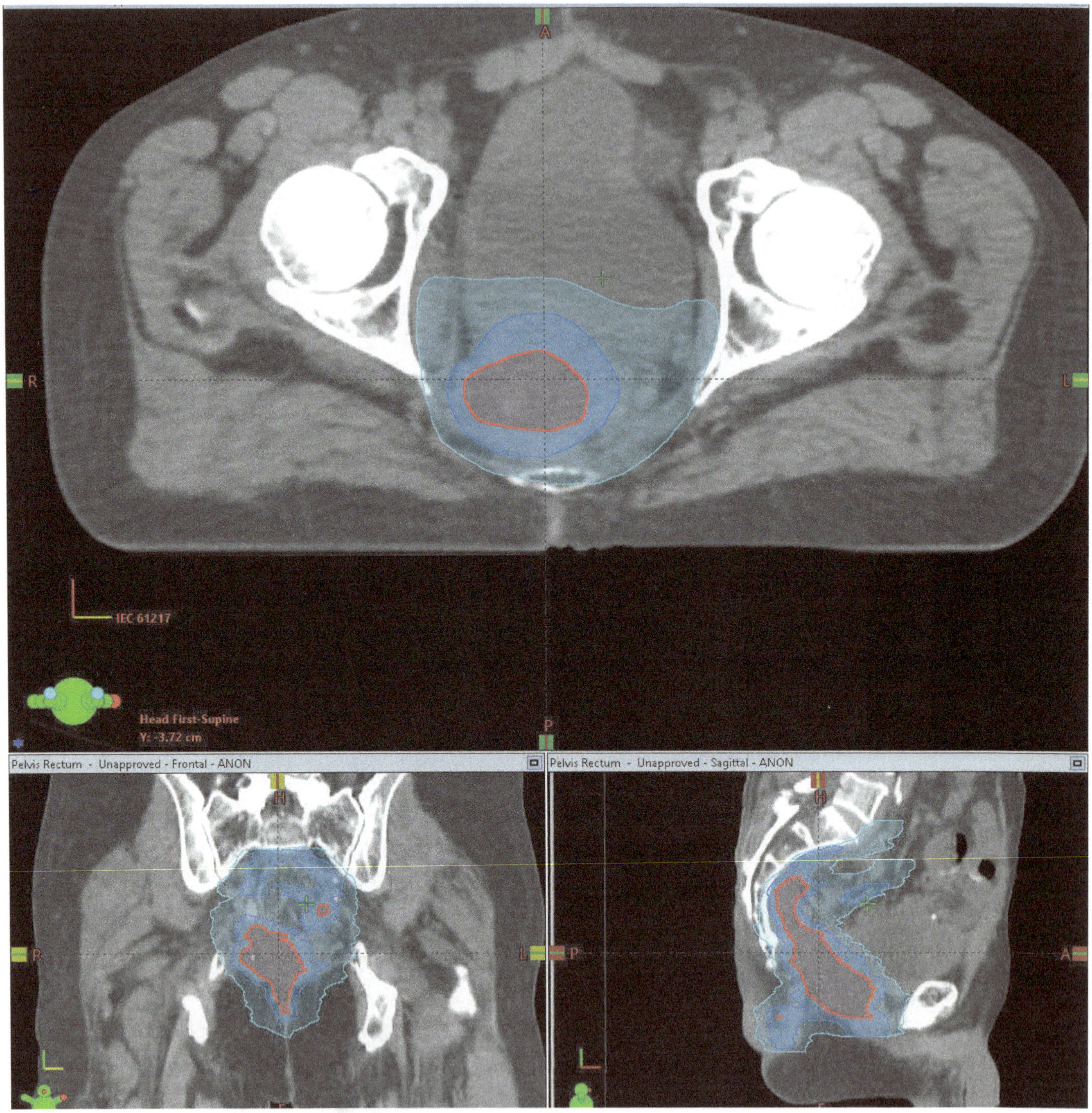

Figure 8.77 Target contours: PTV 6000 (red), PTV 5000 (colorwash dark blue), and PTV 3000 (colorwash green).

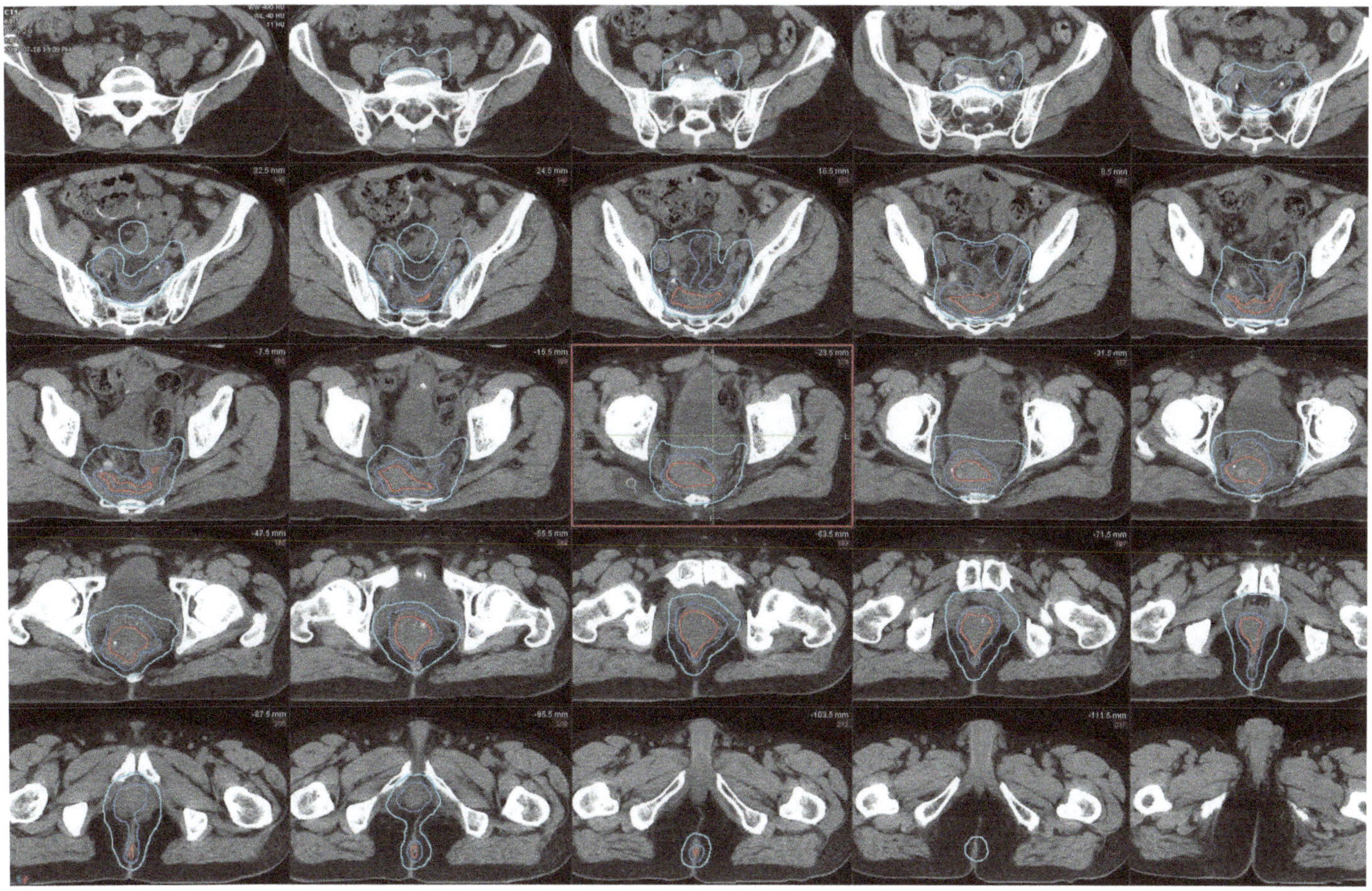

Figure 8.78 Target contours: PTV 6000 (red), PTV 5000 (colorwash dark blue), and PTV 3000 (colorwash green) in axial slices.

Beam Arrangement

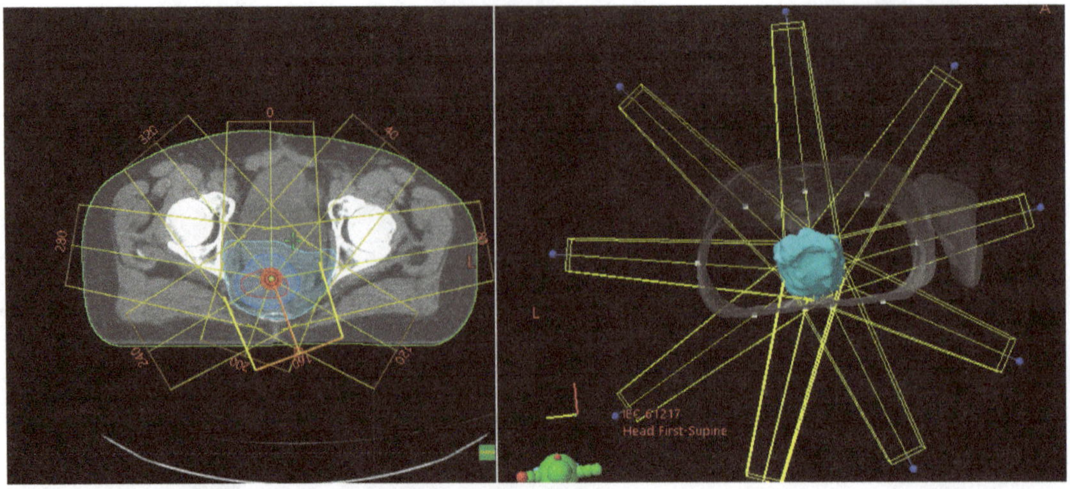

Figure 8.79 IMRT beam arrangement with 9 fields.

Gantry Angle	Couch	Collimator	Energy (MV)
200	0	0	10
240	0	0	10
280	0	0	10
320	0	0	10
0	0	0	10
40	0	0	10
80	0	0	10
120	0	0	10
160	0	0	10

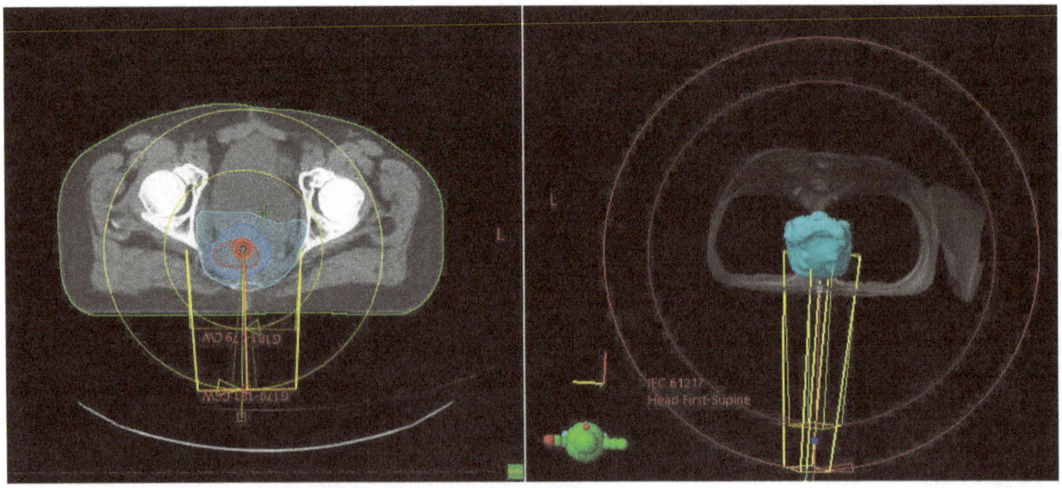

Figure 8.80 VMAT beam arrangement with 2 arcs.

Gantry Angle	Couch	Collimator	Energy (MV)
181–179 CW	0	30	10
179–181 CCW	0	330	10

Optimization

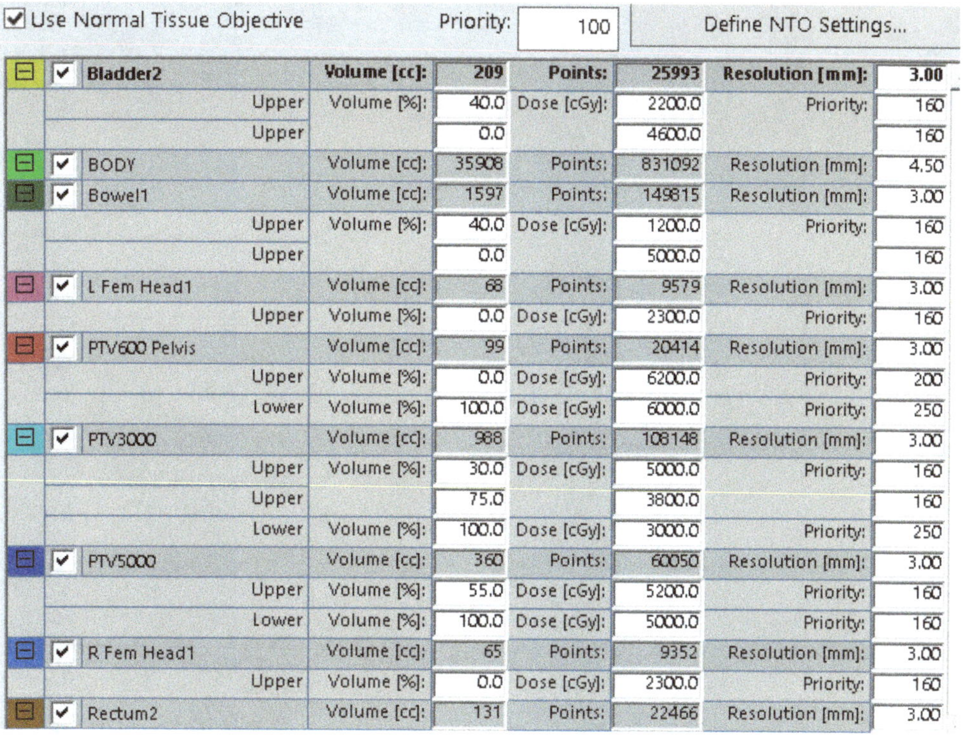

Figure 8.81 Initial optimization parameters.

The optimization will be centered in sparing organs at risk, such as the bladder and bowel, while achieving dose coverage in the target. Dose conformity for the different dose levels is also important. The planner can make use of separate contours for each dose level, as well as ring structures.

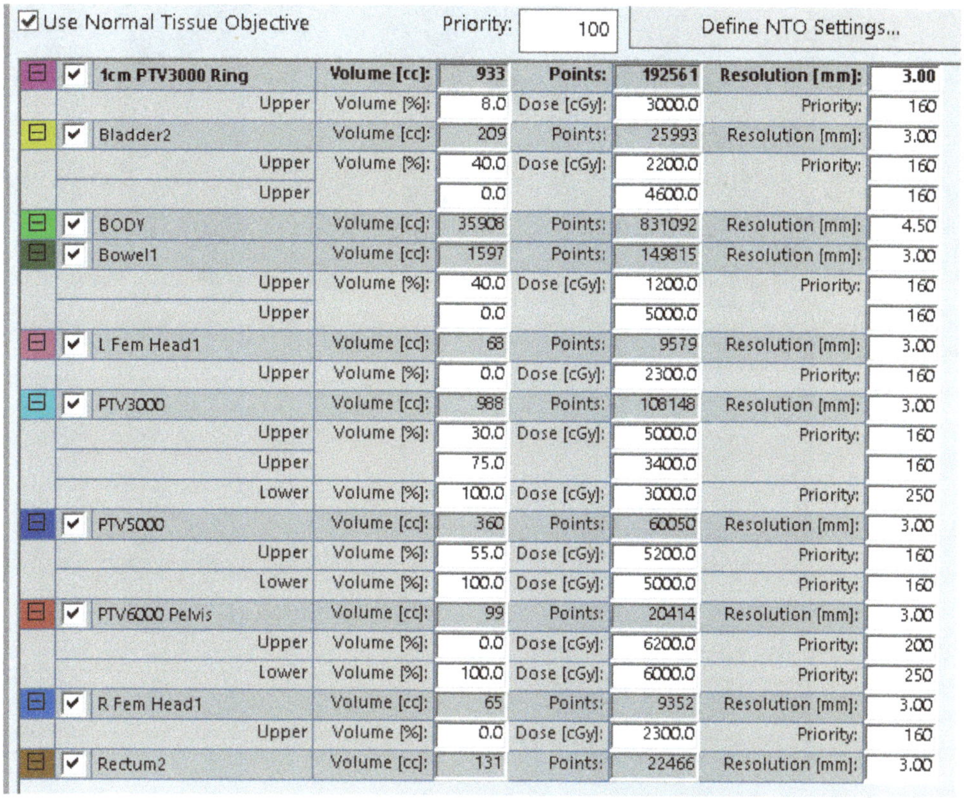

Figure 8.82 Final optimization objectives.

Dose Distribution

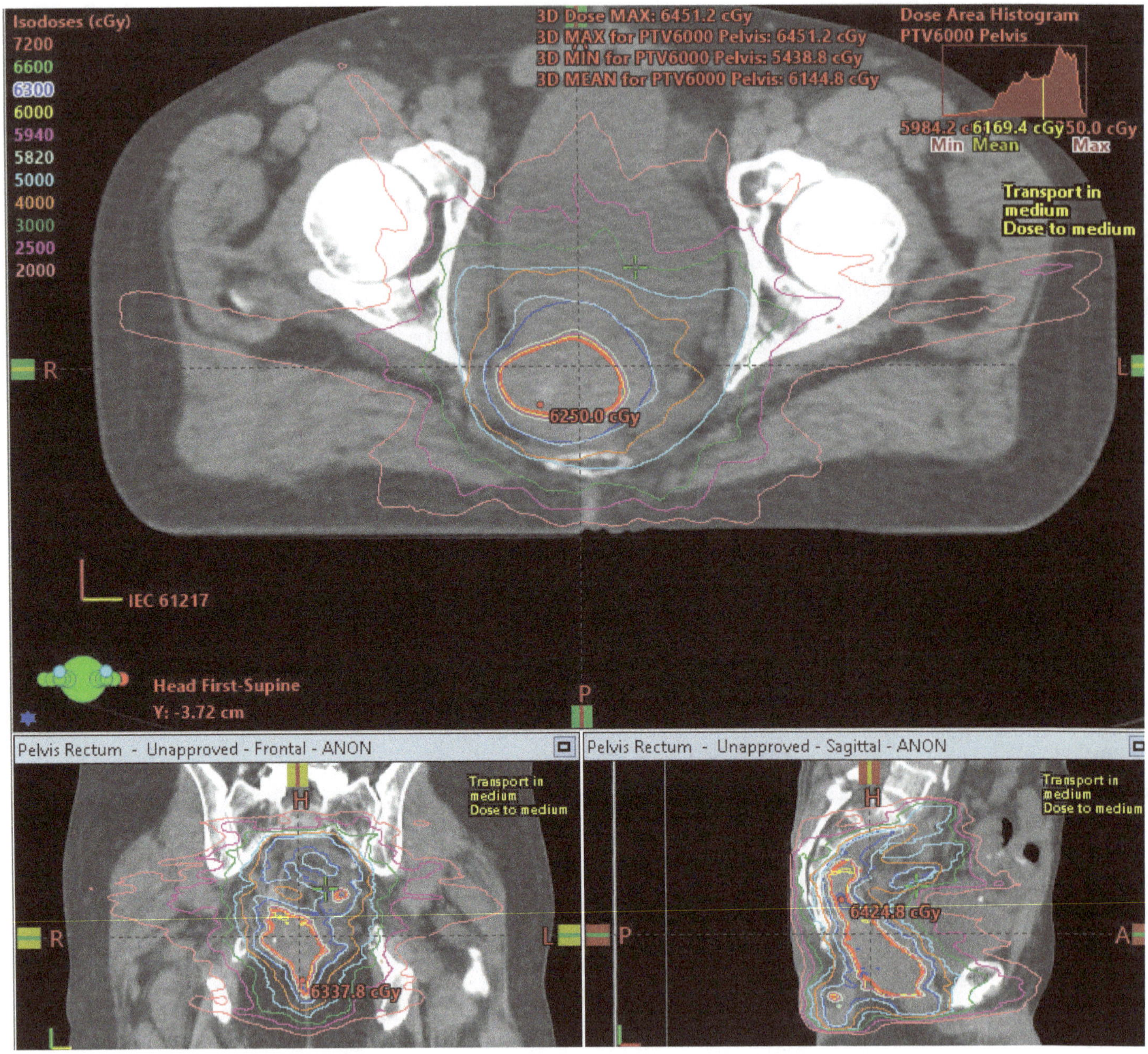

Figure 8.83 Final dose distribution.

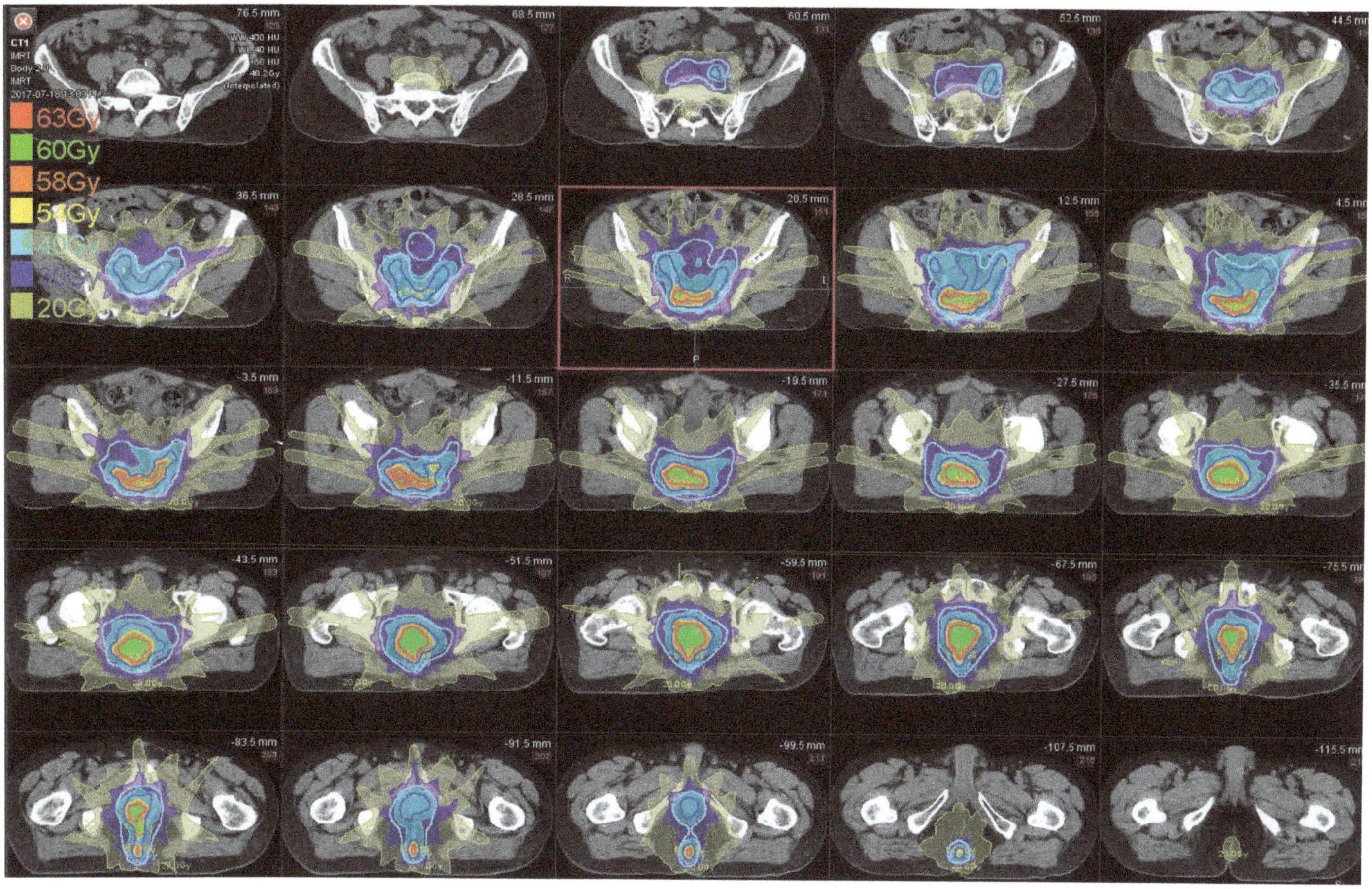

Figure 8.84 Final dose distribution on axial slices.

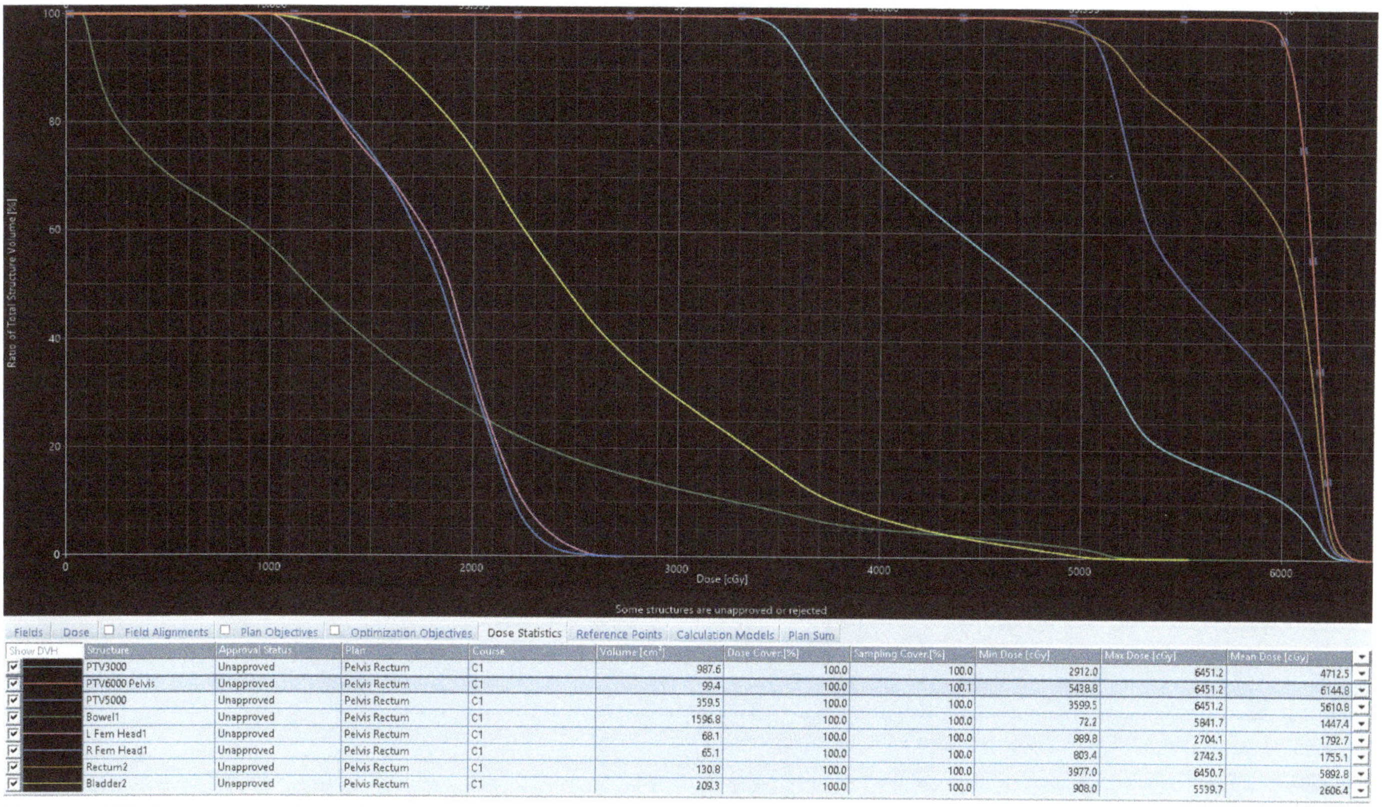

Show DVH	Structure	Approval Status	Plan	Course	Volume [cm³]	Dose Cover.[%]	Sampling Cover.[%]	Min Dose [cGy]	Max Dose [cGy]	Mean Dose [cGy]
☑	PTV3000	Unapproved	Pelvis Rectum	C1	987.6	100.0	100.0	2912.0	6451.2	4712.5
☑	PTV6000 Pelvis	Unapproved	Pelvis Rectum	C1	99.4	100.0	100.1	5438.8	6451.2	6144.8
☑	PTV5000	Unapproved	Pelvis Rectum	C1	359.5	100.0	100.0	3599.5	6451.2	5610.8
☑	Bowel1	Unapproved	Pelvis Rectum	C1	1596.8	100.0	100.0	72.2	5841.7	1447.4
☑	L Fem Head1	Unapproved	Pelvis Rectum	C1	68.1	100.0	100.0	989.8	2704.1	1792.7
☑	R Fem Head1	Unapproved	Pelvis Rectum	C1	65.1	100.0	100.0	803.4	2742.3	1755.1
☑	Rectum2	Unapproved	Pelvis Rectum	C1	130.8	100.0	100.0	3977.0	6450.7	5892.8
☑	Bladder2	Unapproved	Pelvis Rectum	C1	209.3	100.0	100.0	908.0	5539.7	2606.4

Figure 8.85 Final dose volume histogram.

9. RayStation TPS Sample IMRT and VMAT Plans

Case Number rTA01 Thorax and Abdomen: Right Lung Cancer Radiation Treatment

General Information

Brief Clinical History

80-year-old man with non-small-cell lung cancer and an oligometastatic right lung mass. Radiation therapy was to be delivered with concurrent chemotherapy.

Prescription

Right lung:
PTV 6000 = 750 cGy × 8 fractions

Simulation

The simulation was accomplished with a vacuum cushion as an immobilization device. An image fusion was completed with a 4DCT performed at the same time of simulation to evaluate tumor motion and a recent PET/CT for evaluating the extent of the tumor.

NOTE: All sample cases in this chapter are for educational purposes only. Hypofractionated regimens, SRS and SBRT procedures must be tested and implemented clinically in a separate commissioning process from basic IMRT/VMAT implementation. All dose distributions are not suggestions for clinical treatments. Objectives in the optimizer as well as beam arrangements are for illustration/educational purposes only.

Targets

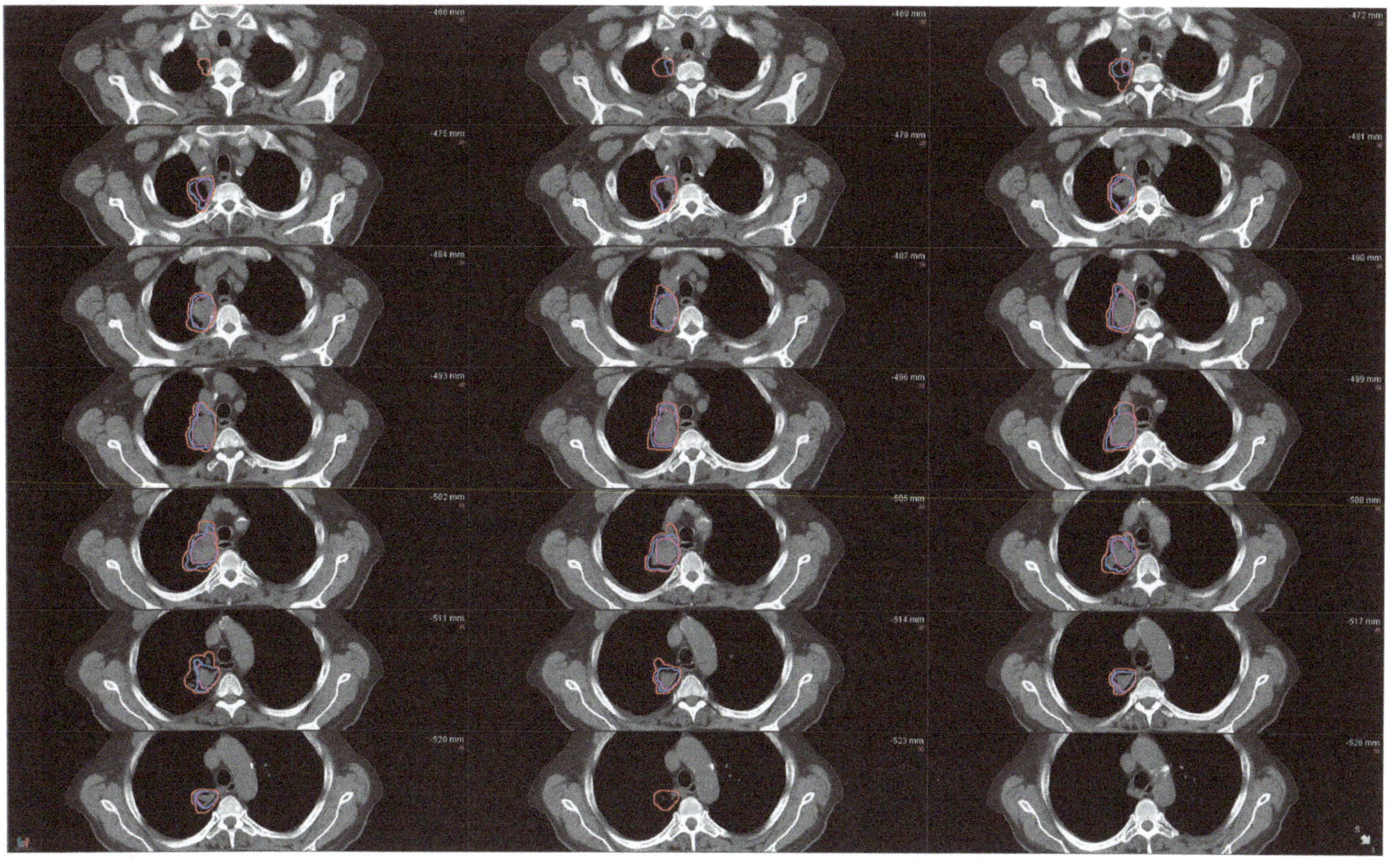

Figure 9.1 The PTV 6000 cGy is in red, ITV in blue, and GTV in purple.

There is only one PTV contour in this case. The GTV and ITV are shown as examples. The extent of the lesion limits dose to organs in the superior thorax, such as the lungs, heart, great vessels, and spinal cord.

Beam Arrangement

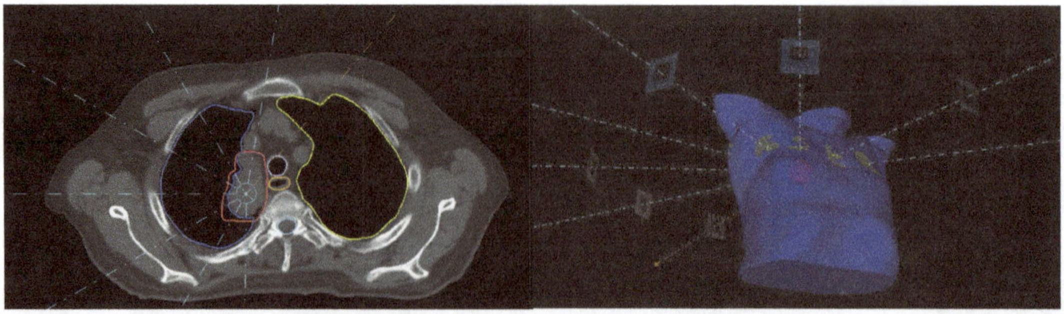

Figure 9.2 IMRT beam arrangement for case rTA01.

Gantry Angle	Couch	Collimator	Energy (MV)
181	0	0	10 FFF
210	0	0	10 FFF
240	0	0	10 FFF
270	0	0	10 FFF
300	0	0	10 FFF
330	0	0	10 FFF
350	0	0	10 FFF
10	0	0	10 FFF
40	0	0	10 FFF

Figures 9.2 and 9.3 show the beam arrangement for IMRT and VMAT plans. The IMRT plan has 8 beams distributed around the right lung to limit dose entry and exit through the left lung. A 10-MV flattening filter-free (FFF) beam has been used to decrease treatment time. The planner can evaluate the use of a non-FFF beam with considerations such as modulation, time, and dose homogeneity. The planner could also choose a 6-MV beam to help with the lateral electronic equilibrium in the part of the tumor close to the lung. In this particular case, a 10 FFF beam disrupts lateral electronic equilibrium in the lung tissue and sparing it as a result.

Figure 9.3 shows the VMAT plan has two partial arcs that span the length of the IMRT fields. The collimator on the VMAT plan has different rotations to prevent dose accumulation from inter-leaf leakage and the tongue-and-groove effect. A 10-MV FFF has been used to decrease treatment time.

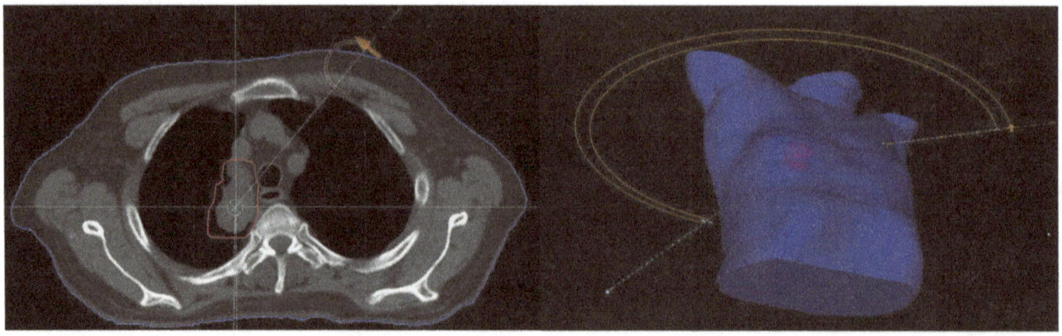

Figure 9.3 VMAT beam arrangement

Gantry Angle	Couch	Collimator	Energy (MV)
181–40 CW	0	30	10 FFF
40–181 CCW	0	330	10 FFF

Optimization

■ Physical Composite Objective				
Min Dose	Plan	PTV60Gy	Min Dose 6000 cGy	30
Max Dose	Plan	PTV60Gy	Max Dose 7200 cGy	1
Dose Fall-Off	Plan	External	Dose Fall-Off [H]6000 cGy [L]0 cGy, Low dose distance 5.00 cm	1
Max Dose	Plan	L Lung	Max Dose 1000 cGy	1
Dose Fall-Off	Plan	R Lung	Dose Fall-Off [H]6000 cGy [L]0 cGy, Low dose distance 5.00 cm	1
Max Dose	Plan	Esophagus	Max Dose 3700 cGy	3
Dose Fall-Off	Plan	Esophagus	Dose Fall-Off [H]6000 cGy [L]0 cGy, Low dose distance 4.00 cm	1
Max DVH	Plan	Canal	Max DVH 2000 cGy to 10% volume	1
Max Dose	Plan	Canal	Max Dose 3300 cGy	1
Max Dose	Plan	Chest Wall	Max Dose 4000 cGy	3
Max DVH	Plan	Chest Wall	Max DVH 2800 cGy to 1% volume	1
Max Dose	Plan	Trachea	Max Dose 4500 cGy	3

Figure 9.4 Initial objectives for case rTA01.

Optimization contours: RayStation's optimization functions can often eliminate the need for several optimization contours, such as rings. In cases where the target volume has significant concavities, rings may be necessary, but the dose fall-off function can frequently be used without the need for rings.

Optimizer: The optimization and segmentation settings allowed for 90 segments. For the first optimization, the planner could focus on achieving dose coverage on the target. In later optimizations, the planner can start to add objectives to reduce the dose to OARs. Note that the dose fall-off objective on the external contour is similar to a ring structure. A convenient feature of RayStation is the option to enter and evaluate a plan based on clinical goals. One can see the physician-defined clinical goals below and how the first optimization did with respect the achieving of those goals.

Plan dose: R Lung (CT 1)	Canal	At most 10.00 % volume at 2640 cGy dose	4.94 %	✓	0 %
Plan dose: R Lung (CT 1)	Canal	At most 3360 cGy dose at 0.0 cm³ volume	3473 cGy	!	0 %
Plan dose: R Lung (CT 1)	Chest Wall	At most 30.0 cm³ volume at 3000 cGy dose	47.5 cm³	!	0 %
Plan dose: R Lung (CT 1)	Esophagus	At most 5.0 cm³ volume at 2160 cGy dose	4.7 cm³	✓	0 %
Plan dose: R Lung (CT 1)	Esophagus	At most 2000 cGy average dose	722 cGy	✓	0 %
Plan dose: R Lung (CT 1)	Esophagus	At most 3840 cGy dose at 0.0 cm³ volume	4799 cGy	!	0 %
Plan dose: R Lung (CT 1)	Great Vessels	At most 10.0 cm³ volume at 5520 cGy dose	0.0 cm³	✓	0 %
Plan dose: R Lung (CT 1)	Great Vessels	At most 6300 cGy dose at 0.0 cm³ volume	5139 cGy	✓	0 %
Plan dose: R Lung (CT 1)	Heart	At most 15.0 cm³ volume at 3440 cGy dose	0.0 cm³	✓	0 %

Figure 9.5 RayStation's Clinical Goals window.

There are various tools with which the planner can evaluate how to add or adjust optimization functions, such as objectives or weights, to better achieve the dose distribution goals.

Function	Constraint	Dose	ROI	Description	Robust	Weight
■ Physical Composite Objective						
Min Dose		Plan	PTV60Gy	Min Dose 6000 cGy		30
Max Dose		Plan	PTV60Gy	Max Dose 7200 cGy		1
Dose Fall-Off		Plan	External	Dose Fall-Off [H]6000 cGy [L]0 cGy, Low dose distance 5.00 cm		1
Max Dose		Plan	L Lung	Max Dose 1000 cGy		1
Dose Fall-Off		Plan	R Lung	Dose Fall-Off [H]6000 cGy [L]0 cGy, Low dose distance 5.00 cm		1
Max Dose		Plan	Esophagus	Max Dose 3000 cGy		15
Dose Fall-Off		Plan	Esophagus	Dose Fall-Off [H]6000 cGy [L]0 cGy, Low dose distance 4.00 cm		4
Max DVH		Plan	Canal	Max DVH 2000 cGy to 10% volume		4
Max Dose		Plan	Canal	Max Dose 2900 cGy		4
Max Dose		Plan	Chest Wall	Max Dose 3600 cGy		6
Max DVH		Plan	Chest Wall	Max DVH 2500 cGy to 1% volume		4
Max Dose		Plan	Trachea	Max Dose 4000 cGy		6

Figure 9.6 Final objectives window (not all objectives shown).

Dose Distribution

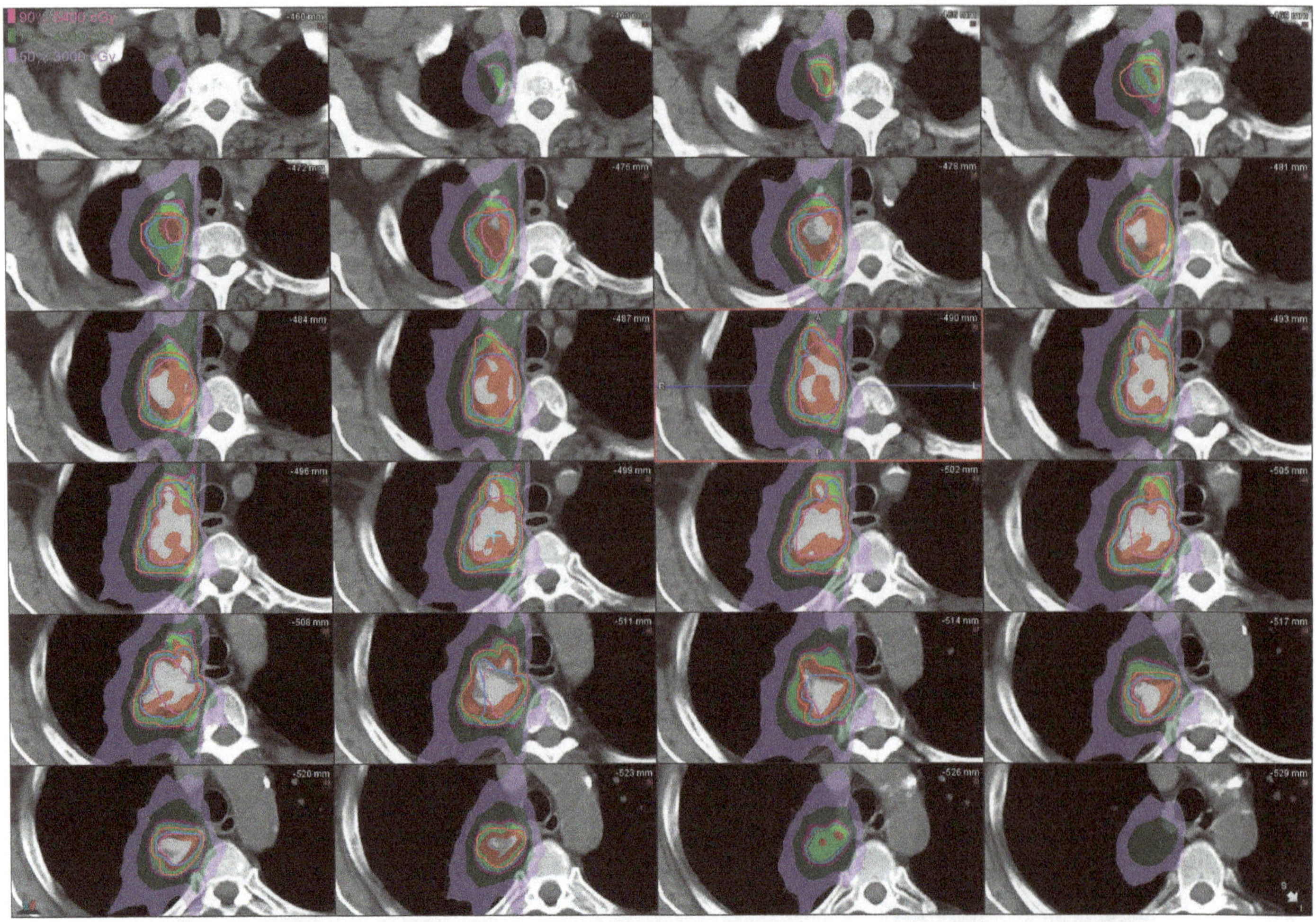

Figure 9.7 Final dose distribution.

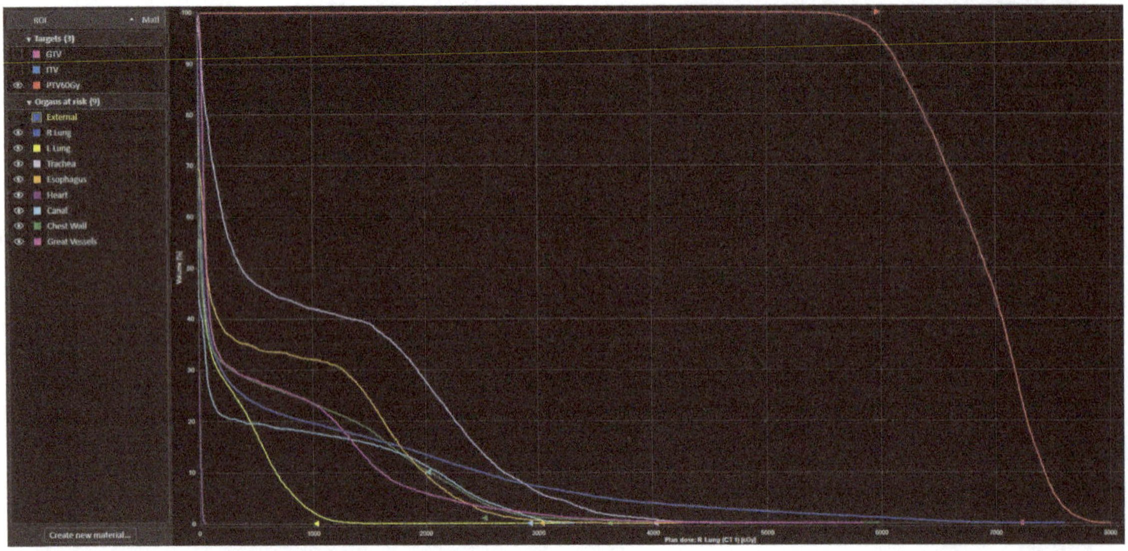

Figure 9.8 Final dose volume histogram.

In this particular case of hypofractionation, the radiation oncologist accepted a maximum dose of 8000 cGy or roughly 130% of the prescription dose. The optimization took advantage of this scenario and minimized the low-dose region around the PTV volume with a short dose fall-off distance.

Case Number rTA02 Thorax and Abdomen: Lower Esophagus Cancer Radiation Treatment

General Information

Brief Clinical History

61-year-old male patient diagnosed with squamous cell carcinoma from unknown primary pathology. He presents a large lymph node in his neck with a cancerous mass in his lower esophagus. A clinical decision was made to treat the esophagus with chemotherapy and radiation therapy.

Prescription

Esophagus:
PTV 5400 cGy = 216 cGy × 25 fractions
PTV 4500 cGy = 180 cGy × 25 fractions

Simulation/Comments

The simulation was accomplished with a vacuum cushion as an immobilization device. A PET/CT was fused to the planning CT to assist the radiation oncologist in targeting and evaluating the extent of the lesion.

Targets

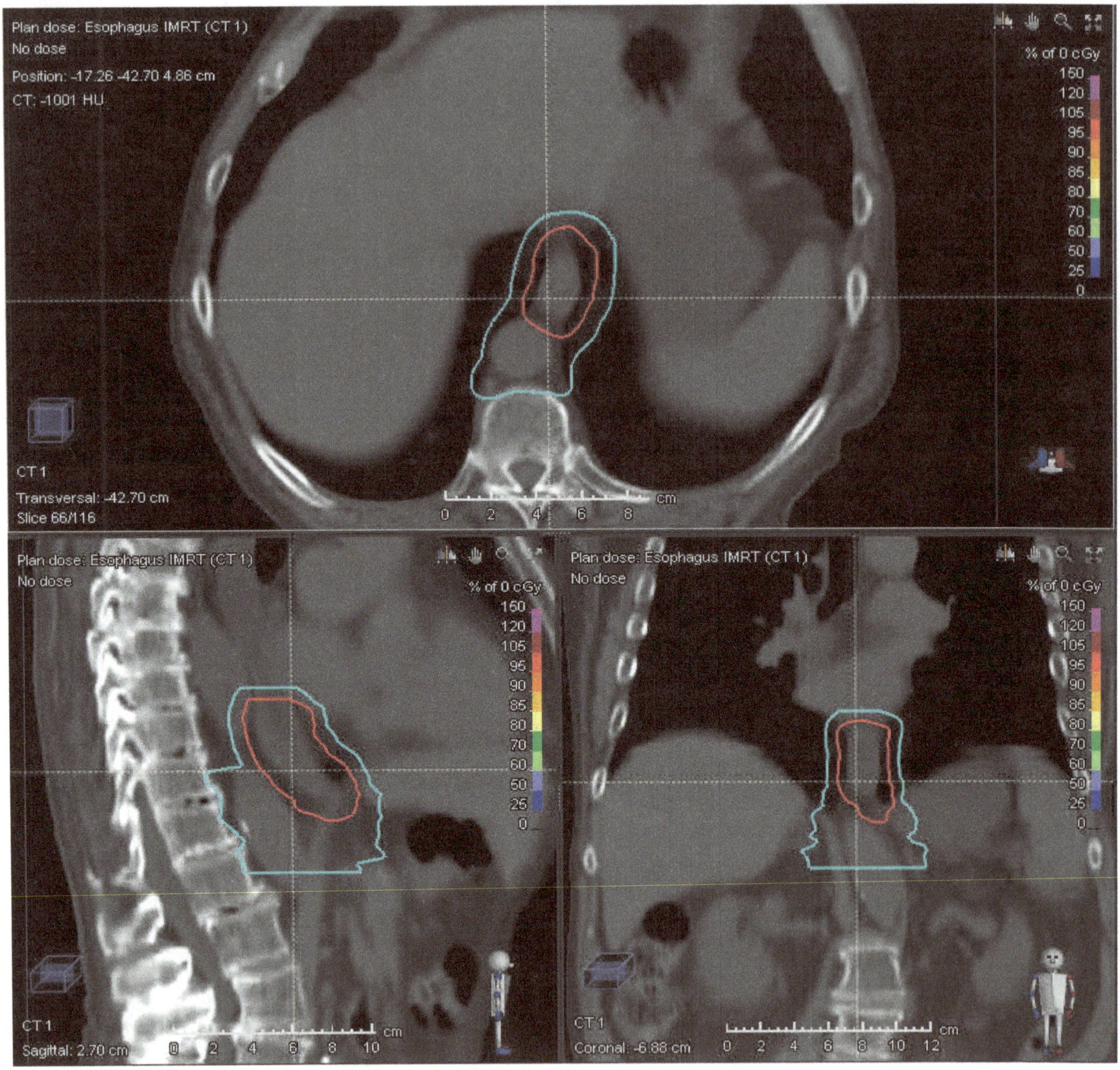

Figure 9.9 PTV 5400 (red) and PTV 4500 (cyan).

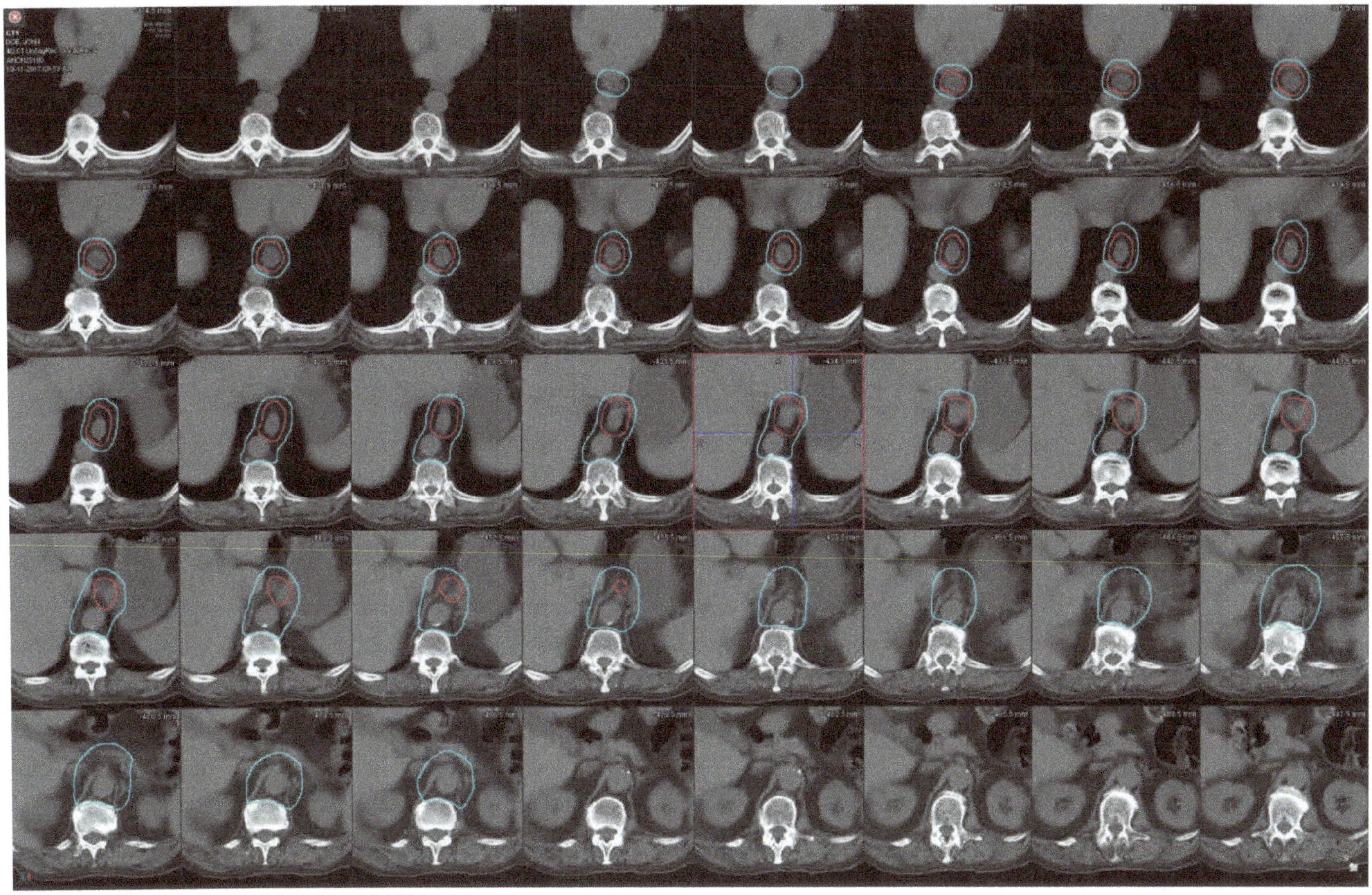

Figure 9.10 PTV 5400 (red) and PTV 4500 (blue).

Beam Arrangement

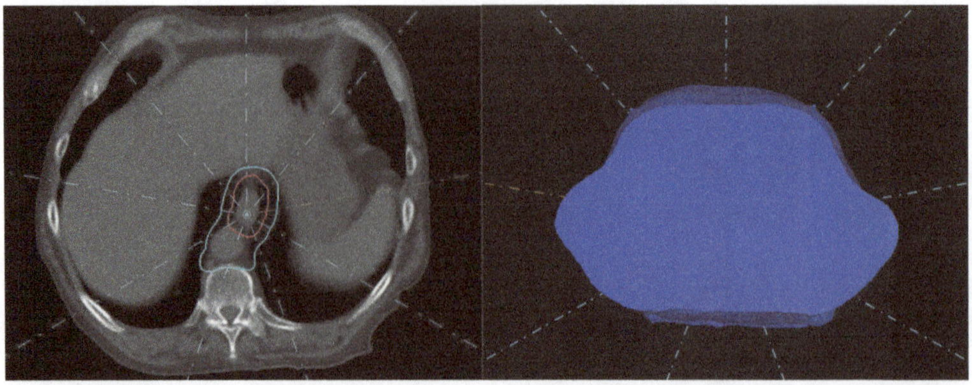

Figure 9.11 IMRT beam arrangement.

Gantry Angle	Couch	Collimator	Energy (MV)
200	0	0	10
240	0	0	10
280	0	0	10
320	0	0	10
0	0	0	10
40	0	0	10
80	0	0	10
120	0	0	10
160	0	0	10

A 9-field beam arrangement with beams evenly distributed every 40 degrees can be used due to the central location of the tumor. The planner could also choose to use a 7-field arrangement to decrease the treatment time, but 9 fields maximizes the modulation and dose sparing.

For VMAT, one or two 360-degree arcs can be used, as shown in Figure 9.12.

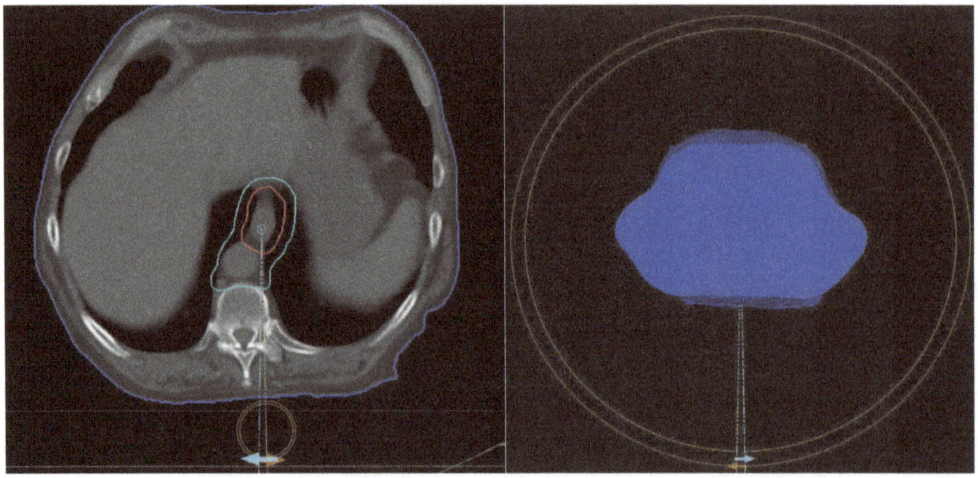

Figure 9.12 VMAT beam arrangement.

Gantry Angle	Couch	Collimator	Energy (MV)
G181–179 CW	0	30	10
G179–181 CCW	0	330	10

Optimization

Dose	ROI/POI	Clinical goal	Value	Result	% outside grid
Plan dose: Esophagus I...	Bowel Bag	At most 120.0 cm³ volume at 4400 cGy dose		N/A	0 %
Plan dose: Esophagus I...	Bowel Bag	At most 5000 cGy dose at 0.0 cm³ volume		N/A	0 %
Plan dose: Esophagus I...	Duodenum	At most 5.0 cm³ volume at 4000 cGy dose		N/A	0 %
Plan dose: Esophagus I...	Duodenum	At most 4400 cGy dose at 0.0 cm³ volume		N/A	0 %
Plan dose: Esophagus I...	Esopagus	At most 5.0 cm³ volume at 5500 cGy dose		N/A	0 %
Plan dose: Esophagus I...	Esopagus	At most 5800 cGy dose at 0.0 cm³ volume		N/A	0 %
Plan dose: Esophagus I...	Heart	At most 15.0 cm³ volume at 4600 cGy dose		N/A	0 %
Plan dose: Esophagus I...	Heart	At most 5200 cGy dose at 0.0 cm³ volume		N/A	0 %
Plan dose: Esophagus I...	Kidneys	At most 260.0 cm³ volume at 2000 cGy dose		N/A	0 %
Plan dose: Esophagus I...	Liver	At most 900.0 cm³ volume at 2800 cGy dose		N/A	0 %
Plan dose: Esophagus I...	PTV4500	At least 95.00 % volume at 4500 cGy dose		N/A	0 %
Plan dose: Esophagus I...	PTV5400	At least 95.00 % volume at 5400 cGy dose		N/A	0 %
Plan dose: Esophagus I...	Spinal Canal	At most 5.0 cm³ volume at 4200 cGy dose		N/A	0 %
Plan dose: Esophagus I...	Spinal Canal	At most 4600 cGy dose at 0.0 cm³ volume		N/A	0 %
Plan dose: Esophagus I...	Stomach	At most 50.0 cm³ volume at 4400 cGy dose		N/A	0 %
Plan dose: Esophagus I...	Stomach	At most 5200 cGy dose at 0.0 cm³ volume		N/A	0 %

Figure 9.13 Clinical goals for plan rTA02.

In order to achieve these objectives with the 9-field IMRT arrangement, the planner allowed up to 90 segments in the plan and set the following initial dose functions. The number of segments can offer a dosimetric advantage in terms of modulation. However, the planner must be aware to avoid extending the delivery time.

Function	Constraint	Dose	ROI	Description	Robust	Weight
Physical Composite Objective						
Min Dose		Plan	PTV5400	Min Dose 5400 cGy		30
Min Dose		Plan	PTV4500	Min Dose 4500 cGy		30
Dose Fall-Off		Plan	PTV4500	Dose Fall-Off [H]5400 cGy [L]4500 cGy, Low dose distance 0.10 cm		1
Dose Fall-Off		Plan	External	Dose Fall-Off [H]5400 cGy [L]0 cGy, Low dose distance 10.00 cm		1
Dose Fall-Off		Plan	Spinal Canal	Dose Fall-Off [H]5400 cGy [L]0 cGy, Low dose distance 3.00 cm		1
Max Dose		Plan	Spinal Canal	Max Dose 1800 cGy		1
Dose Fall-Off		Plan	Stomach	Dose Fall-Off [H]5400 cGy [L]0 cGy, Low dose distance 10.00 cm		1
Dose Fall-Off		Plan	Kidneys	Dose Fall-Off [H]5400 cGy [L]0 cGy, Low dose distance 5.00 cm		1
Max Dose		Plan	PTV5400	Max Dose 5700 cGy		1
Max Dose		Plan	Heart	Max Dose 5450 cGy		1

Figure 9.14 Initial optimization objectives.

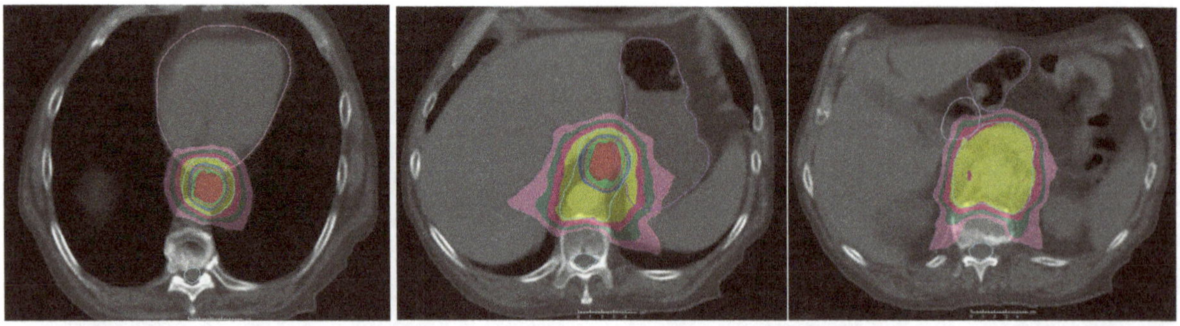

Figure 9.15 From left to right, the PTV overlap with the heart and esophagus can be observed. On the far right, the duodenum is directly adjacent to but not overlapping with the PTV.

After the first optimization, the user can evaluate the DVH against the clinical goals. The esophagus receives prescription dose as it is the primary treatment site. The radiation oncologist allowed for prescription dose on this OAR with minimal hot spots.

Function	Constraint	Dose	ROI	Description	Robust	Weight
Physical Composite Objective						
Min Dose		Plan	PTV5400	Min Dose 5400 cGy		30
Min Dose		Plan	PTV4500	Min Dose 4500 cGy		30
Dose Fall-Off		Plan	PTV4500	Dose Fall-Off [H]5400 cGy [L]4500 cGy, Low dose distance 0.10 cm		5
Dose Fall-Off		Plan	External	Dose Fall-Off [H]5400 cGy [L]0 cGy, Low dose distance 10.00 cm		1
Dose Fall-Off		Plan	Spinal Canal	Dose Fall-Off [H]5400 cGy [L]0 cGy, Low dose distance 3.00 cm		1
Max Dose		Plan	Spinal Canal	Max Dose 1800 cGy		1
Dose Fall-Off		Plan	Stomach	Dose Fall-Off [H]5400 cGy [L]0 cGy, Low dose distance 10.00 cm		1
Dose Fall-Off		Plan	Kidneys	Dose Fall-Off [H]5400 cGy [L]0 cGy, Low dose distance 5.00 cm		1
Max Dose		Plan	PTV5400	Max Dose 5550 cGy		1
Max Dose		Plan	Heart	Max Dose 5450 cGy		1
Max Dose		Plan	Duodenum	Max Dose 4000 cGy		3
Max Dose		Plan	Stomach	Max Dose 5450 cGy		5

Figure 9.16 Final optimization objectives (not all objectives are pictured).

Dose Distribution

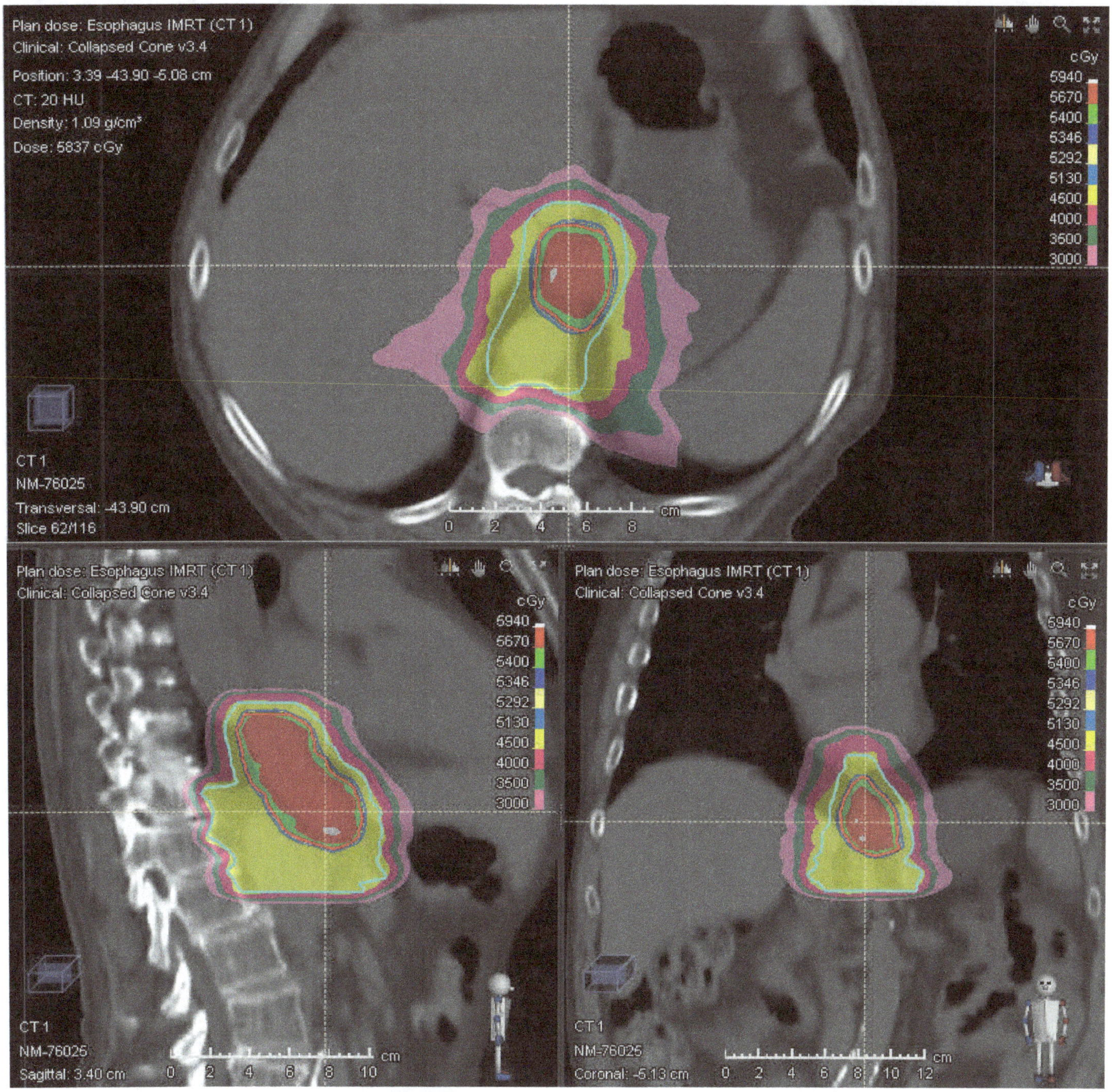

Figure 9.17 Final dose distribution for case rTA02.

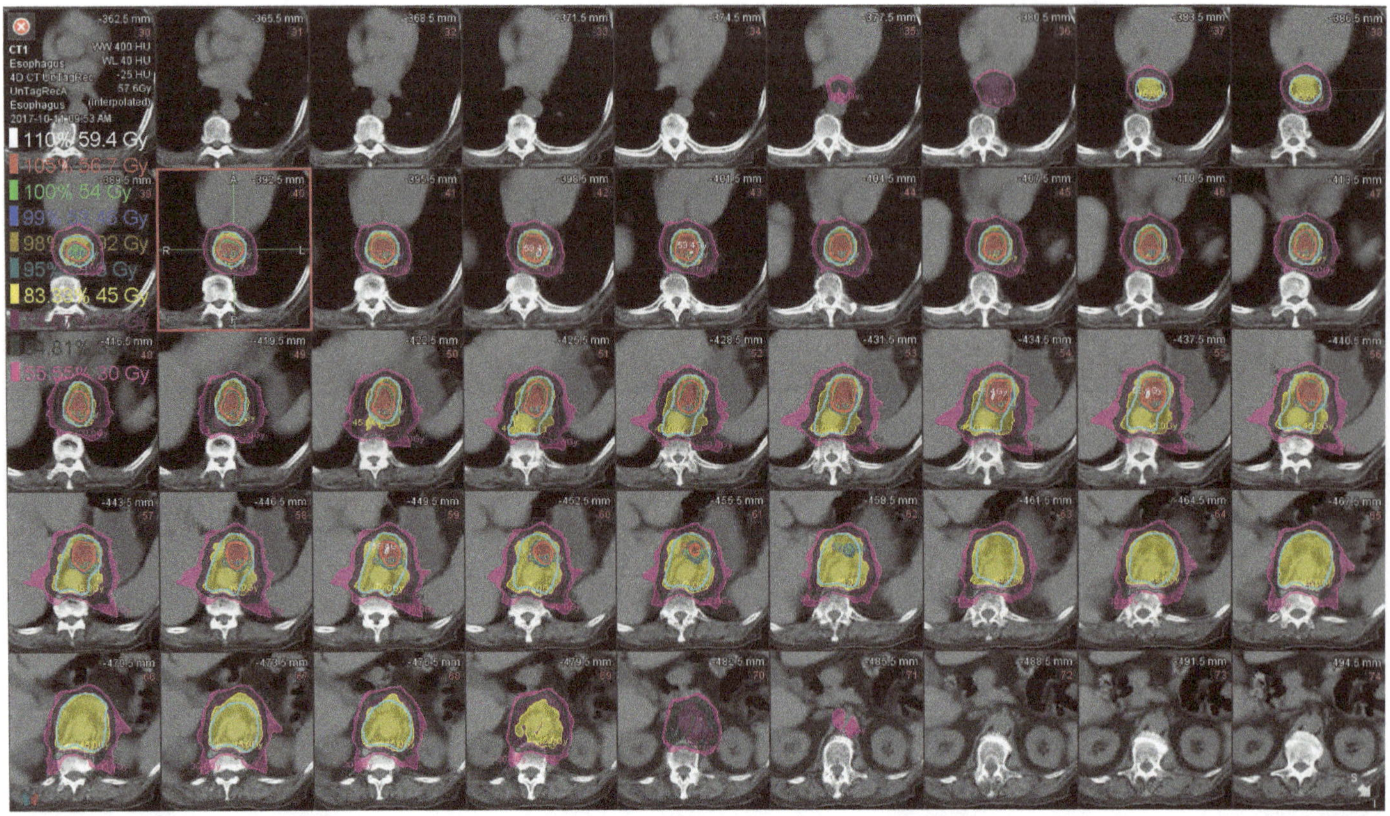

Figure 9.18 Final dose distribution for case rTA02 on axial slices.

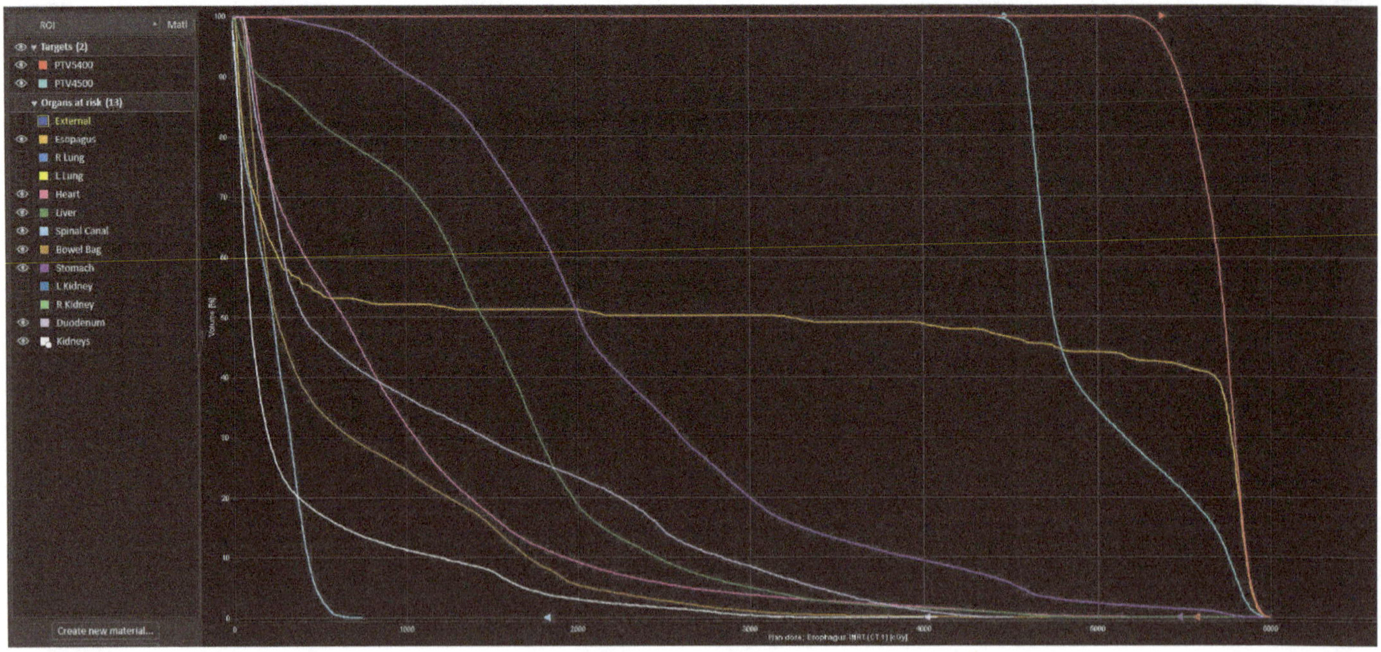

Figure 9.19 Final dose volume histogram.

Case Number rTA03 Thorax and Abdomen: T7 Vertebral Body Radiation Treatment

General Information

Brief Clinical History

72-year-old man with multiple bone metastasis. He was feeling pain in his spine, and PET/CT revealed an avid metastasis in his T7 vertebral body. The radiation oncologist decided to treat with SBRT to the vertebral bodies that were currently causing discomfort.

Prescription

T7 vertebral body:
PTV 2700 = 900 cGy × 3 fractions
PTV 2100 = 700 cGy × 3 fractions

Simulation/Comments

The simulation was accomplished with a vacuum cushion as an immobilization device positioned beneath the patient's thorax up to his head. The patient had his arms above his head and a small knee cushion for comfort. An image fusion was completed with PET/CT and MRI for target and spinal cord delineation.

Targets

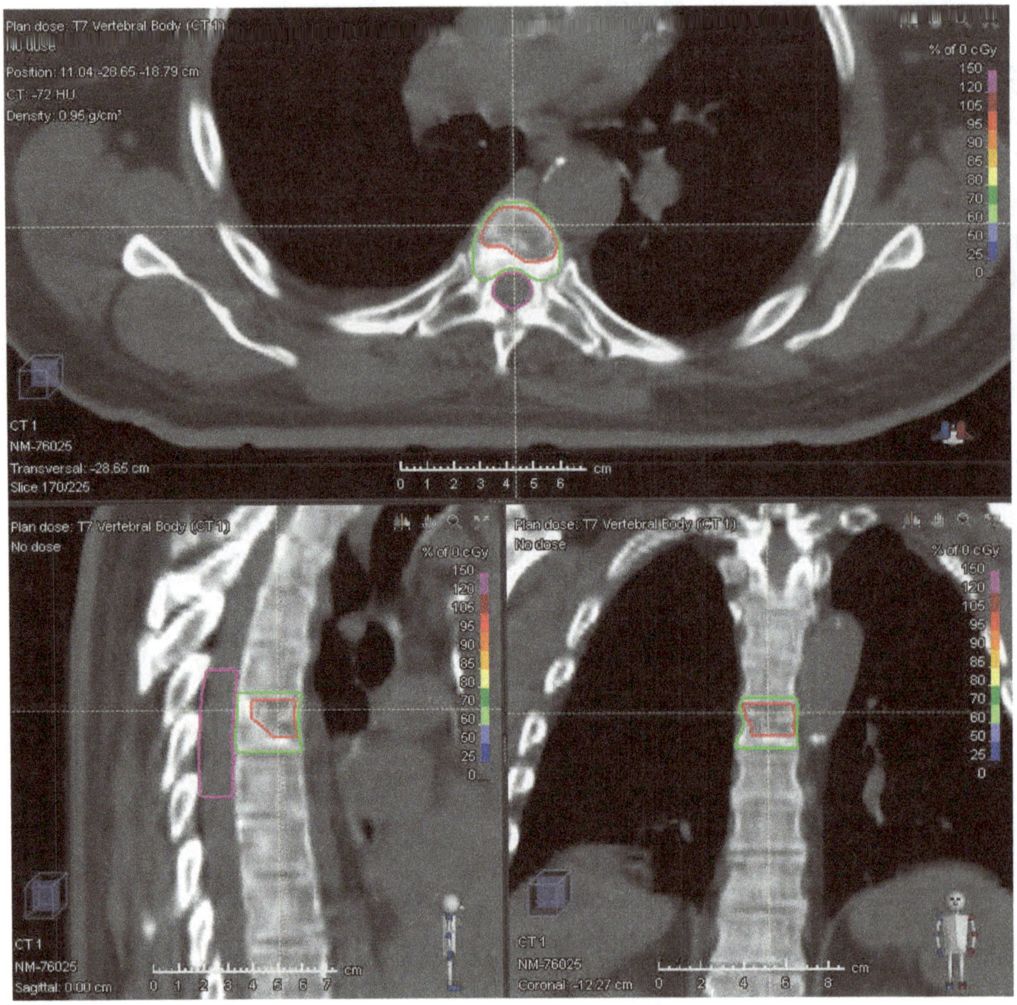

Figure 9.20 PTV 2700 (red) and PTV 2100 (green).

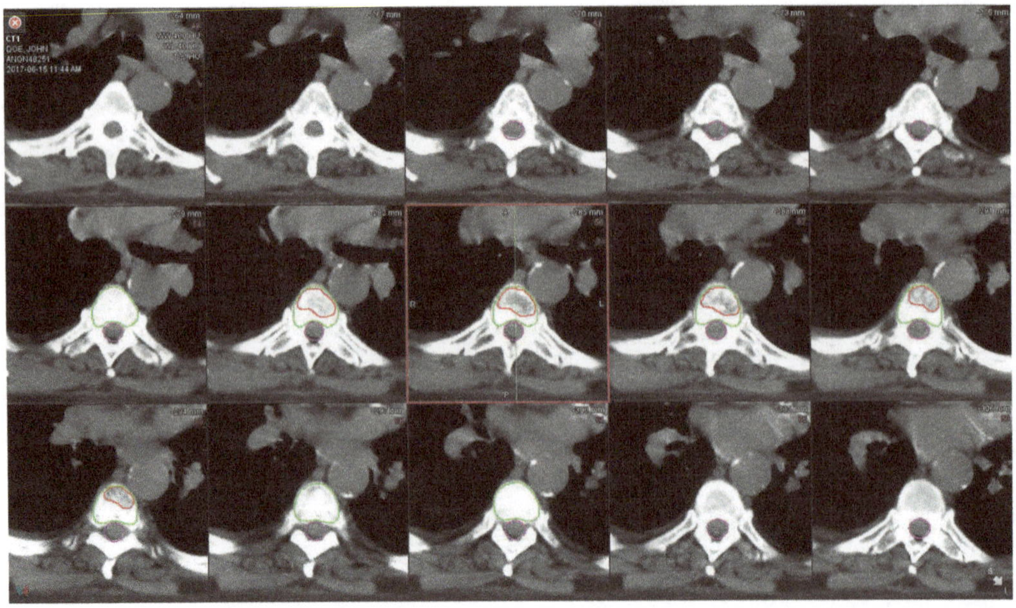

Figure 9.21 PTV 2700 (red) and PTV 2100 (green) drawn by the radiation oncologist. The purple spinal cord sub-volume contour was drawn on MRI to accurately delineate the spinal cord. This sub-volume extends 1 centimeter superior and inferior to the PTV contour.

Beam Arrangement

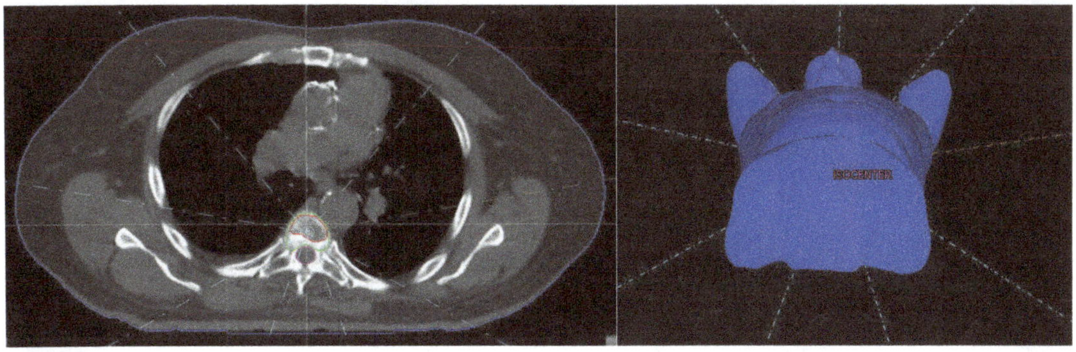

Figure 9.22 The IMRT beam arrangement for the T7 vertebral body treatment using 9 evenly distributed fields every 40 degrees.

Given the proximity of the target with the spinal cord, a 9-beam arrangement can be chosen to maximize sparing of the cord.

Gantry Angle	Couch	Collimator	Energy (MV)
200	0	0	10 FFF
240	0	0	10 FFF
280	0	0	10 FFF
320	0	0	10 FFF
0	0	0	10 FFF
40	0	0	10 FFF
80	0	0	10 FFF
120	0	0	10 FFF
160	0	0	10 FFF

For this example the planner has used a 2-full-arc VMAT plan. Two arcs back and forth will allow sufficient modulation and sparing of the cord while decreasing delivery time.

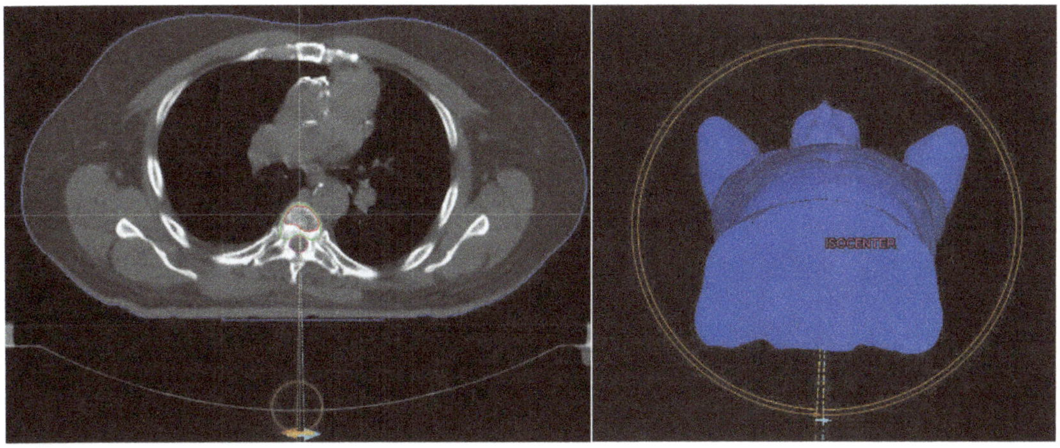

Figure 9.23 VMAT beam arrangement.

Gantry Angle	Couch	Collimator	Energy (MV)
G181–179 CW	0	30	10 FFF
G179–181 CCW	0	330	10 FFF

Optimization

Dose	ROI/POI	Clinical goal	Value	Result	% outside grid
Plan dose: T7 Vertebra...	Bilateral Lungs	At most 1250.0 cm³ volume at 1050 cGy dose		N/A	0 %
Plan dose: T7 Vertebra...	Bilateral Lungs	At most 1800.0 cm³ volume at 1140 cGy dose		N/A	0 %
Plan dose: T7 Vertebra...	Cord Subvolume	At most 10.00 % volume at 1800 cGy dose		N/A	0 %
Plan dose: T7 Vertebra...	Cord Subvolume	At most 2250 cGy dose at 0.0 cm³ volume		N/A	0 %
Plan dose: T7 Vertebra...	Esophagus	At most 5.0 cm³ volume at 1770 cGy dose		N/A	0 %
Plan dose: T7 Vertebra...	Esophagus	At most 2520 cGy dose at 0.0 cm³ volume		N/A	0 %
Plan dose: T7 Vertebra...	Heart	At most 15.0 cm³ volume at 2400 cGy dose		N/A	0 %
Plan dose: T7 Vertebra...	Heart	At most 3000 cGy dose at 0.0 cm³ volume		N/A	0 %
Plan dose: T7 Vertebra...	T7_PTV 2100	At least 95.00 % volume at 2100 cGy dose		N/A	0 %
Plan dose: T7 Vertebra...	T7_PTV 2700	At least 95.00 % volume at 2700 cGy dose		N/A	0 %
Plan dose: T7 Vertebra...	Trachea	At most 5.0 cm³ volume at 2580 cGy dose		N/A	0 %
Plan dose: T7 Vertebra...	Trachea	At most 3000 cGy dose at 0.0 cm³ volume		N/A	0 %

Figure 9.24 Clinical goals for plan rTA03.

Figure 9.24 shows the physician's clinical goals for the planner to achieve for this case. To achieve these objectives with the two-arc VMAT, the user set the Gantry Spacing to 2 degrees and maximum delivery time to 60 seconds (the minimum amount of time required for the gantry to perform a full rotation). The following initial optimization functions were set achieve a steep dose gradient and limit dose to the spinal cord. In this SBRT case, the radiation oncologist allowed a dose maximum of 130% of the prescription dose.

Function	Constraint	Dose	ROI	Description	Robust	Weight	Value
Physical Composite Objective							
Min Dose		Plan	T7_PTV 2700	Min Dose 2700 cGy		30	
Max Dose		Plan	T7_PTV 2700	Max Dose 3500 cGy		1	
Dose Fall-Off		Plan	External	Dose Fall-Off [H]2700 cGy [L]0 cGy, Low dose distance 10.00 cm		1	
Max DVH		Plan	Cord Subvolume	Max DVH 800 cGy to 8% volume		10	
Max Dose		Plan	Cord Subvolume	Max Dose 1600 cGy		10	
Min Dose		Plan	T7_PTV 2100	Min Dose 2100 cGy		30	
Dose Fall-Off		Plan	T7_PTV 2100	Dose Fall-Off [H]2700 cGy [L]2100 cGy, Low dose distance 0.10 cm		1	

Figure 9.25 Initial list of optimization objectives.

After the first optimization, the user can evaluate the DVH against the clinical goals. The objectives and their weights can be adjusted to meet the dose goals.

Function	Constraint	Dose	ROI	Description	Robust	Weight	Value
Physical Composite Objective							
Min Dose		Plan	T7_PTV 2700	Min Dose 2700 cGy		30	
Max Dose		Plan	T7_PTV 2700	Max Dose 3500 cGy		1	
Dose Fall-Off		Plan	External	Dose Fall-Off [H]2700 cGy [L]0 cGy, Low dose distance 10.00 cm		1	
Max DVH		Plan	Cord Subvolume	Max DVH 800 cGy to 8% volume		10	
Max Dose		Plan	Cord Subvolume	Max Dose 1600 cGy		10	
Min Dose		Plan	T7_PTV 2100	Min Dose 2100 cGy		30	
Dose Fall-Off		Plan	T7_PTV 2100	Dose Fall-Off [H]2700 cGy [L]2100 cGy, Low dose distance 0.10 cm		1	
Min Dose		Plan	PTV2100 Push	Min Dose 2100 cGy		35	
Max DVH		Plan	PTV 2100 1cm Hot Ring	Max DVH 2100 cGy to 3% volume		20	

Figure 9.26 Final list of optimization objectives (not all objectives pictured).

RAYSTATION SAMPLES

Dose Distribution

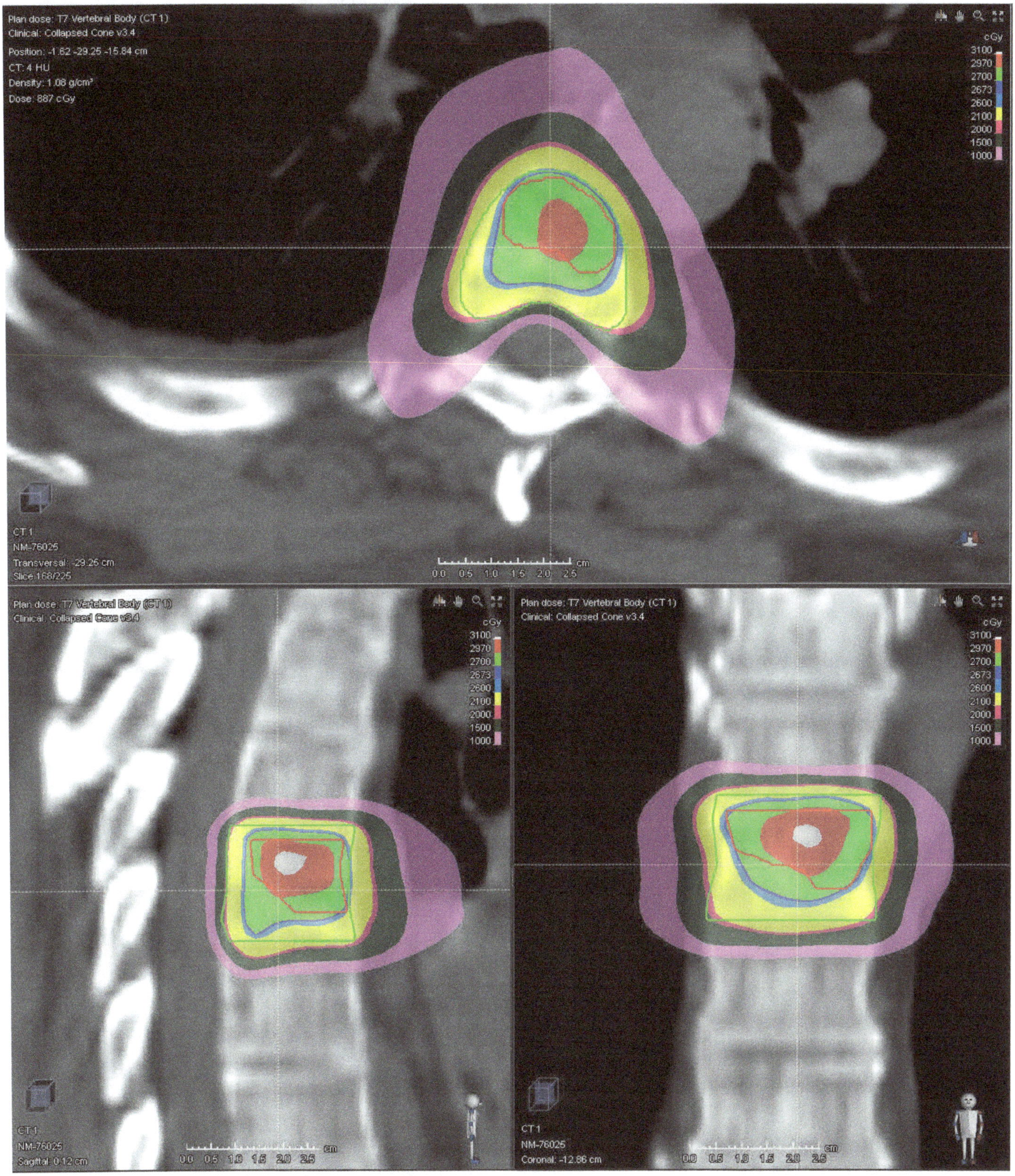

Figure 9.27 Final dose distribution for case rTA03.

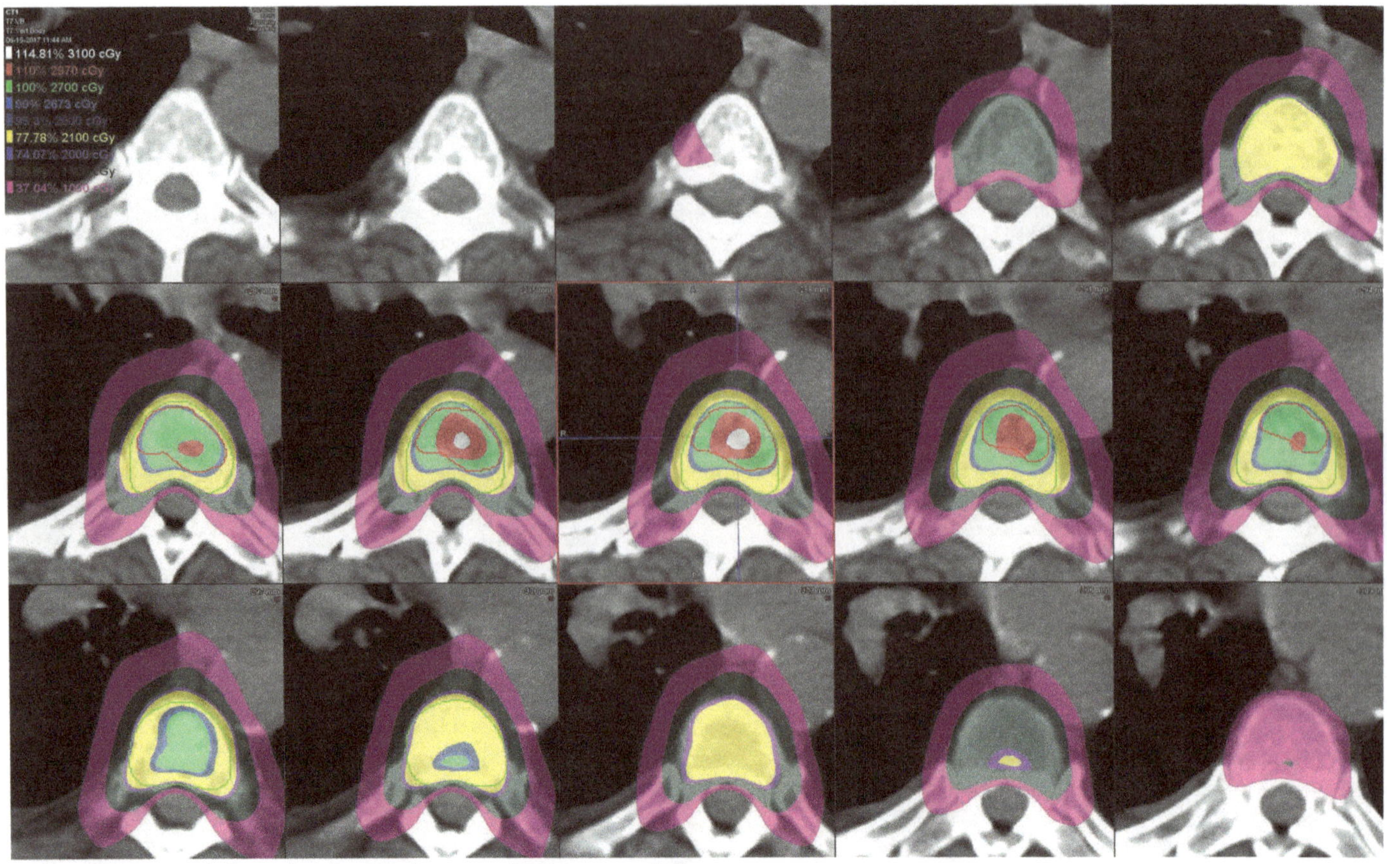

Figure 9.28 Final dose distribution for case rTA03 in axial slices.

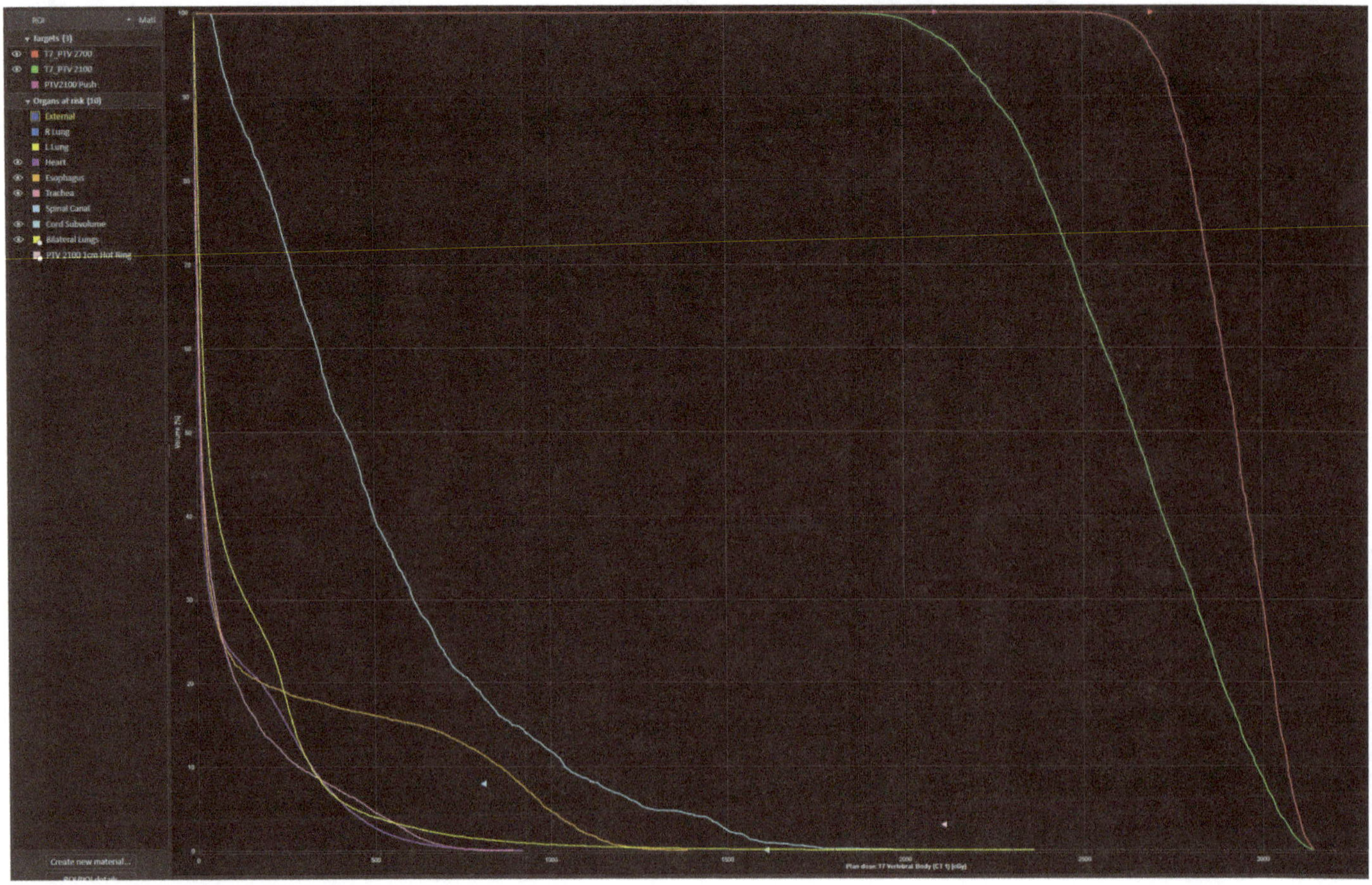

Figure 9.29 Final dose volume histogram.

Case Number rHN01 Head and Neck: Bilateral Neck Radiation Treatment

General Information

Brief Clinical History

59-year-old man diagnosed with nasopharyngeal primary cancer and bilateral neck lymphadenopathy.

Prescription

Bilateral neck nodes:
PTV 7000 = 250 cGy × 28 fractions
PTV 5610 = 200.4 cGy × 28 fractions

Simulation/Comments

The simulation was accomplished with the patient on head and neck indexed immobilization board, thermoplastic mask, and moldable pillow. Fusion of images was performed with a recent PET/CT for evaluation of the extent of the lesion and targeting by the radiation oncologist.

Targets

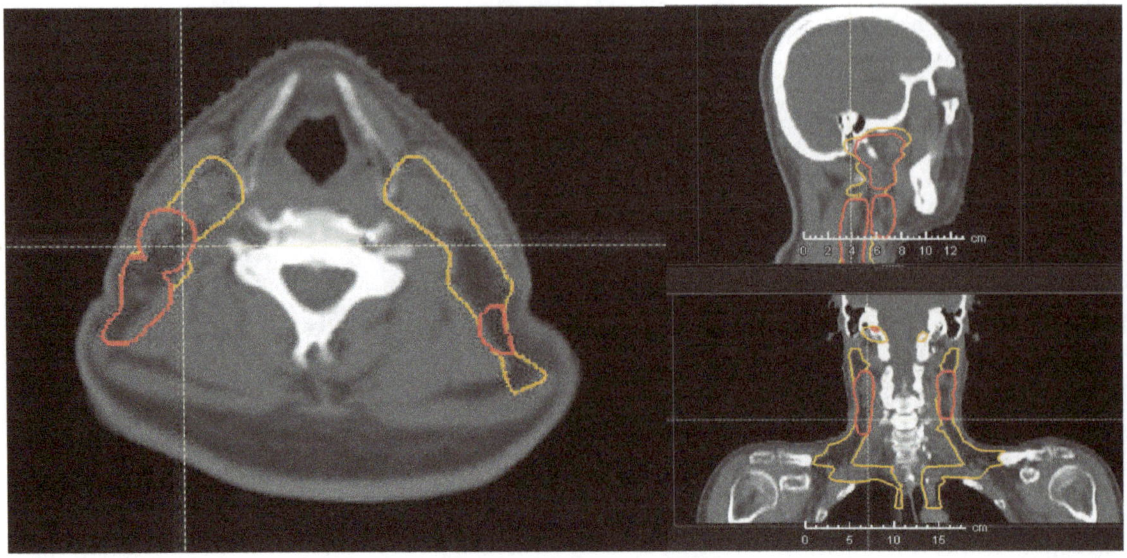

Figure 9.30 PTV contours: PTV 5610 (orange) and PTV 7000 (red).

Figures 9.30 and 9.31 show the PTV contours (PTV 5610 and PTV 7000) drawn by the radiation oncologist.

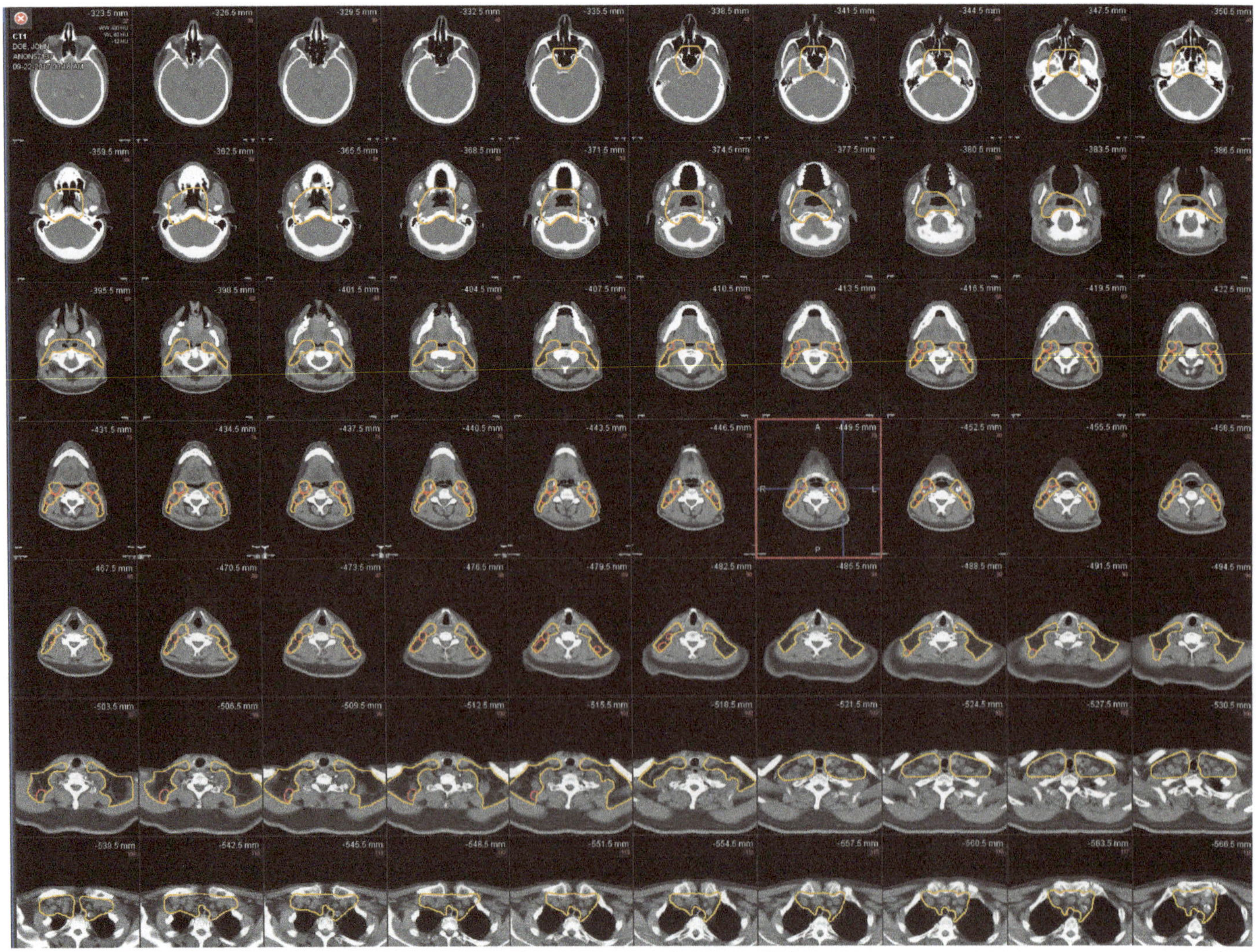

Figure 9.31 PTV contours: PTV 5610 (orange) and PTV 7000 (red) on axial slices.

Beam Arrangement

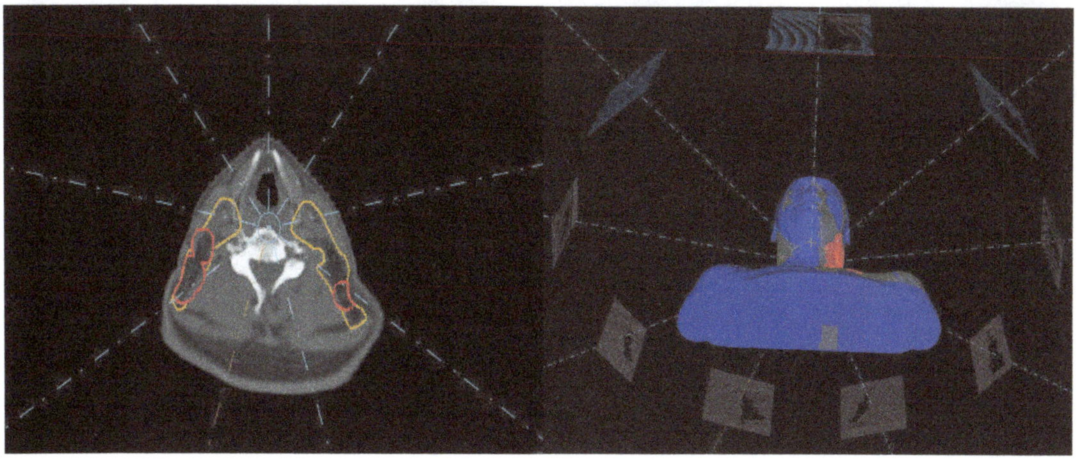

Figure 9.32 IMRT beam arrangement.

Gantry Angle	Couch	Collimator	Energy (MV)
200	0	0	6
240	0	0	6
280	0	0	6
320	0	0	6
0	0	0	6
40	0	0	6
80	0	0	6
120	0	0	6
160	0	0	6

The beam arrangement for IMRT can be seen in Figure 9.32. Since the target contour is distributed from the left to the right side of the patient, the planner can place the isocenter in the center of the PTV and evenly distribute the beams around the patient. If the PTV favored a particular side, the beam distribution would be placed to limit dose to the healthy contralateral neck.

VMAT beam arrangement can be seen in Figure 9.33. In this case, the planner has split the beams to form 4 arcs. This beam arrangement takes into consideration the travel distance limitation on the MLCs by limiting the maximum field size. Collimator and jaw arrangements can be seen in Figure 9.34. The planner is encouraged to try 2 arcs to begin the optimization and add more as the complexity of cases increases.

Gantry Angle	Couch	Collimator	Energy (MV)
G181–179 CW 1	0	30	6
G179–181 CCW 1	0	330	6
G181–179 CW 2	0	30	6
G179–181 CCW 2	0	330	6

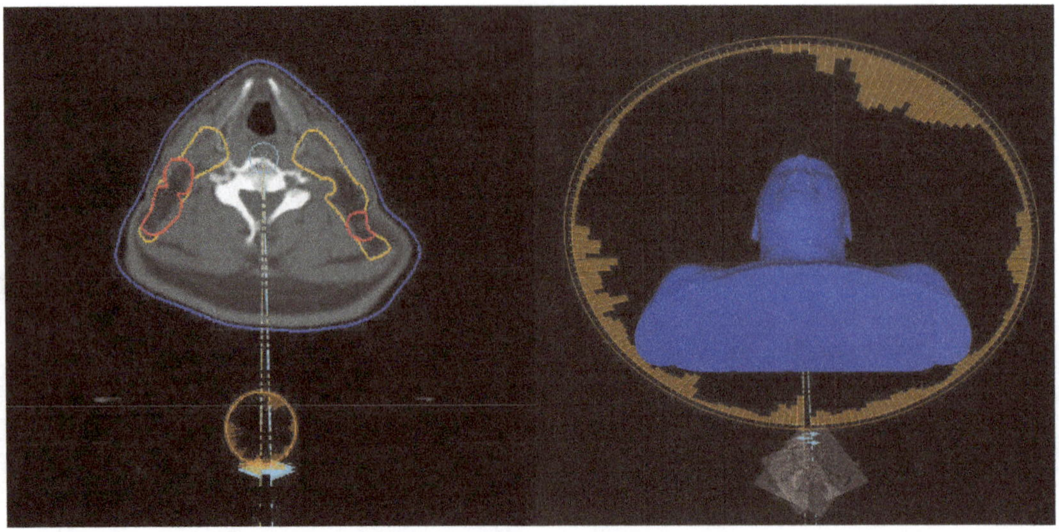

Figure 9.33 VMAT Beam arrangement.

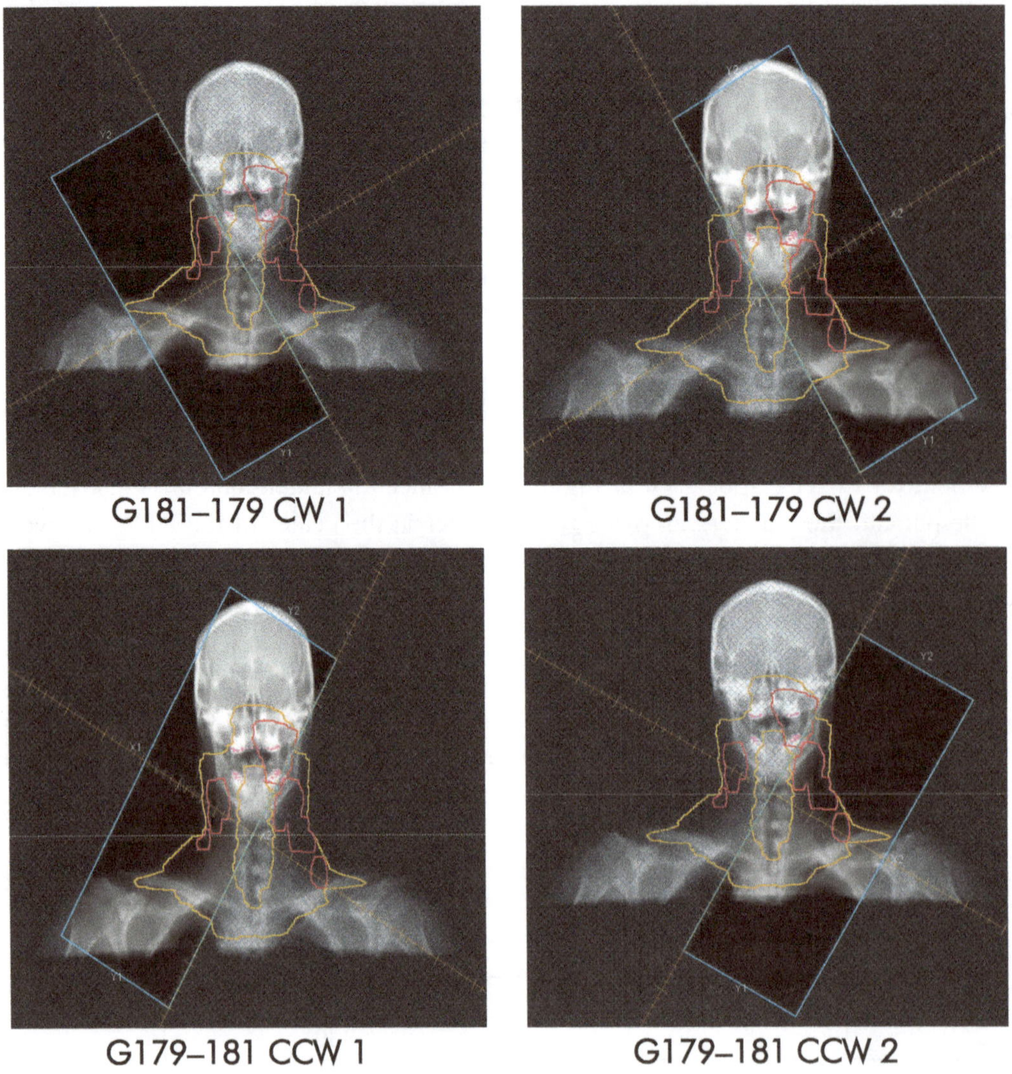

Figure 9.34 Jaw and collimator configuration for arcs in case rHN01. The jaw settings are set at the maximum allowed field dimensions for the purpose of jaw tracking. During optimization, if the jaw tracking is active, the jaw will be limited to the aperture of the leaves, which will be limited to the target volumes. If the machine does not have jaw-tracking capabilities, the planner can just set the jaws to cover the treatment volume.

Optimization

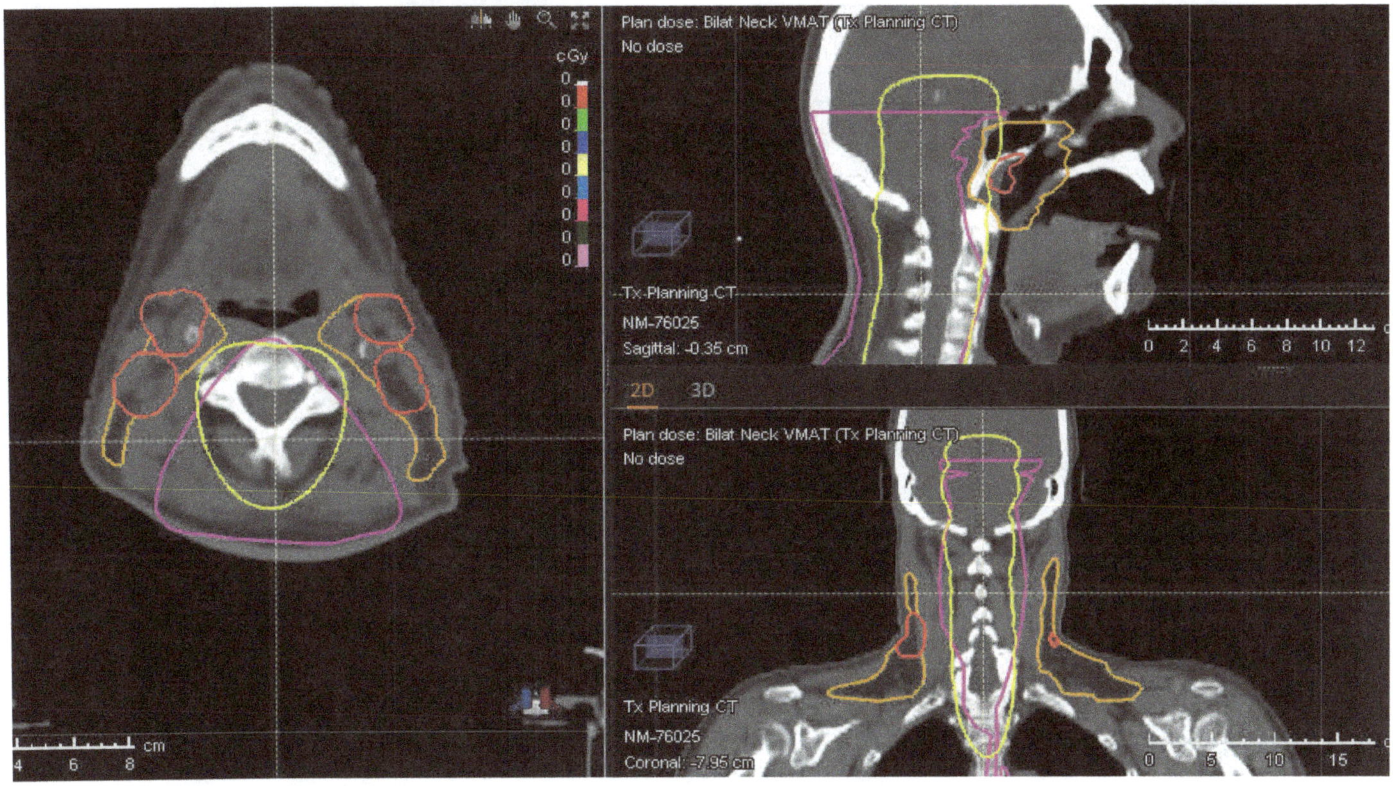

Figure 9.35 Dose avoidance contours: spinal canal and brainstem expansion contour (yellow) and posterior avoidance contour (purple).

Organs at risk for this case include: brainstem, cochleae, optic chiasm, optic nerves, oral cavity, parotid glands, brain, eye globes, larynx, esophagus, trachea, spinal canal, and mandible. Additionally, the dose should be limited to the tissue along the posterior of the neck. This can be done by drawing another contour and giving a margin on the spinal canal and brainstem.

The planner can start setting up objectives to cover the PTV and then modifying and adding objectives for organs at risk after the first optimization. Another option is that of using an existing template with dose objectives.

Physical Composite Objective				
Max EUD	Plan	parotid L	Max EUD 2500 cGy, Parameter A 1	1
Max EUD	Plan	cochlea L	Max EUD 3400 cGy, Parameter A 1	1
Max EUD	Plan	cochlea R	Max EUD 3400 cGy, Parameter A 1	1
Max EUD	Plan	oral cavity	Max EUD 2800 cGy, Parameter A 1	1
Max Dose	Plan	Optics	Max Dose 2600 cGy	1
Max Dose	Plan	Spinal Canal	Max Dose 4000 cGy	1
Max Dose	Plan	Mandible	Max Dose 6500 cGy	1
Dose Fall-Off	Plan	Posterior Avoid	Dose Fall-Off [H]7000 cGy [L]1000 cGy, Low dose distance 12.00 cm	1
Dose Fall-Off	Plan	Spinal Canal	Dose Fall-Off [H]7000 cGy [L]0 cGy, Low dose distance 7.00 cm	1
Max Dose	Plan	cochlea L	Max Dose 3800 cGy	1
Max Dose	Plan	cochlea R	Max Dose 3800 cGy	1
Max Dose	Plan	PTV5610	Max Dose 7200 cGy	10
Dose Fall-Off	Plan	External	Dose Fall-Off [H]7000 cGy [L]1000 cGy, Low dose distance 10.00 cm	1
Max Dose	Plan	brainstem	Max Dose 3000 cGy	1
Max Dose	Plan	PTV7000 nodes	Max Dose 7200 cGy	20
Max EUD	Plan	parotid R	Max EUD 2500 cGy, Parameter A 1	1
Dose Fall-Off	Plan	brainstem	Dose Fall-Off [H]7000 cGy [L]0 cGy, Low dose distance 3.00 cm	1
Min Dose	Plan	PTV7000 nodes	Min Dose 7000 cGy	30
Max Dose	Plan	PTV7000	Max Dose 7200 cGy	20
Dose Fall-Off	Plan	PTV5610	Dose Fall-Off [H]7000 cGy [L]5600 cGy, Low dose distance 1.00 cm	2
Min Dose	Plan	PTV5610	Min Dose 5610 cGy	30
Min Dose	Plan	PTV7000	Min Dose 7000 cGy	30

Figure 9.36 Sample initial objectives for case rHN01.

Plan dose: Bilat Neck...	brainstem	At most 2.0 cm³ volume at 5000 cGy dose
Plan dose: Bilat Neck...	brainstem	At most 2000 cGy average dose
Plan dose: Bilat Neck...	brainstem	At most 6000 cGy dose at 0.0 cm³ volume
Plan dose: Bilat Neck...	cochlea L	At most 0.5 cm³ volume at 3600 cGy dose
Plan dose: Bilat Neck...	cochlea L	At most 3600 cGy average dose
Plan dose: Bilat Neck...	cochlea L	At most 4000 cGy dose at 0.0 cm³ volume
Plan dose: Bilat Neck...	cochlea R	At most 0.5 cm³ volume at 3600 cGy dose
Plan dose: Bilat Neck...	cochlea R	At most 3600 cGy average dose
Plan dose: Bilat Neck...	cochlea R	At most 4000 cGy dose at 0.0 cm³ volume
Plan dose: Bilat Neck...	Esophagus	At most 5.0 cm³ volume at 5500 cGy dose
Plan dose: Bilat Neck...	Esophagus	At most 4500 cGy average dose
Plan dose: Bilat Neck...	Esophagus	At most 6000 cGy dose at 0.0 cm³ volume
Plan dose: Bilat Neck...	globe L	At most 3800 cGy average dose
Plan dose: Bilat Neck...	globe L	At most 4500 cGy dose at 0.0 cm³ volume
Plan dose: Bilat Neck...	globe R	At most 3800 cGy average dose
Plan dose: Bilat Neck...	globe R	At most 4500 cGy dose at 0.0 cm³ volume
Plan dose: Bilat Neck...	Larynx	At most 4500 cGy average dose
Plan dose: Bilat Neck...	Mandible	At most 7000 cGy dose at 0.0 cm³ volume
Plan dose: Bilat Neck...	optic chiasm	At most 3000 cGy average dose
Plan dose: Bilat Neck...	optic chiasm	At most 5400 cGy dose at 0.0 cm³ volume
Plan dose: Bilat Neck...	optic nerve L	At most 3000 cGy dose at 0.0 cm³ volume
Plan dose: Bilat Neck...	optic nerve R	At most 3000 cGy dose at 0.0 cm³ volume
Plan dose: Bilat Neck...	oral cavity	At most 3000 cGy average dose
Plan dose: Bilat Neck...	parotid L	At most 2600 cGy average dose
Plan dose: Bilat Neck...	parotid R	At most 2600 cGy average dose
Plan dose: Bilat Neck...	PTV5610	At least 95.00 % volume at 5610 cGy dose

Figure 9.37 Sample list of clinical objectives to be achieved in case rHN01. Not meeting these objectives is at the discretion of the radiation oncologist.

Dose Distribution

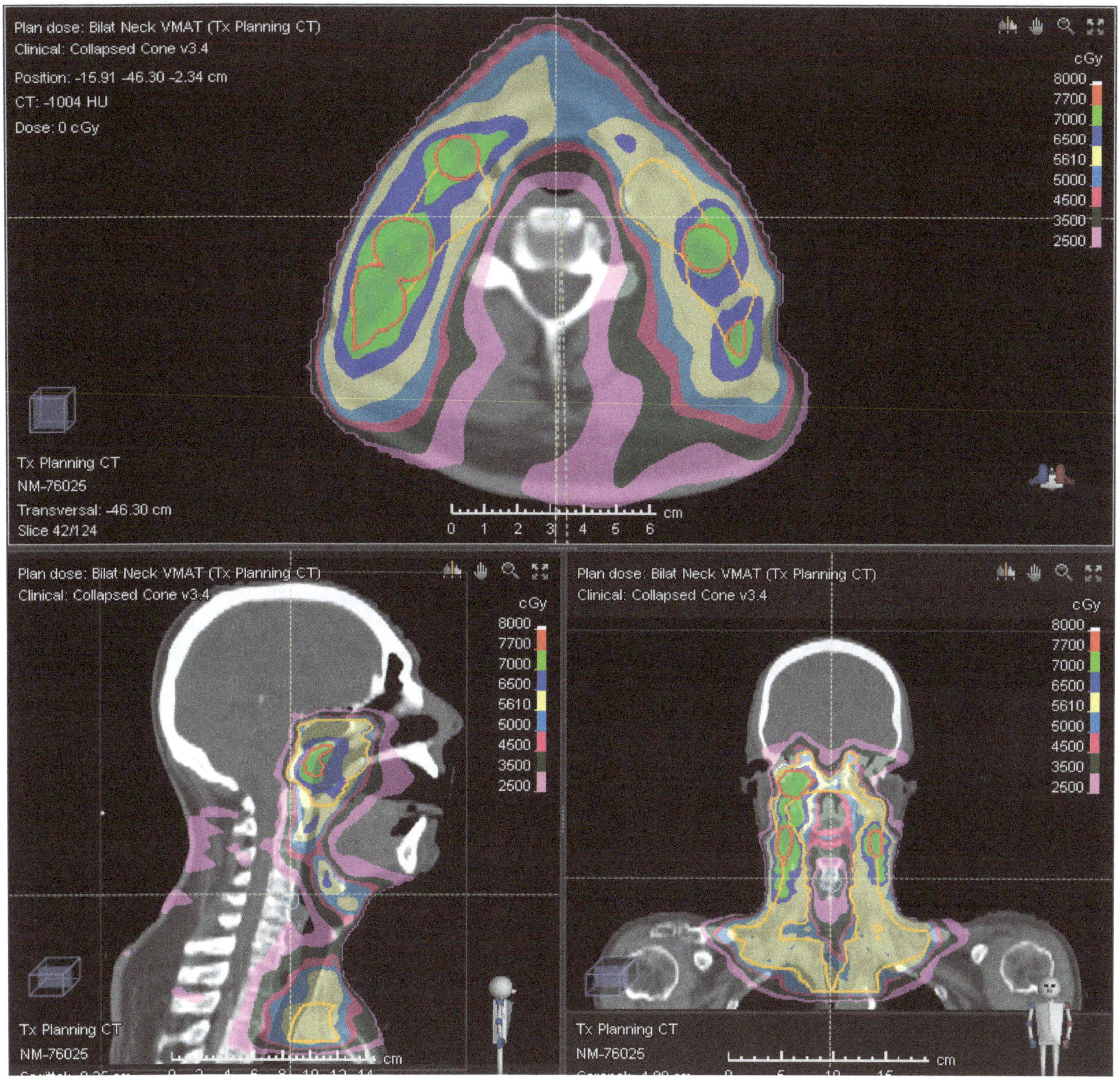

Figure 9.38 Final dose distribution for case rHN01.

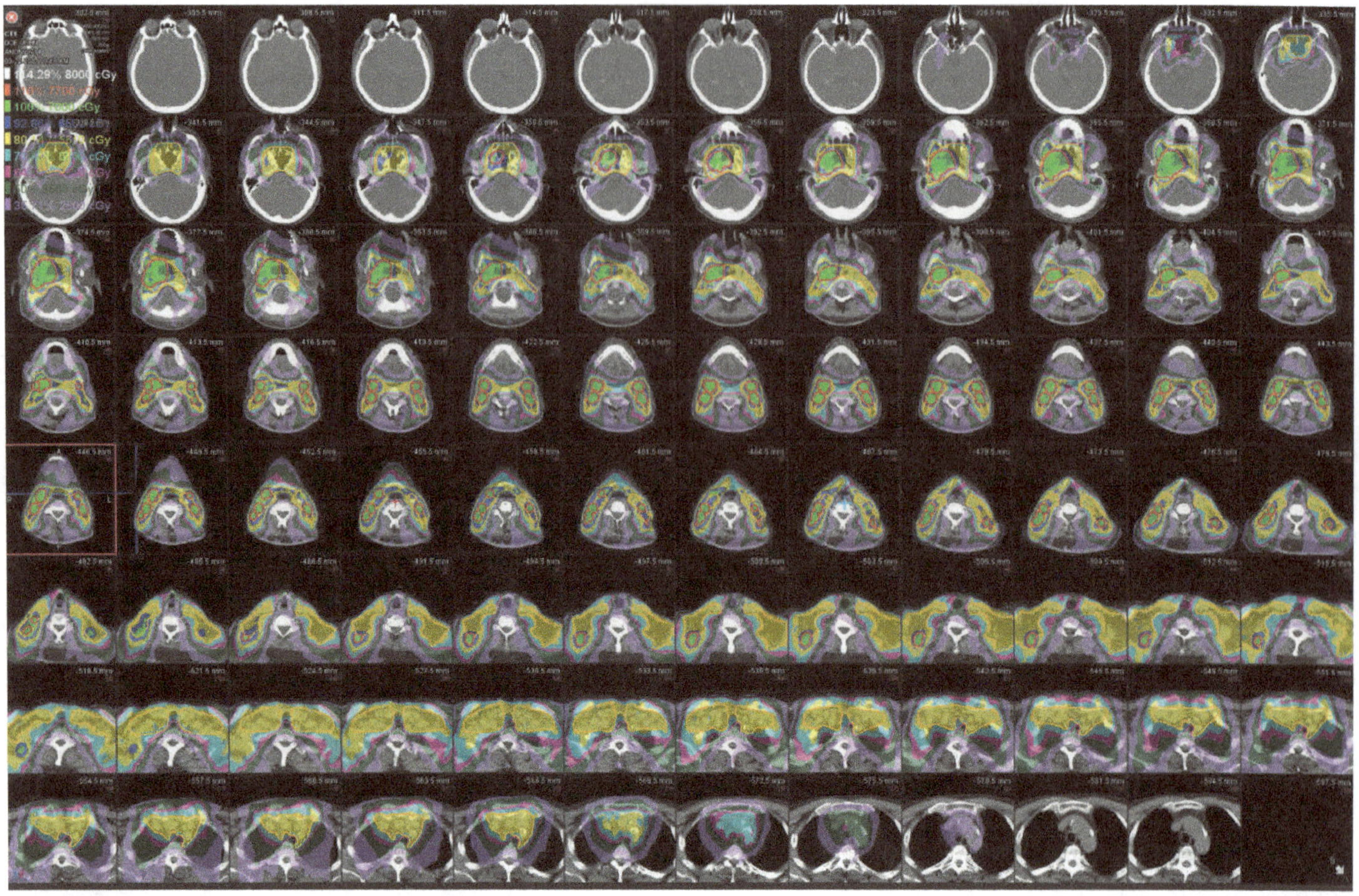

Figure 9.39 Final dose distribution for case rHN01 in axial slices.

Case Number rHN02 Head and Neck: Right Neck Radiation Treatment

General Information

Brief Clinical History

75-year-old woman diagnosed with diffuse large B-cell lymphoma and prescribed radiation therapy to bulky disease in the right neck.

Prescription

Right neck:
PTV 4500 = 300 cGy × 15 fractions

Simulation/Comments

The simulation was accomplished with the patient on head and neck index/immobilization board, thermoplastic mask, moldable pillow, and head rest. Fusion of images was performed with a recent PET/CT for evaluation of the extent of the lesion and targeting by the radiation oncologist.

Targets

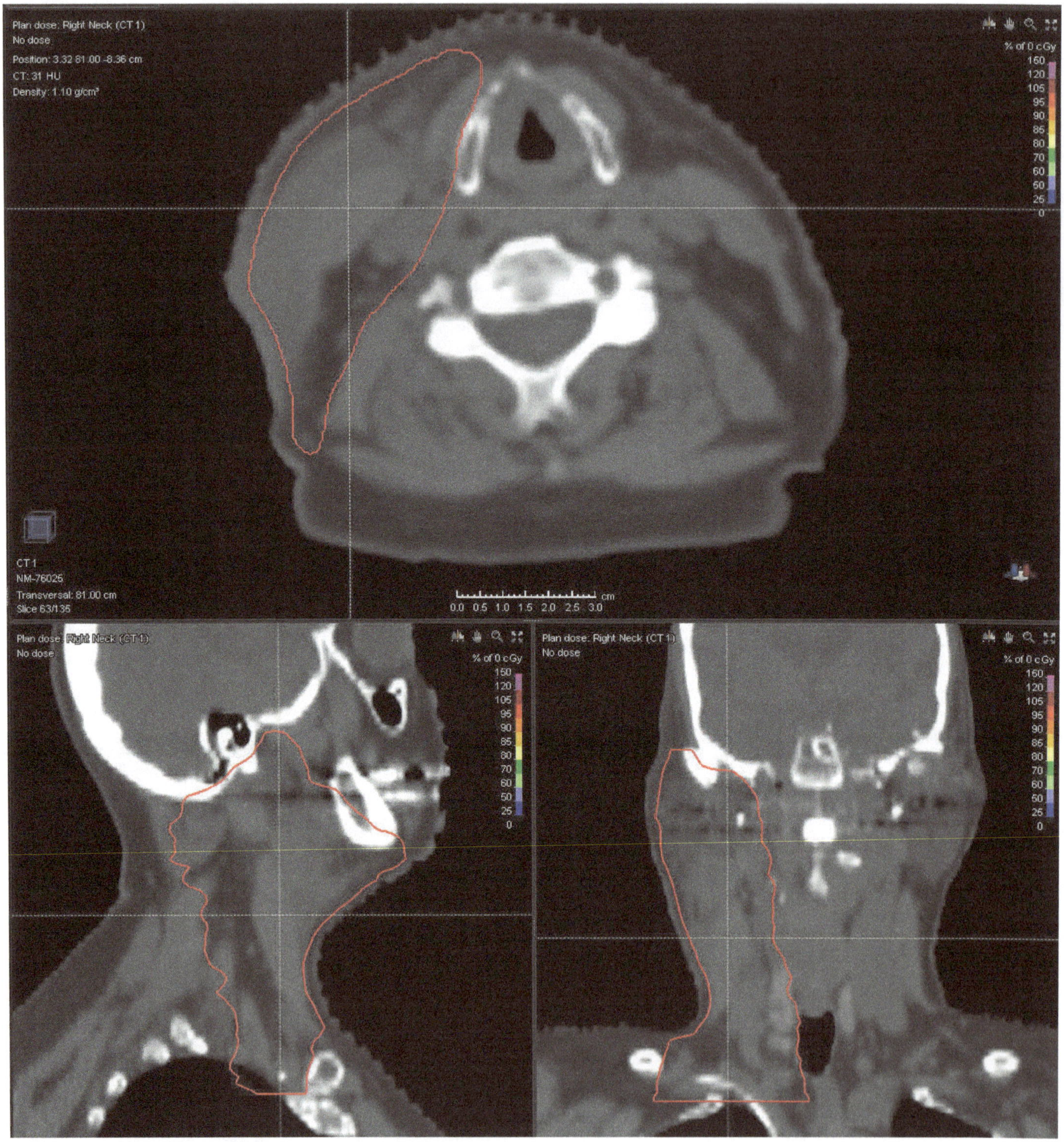

Figure 9.40 Prescription contour PTV 4500 (red contour).

Figures 9.40 and 9.41 show the PTV and GTV contours drawn by the radiation oncologist.

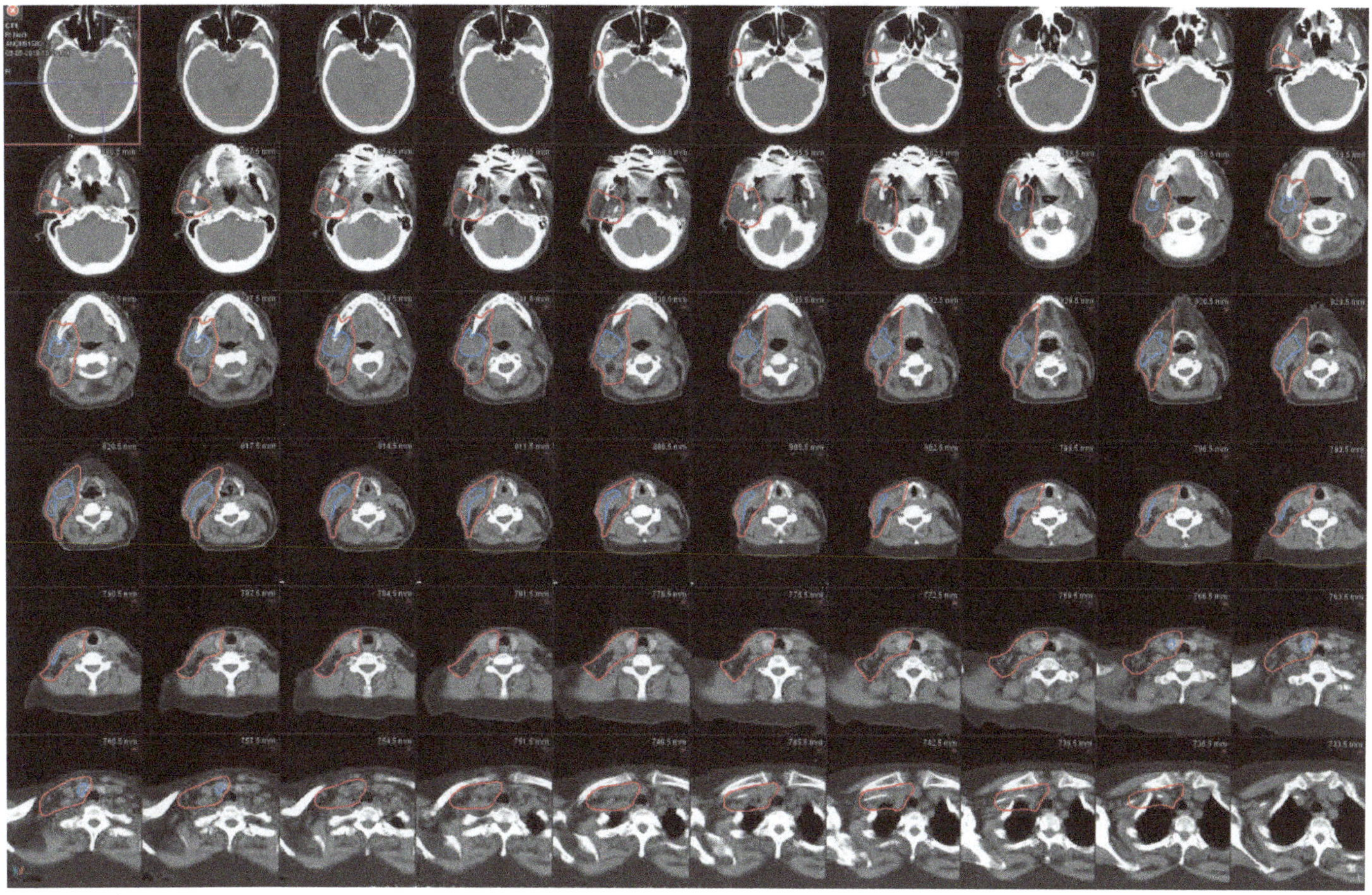

Figure 9.41 Prescription contour PTV 4500 (red) and GTV (blue) in axial slices.

Beam Arrangement

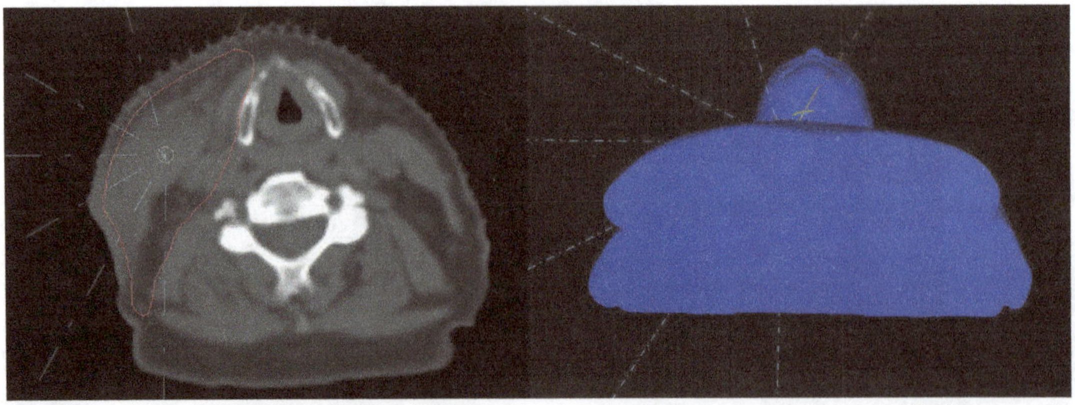

Figure 9.42 IMRT beam arrangement.

Gantry Angle	Couch	Collimator	Energy (MV)
181	0	0	6
210	0	0	6
240	0	0	6
270	0	0	6
300	0	0	6
345	0	0	6
30	0	0	6

The beam arrangement for IMRT can be seen in Figure 9.42. The target contour favors the right side of the patient and the beam arrangement spares the contralateral neck tissue.

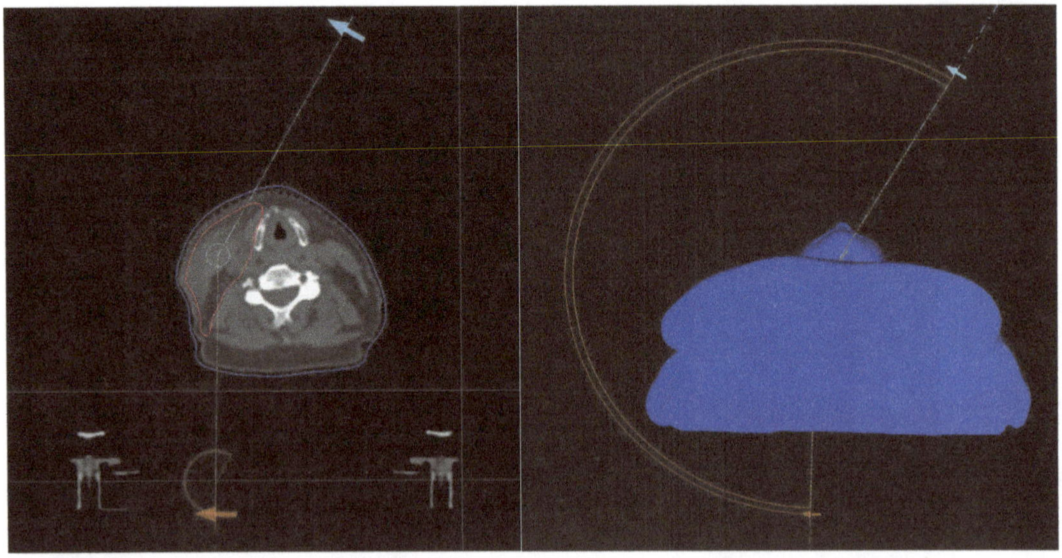

Figure 9.43 VMAT beam arrangement.

Gantry Angle	Couch	Collimator	Energy (MV)
G181–30 CW	0	30	6
G30–181 CCW	0	330	6

The beam arrangement for VMAT is shown in Figure 9.43. The beams span the same length as the IMRT beam arrangement.

Optimization

Dose	ROI/POI	Clinical goal	Value	Result	% outside grid
Plan dose: Right Neck I...	Brainstem	At most 4800 cGy dose at 0.0 cm³ volume		N/A	0 %
Plan dose: Right Neck I...	Canal	At most 3600 cGy dose at 0.0 cm³ volume		N/A	0 %
Plan dose: Right Neck I...	Esophagus	At most 5.0 cm³ volume at 5130 cGy dose		N/A	0 %
Plan dose: Right Neck I...	Esophagus	At most 5530 cGy dose at 0.0 cm³ volume		N/A	0 %
Plan dose: Right Neck I...	Larynx	At most 3.0 cm³ volume at 3750 cGy dose		N/A	0 %
Plan dose: Right Neck I...	Larynx	At most 2500 cGy average dose		N/A	0 %
Plan dose: Right Neck I...	Larynx	At most 4000 cGy dose at 0.0 cm³ volume		N/A	0 %
Plan dose: Right Neck I...	Lt Cochlea	At most 3300 cGy dose at 0.0 cm³ volume		N/A	0 %
Plan dose: Right Neck I...	Lt Parotid	At most 7.0 cm³ volume at 1800 cGy dose		N/A	0 %
Plan dose: Right Neck I...	Lt Parotid	At most 2250 cGy average dose		N/A	0 %
Plan dose: Right Neck I...	Lt Parotid	At most 2700 cGy dose at 0.0 cm³ volume		N/A	0 %
Plan dose: Right Neck I...	Lt Salivary Gland	At most 2250 cGy average dose		N/A	0 %
Plan dose: Right Neck I...	Lt Salivary Gland	At most 2700 cGy dose at 0.0 cm³ volume		N/A	0 %
Plan dose: Right Neck I...	Oral Cavity Avoidance	At most 2000 cGy average dose		N/A	0 %
Plan dose: Right Neck I...	Oral Cavity Avoidance	At most 3000 cGy dose at 0.0 cm³ volume		N/A	0 %
Plan dose: Right Neck I...	PTV4500	At least 95.00 % volume at 4500 cGy dose		N/A	0 %
Plan dose: Right Neck I...	PTV4500	At most 4950 cGy dose at 0.04 % volume		N/A	0 %
Plan dose: Right Neck I...	Rt Cochlea	At most 3300 cGy dose at 0.0 cm³ volume		N/A	0 %

Figure 9.44 Sample clinical goals for plan rHN02.

The planner has used the dose fall-off functions to minimize dose outside of the target.

Function	Constraint	Dose	ROI	Description	Robust	Weight	Value
Physical Composite Objective							0.0017
Min Dose		Plan	PTV4500	Min Dose 4500 cGy		30	1.7278E-4
Max Dose		Plan	PTV4500	Max Dose 4700 cGy		1	6.8038E-5
Dose Fall-Off		Plan	External	Dose Fall-Off [H]4500 cGy [L]0 cGy, Low dose distance 10.00 cm		1	0.0010
Max Dose		Plan	Canal	Max Dose 1800 cGy		1	8.9461E-6
Dose Fall-Off		Plan	Canal	Dose Fall-Off [H]4500 cGy [L]0 cGy, Low dose distance 5.00 cm		1	2.3413E-4
Dose Fall-Off		Plan	Brainstem	Dose Fall-Off [H]4500 cGy [L]0 cGy, Low dose distance 5.00 cm		1	2.2823E-4
Max Dose		Plan	Brainstem	Max Dose 1800 cGy		1	0.0000
Max EUD		Plan	Lt Parotid	Max EUD 1800 cGy, Parameter A 1		1	0.0000
Max EUD		Plan	Lt Salivary Gland	Max EUD 1800 cGy, Parameter A 1		1	0.0000
Max EUD		Plan	Oral Cavity	Max EUD 1800 cGy, Parameter A 1		1	0.0000
Max EUD		Plan	Rt Cochlea	Max EUD 1800 cGy, Parameter A 1		1	0.0000
Max EUD		Plan	Larynx	Max EUD 1800 cGy, Parameter A 1		1	0.0000

Figure 9.45 Initial optimization parameters.

The planner can start setting up objectives to cover the PTV, modifying and adding objectives for organs at risk after the first optimization to meet the clinical goals. Failing to meet criteria is at the discretion of the Radiation Oncologist and the level of dose coverage requested.

Dose Distribution

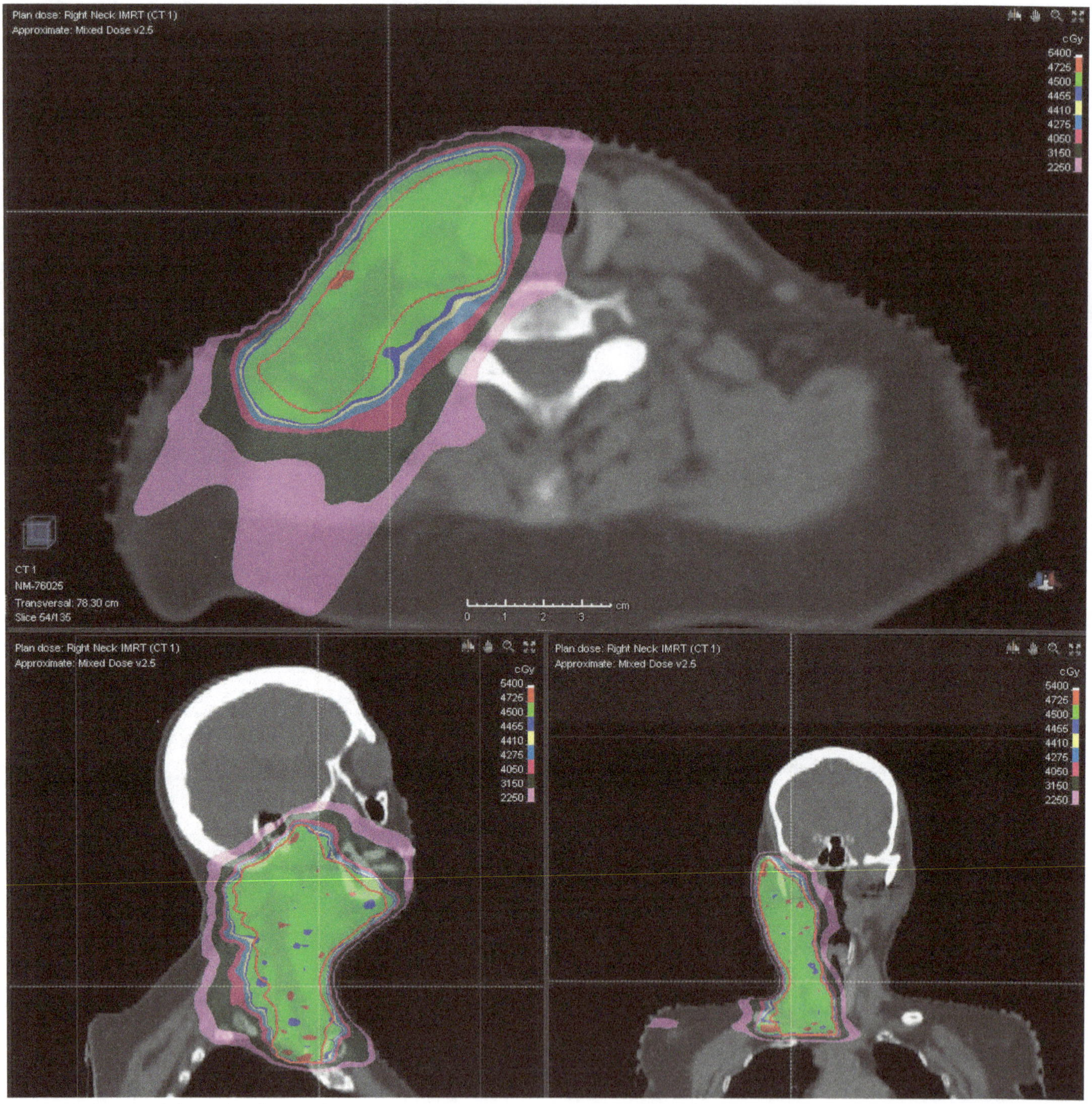

Figure 9.46 Final dose distribution for case rHN02.

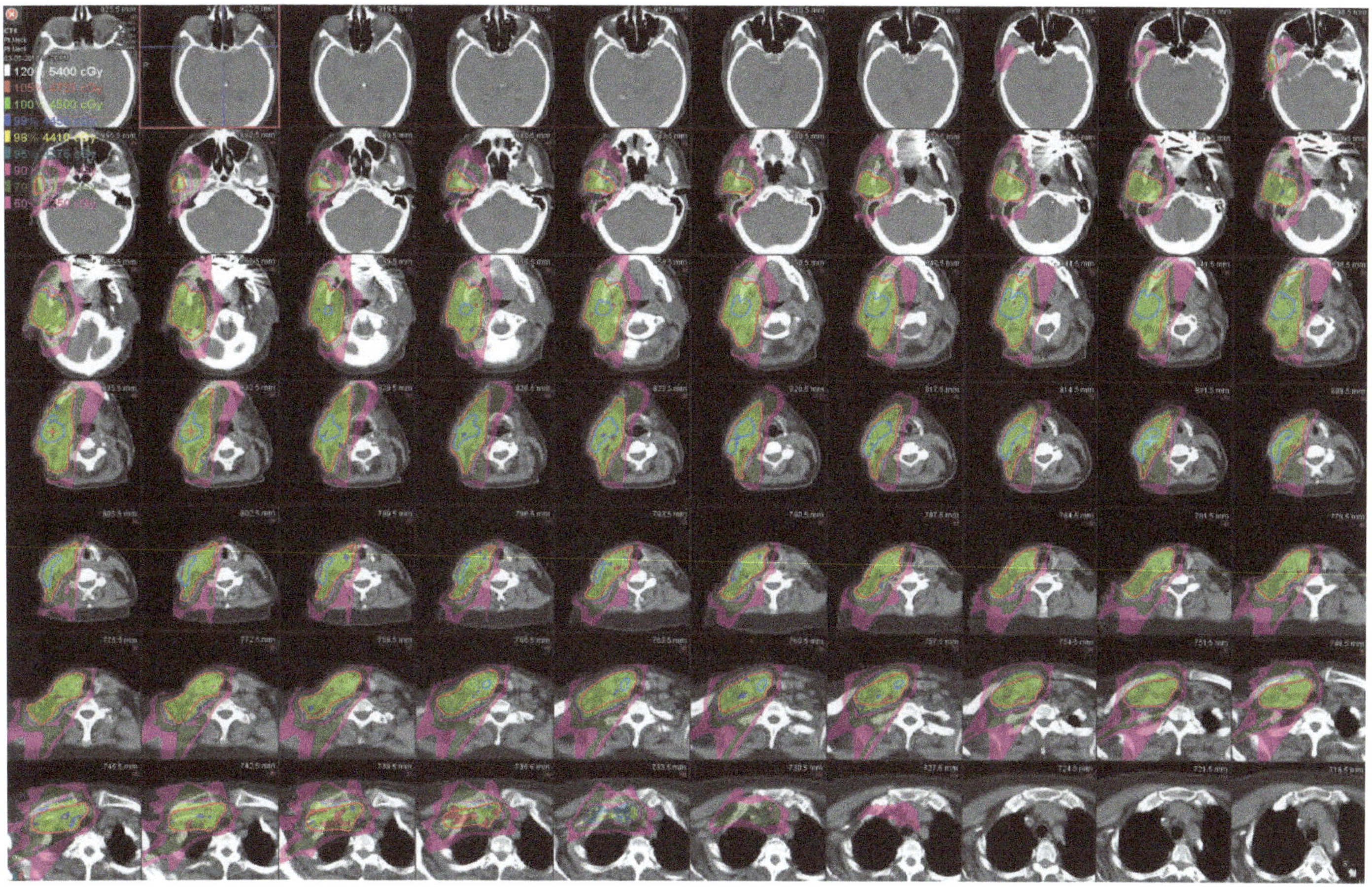

Figure 9.47 Final dose distribution for case rHN02 in axial slices.

Case Number rHN03 Head and Neck: Left Parotid Radiation Treatment

General Information

Brief Clinical History

65-year-old female diagnosed with T1N0 squamous cell carcinoma of the left parotid. Prescribed definitive radiation therapy to the parotid and elective lymph nodes.

Prescription

Left neck:
PTV 6250 = 250 cGy × 25 fractions
PTV 4500 = 180 cGy × 25 fractions

Simulation/Comments

The simulation was accomplished with the patient on head and neck index/immobilization board, thermoplastic mask, moldable pillow, and headrest. Fusion of images was performed with a recent PET/CT for evaluation of the extent of the lesion and targeting by the radiation oncologist.

Targets

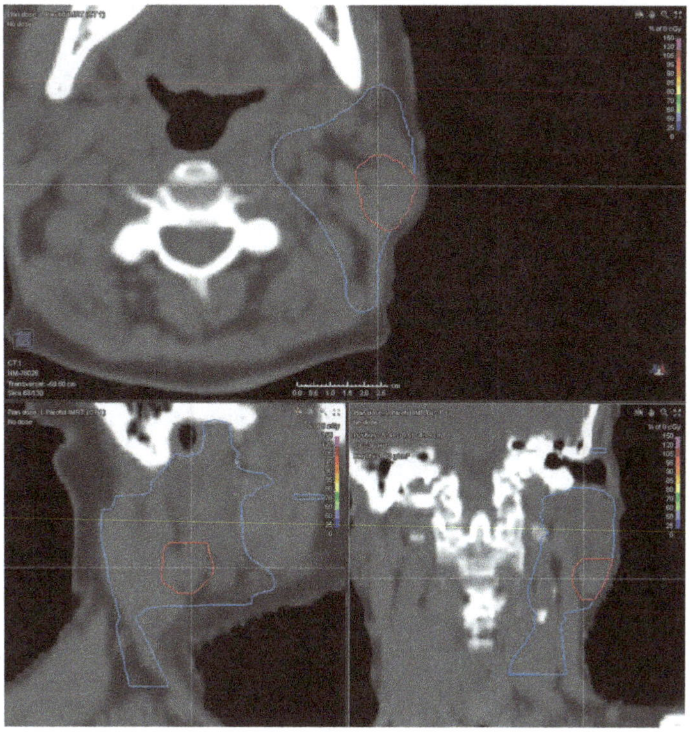

Figure 9.48 Target contours: PTV 6250 (red) and PTV 4500 (blue).

Figures 9.48 and 9.49 show the physician drawn targets for the left parotid treatment volume.

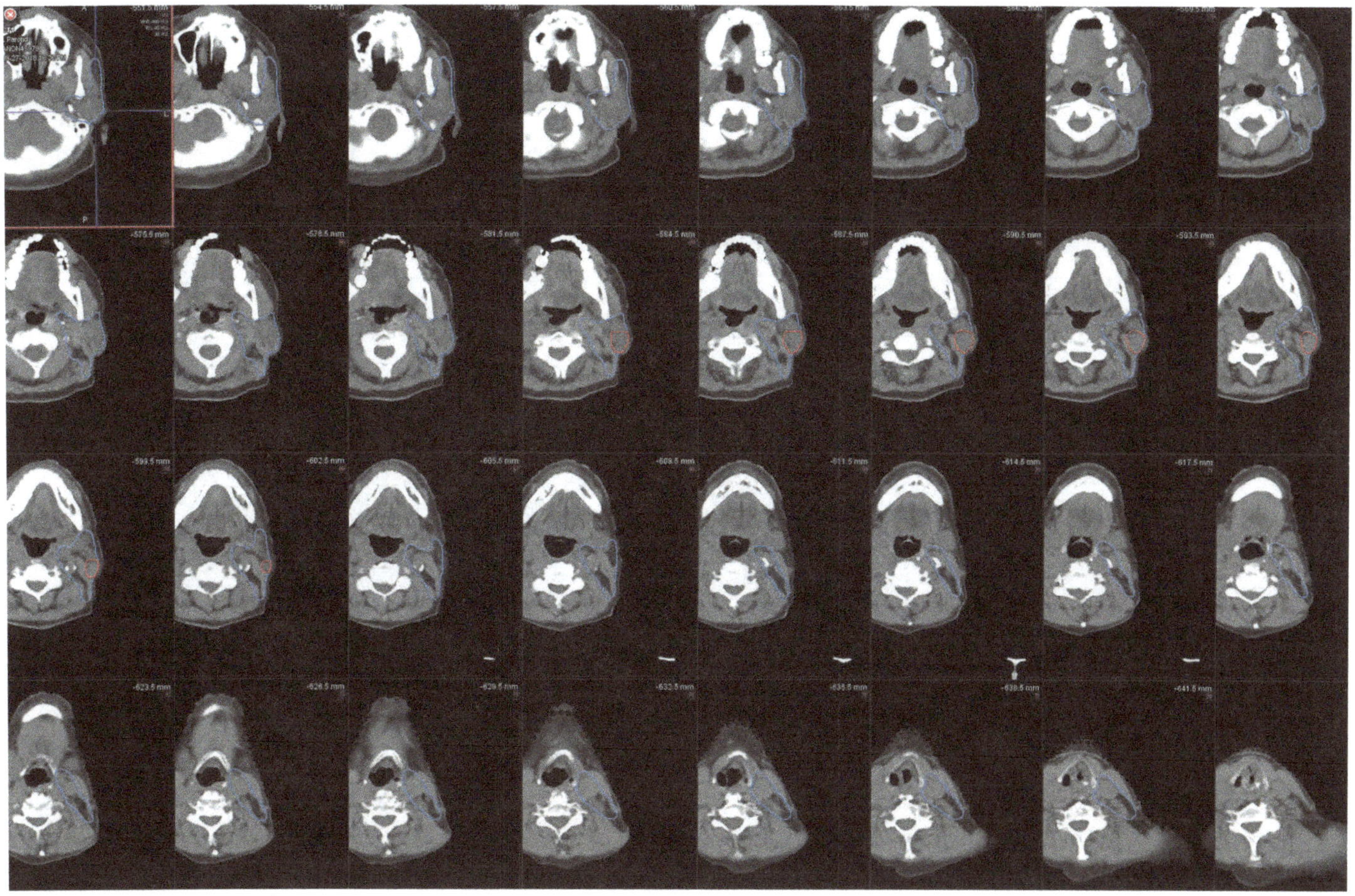

Figure 9.49 Target contours: PTV 6250 (red) and PTV 4500 (blue) in axial slices.

Beam Arrangement

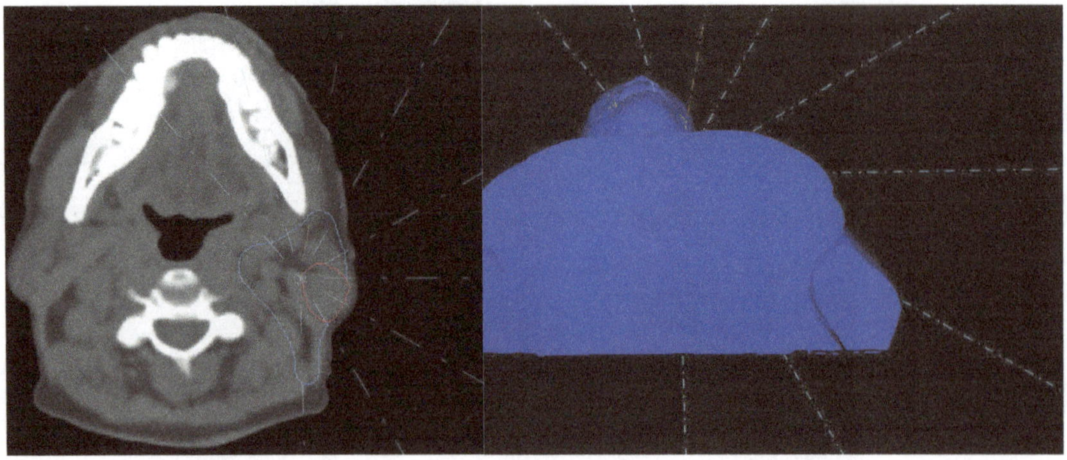

Figure 9.50 IMRT beam arrangement.

Gantry Angle	Couch	Collimator	Energy (MV)
179	0	0	6
150	0	0	6
120	0	0	6
90	0	0	6
60	0	0	6
30	0	0	6
10	0	0	6
345	0	0	6
320	0	0	6

The beam arrangement for IMRT and VMAT can be seen in figures 9.50 and 9.51, respectively. The target favors the left side of the neck and the beam arrangement is placed to minimize dose to the contralateral neck.

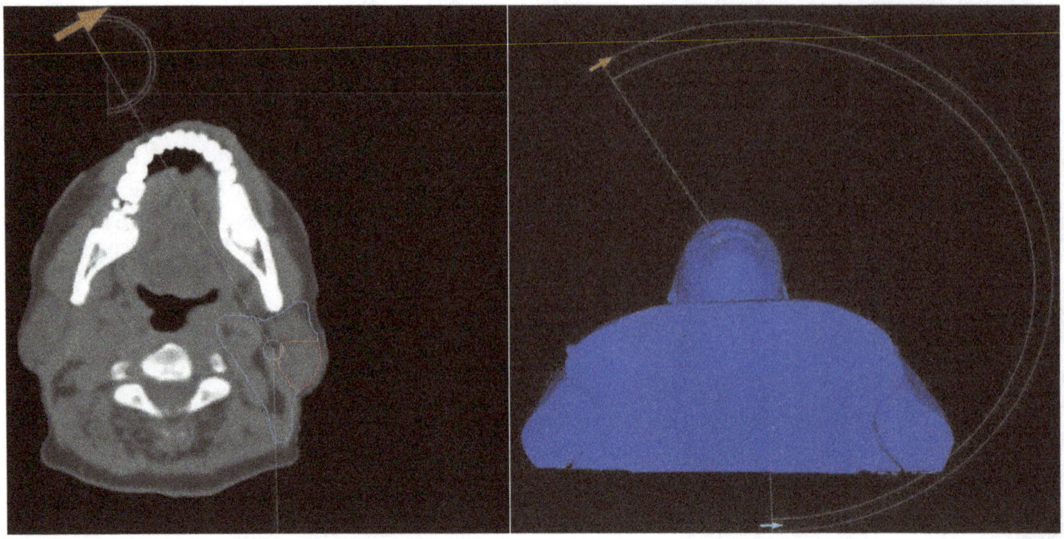

Figure 9.51 VMAT beam arrangement.

Gantry Angle	Couch	Collimator	Energy (MV)
G179–330 CCW	0	30	6
G330–171 CW	0	330	6

Optimization

Dose	ROI/POI	Clinical goal	Value	Result	% outside grid
Plan dose: L Parotid I...	Brainstem	At most 2.0 cm³ volume at 4400 cGy dose		N/A	0 %
Plan dose: L Parotid I...	Brainstem	At most 5200 cGy dose at 0.0 cm³ volume		N/A	0 %
Plan dose: L Parotid I...	Canal	At most 5.0 cm³ volume at 4200 cGy dose		N/A	0 %
Plan dose: L Parotid I...	Canal	At most 4600 cGy dose at 0.0 cm³ volume		N/A	0 %
Plan dose: L Parotid I...	Larynx	At most 3.0 cm³ volume at 3800 cGy dose		N/A	0 %
Plan dose: L Parotid I...	Larynx	At most 4400 cGy dose at 0.0 cm³ volume		N/A	0 %
Plan dose: L Parotid I...	Lt Cochlea	At most 0.5 cm³ volume at 3200 cGy dose		N/A	0 %
Plan dose: L Parotid I...	Lt Cochlea	At most 3600 cGy dose at 0.0 cm³ volume		N/A	0 %
Plan dose: L Parotid I...	Lt SMG	At most 2400 cGy average dose		N/A	0 %
Plan dose: L Parotid I...	PTV4500	At least 95.00 % volume at 4500 cGy dose		N/A	0 %
Plan dose: L Parotid I...	PTV6250	At least 95.00 % volume at 6250 cGy dose		N/A	0 %

Figure 9.52 Clinical goals set by the radiation oncologist.

The planner can start setting up objectives to cover the PTV, modifying and adding objectives for organs at risk after the first optimization to meet the clinical goals. Failing to meet criteria is at the discretion of the radiation oncologist and the level of dose coverage requested.

Function	Constraint	Dose	ROI	Description	Robust	Weight	Value
Physical Composite Objective							
Min Dose		Plan	PTV4500	Min Dose 4500 cGy		30	
Dose Fall-Off		Plan	PTV4500	Dose Fall-Off [H]6250 cGy [L]4500 cGy, Low dose distance 0.10 cm		1	
Min Dose		Plan	PTV6250	Min Dose 6250 cGy		30	
Max Dose		Plan	PTV6250	Max Dose 6400 cGy		1	
Dose Fall-Off		Plan	External	Dose Fall-Off [H]6250 cGy [L]0 cGy, Low dose distance 10.00 cm		1	
Max Dose		Plan	Canal	Max Dose 2000 cGy		1	
Dose Fall-Off		Plan	Canal	Dose Fall-Off [H]6250 cGy [L]0 cGy, Low dose distance 5.00 cm		1	
Max DVH		Plan	Lt SMG	Max DVH 4000 cGy to 18% volume		10	
Max DVH		Plan	Lt SMG	Max DVH 2000 cGy to 40% volume		10	
Max Dose		Plan	Skin 5mm	Max Dose 6400 cGy		5	
Max Dose		Plan	Larynx	Max Dose 4300 cGy		1	

Figure 9.53 Initial optimization parameters.

RAYSTATION SAMPLES

Dose Distribution

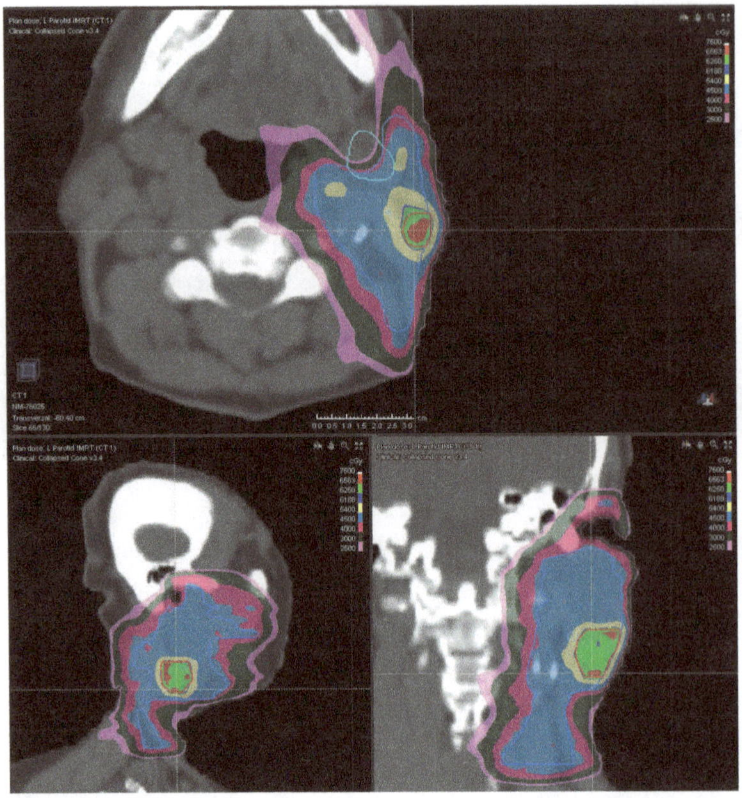

Figure 9.54 Final dose distribution for plan rHN03.

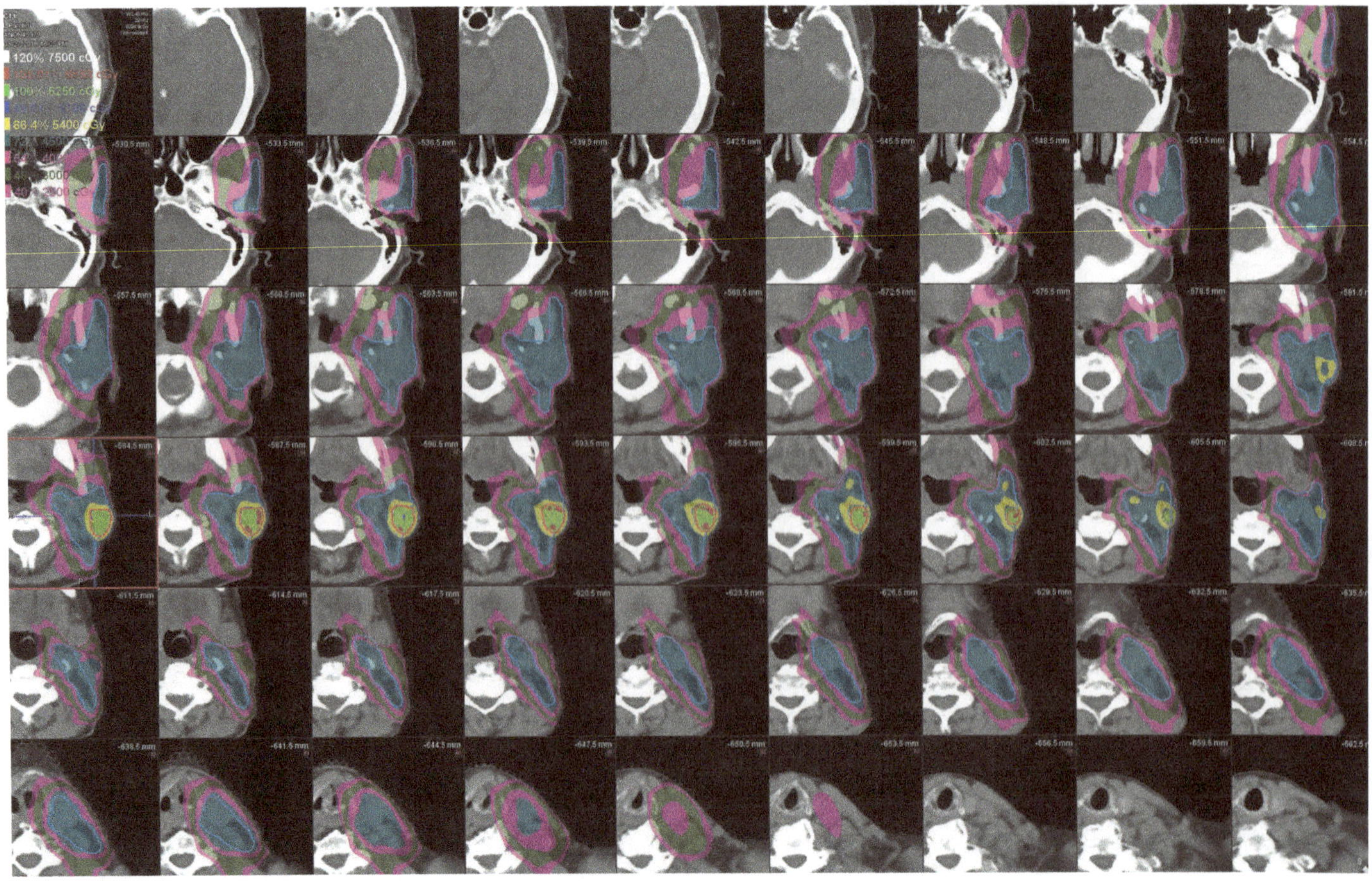

Figure 9.55 Final dose distribution for plan rHN03 in axial slices.

Case Number rP01 Pelvis: Prostate + Lymph Nodes Treatment

General Information

Brief Clinical History

70-year-old man diagnosed with biochemically recurrent prostate adenocarcinoma.

Prescription

Prostate and nodes:
PTV 6600 = 275 cGy × 24 fractions
PTV 4000 = 166.7 cGy × 24 fractions

Simulation/Comments

The simulation was accomplished with the patient in the supine position with a vacuum cushion underneath the pelvis. Fusion of images was performed with a recent MRI to evaluate the extent of the target by the radiation oncologist.

Targets

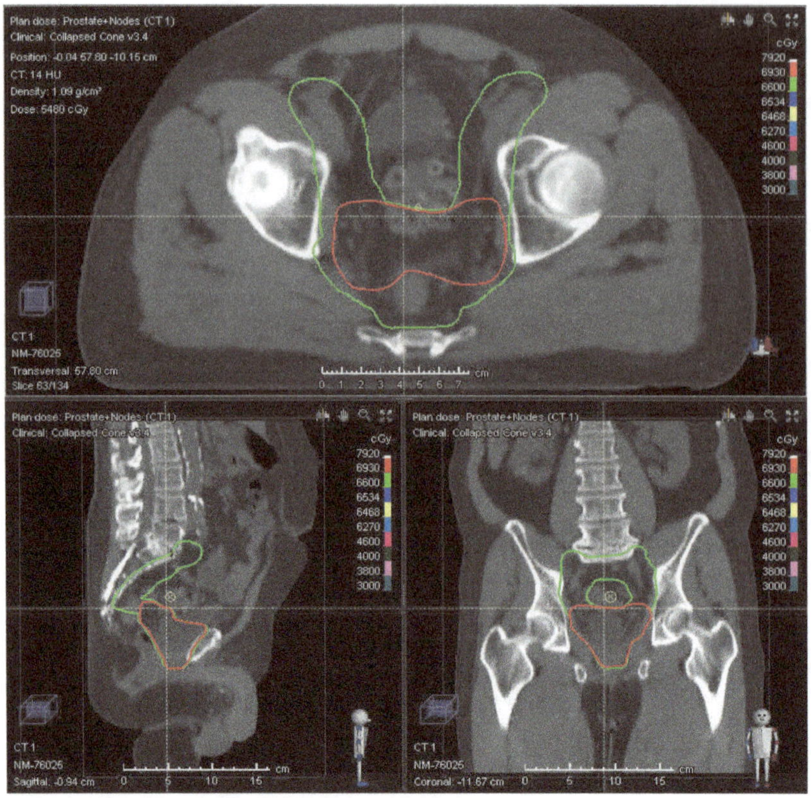

Figure 9.56 Target contours: PTV 6600 (red) and PTV 4000 (green).

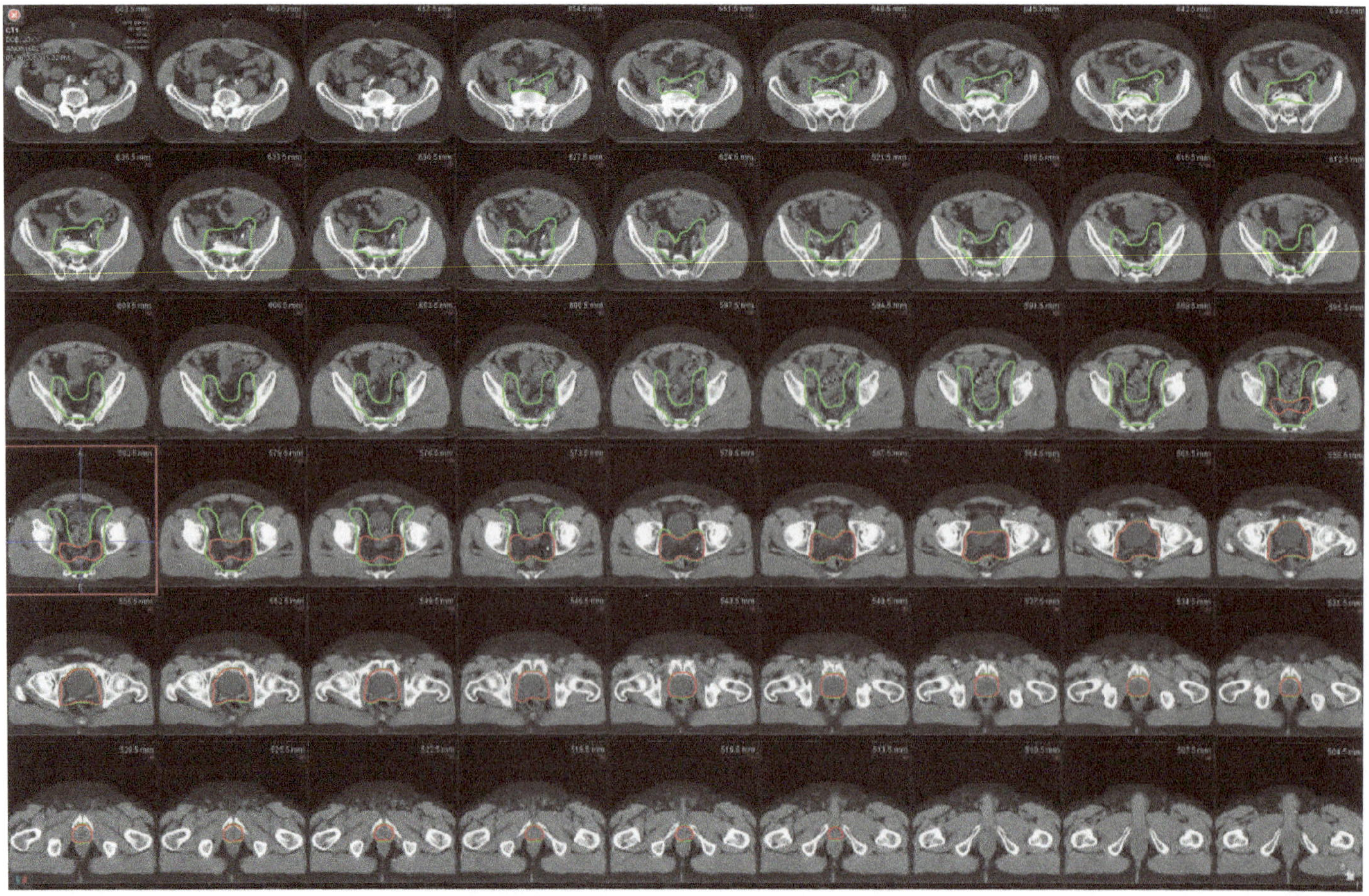

Figure 9.57 Target contours: PTV 6600 (red) and PTV 4000 (green) in axial slices.

Beam Arrangement

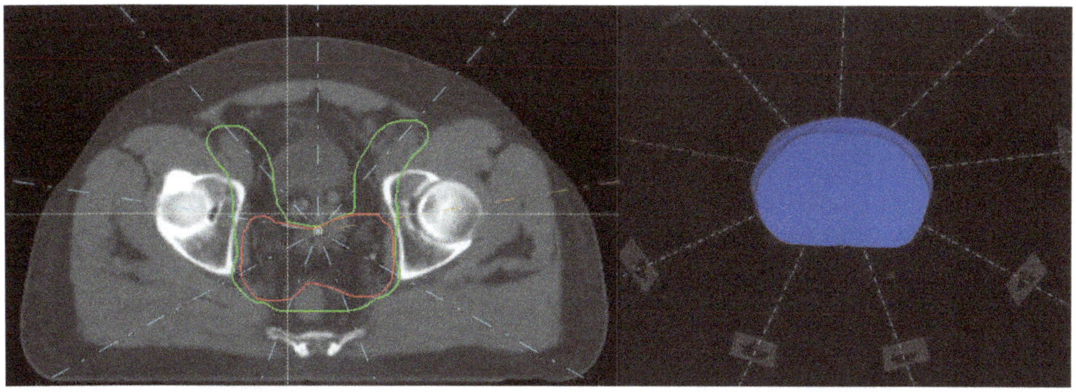

Figure 9.58 IMRT beam arrangement for case rP01.

Gantry Angle	Couch	Collimator	Energy (MV)
200	0	0	10
240	0	0	10
280	0	0	10
320	0	0	10
0	0	0	10
40	0	0	10
80	0	0	10
120	0	0	10
160	0	0	10

Figures 9.58 and 9.59 show the IMRT and VMAT beam arrangements, respectively. The IMRT plan utilizes 9 evenly distributed fields spaced every 40 degrees. A 7-field plan could also have been used. However, 9 fields offer more modulation to better spare nearby critical structures. Prostate plans with nodes often have a fairly large treatment volume, and the planner must verify that the entirety of the target volumes is within the machine's capabilities.

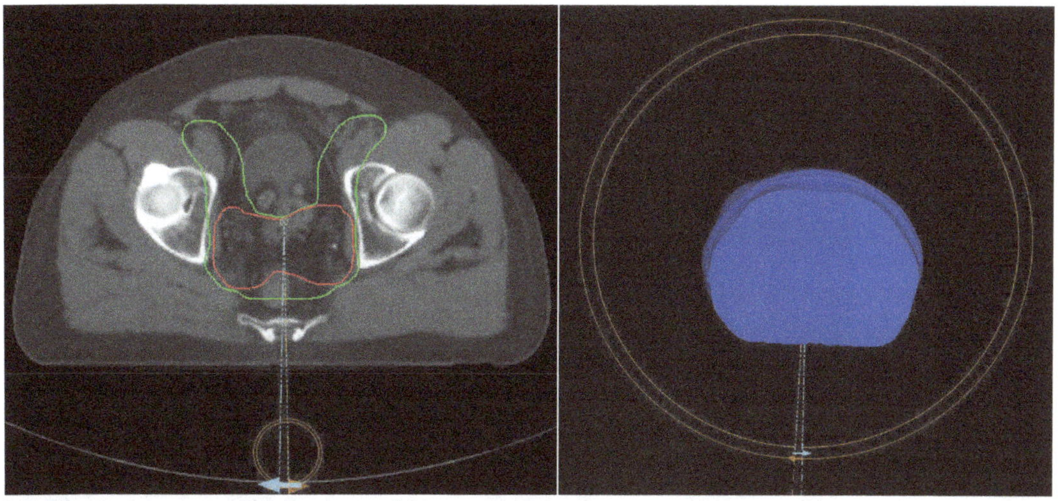

Figure 9.59 VMAT beam arrangement for case rP01.

Gantry Angle	Couch	Collimator	Energy (MV)
G181–179 CW	0	30	10
G179–181 CCW	0	330	10

Optimization

Priority	Dose	ROI/POI	Clinical goal	Value	Result	% outside grid
	Plan dose: Prostate+N...	Bladder	At most 90.0 cm³ volume at 6000 cGy dose		N/A	0 %
	Plan dose: Prostate+N...	Bladder	At most 125.0 cm³ volume at 5600 cGy dose		N/A	0 %
	Plan dose: Prostate+N...	Bladder	At most 6600 cGy dose at 0.0 cm³ volume		N/A	0 %
	Plan dose: Prostate+N...	Bowel	At most 5.0 cm³ volume at 4000 cGy dose		N/A	0 %
	Plan dose: Prostate+N...	Bowel	At most 4400 cGy dose at 0.0 cm³ volume		N/A	0 %
	Plan dose: Prostate+N...	LT Femoral Head	At most 10.0 cm³ volume at 4400 cGy dose		N/A	0 %
	Plan dose: Prostate+N...	LT Femoral Head	At most 5000 cGy dose at 0.0 cm³ volume		N/A	0 %
	Plan dose: Prostate+N...	PTV4000	At least 95.00 % volume at 4000 cGy dose		N/A	0 %
	Plan dose: Prostate+N...	PTV6600	At least 95.00 % volume at 6600 cGy dose		N/A	0 %
	Plan dose: Prostate+N...	Rectum	At most 10.0 cm³ volume at 6400 cGy dose		N/A	0 %
	Plan dose: Prostate+N...	Rectum	At most 20.0 cm³ volume at 6000 cGy dose		N/A	0 %
	Plan dose: Prostate+N...	Rectum	At most 30.0 cm³ volume at 5600 cGy dose		N/A	0 %
	Plan dose: Prostate+N...	Rectum	At most 40.0 cm³ volume at 5200 cGy dose		N/A	0 %
	Plan dose: Prostate+N...	Rectum	At most 6600 cGy dose at 0.0 cm³ volume		N/A	0 %
	Plan dose: Prostate+N...	RT Femoral Head	At most 10.0 cm³ volume at 4400 cGy dose		N/A	0 %
	Plan dose: Prostate+N...	RT Femoral Head	At most 5000 cGy dose at 0.0 cm³ volume		N/A	0 %

Figure 9.60 Clinical goals set by the radiation oncologist.

The radiation oncologist can assess areas of overlap of prescription dose and organs at risk in terms of coverage. One common case is the sparing of the rectum structure. The planner can use an expansion of contour to further prevent high dose in this structure.

Function	Constraint	Dose	ROI	Description	Robust	Weight
Physical Composite Objective						
Min Dose		Plan	PTV6600	Min Dose 6600 cGy		30
Max Dose		Plan	PTV6600	Max Dose 6700 cGy		28
Dose Fall-Off		Plan	External	Dose Fall-Off [H]6600 cGy [L]0 cGy, Low dose distance 1.00 cm		1
Dose Fall-Off		Plan	Rectum	Dose Fall-Off [H]6600 cGy [L]0 cGy, Low dose distance 2.00 cm		1
Min Dose		Plan	PTV4000	Min Dose 4000 cGy		30
Dose Fall-Off		Plan	PTV4000	Dose Fall-Off [H]6600 cGy [L]4000 cGy, Low dose distance 0.10 cm		1
Dose Fall-Off		Plan	External	Dose Fall-Off [H]6600 cGy [L]0 cGy, Low dose distance 10.00 cm		1
Dose Fall-Off		Plan	Bladder	Dose Fall-Off [H]6600 cGy [L]0 cGy, Low dose distance 5.00 cm		1
Max Dose		Plan	Rectum	Max Dose 6700 cGy		10
Dose Fall-Off		Plan	Bowel	Dose Fall-Off [H]6600 cGy [L]0 cGy, Low dose distance 10.00 cm		1
Max Dose		Plan	Bladder	Max Dose 6700 cGy		10
Max Dose		Plan	Rectum+5mm	Max Dose 6650 cGy		1

Figure 9.61 Initial objectives for optimization. Initial optimization objectives for this plan aim to minimize hot spots within the rectum and bladder without losing coverage to the PTVs.

Function	Constraint	Dose	ROI	Description	Robust	Weight
Physical Composite Objective						
Min Dose		Plan	PTV6600	Min Dose 6600 cGy		30
Max Dose		Plan	PTV6600	Max Dose 6700 cGy		28
Dose Fall-Off		Plan	External	Dose Fall-Off [H]6600 cGy [L]0 cGy, Low dose distance 1.00 cm		1
Dose Fall-Off		Plan	Rectum	Dose Fall-Off [H]6600 cGy [L]0 cGy, Low dose distance 2.00 cm		1
Min Dose		Plan	PTV4000	Min Dose 4000 cGy		30
Dose Fall-Off		Plan	PTV4000	Dose Fall-Off [H]6600 cGy [L]4000 cGy, Low dose distance 0.10 cm		1
Dose Fall-Off		Plan	External	Dose Fall-Off [H]6600 cGy [L]0 cGy, Low dose distance 10.00 cm		1
Dose Fall-Off		Plan	Bladder	Dose Fall-Off [H]6600 cGy [L]0 cGy, Low dose distance 5.00 cm		1
Max Dose		Plan	Rectum	Max Dose 6700 cGy		10
Dose Fall-Off		Plan	Bowel	Dose Fall-Off [H]6600 cGy [L]0 cGy, Low dose distance 10.00 cm		1
Max Dose		Plan	Bladder	Max Dose 6700 cGy		10
Max Dose		Plan	Rectum+5mm	Max Dose 6650 cGy		12
Max DVH		Plan	Ring - 5mm PTV6600	Max DVH 6600 cGy to 1% volume		1
Max Dose		Plan	Ring - 5mm PTV6600	Max Dose 6610 cGy		10
Uniformity Constraint	★	Plan	PTV6600	Uniformity, relative standard deviation 1%		

Figure 9.62 Final objectives for optimization (not all objectives are pictured).

Dose Distribution

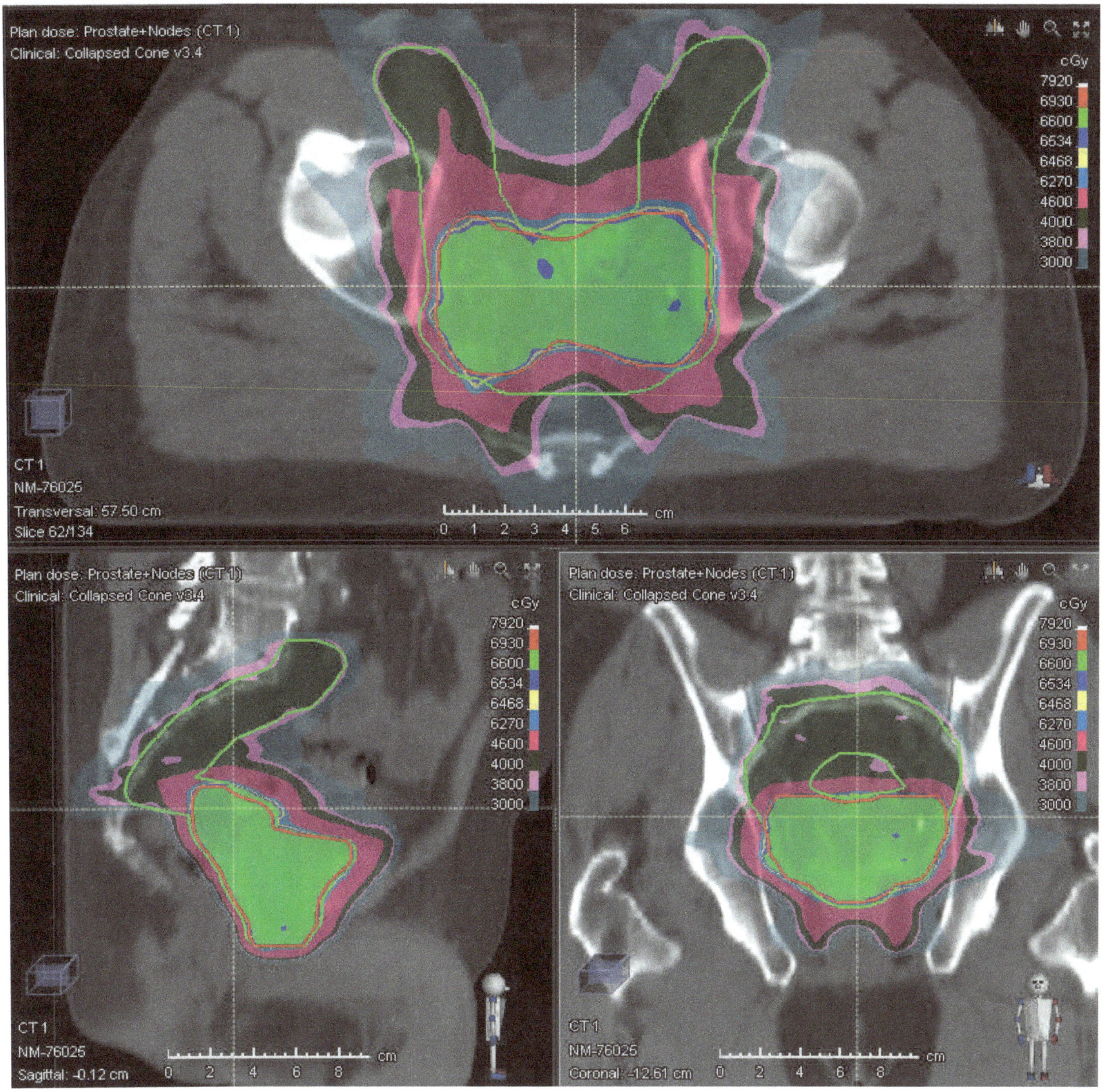

Figure 9.63 Final dose distribution for case rP01.

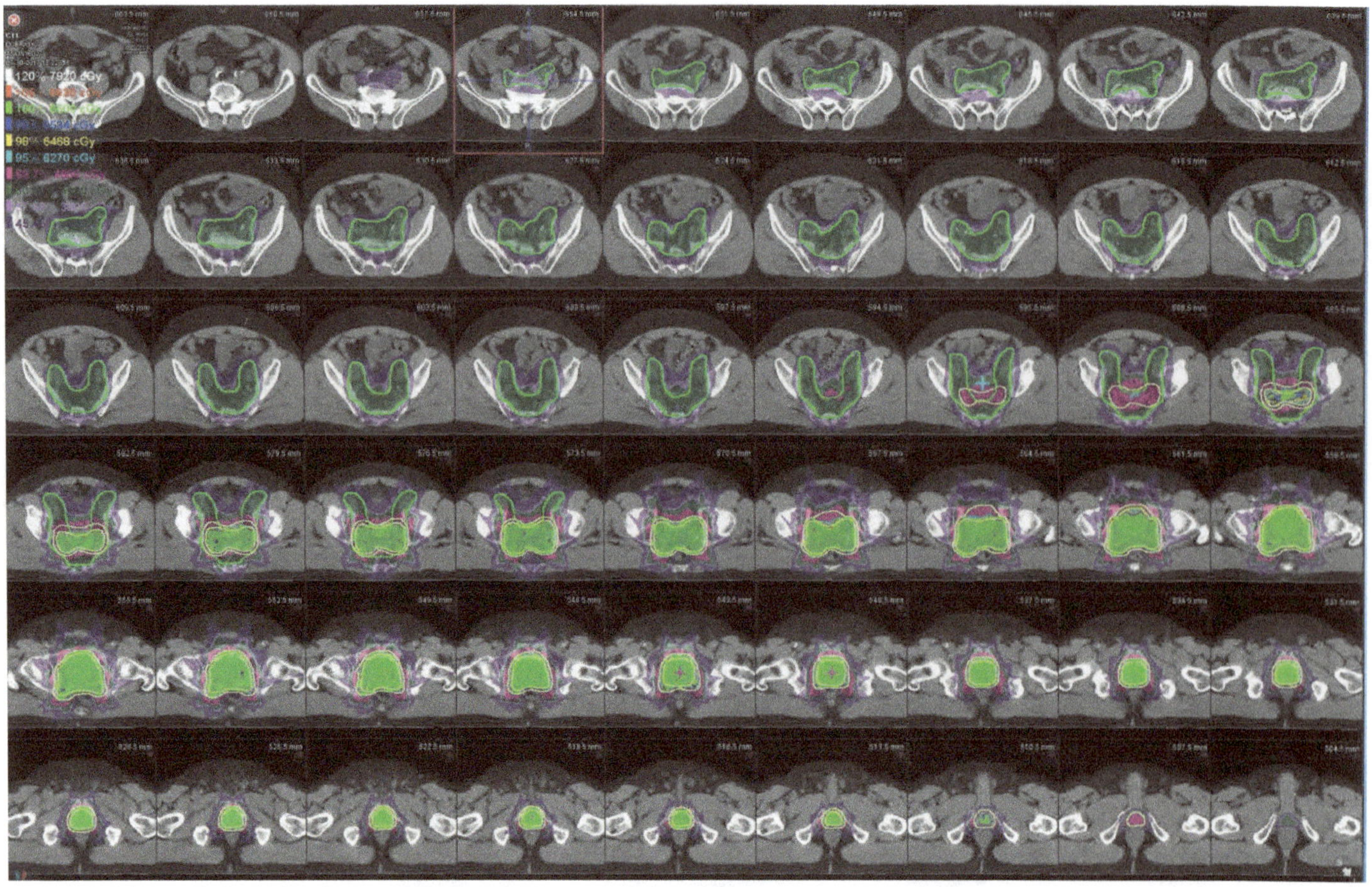

Figure 9.64 Final dose distribution for case rP01 in axial slices.

Case Number rP02 Pelvis: Left Acetabulum Treatment

General Information

Brief Clinical History

57-year-old female diagnosed with multiple bone metastasis prescribed radiation therapy to symptomatic sites.

Prescription

Left acetabulum:
PTV 3000 = 300 cGy × 10 fractions
PTV 2100 = 210 cGy × 10 fractions

Simulation/Comments

The simulation was accomplished with the patient in the supine position with a vacuum cushion underneath their pelvis and their hands on their chest. Fusion of images was performed with a recent PET/CT to evaluate the extent of the target by the radiation oncologist.

Targets

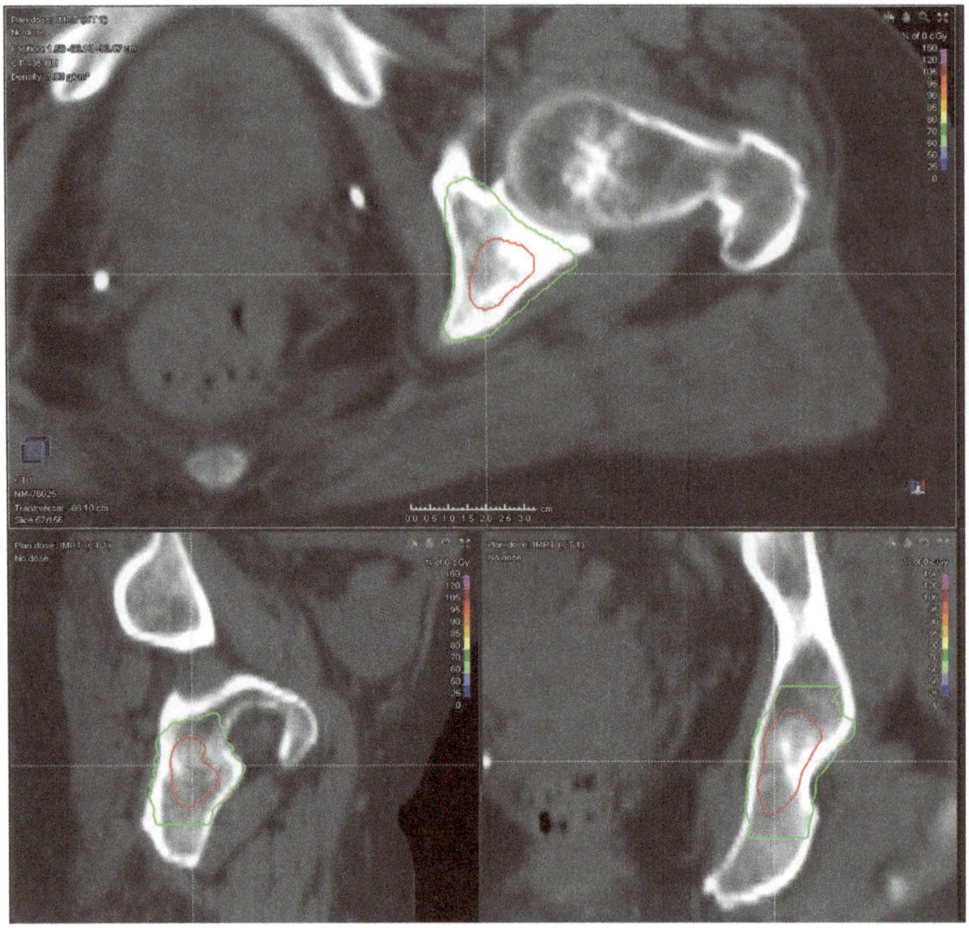

Figure 9.65 Target contours: PTV 3000 (red) and PTV 2100 (green).

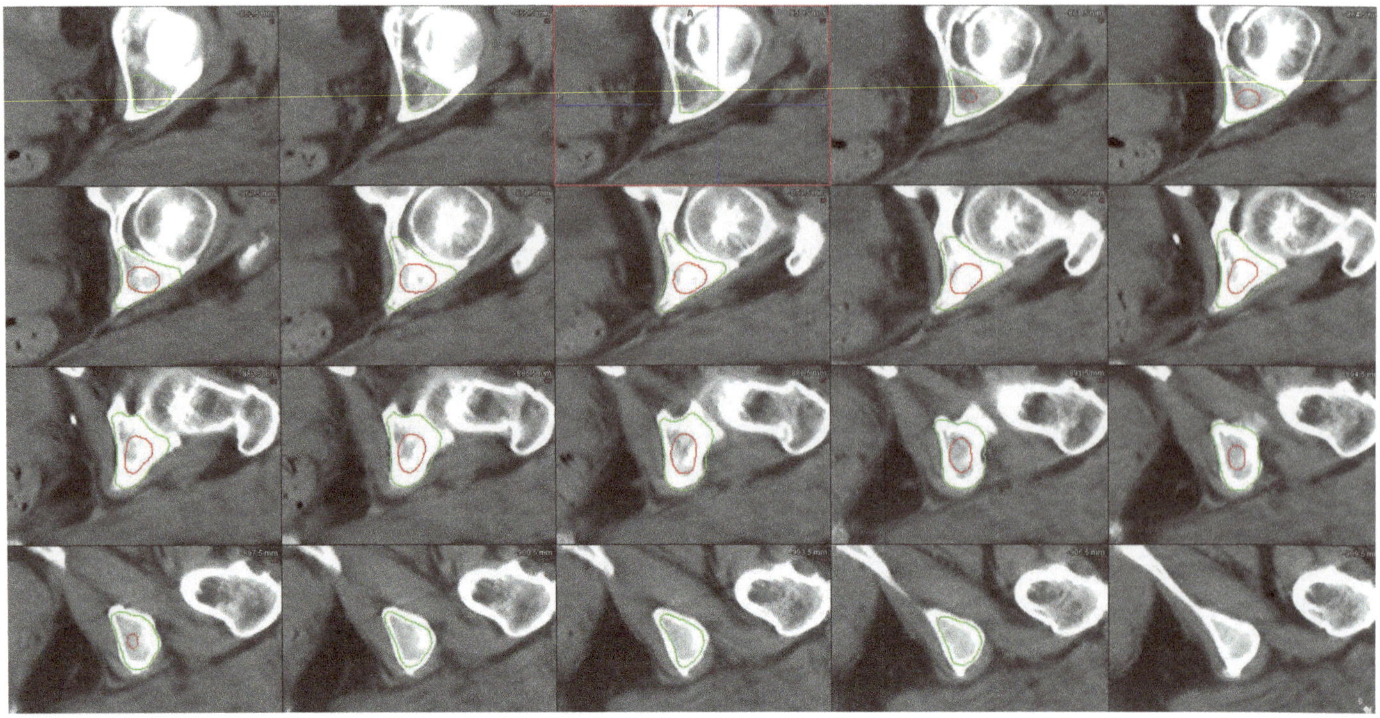

Figure 9.66 Target contours: PTV 3000 (red) and PTV 2100 (green).

Beam Arrangement

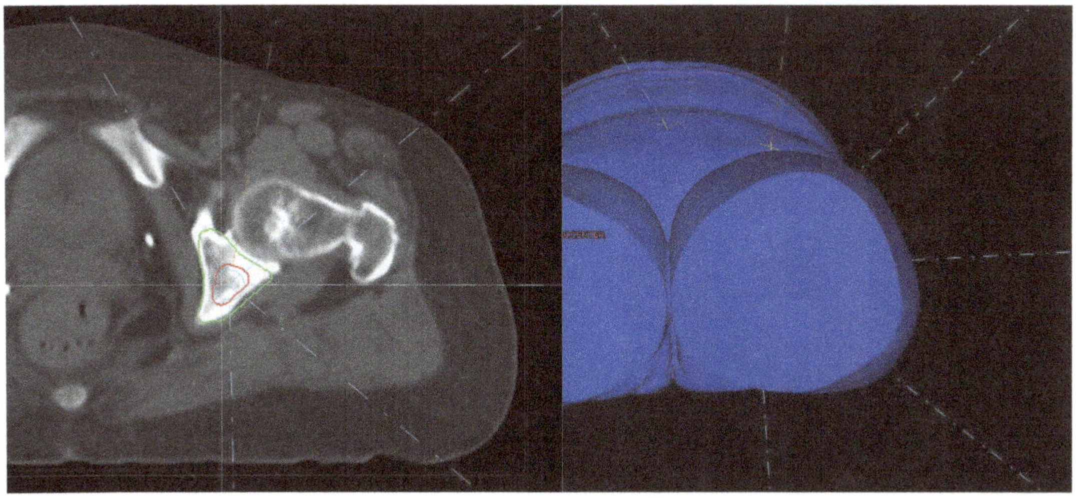

Figure 9.67 IMRT beam arrangement. The 10-MV FFF energy increases dose rate and decrease delivery time.

Gantry Angle	Couch	Collimator	Energy (MV)
179	0	0	10 FFF
130	0	0	10 FFF
90	0	0	10 FFF
50	0	0	10 FFF
10	0	0	10 FFF
330	0	0	10 FFF

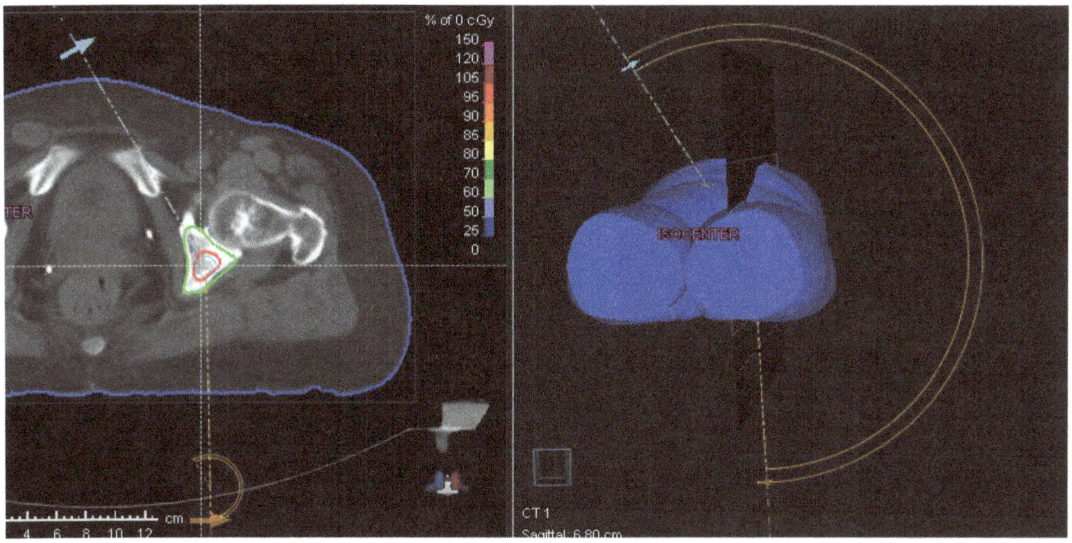

Figure 9.68 VMAT beam arrangement.

Gantry Angle	Couch	Collimator	Energy (MV)
330–179 CW	0	30	10 FFF
179–330 CCW	0	330	10 FFF

Optimization

Dose	ROI/POI	Clinical goal	Value	Result	% outside grid
Plan dose: IMRT (CT 1)	Bladder	At most 15.0 cm³ volume at 1700 cGy dose		N/A	0 %
Plan dose: IMRT (CT 1)	Bladder	At most 3300 cGy dose at 0.0 cm³ volume		N/A	0 %
Plan dose: IMRT (CT 1)	Bowel Bag	At most 20.0 cm³ volume at 2400 cGy dose		N/A	0 %
Plan dose: IMRT (CT 1)	Bowel Bag	At most 3000 cGy dose at 0.0 cm³ volume		N/A	0 %
Plan dose: IMRT (CT 1)	Cauda	At most 5.0 cm³ volume at 2190 cGy dose		N/A	0 %
Plan dose: IMRT (CT 1)	Cauda	At most 2550 cGy dose at 0.0 cm³ volume		N/A	0 %
Plan dose: IMRT (CT 1)	L Femoral Head	At most 10.0 cm³ volume at 2400 cGy dose		N/A	0 %
Plan dose: IMRT (CT 1)	PTV2100	At least 95.00 % volume at 2100 cGy dose		N/A	0 %
Plan dose: IMRT (CT 1)	PTV3000	At least 95.00 % volume at 3000 cGy dose		N/A	0 %
Plan dose: IMRT (CT 1)	R Femoral Head	At most 10.0 cm³ volume at 2400 cGy dose		N/A	0 %
Plan dose: IMRT (CT 1)	Rectum	At most 20.0 cm³ volume at 2750 cGy dose		N/A	0 %
Plan dose: IMRT (CT 1)	Rectum	At most 3500 cGy dose at 0.0 cm³ volume		N/A	0 %

Figure 9.69 Clinical goals for case rP02.

The maximum dose for the lesion was set to 135% of the prescription dose by the radiation oncologist. This is easily achievable in such small target volume.

Function	Constraint	Dose	ROI	Description	Robust	Weight	Value
Physical Composite Objective							
Min Dose		Plan	PTV2100	Min Dose 2100 cGy		30	
Min Dose		Plan	PTV3000	Min Dose 3000 cGy		30	
Dose Fall-Off		Plan	PTV2100	Dose Fall-Off [H]3000 cGy [L]2100 cGy, Low dose distance 0.10 cm		1	
Max Dose		Plan	PTV3000	Max Dose 4100 cGy		1	
Dose Fall-Off		Plan	External	Dose Fall-Off [H]3000 cGy [L]0 cGy, Low dose distance 10.00 cm		1	
Dose Fall-Off		Plan	Rectum	Dose Fall-Off [H]3000 cGy [L]0 cGy, Low dose distance 5.00 cm		1	
Dose Fall-Off		Plan	Bladder	Dose Fall-Off [H]3000 cGy [L]0 cGy, Low dose distance 8.00 cm		1	
Dose Fall-Off		Plan	L Femoral Head	Dose Fall-Off [H]3000 cGy [L]0 cGy, Low dose distance 3.00 cm		1	
Max Dose		Plan	L Femoral Head	Max Dose 2000 cGy		1	

Figure 9.70 Initial optimization parameters for case rP02.

Function	Constraint	Dose	ROI	Description	Robust	Weight	Value
Physical Composite Objective							
Min Dose		Plan	PTV2100	Min Dose 2100 cGy		30	
Min Dose		Plan	PTV3000	Min Dose 3000 cGy		30	
Dose Fall-Off		Plan	PTV2100	Dose Fall-Off [H]3000 cGy [L]2100 cGy, Low dose distance 0.10 cm		1	
Max Dose		Plan	PTV3000	Max Dose 4100 cGy		1	
Dose Fall-Off		Plan	External	Dose Fall-Off [H]3000 cGy [L]0 cGy, Low dose distance 10.00 cm		1	
Dose Fall-Off		Plan	Rectum	Dose Fall-Off [H]3000 cGy [L]0 cGy, Low dose distance 5.00 cm		1	
Dose Fall-Off		Plan	Bladder	Dose Fall-Off [H]3000 cGy [L]0 cGy, Low dose distance 8.00 cm		1	
Dose Fall-Off		Plan	L Femoral Head	Dose Fall-Off [H]3000 cGy [L]0 cGy, Low dose distance 3.00 cm		1	
Max Dose		Plan	L Femoral Head	Max Dose 2000 cGy		1	
Max DVH		Plan	PTV2100 1cm Ring	Max DVH 2100 cGy to 3% volume		10	

Figure 9.71 Final optimization parameters for case rP02.

Dose Distribution

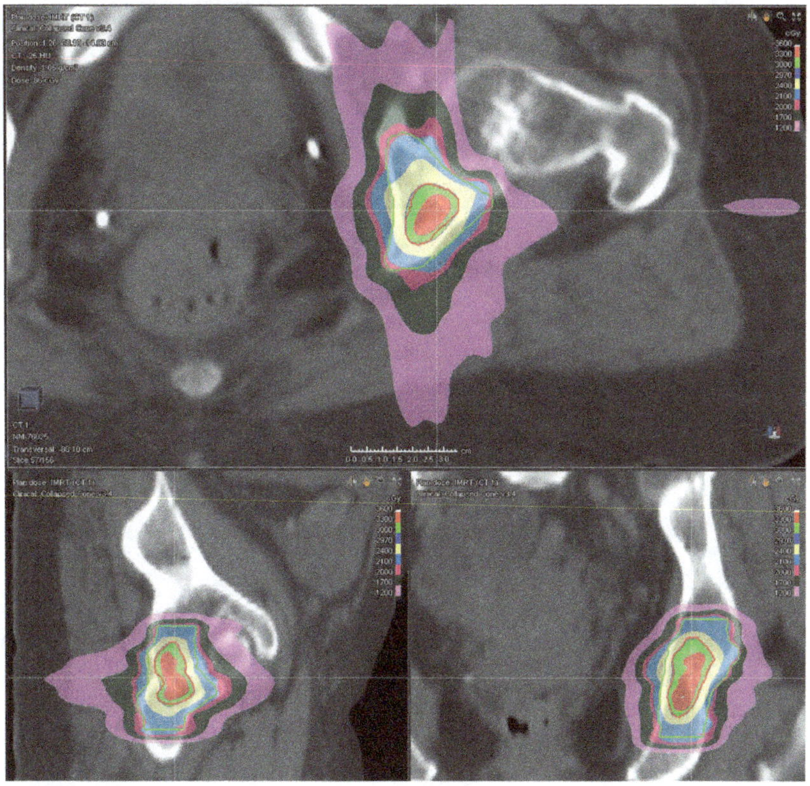

Figure 9.72 Final dose distribution for case rP02.

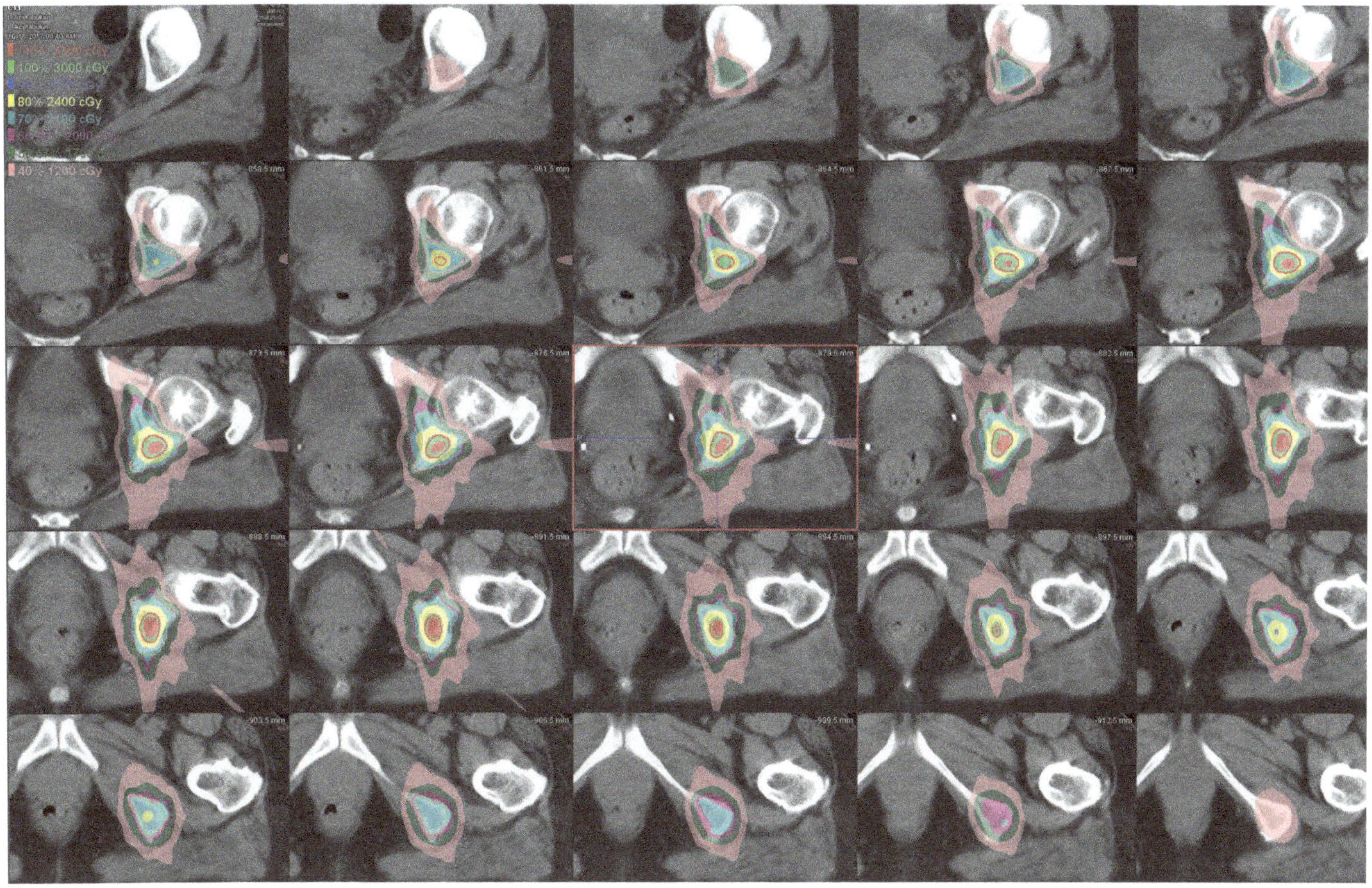

Figure 9.73 Final dose distribution for case rP02 in axial slices.

Dose	ROI/POI	Clinical goal	Value	Result	% outside grid
Plan dose: IMRT (CT 1)	Bladder	At most 15.0 cm³ volume at 1700 cGy dose	0.0 cm³	✓	0 %
Plan dose: IMRT (CT 1)	Bladder	At most 3300 cGy dose at 0.0 cm³ volume	621 cGy	✓	0 %
Plan dose: IMRT (CT 1)	Bowel Bag	At most 20.0 cm³ volume at 2400 cGy dose	0.0 cm³	✓	0 %
Plan dose: IMRT (CT 1)	Bowel Bag	At most 3000 cGy dose at 0.0 cm³ volume	807 cGy	✓	0 %
Plan dose: IMRT (CT 1)	Cauda	At most 5.0 cm³ volume at 2190 cGy dose	0.0 cm³	✓	0 %
Plan dose: IMRT (CT 1)	Cauda	At most 2550 cGy dose at 0.0 cm³ volume	2 cGy	✓	0 %
Plan dose: IMRT (CT 1)	L Femoral Head	At most 10.0 cm³ volume at 2400 cGy dose	0.0 cm³	✓	0 %
Plan dose: IMRT (CT 1)	PTV2100	At least 95.00 % volume at 2100 cGy dose	96.71 %	✓	0 %
Plan dose: IMRT (CT 1)	PTV3000	At least 95.00 % volume at 3000 cGy dose	95.00 %	✓	0 %
Plan dose: IMRT (CT 1)	R Femoral Head	At most 10.0 cm³ volume at 2400 cGy dose	0.0 cm³	✓	0 %
Plan dose: IMRT (CT 1)	Rectum	At most 20.0 cm³ volume at 2750 cGy dose	0.0 cm³	✓	0 %
Plan dose: IMRT (CT 1)	Rectum	At most 3500 cGy dose at 0.0 cm³ volume	426 cGy	✓	0 %

Figure 9.74 Clinical goals met for case rP02.

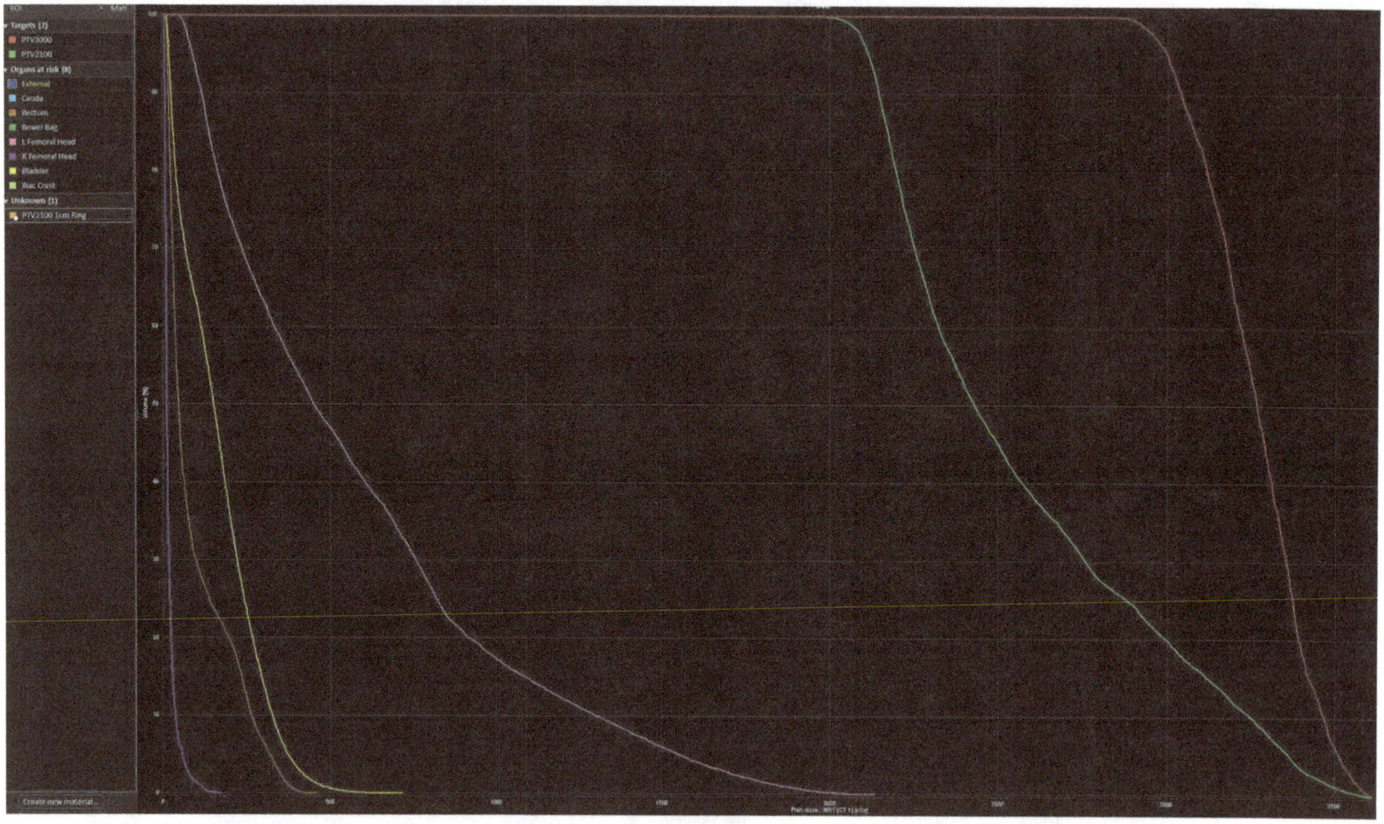

Figure 9.75 Final DVH for case rP02.

Case Number rP03 Pelvis: Prostate Bed Treatment

General Information

Brief Clinical History

68-year-old man received a radical prostatectomy and bilateral extended pelvic lymph node dissection. The patient was prescribed adjuvant IMRT to the prostate bed.

Prescription

Prostate and nodes:
PTV 6600 = 200 cGy × 33 fractions

Simulation/Comments

The simulation was accomplished with the patient in the supine position with a vacuum cushion underneath his pelvis.

Targets

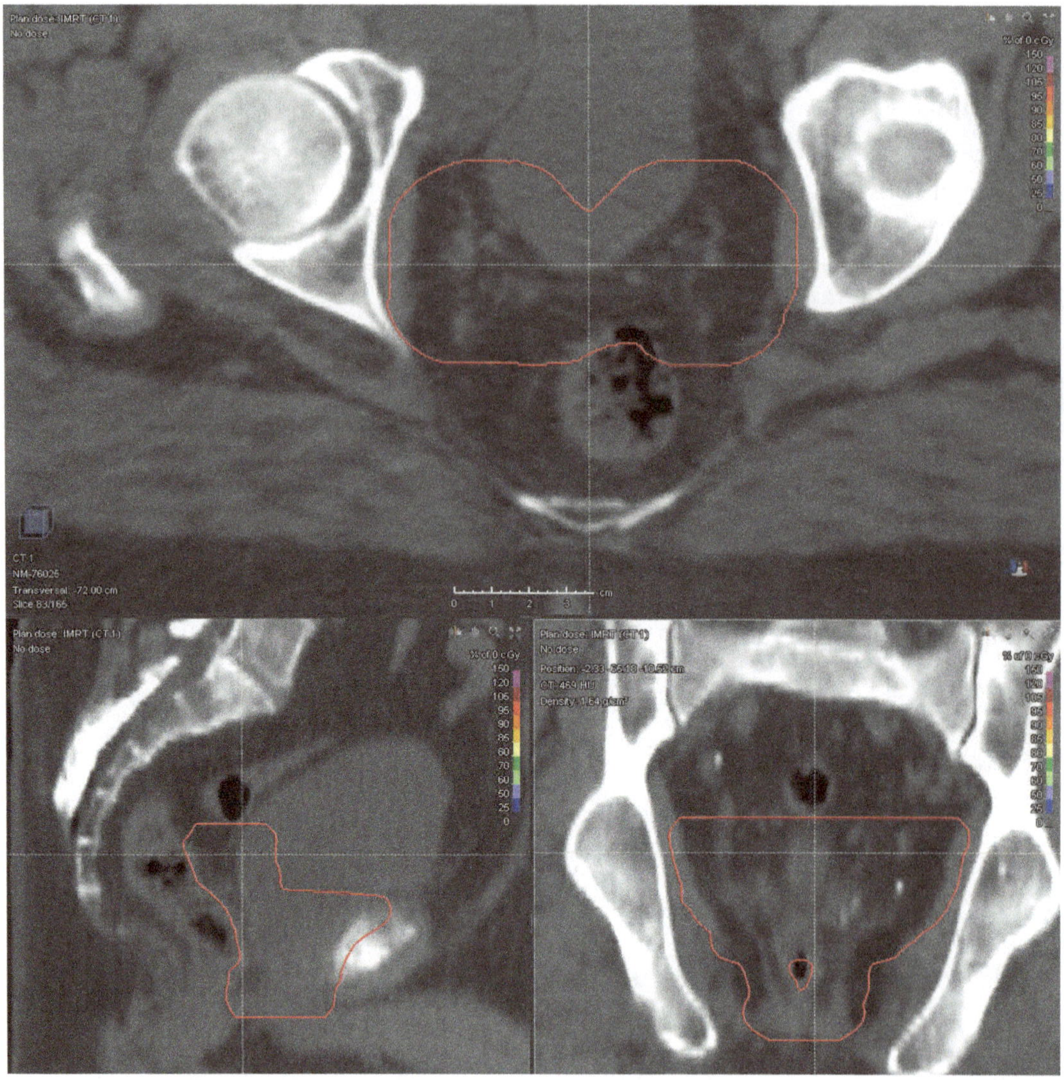

Figure 9.76 Treatment target: PTV 6600 (red).

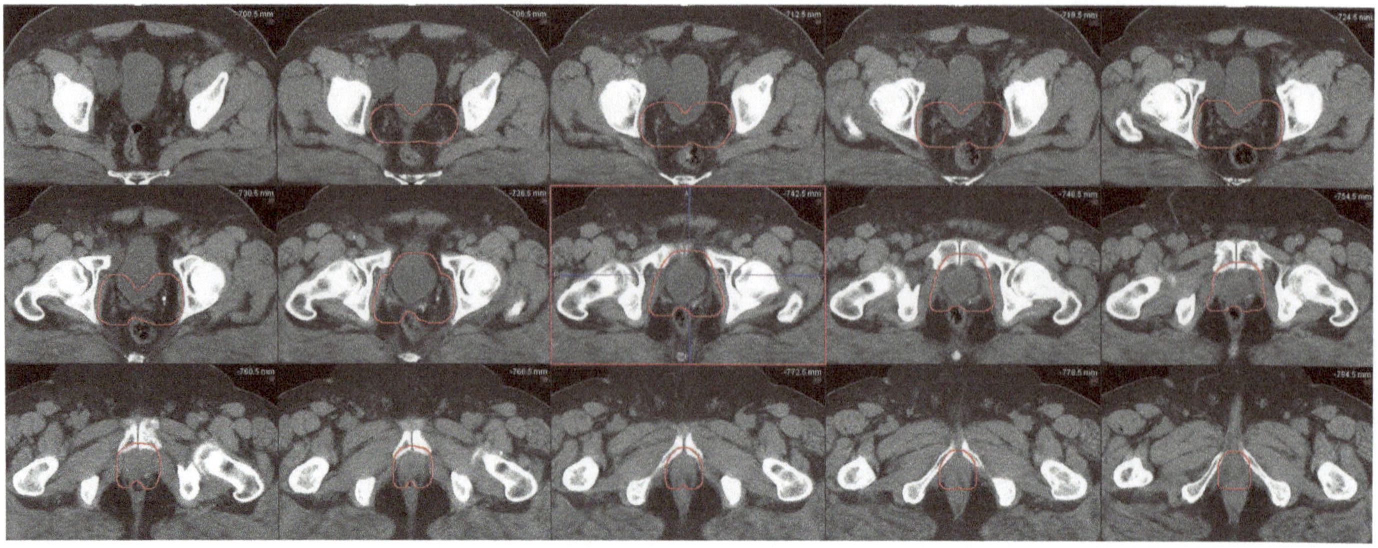

Figure 9.77 Treatment target: PTV 6600 (red) in axial slices.

Beam Arrangement

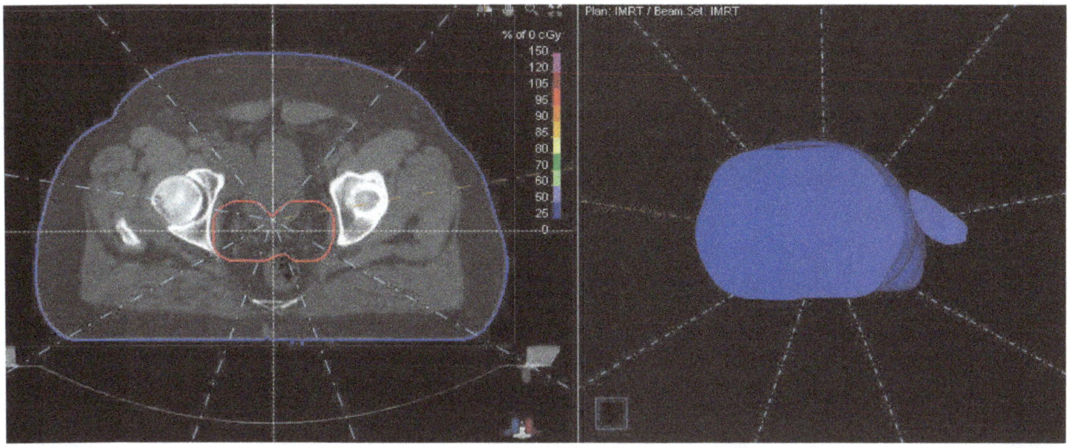

Figure 9.78 Nine-field IMRT arrangement for prostate bed irradiation.

Gantry Angle	Couch	Collimator	Energy (MV)
200	0	0	10
240	0	0	10
280	0	0	10
320	0	0	10
0	0	0	10
40	0	0	10
80	0	0	10
120	0	0	10
160	0	0	10

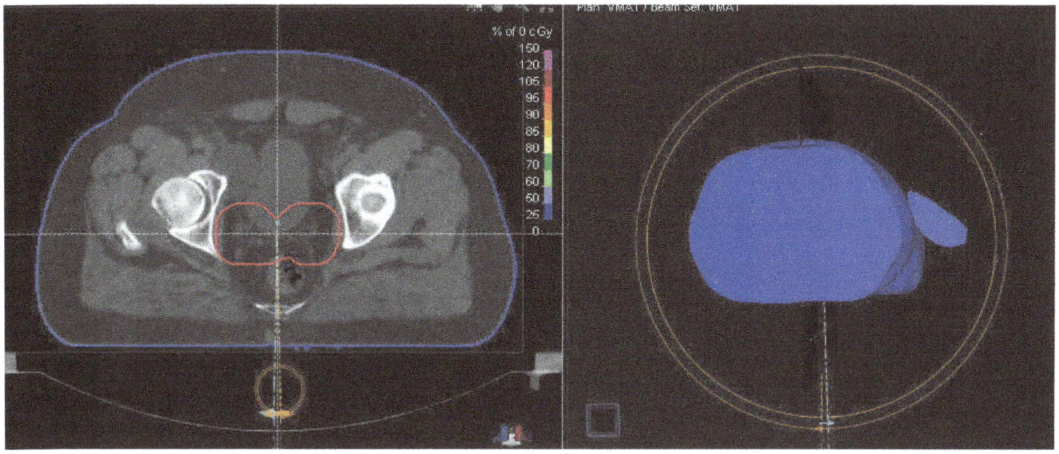

Figure 9.79 Two-arc VMAT beam arrangement.

Gantry Angle	Couch	Collimator	Energy (MV)
181–179 CW	0	30	10
179–181 CCW	0	330	10

Optimization

Dose	ROI/POI	Clinical goal	Value	Result	% outside grid
Plan dose: VMAT (CT 1)	Bladder	At most 90.0 cm³ volume at 7000 cGy dose		N/A	0 %
Plan dose: VMAT (CT 1)	Bladder	At most 150.0 cm³ volume at 6500 cGy dose		N/A	0 %
Plan dose: VMAT (CT 1)	Bladder	At most 7900 cGy dose at 0.0 cm³ volume		N/A	0 %
Plan dose: VMAT (CT 1)	Bowels	At most 5.0 cm³ volume at 4200 cGy dose		N/A	0 %
Plan dose: VMAT (CT 1)	Bowels	At most 5000 cGy dose at 0.0 cm³ volume		N/A	0 %
Plan dose: VMAT (CT 1)	L Femur head	At most 10.0 cm³ volume at 4800 cGy dose		N/A	0 %
Plan dose: VMAT (CT 1)	L Femur head	At most 5600 cGy dose at 0.0 cm³ volume		N/A	0 %
Plan dose: VMAT (CT 1)	PTV6600	At least 95.00 % volume at 6600 cGy dose		N/A	0 %
Plan dose: VMAT (CT 1)	R Femur head	At most 10.0 cm³ volume at 4800 cGy dose		N/A	0 %
Plan dose: VMAT (CT 1)	R Femur head	At most 5600 cGy dose at 0.0 cm³ volume		N/A	0 %
Plan dose: VMAT (CT 1)	Rectum	At most 10.0 cm³ volume at 7500 cGy dose		N/A	0 %
Plan dose: VMAT (CT 1)	Rectum	At most 20.0 cm³ volume at 7000 cGy dose		N/A	0 %
Plan dose: VMAT (CT 1)	Rectum	At most 30.0 cm³ volume at 6500 cGy dose		N/A	0 %
Plan dose: VMAT (CT 1)	Rectum	At most 40.0 cm³ volume at 6000 cGy dose		N/A	0 %
Plan dose: VMAT (CT 1)	Rectum	At most 7900 cGy dose at 0.0 cm³ volume		N/A	0 %

Figure 9.80 Clinical goals for case rP03.

Function	Constraint	Dose	ROI	Description	Robust	Weight
Physical Composite Objective						
Min Dose		Plan	PTV6600	Min Dose 6600 cGy		30
Max Dose		Plan	PTV6600	Max Dose 6900 cGy		5
Dose Fall-Off		Plan	External	Dose Fall-Off [H]6600 cGy [L]0 cGy, Low dose distance 10.00 cm		1
Dose Fall-Off		Plan	Rectum	Dose Fall-Off [H]6600 cGy [L]0 cGy, Low dose distance 3.00 cm		1
Dose Fall-Off		Plan	Bladder	Dose Fall-Off [H]6600 cGy [L]0 cGy, Low dose distance 5.00 cm		1
Max Dose		Plan	Bladder	Max Dose 6750 cGy		1
Max Dose		Plan	Rectum+5mm	Max Dose 6700 cGy		1

Figure 9.81 Initial optimization criteria for case rP03.

Function	Constraint	Dose	ROI	Description	Robust	Weight
Physical Composite Objective						
Min Dose		Plan	PTV6600	Min Dose 6600 cGy		30
Max Dose		Plan	PTV6600	Max Dose 6900 cGy		5
Dose Fall-Off		Plan	External	Dose Fall-Off [H]6600 cGy [L]0 cGy, Low dose distance 10.00 cm		1
Dose Fall-Off		Plan	Rectum	Dose Fall-Off [H]6600 cGy [L]0 cGy, Low dose distance 2.00 cm		1
Dose Fall-Off		Plan	Bladder	Dose Fall-Off [H]6600 cGy [L]0 cGy, Low dose distance 5.00 cm		1
Max Dose		Plan	Bladder	Max Dose 6750 cGy		1
Max Dose		Plan	Rectum+5mm	Max Dose 6700 cGy		1
Uniformity Constraint	★	Plan	PTV6600	Uniformity, relative standard deviation 2%		

Figure 9.82 Final optimization criteria for case rP03.

Dose Distribution

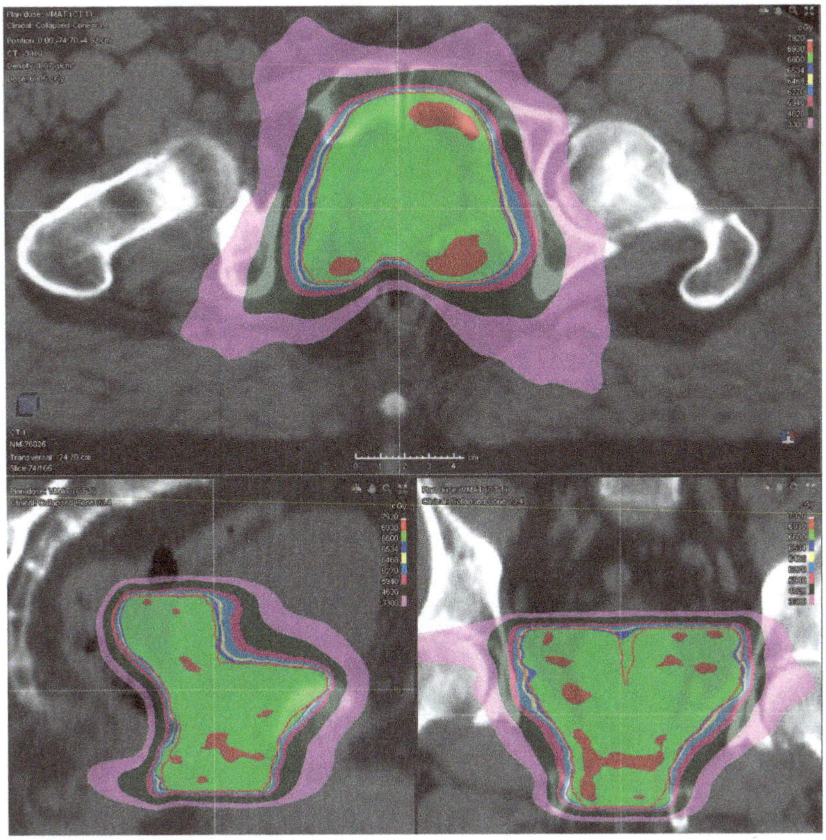

Figure 9.83 Final dose distribution for case rP03.

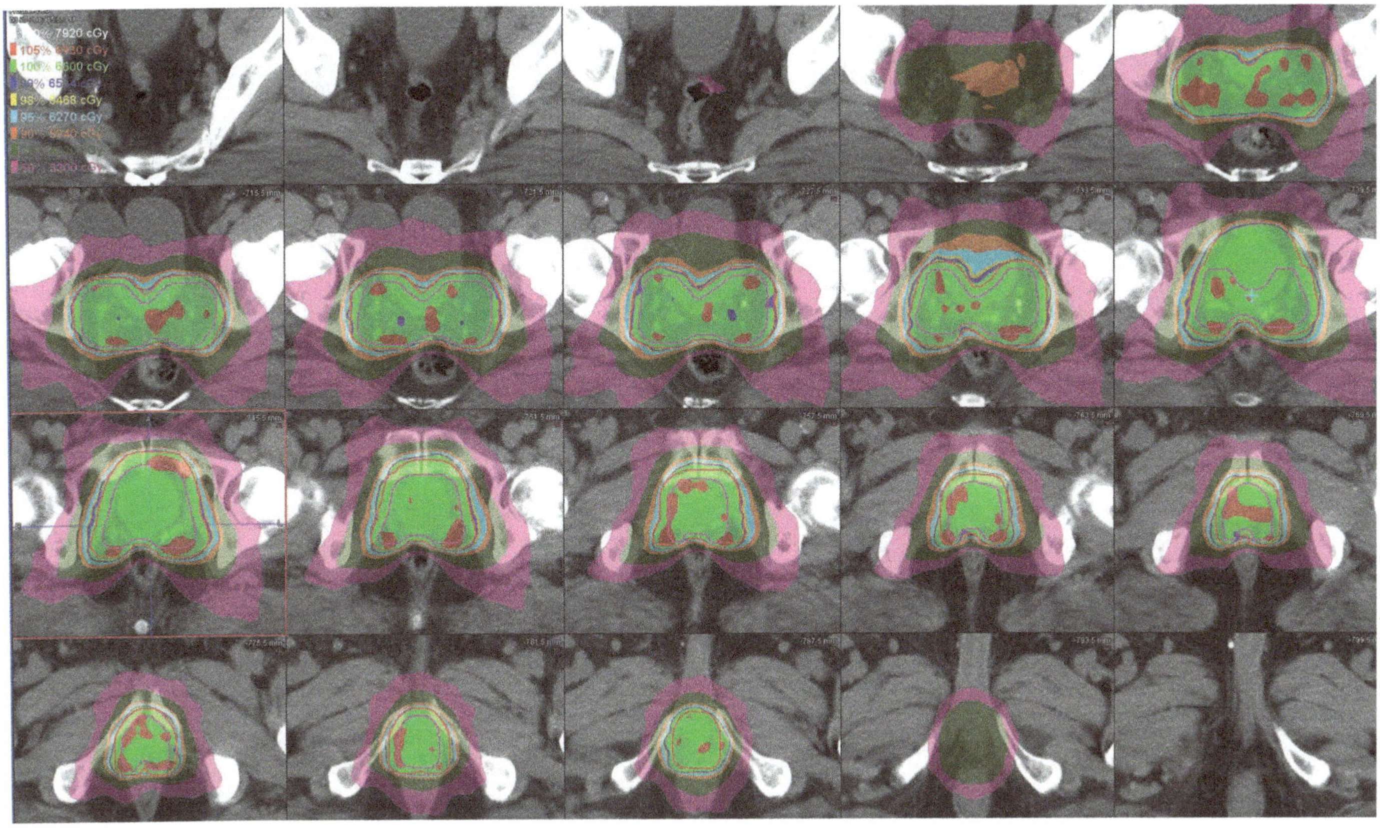

Figure 9.84 Final dose distribution for case rP03 in axial slices.

Dose	ROI/POI	Clinical goal	Value	Result	% outside grid
Plan dose: VMAT (CT 1)	Bladder	At most 90.0 cm³ volume at 7000 cGy dose	0.1 cm³	✓	0 %
Plan dose: VMAT (CT 1)	Bladder	At most 150.0 cm³ volume at 6500 cGy dose	68.4 cm³	✓	0 %
Plan dose: VMAT (CT 1)	Bladder	At most 7900 cGy dose at 0.0 cm³ volume	7027 cGy	✓	0 %
Plan dose: VMAT (CT 1)	Bowels	At most 5.0 cm³ volume at 4200 cGy dose	2.1 cm³	✓	0 %
Plan dose: VMAT (CT 1)	Bowels	At most 5000 cGy dose at 0.0 cm³ volume	6675 cGy	!	0 %
Plan dose: VMAT (CT 1)	L Femur head	At most 10.0 cm³ volume at 4800 cGy dose	0.0 cm³	✓	0 %
Plan dose: VMAT (CT 1)	L Femur head	At most 5600 cGy dose at 0.0 cm³ volume	4543 cGy	✓	0 %
Plan dose: VMAT (CT 1)	PTV6600	At least 95.00 % volume at 6600 cGy dose	95.00 %	✓	0 %
Plan dose: VMAT (CT 1)	R Femur head	At most 10.0 cm³ volume at 4800 cGy dose	0.0 cm³	✓	0 %
Plan dose: VMAT (CT 1)	R Femur head	At most 5600 cGy dose at 0.0 cm³ volume	4338 cGy	✓	0 %
Plan dose: VMAT (CT 1)	Rectum	At most 10.0 cm³ volume at 7500 cGy dose	0.0 cm³	✓	0 %
Plan dose: VMAT (CT 1)	Rectum	At most 20.0 cm³ volume at 7000 cGy dose	0.0 cm³	✓	0 %
Plan dose: VMAT (CT 1)	Rectum	At most 30.0 cm³ volume at 6500 cGy dose	8.7 cm³	✓	0 %
Plan dose: VMAT (CT 1)	Rectum	At most 40.0 cm³ volume at 6000 cGy dose	13.6 cm³	✓	0 %
Plan dose: VMAT (CT 1)	Rectum	At most 7900 cGy dose at 0.0 cm³ volume	6952 cGy	✓	0 %

Figure 9.85 Final clinical goals list.

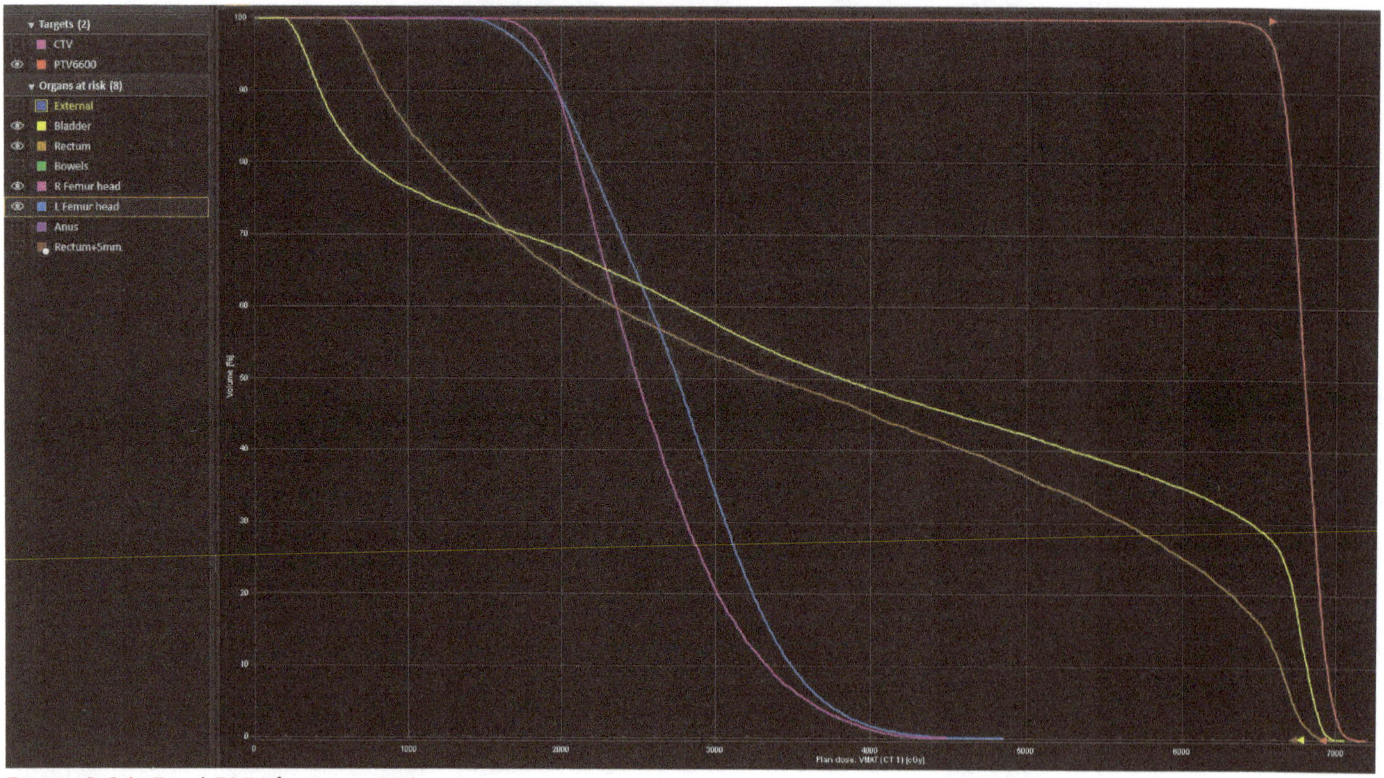

Figure 9.86 Final DVH for case rP03.